Biology of the Cell
An Evolutionary Approach

WILLIAM DE WITT
Williams College

1977 W.B. SAUNDERS COMPANY • Philadelphia • London • Toronto

W. B. Saunders Company: West Washington Square
Philadelphia, Pa. 19105

1 St. Anne's Road
Eastbourne, East Sussex BN21 3UN, England

833 Oxford Street
Toronto, Ontario M8Z 5T9, Canada

> **Library of Congress Cataloging in Publication Data**
>
> De Witt, William
>
> Biology of the cell: an evolutionary approach
>
> Includes index.
>
> 1. Cytology.   2. Evolution.   I. Title. [DNLM: 1. Cells. QH581.2 D523b]
>
> QH581.2.D48   574.8'7   75-21146
>
> ISBN 0-7216-3045-6

Cover photography of a sugar crystal by Phillip A. Harrington.

Biology of the Cell: An Evolutionary Approach   ISBN 0-7216-3045-6

© 1977 by W. B. Saunders Company. Copyright under the International Copyright Union. All rights reserved. This book is protected by copyright. No part of it may be reproduced, stored in a retrieval system, or transmitted in any form or by any means, electronic, mechanical, photocopying, recording, or otherwise, without written permission from the publisher. Made in the United States of America. Press of W. B. Saunders Company. Library of Congress Catalog card number 75-21146.

Last digit is the print number: 9   8   7   6   5   4   3   2   1

To Chunk, Fusky *and* Snout,
and to Billy, Josephine *and* Thomas —
for their help.

Preface

The study of cell biology is often limited to detailed and interminable descriptions of cell structure and function. Although such an approach is undoubtedly valid and provides the student with a wealth of necessary information, it is not without problems. One of these is that cell biology comes across as rather uninteresting and tedious, and it is often difficult, particularly for the beginning student, to deal with this large body of information in any meaningful way. In short, the student does not come to appreciate how the living cell operates. Perhaps an even more serious problem with the "catalogue" approach to cell biology is that it frequently obscures the dynamic aspect of cellular structure and function—that quality which determines the very essence of living systems.

One way to overcome these problems is to study the cell from an evolutionary perspective. Theories on the origin and evolutionary development of the living cell can provide a useful framework on which to organize the massive amounts of information that we call cell biology. By perceiving cell structure and function as evolutionary solutions to environmental challenges, the student can develop a meaningful synthesis of what appeared earlier as unrelated facts.

This textbook is designed for students who have a minimal background in biology and chemistry; it may be used before, as a part of, or after a course in general introductory biology. The first five chapters introduce the evolutionary theme and provide the background in biology and chemistry that is needed in later chapters. Chapters 6 through 9 discuss possible mechanisms for the origin of the first cells, while providing information on the chemical composition and organizational design of present-day cells. Chapters 10 and 11 examine the structure of prokaryotic and eukaryotic cell types and speculate on their evolution. Chapters 12 through 18 consider metabolic subsystems of cells in relation to their possible development during evolution. In order to carry the evolutionary approach into the laboratory, Eleanor Brown and I have written a companion laboratory manual, *Biology of the Cell: Laboratory Explorations,* which contains a number of simple experiments illustrating possible mechanisms for the origin and development of cell structure and function.

I am appreciative to William Grant for reading and criticizing several chapters; to my guinea pigs in Biology 102 at Williams College for providing invaluable feedback; to Judy Counter and Mary Lou DeWitt, who typed the manuscript; to the helpful staff at W.B. Saunders Company; and to Williams College, for a sabbatical leave that took me to Santa Fe, New Mexico, to write unhassled in our sunny arroyo.

WILLIAM DE WITT

Contents

Chapter 1
THE ORIGIN AND FUNCTION OF LIVING SYSTEMS 1

Chapter 2
ANCESTRAL ORGANISMS 18

Chapter 3
LIFE AND THE FITNESS OF CHEMICAL ELEMENTS 39

Chapter 4
THE SPECIAL ROLES OF WATER AND CARBON DIOXIDE 63

Chapter 5
CELLULAR ENERGETICS 83

Chapter 6
THE PRIMITIVE EARTH AND ITS ENVIRONMENT 113

Chapter 7
THE CHEMICAL NATURE OF BIOLOGICAL SYSTEMS 138

Chapter 8
PREBIOTIC SYNTHESIS OF ORGANIC COMPOUNDS 191

Chapter 9
FROM MOLECULES TO CELLS 217

Chapter 10
THE STRUCTURE OF PROKARYOTIC CELLS.. 234

Chapter 11
THE STRUCTURE AND ORIGIN OF EUKARYOTIC CELLS...................................... 260

Chapter 12
THE EVOLUTION OF CELLULAR METABOLISM.. 317

Chapter 13
THE METABOLISM OF ANAEROBIC HETEROTROPHS... 340

Chapter 14
PHOTOSYNTHESIS AND THE BUILDUP OF OXYGEN IN THE ATMOSPHERE............... 358

Chapter 15
AEROBIC METABOLISM 383

Chapter 16
MEMBRANE STRUCTURE AND THE TRANSPORT OF MATERIALS........................... 406

Chapter 17
CELL DIVISION 443

Chapter 18
THE GENETIC CODE: FUNCTION AND EVOLUTION 498

Chapter 19
EPILOGUE 534

Appendix I
MICROSCOPY 535

Contents

Appendix II
SOME PRACTICAL RULES OF BONDING 541

Appendix III
THE pH SCALE 544

Appendix IV
MECHANISMS OF OXIDATIVE PHOSPHORYLATION... 546

Index 553

CHAPTER 1

The Origin and Function of Living Systems

It is a unifying principle of biology that all living organisms are composed of cells and cell products. This concept, the *cell theory,* was originally proposed independently by two German biologists, Matthias Schleiden and Theodor Schwann. Schleiden first reached this conclusion in 1838 from his studies on plant cells; a year later Schwann, working with animal cells, reported more extensive results in which he described the cell as the structural unit of living organisms.

Although this generalization may seem obvious today, the enormous morphological diversity of living forms tended to obscure the fact that all living organisms are composed of fundamental units of biological organization. Thus, it is not surprising that the cell theory was suggested long after cell structure had been described by early microscopists and nearly two hundred years after Robert Hooke first examined thin slices of cork under the microscope, observing empty spaces delineated by heavy structural walls. He described the spaces as "cellulae," from the medieval Latin root *cella,* meaning small room. Although Hooke was in fact only observing the space once occupied by living material, the term *cell* has been used since to describe the fundamental unit of life (Fig. 1–1).

A study of the cell is perhaps best approached by attempting to trace the development of cellular pattern during the origins of life on earth, for the physical and chemical organization of the cell is best understood when viewed from a historical perspective. The living cell is an extremely complex entity capable of performing intricate functions with astonishing ease. In fact, its structure and function are so closely interwoven that it becomes at best difficult, and at worst impossible, to understand one without understanding the other. This is so precisely because the structure of the cell has developed through time concomitantly with its function.

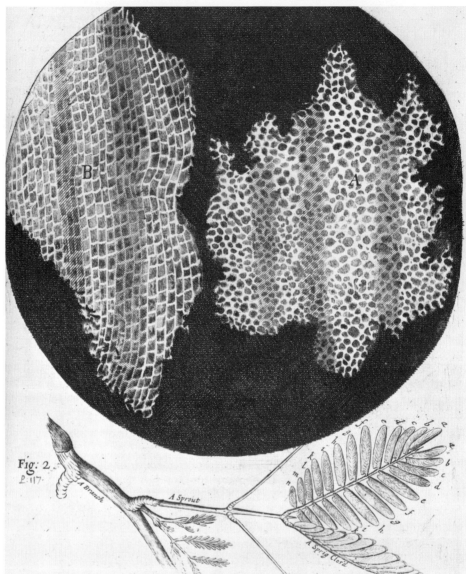

Figure 1-1 The structure of cork as seen by Robert Hooke. This reproduction of a plate from his book *Micrographia*, published in 1665, is furnished through the courtesy of Chapin Library, Williams College.

BIOLOGICAL EVOLUTION

Biologists describe and explain the development of living forms by the theory of evolution, which was originally proposed independently by Charles Darwin and Alfred Russel Wallace in 1858, and was further documented in the following year with the publication of Darwin's *On the Origin of Species.* The theory is summarized by Darwin in his introduction to the book:

> As many more individuals of each species are born than can possibly survive; and as, consequently, there is a frequently recurring struggle for existence, it follows that any being, if it vary in any manner profitable to itself, under the complex and some-

Biological Evolution

times varying conditions of life, will have a better chance of surviving and thus be *naturally selected*. From the strong principle of inheritance, any selected variety will tend to propagate its new and modified form.

Although Darwin did not explain how individual variation occurred, it has since been attributed to random change in the inherited characteristics of an individual. Consequently, we now consider evolution as a two-step process, involving the *chance* production of hereditary variation among the members of a species, followed by the natural selection of those individuals who are *reproductively* more fit and will thus, on the average, contribute to the next generation a greater proportion of individuals that reach reproductive age than do forms which are less adapted to the environment. By attributing variation to random hereditary change, modern evolutionary theory as we now understand it differs fundamentally from previous evolutionary theories, which held that the environment acted directly on an individual to produce the observed variation.

Differential reproduction arising from randomly occurring hereditary variation is, then, the key to evolutionary change. The point cannot be overemphasized, for it is a failure to understand this central concept that has resulted in countless misinterpretations of the evolutionary process (Fig. 1–2).

The consequence of evolutionary theory is that we now believe that all living species have originated from a single common ancestral organism (or from several *extremely similar* ancestral organisms) by a process involving the gradual accumulation of selectively advantageous heritable variation. We are able to infer this

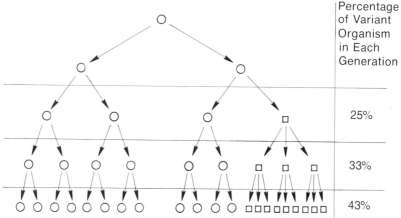

Illustration of evolutionary change by differential reproduction. One type of organism, represented by circles, leaves to the next generation *on the average* two individuals that reach reproductive age. A variant type, represented by squares, arises by chance and is, for any of a number of reasons, reproductively more fit, leaving to the next generation *on the average* three individuals that reach reproductive age. In succeeding generations, this new type will form an increasingly large proportion of the population. Competition for available resources between the original and variant organisms may lead to the extinction of the original form. If the two do not compete, they may coexist successfully.

Figure 1–2

evolutionary continuity not because it has been proved by definitive experimentation but rather because the basic structural and functional organization of all living forms is so similar that it would be of the very highest improbability that systems so identical would have arisen independently of one another.

The assumed relationship between all organisms has led biologists to try to reconstruct the evolutionary history, or *phylogeny*, of living organisms. Although such reconstructions are admittedly speculative, they are exceedingly useful because they categorize living organisms according to structural and functional similarities. One such scheme, which we will use in our discussions throughout this book, is shown in Figure 1–3. For various reasons that we will come to understand, the living world has been divided into five major groups, or *kingdoms*: Monera (the prokaryotes), Protista, Fungi, Plantae, and Animalia. As indicated by this scheme, all organisms are derived from a common ancestral organism, the first living cell. In addition, the available evidence suggests that all organisms

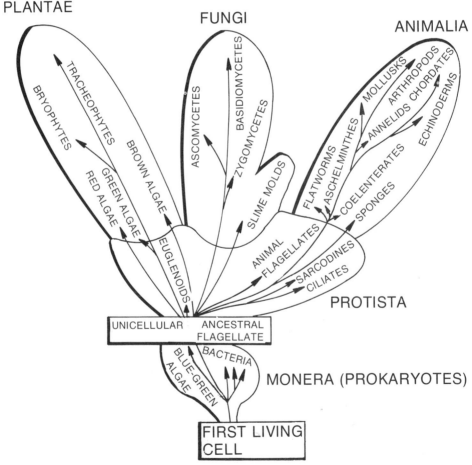

Figure 1–3 A proposed scheme indicating the evolutionary relationship between organisms. (Modified from R.H. Whittaker [1969]. *Science, 163*:157.)

except the prokaryotes (bacteria and blue-green algae) are derived from some common, but unidentified, unicellular flagellated ancestor. How this latter ancestral organism may have evolved from a prokaryotic one is currently the subject of much debate in the field of biology.

UNICELLULAR AND MULTICELLULAR ORGANISMS

Some organisms, such as bacteria and protozoa, are *unicellular* (single-celled) and others are *multicellular.* Multicellular organisms range from the very simplest, which are composed of individual cells of similar structure and function that show little coordination in activity, to the most complex, in which individual cells not only are specialized both structurally and functionally to perform particular tasks but also are finely coordinated with one another in activity. Such coordination may occur on a variety of levels. Cells with specialized structure and function may be organized into *tissues,* such as the cells composing the muscle of the heart. One or more tissues may, in turn, be coordinated as a structural and functional unit, termed an *organ.* The heart, for example, is an organ whose function is to pump blood. A group of organs may form a *system* responsible for carrying out one general function, such as respiration, circulation, or digestion. The individual systems are, of course, interdependent, and are integrated in a complicated manner to assure the survival and reproduction of the organism. This interdependence becomes obvious if we attempt to discuss one particular system of an organism in isolation from the others. The digestive system in a mammal, for example, requires a functioning circulatory system to carry nutrients, supplied by the digestive system, to the cells of the body, just as the cells of the heart and other components of the circulatory system are dependent on the digestive, respiratory, and other systems for obtaining nutrients, oxygen, and other necessities. The degree of cellular integration will, of course, vary with the complexity of the organism. In the final analysis, it is this integration of cellular activity that determines the survival of the individual: the multicellular organism is clearly more than the sum of its parts. While our discussions in this book will focus on the properties of the individual cell, it is important that we never lose sight of this fact.

The great diversity in form and function that we see among living organisms may ultimately be traced, on one hand, to variability in cellular structure and function, and on the other, to variability in the way individual cells are integrated to form an organism. Both types of variation have arisen during the historical development of living forms. It is, however, true that although there is considerable structural and functional variability between both cells of different organisms and different types of cells in the same organism, the fundamental similarities between all cells are far more striking than the differences.

CHARACTERISTICS OF LIVING SYSTEMS

CELLULAR ORGANIZATION

What are some of the common features of living systems that allow us to assume similar ancestry to all present-day forms of life? As we mentioned previously, all organisms are composed of cells and thus share a *common cellular organization.*

All cells, regardless of other structural details, are bounded by a delicate *cell* or *plasma membrane* of similar chemical composition. By controlling the passage of nutrients and other essential substances into the cell as well as the passage of waste products and other substances out of the cell, the plasma membrane functions as a highly discriminating chemical barrier. It thus serves to maintain the physical and chemical integrity of the cell.

METABOLISM

All cells *metabolize.* Their *metabolic activities* include all the chemical processes by which cellular material is produced, maintained, and broken down, and by which energy is supplied for these functions. The cell is, in a sense, a chemical factory: chemical substances obtained as nutrients from the external environment are converted by chemical means into other substances. Often these chemical transformations are not simple, and may involve a series of chemical reactions (*metabolic pathways*) by which a nutrient (A) is converted into one or more *metabolic intermediates* (B–E) before the *end product* (F) is formed:

$$A \rightarrow B \rightarrow C \rightarrow D \rightarrow E \rightarrow F$$

Some of these metabolic transformations, termed *anabolic,* require energy and involve the manufacture of highly complex cellular components from simpler nutrient substances. Others are *catabolic* transformations that involve the breakdown of complex molecules into simpler ones, often with the release of energy that may be used to drive anabolic pathways as well as other energy-requiring processes in the cell (Fig. 1–4).

All cells constantly expend energy, which they require for the synthesis of cellular constituents and for other cellular processes; these energy requirements are met in various ways. Green plants, some protists, and some prokaryotes are able to synthesize all the organic (carbon-containing) compounds they need by using carbon dioxide (CO_2) as their only source of carbon. These organisms, which are called *autotrophs,* are of two main types. *Photosynthetic autotrophs* utilize the energy of solar radiation to produce organic compounds. Other autotrophs can extract usable energy from simple inorganic compounds and are called *chemosynthetic autotrophs.* All other organisms, including animals, fungi, most protists,

Characteristics of Living Systems

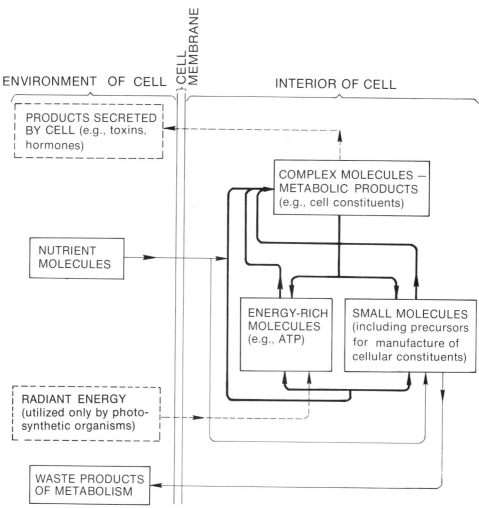

A diagrammatic representation of the general features of cellular metabolism. Dotted lines and boxes indicate processes that occur in some cells, but not in all. Energy-rich molecules may be used for processes other than the manufacture of metabolic products, including transport of substances into and out of cells, muscular contraction, and many others. (After A.H. Rose [1965]. *Chemical Microbiology.* London, Butterworth & Co. [Publishers] Ltd., p. 80.)

Figure 1-4

and most prokaryotes, are *heterotrophs;* they are unable to synthesize from carbon dioxide all the organic compounds they need, and must obtain as nutrient molecules from their environment one or more organic compounds from which they are able to extract energy and synthesize other organic compounds. Since many of the organic compounds essential to heterotrophs are manufactured from carbon dioxide only by autotrophs, heterotrophic organisms are ultimately dependent upon autotrophic organisms for survival.

A universal characteristic of cellular metabolism is that it is performed with unusual efficiency and economy. The number of different chemical compounds present in a living cell is surprisingly small considering both its complexity and its functional capabilities. The cell achieves this economy in a number of ways. A single chemical compound may be used by the cell for more than one function. Thus, for example, a particular compound may be broken down as a source of energy for the cell under one set of metabolic conditions, and under another set of conditions may be used as a starting material for the manufacture of a certain cellular product. Often complex cellular constituents are degraded by the cell when they are no longer needed, producing energy and simpler substances which may be reused by the cell for manufacturing entirely different substances. This recycling of cellular substances is apparent in Figure 1–5, which shows the interrelationships between a large number of the metabolic pathways found in living cells.

All types of cells perform countless metabolic processes using similar, and often identical, metabolic pathways. In fact, it is this uniformity in the metabolic activities of all cells, from the simplest to the most complex, that is perhaps the most astonishing feature of living systems. These similarities are perhaps best illustrated by examining some general features of the functional organization of cells.

CHEMICAL COMPOSITION AND FUNCTIONAL ORGANIZATION

In our everyday experience, we are likely to consider the attributes of life so unique that it is perhaps easy to forget that cells, like all things, are composed of chemical substances. In addition to a few inorganic (non-carbon-containing) substances and some small organic molecules dissolved in a watery matrix, the cells of all organisms are composed primarily of four classes of organic substances: *nucleic acids, proteins, carbohydrates,* and *lipids.* Indeed, not only do all cells have a similar chemical composition but they also use the same types of compounds for similar functions.

Nucleic Acids and Proteins

The hereditary information of all cells is stored in *deoxyribonucleic acid* (DNA). The DNA contains the information necessary to determine not only the structural and functional pattern of the cell itself, but also the general characteristics of the whole organism of which the cell is a part. The DNA molecule is a polymer (poly = many; meros = part) composed of a large number of subunits called *nucleotides.* There are four different kinds of nucleotides, and the linear sequence of these nucleotides in the DNA molecule determines a code which provides instructions for the synthesis of all cellular *proteins.* Proteins are themselves polymers of *amino acids;* there are twenty different amino acid "building blocks" found in nat-

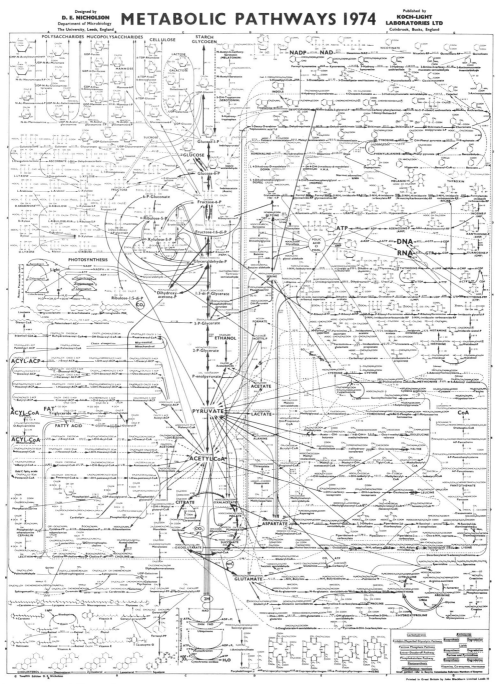

A summary of the main cellular metabolic pathways. All reactions shown do not necessarily occur in every type of cell, but the complex interrelationship of pathways is characteristic of all living cells. (*From* D. E. Nicholson [1974]. *Metabolic Pathways*, 12th Ed. Koch-Light Laboratories, Inc., Colnbruck, Bucks, England. Copyright © 1974 by D.E. Nicholson. Distributed in the United States by General Biochemicals, Inc., Chagrin Falls, Ohio. Used by permission.)

Figure 1-5

urally occurring proteins. The characteristic chemical properties of any protein are determined ultimately by its amino acid sequence.

During the process of protein synthesis, the cell uses another type of nucleic acid, ribonucleic acid (RNA), to decode the nucleotide sequence in DNA into specific sequences of amino acids in proteins. Proteins are actually put together on small spherical bodies called ribosomes, which contain approximately 40 per cent protein and 60 per cent RNA. The purpose here is not to describe in detail the process of cellular protein synthesis, but to point out that the entire process involves an extremely complex series of metabolic reactions *which are virtually identical in all cell types.* This similarity includes the way the DNA code is set up, the way the coded information is decoded, and the way the amino acids are put together to form proteins.

Although some proteins have structural or other functions, many are *enzymes.* These are *biological catalysts,* each of which promotes a specific chemical reaction. Without enzymes, most metabolic reactions could not occur (at least at speeds rapid enough to be useful) in the chemical environment of the living cell. The enzymes present in a cell thus determine *the sum total of its chemical capabilities,* and it is therefore the specific proteins synthesized by a cell that endow it with its structural and functional pattern.

The specific nucleotide sequence in the DNA molecule may, on occasion, become altered, or *mutated.* Since the DNA code is determined by the nucleotide sequence in the DNA molecule, such mutations may result in the production of altered proteins possessing chemical properties entirely different from those of the original proteins. If a mutation occurs in a cell that will eventually give rise to a new organism, then the mutation is transmitted to the next generation. It is by this mechanism that hereditary variation, so important from an evolutionary standpoint, arises. Indeed, the ability to reproduce with precision but with the additional capacity for hereditary variation is a characteristic property of all living forms.

Carbohydrates

Much of the cell's energy is provided by carbohydrates, of which *glucose* is the most important. When glucose is chemically degraded by the cell, the energy released from this process is stored in another chemical compound, adenosine triphosphate (ATP). The energy in ATP is subsequently used for driving the energy-requiring reactions of the cell, such as protein synthesis. ATP is used by all cells as an energy-storing molecule, and at least part of the complex metabolic pathway of glucose degradation is identical in every living cell.

Lipids

Lipids are important not only as a major constituent, along with proteins, of all cell membranes, but also as a nutrient source. In addition, lipid compounds serve a variety of different functions in cellular metabolism. Some of the *vitamins* are lipids; these are essential for certain metabolic reactions. Many animal hormones, which serve as regulators of physiological function, are also lipid in nature.

THE DESIGN OF CELLULAR METABOLISM

These basic design features — the storage of information in DNA molecules, the retrieval of this information in the form of specific patterns of protein synthesis, the use of proteins as enzymes which promote the metabolic reactions of the cell, the storage of energy in ATP molecules, and the physical separation of the entire system in a morphological entity bounded by a highly discriminating chemical barrier — are universal in all cell types.

All cells face the same fundamental problems in their struggle for existence; the fact that they solve these problems in similar ways is strong evidence for common ancestry. It would be quite difficult to account for the establishment of such complex, highly integrated, efficient, and diverse but similar systems by a number of independent chance occurrences. It thus seems reasonable to consider that the development of these systems is a result of the process of evolution acting upon a common ancestral organism (or upon several ancestral organisms of remarkably similar biochemical design).

LIVING VERSUS NON-LIVING

Before leaving our discussion of the characteristics of living systems, it is perhaps worth commenting briefly on one confusing point. We have defined living systems in such a way as to exclude viruses from the living world. We will, however, have occasion to discuss them at length, as they possess many properties that are inherently interesting to cell biologists and that help us to understand many cellular processes. As we shall see, viruses are not composed of cells, nor do they metabolize independently. They are cellular parasites and as such are dependent upon their host cells for supplying the raw materials and providing the metabolic machinery necessary for their reproduction. They are clearly borderline systems, and for our purposes we have arbitrarily chosen *not* to include them within our definition of living systems. It should be emphasized, however, that there is not always a clear distinction between the living and the non-living. Although the vast majority of scientists would not dispute the contention that a tree is living and a rock is non-living, it is not always possible to obtain agreement when considering entities that have some characteristics we normally attribute to living systems, but other characteristics that we consider attributes of the non-living world. Asking if such a borderline system is alive is much like wondering whether someone is happy—the ultimate judgment will depend on individual interpretation.

THEORIES ON THE ORIGIN OF LIFE

From earliest times, man's interest in the difference between the living and the inanimate has led to considerable speculation and controversy regarding the origins of life on earth. A number of ideas have been advanced over the centuries.

SPONTANEOUS GENERATION

Greek philosophers and scientists were apparently the first to attribute the origin of living forms to a natural phenomenon rather than to a creative act of God. They formulated the theory of *spontaneous generation,* which contended that living organisms, even those of a high order of complexity, arose spontaneously from nonliving substances. It was generally believed that a wide variety of plants and animals were of spontaneous origin, and in ancient writings it is not uncommon to encounter suggestions that maggots originated from rotting meat, or mice from rags, or frogs from the action of sunshine on mud. Even Aristotle, perhaps the greatest collector of knowledge of the ancient world, concluded in one of his many treatises on the origin of living forms: "Certain fishes come spontaneously into existence not being derived from eggs or from copulation. Such fish arise all from one of two sources, from mud, or from sand and decayed matter that rises thence as a scum." Such speculations by leaders in scientific thought were conceivable only because so little was known about the relationship of living organisms to one another and to the inanimate world (Fig. 1–6).

The theory of spontaneous generation remained essentially unchallenged until the middle of the seventeenth century, when Francesco Redi (1626–1679) performed an experiment of brilliantly simple design which effectively disproved the notion that maggots arose spontaneously from rotting meat. By protecting meat with a fine gauze covering, Redi was able to show that maggots were the result of eggs deposited by flies.

Old beliefs are difficult to change, and it is not surprising that Redi's experiments did not immediately dispel all ideas of spontaneous generation. Intense controversy ensued for the next two centuries, particularly with regard to the origin of microorganisms, which were discovered in 1677 by the great Dutch microscopist Anton van Leeuwenhoek (1632–1723). Although people were at least partially willing to give up their belief in the spontaneous origin of highly complex organisms such as insects, they held firmly to the idea that the minute forms of life arose by spontaneous generation. Toward the end of the eighteenth century, Lazaro Spallanzani (1729–1799) was able to prove that neither bacteria nor protozoa

Theories on the Origin of Life

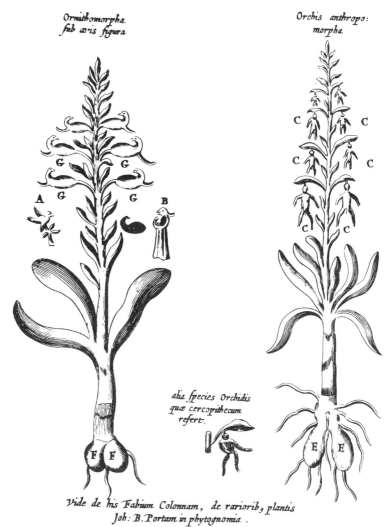

Figure 1-6

The early lack of appreciation for the relationships among organisms is illustrated by this figure depicting doves and little men arising as flowers on orchids, which appeared in 1665. (Figure taken from the *Mundus Subterraneus* of Athanasius. Kircher, Amsterdam, 1665.)

would develop in liquids that had been boiled for extended periods of time and kept in closed containers. However, not only did the experiments of several well-known scientists contradict Spallanzani's results, but some scientists contended that the process of spontaneous generation required a free supply of air and could not occur in sealed containers. The obstacle in disproving this latter contention was the fact that microorganisms are present in air, and consequently it was exceedingly difficult to design an experiment that would allow air to enter a container of sterile nutrient medium while excluding airborne microorganisms. The issue was not finally resolved until 1861, when Louis Pasteur, the noted French biologist and chemist (1822–1895), published the results of a now classic experiment.

Pasteur added a nutrient liquid that would support the growth of microorganisms to a flask and then, by heating the top of the flask, drew it out into a long S-shaped open tube. He boiled the liquid in the flask for several minutes in order to kill any organisms that might be present and, after allowing it to cool, placed it in an incubator at a temperature that would encourage the growth of microorganisms. Even after many months, no organisms developed in the flask. However, when Pasteur removed the curved neck of the flask, numerous microorganisms began to appear within short intervals of time. The S-shaped neck had served as an effective filter to exclude contaminating organisms from the flask by trapping them on its curved and moist surface (Fig. 1–7).*

---

*Fortunately, Pasteur by chance did not use a nutrient medium that would support the growth of heat-resistant spores.

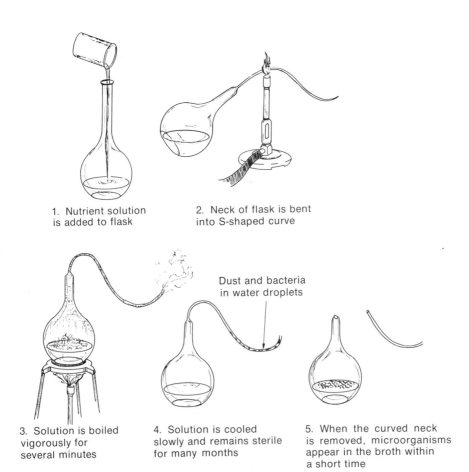

1. Nutrient solution is added to flask

2. Neck of flask is bent into S-shaped curve

3. Solution is boiled vigorously for several minutes

4. Solution is cooled slowly and remains sterile for many months

Dust and bacteria in water droplets

5. When the curved neck is removed, microorganisms appear in the broth within a short time

Figure 1–7 The basic design of Pasteur's experiment disproving the spontaneous origin of microorganisms. (*From* R.S. Young [1966], *Extraterrestrial Biology.* New York, Holt, Rinehart, and Winston, Inc.)

Theories on the Origin of Life

By a series of experiments similar to this, Pasteur proved that the presence of microorganisms was always due either to growth of those already present in the nutrient medium or to contamination from the air. Pasteur's results were so conclusive that the doctrine *omne vivum ex vivo*—all living things come from living things—became universally accepted by the scientific community.*

PANSPERMIA

Pasteur's experiments, while silencing the proponents of spontaneous generation, did not explain the origin of the first living organism. Richter attempted to resolve this dilemma by proposing in 1865 the *theory of panspermia* (or *cosmozoa*), which held that spores of microorganisms were carried to the primitive earth on meteorites originating from cosmic bodies on which life already existed. Finding a favorable environment, the spores allegedly gave rise to an ancestral organism from which all forms of life subsequently evolved.

This theory is certainly difficult to disprove. A spore of terrestrial type, embedded within a large meteorite, might be effectively shielded both from the damaging effects of cosmic radiation and from the extremes of heat which a meteorite would encounter when entering the earth's upper atmosphere. Indeed, the idea would be worthy of more serious consideration if it offered us some additional clues as to the ultimate origin of living systems. Unfortunately, it merely provides for a change in the time and place of life's origin without shedding further light on the problem.

MECHANISTIC THEORY

Ernst Haeckel, the German biologist and philosopher, rejected the notion of panspermia as an explanation for the origin of life on earth. In 1866, shortly after the publication of Darwin's work, Haeckel proposed what is now known as the *mechanistic theory* of life's origin. He contended that at *one particular time,* or at most several times, in the early history of the earth, conditions were favorable for the formation of the simplest living entity from inanimate matter. The theory does not explain how this process occurred, but suggests that once this first living organism was formed, it began to evolve by natural selection into present-day forms of life. Most biologists now consider this explanation very unlikely because of the high improbability of forming a complete, albeit primitive, living organism in a single, chance event. The theory requires that all of

---

*It is fascinating, however, that spontaneous generation has proved so appealing as an explanation for the origin of living forms that it persists even into the present time. In many farming communities, worms of various types are often thought to arise spontaneously; the sudden appearance in water troughs of long, slender hairworms gives rise to the common belief that they are metamorphosed horsehairs.

the organism's molecular constituents, some of which we know to be extremely complex, happened to accumulate by chance at one time and in one place on the earth's surface, and were then properly assembled, again by chance, into a living form.

MATERIALISTIC THEORY

Biologists now favor the *materialistic theory*, proposed initially in 1924 by the Russian biologist A.I. Oparin and independently several years later by the British biologist J.B.S. Haldane. Oparin and Haldane suggested that the most primitive living organism originated from non-living matter by a gradual process of chemical evolution. Accordingly, the development of the first living organism may be thought of as simply one event in a long and continuous evolution of organic material.

This evolutionary process is thought to have begun with the formation of the earth approximately 4.6 billion years ago. The atmosphere of the primitive earth was probably a reducing one; that is, it contained no free oxygen, but consisted primarily of methane, ammonia, hydrogen, and water vapor, along with smaller amounts of other gases. It is assumed that simple organic compounds were formed from these gases, using energy sources available in the Precambrian environment. These included ultraviolet light from the sun, electric discharge from thunderstorms, and heat from volcanic activity, among others. As simple organic compounds became concentrated in the primitive oceans, chemical substances of ever-increasing complexity were formed. These substances gradually achieved a capacity for self-reproduction and a loose structural organization. Eventually, the first cellular system evolved, signaling the beginning of biological evolution. Once living organisms had evolved, this process could have been repeated only in isolated environments, for any newly developing forms would quickly be eliminated by more efficient competitors.

Although we do not know the precise details, there is good evidence to suggest that cellular life may well have arisen in this way. It is the purpose of this book to examine the development of cellular structure and function by tracing a probable sequence of events that occurred during the evolution of cellular systems from non-living molecules.

REFERENCES: CHAPTER 1

Brock, T.D. (1961). *Milestones in Microbiology.* Prentice-Hall, Inc., Englewood Cliffs, N.J.
Cohen, M.R., and I.E. Drabkin (1948). *A Source Book in Greek Science.* McGraw-Hill Book Co., New York.
Dampier, W. (1936). *A History of Science and its Relation with Philosophy and Religion.* Macmillan, Inc., New York.
Darwin, C. (1859). *On the Origin of Species by Means of Natural Selection.* John Murray, London.

References: Chapter 1

Delbruck, M. (1966). In *Phage and the Origins of Molecular Biology.* J. Cairns, G.S. Stent, and J.D. Watson, Eds. Cold Spring Harbor Laboratory Quantitative Biology.
Edsall, J.T., and J. Wyman (1958). *Biophysical Chemistry.* Academic Press, Inc., New York.
Fiske, J. (1892). *Outlines of Cosmic Philosophy.* Riverside Press.
Florkin, M. (1960). *Aspects of the Origin of Life.* Pergamon Press, Inc., Elmsford, N.Y.
Fox, C.F. (1972). The structure of cell membranes. *Scientific American,* 226:30.
Fox, S.W. (1965). *The Origins of Prebiological Systems.* Academic Press, Inc., New York.
Gardner, E.J. (1960). *History of Science.* Burgess Publishing Co., Minneapolis, Minnesota.
Haldane, J.B.S. (1929). The origin of life. *Rationalist Ann.* p. 3.
Howland, J.L. (1968). *Introduction to Cell Physiology: Information and Control.* The Macmillan Co., New York.
Oparin, A.I. (1938). *The Origin of Life.* Macmillan, Inc., New York.
Oparin, A.I. (1968). *Genesis and Evolutionary Development of Life.* Academic Press, Inc., New York.
Orgel, J.E. (1973). *The Origins of Life: Molecules and Natural Selection.* John Wiley and Sons, New York.
Rose, A.H. (1965). *Chemical Microbiology.* Butterworth and Co., Ltd., London.
Schrödinger, E. (1944). *What Is Life?* Cambridge University Press, Cambridge, England.
Stanier, R.Y., M. Doudoroff, and E.A. Adelberg (1963). *The Microbial World.* Prentice-Hall, Inc., Englewood Cliffs, New Jersey.
Young, R.S. (1966). *Extraterrestrial Biology.* Holt, Rinehart and Winston, Inc., New York.

CHAPTER 2

Ancestral Organisms

In order to trace and to understand the evolutionary process by which we believe living systems arose on earth, it is helpful to examine the organizational plan of the simplest present-day living organisms. We have good reason to believe that these organisms are members of groups that probably have not changed significantly since their origin during early stages of biological evolution. Hence it is not unreasonable to assume that their general design, although undoubtedly more complex, contains basic features similar to those of the first cells. Evidence to support this assumption comes from an examination of patterns of cellular organization and from the fossil record.

CELLULAR ORGANIZATION

Advances in our knowledge of cell structure have largely depended on the improvement of a tool: the microscope (see Appendix I). Although the function of a microscope is to magnify an object so that it becomes visible, the usefulness of a microscope is not dependent primarily upon the magnification it can obtain, but rather upon its *resolving power;* that is, its ability to separate clearly the individual parts of an image. In practice, the resolving power of a microscope defines the smallest object that can be seen distinctly separate from adjacent objects its own size.

The light microscope, which was the main instrument used for the examination of cell structure until the development of the electron microscope several decades ago, is unable to resolve clearly an object smaller than about 0.2 μm (200 nm)* in diameter. Consequently, the light microscope does not possess sufficient sensitivity to provide an accurate picture of internal cellular detail. Bacterial cells, for example, range in size from about 0.2 to 3.0 μm; the smallest of these is just barely detectable in the light microscope. Even animal and plant tissue cells, which are considerably larger

---

*1000 nanometers (nm) = 1 micrometer (μm) = 1 $\times$ 10^{-6} meter (m).

Cellular Organization

and range in size between roughly 10 and 40 μm, possess considerable internal structure that is not discernible by light microscopy. For example, internal cell membranes, which are approximately 7.5 nm thick, and many other small cellular structures, are not visible in the light microscope.

Because of the limited resolving power of the light microscope, cell biologists long believed that all cells were constructed along identical lines. They pictured the cell as a sack of living substance that consisted of a membrane-bound entity containing a watery fluid, termed *cytoplasm (cyto* = cell; plasm = substance), in which various free-floating cell structures were suspended (Fig. 2–1).

In recent years, the electron microscope has greatly changed our view of cellular structure. By using a beam of electrons instead of visible light to illuminate a specimen, an electron microscope is able to obtain a resolution of 0.4 to 0.5 nm. With this greatly improved resolution, it has now become apparent that all living organisms can be divided into two distinct groups on the basis of their cellular organization. The cells of all bacteria and a single group of algae, the blue-green algae, possess the simplest structure, and are termed *prokaryotic* (pro = primitive; karyon = nucleus). All other organisms have a *eukaryotic* (eu = true; karyon = nucleus) cell structure, which is the more complex cell type.

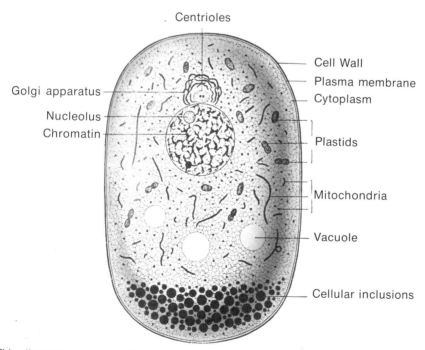

Figure 2–1 This diagram of a generalized cell, originally drawn by E.B. Wilson around 1900, appeared in a well-known cell biology text as late as 1954. It illustrates the conception of cellular structure before the use of the electron microscope. (*From* DeRobertis, Nowinski, and Saez [1954] *General Cytology,* Philadelphia, W.B. Saunders Co., p. 66.)

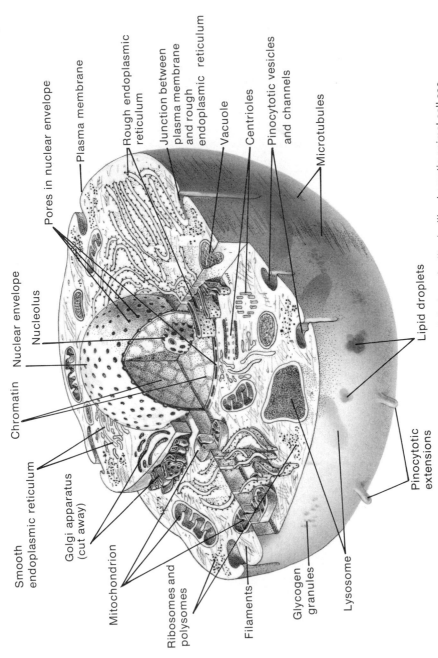

Figure 2-2 An artist's conception of the general design of a "typical" eukaryotic animal cell sectioned in a number of planes. It is shown in order to illustrate the relationship between various cellular components likely to be found in eukaryotic cells. Note particularly the defined nucleus and nuclear envelope, the membrane-bound organelles (such as mitochondria and lysosomes) and the extensive membrane systems. (Modified from S.W. Hurry [1964], *The Microstructure of Cells*, Boston, The Houghton Mifflin Co.)

Cellular Organization

STRUCTURE OF EUKARYOTIC CELLS

Although we have pointed out that there is a great deal of structural variability among different cell types, even those in the same organism, it is possible to make some broad generalizations concerning the structure of a "typical" eukaryotic cell. We must keep in mind, however, that even a generalized description of structural detail does not necessarily apply without exception to all eukaryotic cell types. For example, the circulating human red blood cell does not possess a nucleus or any other significant internal structure. It consists essentially of a cell membrane enclosing an aqueous cytoplasm, in which the oxygen-carrying protein hemoglobin and other chemical substances essential for metabolism are dissolved.

The typical eukaryotic cell (Fig. 2-2) consists of a *nucleus* and surrounding *cytoplasm* and is bounded by a limiting *plasma membrane*. In most eukaryotic plants, but not in animals, a *cell wall* composed of cellulose lies exterior to the plasma membrane and helps to keep the cell rigid. The nucleus of all eukaryotic cells is separated from the cytoplasm by a *nuclear envelope* consisting of two concentric membranes. The hereditary information of the cell, which is coded in large molecules of DNA, is contained in the nucleus and is packaged in structures called *chromosomes,* which are visible during cell division. Chromosomes are composed of protein and some RNA in addition to DNA. During cell division, the chromosomes are duplicated and then partitioned equally to both resultant cells in an intricate process termed *mitosis.* This assures that each cell obtains a complete copy of the hereditary material of the original cell.

The cytoplasm of eukaryotic cells typically contains a variety of subcellular structures, each of which is responsible for carrying out certain specialized functions of the cell. Some of the subcellular structures are bounded by limiting membranes and are known as cytoplasmic organelles. These include, among others: *mitochondria,* which supply most of the cell's energy by breaking down nutrient molecules and using the energy released in the process to synthesize ATP, the "energy currency" of the cell; *chloroplasts,* which are present in green plants and are the site of photosynthesis; and *lysosomes,* which contain enzymes capable of degrading nucleic acids, proteins, carbohydrates, and lipids, and which are important in digestive and other processes in the cell.

In addition to membrane-bound organelles, eukaryotic cells usually possess *Golgi bodies, endoplasmic reticulum,* and other extensive internal *membrane systems.* These serve as sites for many of the metabolic reactions of the cell and as internal channels for the passage and storage of materials within the cell. *Ribosomes,* important in the process of protein synthesis, may lie singly or in clusters in the cytoplasm, or may be attached to membranes of the endoplasmic reticulum.

It should be apparent from the foregoing discussion that the eukaryotic cell, far from being a sack of unstructured cytoplasm, is highly *compartmentalized;* that is, specific metabolic activities of the

STRUCTURE OF PROKARYOTIC CELLS

Prokaryotic cells (Fig. 2–3) are built on simpler lines than eukaryotic ones. Most prokaryotes possess a rigid cell wall that lies exterior to the plasma membrane and determines the characteristic shape of the cell. The prokaryotic cell wall, unlike the cellulose wall found in plants, is composed primarily of carbohydrates and amino acids. The DNA-containing region of the cell, called the *nucleoid,* is not separated from the rest of the cell by a nuclear envelope. There are no chromosomes; when cell division occurs the DNA is simply duplicated and a copy of the original DNA is distributed to each of the resulting cells. No cytoplasmic organelles are present, although internal membrane systems may exist that serve, along with the plasma membrane, as sites for metabolic reactions. Specialized membranous structures called *mesosomes,* which appear to be infoldings of the plasma membrane in the nucleoid region, are found in many prokaryotic cells; it has been proposed that mesosomes may be involved in the process of cell division.

Although prokaryotic cells do not possess some of the specialized structural refinements of eukaryotic cells and hence appear to us to be less compartmentalized and simpler in structure, this should not be interpreted as implying that the prokaryotic cell is in some way deficient in function or less efficient than eukaryotic cell types. Not only do prokaryotic cells share the basic metabolic organization common to other forms of cellular life (see Chap. 1), but there is good evidence from the fossil record to suggest that cells

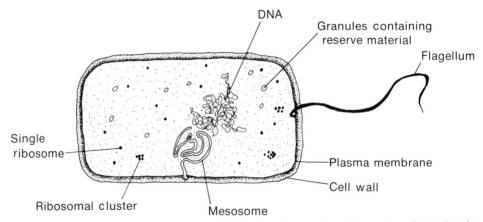

Figure 2-3 A "typical" bacterial prokaryote. A defined nucleus and membrane-bound cytoplasmic organelles are lacking. Blue-green algae and some bacteria may, however, possess extensive internal membrane systems in addition to mesosomes.

similar in morphology to present-day bacteria and blue-green algae have existed since early stages of biological evolution. Prokaryotic cells constitute a diverse group possessing an enormous range of metabolic specializations which have allowed the successful exploitation of widely differing environments. Their continued existence is indicative of the adaptability and efficiency of prokaryotic cell design.

THE FOSSIL RECORD

The fossil record, consisting of remains or impressions of organisms which have been preserved in the earth's crust, tells us something about the types of organisms present at specific times during the past history of the earth.

DATING METHODS

Fossils are dated by determining the age of the rock strata in which they are found. This is often done by a technique called *radioisotope dating* that depends on the predictable nature of radioactive decay. In order to understand the phenomenon of radioactive decay, we must review briefly some elementary aspects of atomic structure.

All matter is composed of particles called *atoms,* of which there are 92 naturally occurring types. Each different type of atom has characteristic chemical properties and is called an element. Thus, hydrogen (H), helium (He), oxygen (O), and carbon (C) all are different types of elements.

An atom is composed primarily of a *nucleus* and one or more negatively charged particles, called *electrons,* that move around the nucleus in cloudlike regions and occupy most of the space taken up by an atom. Almost all the mass of an atom may be found in the nucleus, which consists of *protons* and *neutrons* (except in the case of the lightest element, hydrogen, which contains only one proton and no neutrons). Protons and neutrons are of approximately the same mass; electrons, vastly lighter, possess a mass of about 1/1835 that of a proton. Each proton possesses a positive electrical charge equal in magnitude, but of opposite sign, to that of an electron. Table 2–1 summarizes the properties of the three fundamental particles of an atom.

It is apparent from our description of the fundamental particles that in a neutral atom, the positive charges carried by protons must be exactly balanced by the negative charges of the electrons; hence in a neutral atom the total number of protons and electrons must be equal.

Neutral atoms of each different element possess a characteristic number of protons and electrons. For example, hydrogen atoms possess one proton and one electron; helium atoms possess two protons and two electrons, and so on. The total number of protons

Table 2-1 FUNDAMENTAL ATOMIC PARTICLES

PARTICLE	MASS (RELATIVE TO C^{12}) AMU[a]	CHARGE
Electron	0.00055	−1
Proton	1.00728	+1
Neutron	1.00867	0

[a] AMU = Atomic Mass Units
1 AMU = 1.66035×10^{-24} gram

or electrons in a neutral atom is called its *atomic number,* and is often written as a subscript preceding the chemical symbol for that element. Thus hydrogen and its atomic number are routinely indicated by the symbol $_1$H; helium is written $_2$He, and so forth. Atomic numbers are an important and useful way of classifying elements, since *the total number of electrons in an atom determines to a very large extent its chemical behavior and reactivity.*

In most of the atoms found in living organisms, but not in all, the number of neutrons in the nucleus is equal to the number of protons. The sum of the number of protons and neutrons in the nucleus of an atom is termed its *mass number.* The mass number represents the approximate mass of an atom, since the mass of each proton and neutron is close to one atomic mass unit (AMU; see Table 2-1). The mass of an atom has little effect on its chemical properties, although it does influence such physical properties as diffusion. Mass number is usually indicated by a superscript following the chemical symbol for an element, as in H^1, He^4, C^{12}.

Atoms of the same element may contain different numbers of neutrons in their nuclei. However, because these atoms do not differ in their atomic numbers, they possess essentially identical chemical properties and are called *isotopes* of the element. Thus $_6C^{12}$, $_6C^{13}$ and $_6C^{14}$ are three isotopes of carbon. Occasionally, the lack of equal numbers of neutrons and protons will make the nucleus of an atom unstable, and it will break down spontaneously, releasing particles. These unstable isotopes are termed *radioactive.*

Radioactive isotopes (radioisotopes) undergo spontaneous transformation either into another isotope of the same element or into an entirely different element. The rate of this transformation is predictable and is defined by the half-life of the radioisotope, which is the amount of time required for one half of the radioactive material to decay into its new form. For example, rubidium-87 (Rb^{87}), a radioisotope of rubidium, decays with a half-life of 47 billion years into strontium-87 (Sr^{87}), a non-radioactive, stable form of strontium. This

The Fossil Record

means that fifty per cent of the Rb^{87} present at any given time will be converted into Sr^{87} in 47 billion years.

The decay rate of a radioactive element is so predictable that it can be used as a basis for determining the age of a substance, and the rubidium-strontium system has been employed extensively in geology to date rock samples. This is done by measuring the ratio of the amount of Rb^{87} to the amount of Sr^{87} in the sample. The method, of course, assumes that all of the Sr^{87} produced by radioactive decay of Rb^{87} is still present in the rock, and that there was no Sr^{87} in the rock at the time the rock was formed. In practice, this latter assumption is rarely valid, since Sr^{87} is abundant in mineral matter, but techniques are available for estimating the amount of Sr^{87} originally present when the rock was formed.

The decay of several other naturally occurring radioisotopes has proved particularly useful for determining the age of ancient rocks. Among these are the transformation of potassium-40 (K^{40}) to argon-40 (Ar^{40}), uranium-235 (U^{235}) to lead-207 (Pb^{207}), and uranium-238 (U^{238}) to lead-206 (Pb^{206}).

Rocks form in a number of ways, and not all types of rocks may be accurately dated by radioisotopic methods. These techniques are only applicable to *igneous rocks* such as lava, which have cooled from a molten form, and to *metamorphic rocks,* which have been hardened by recrystallization under the influence of heat and pressure. In both cases, radioisotope dating will determine the age since the rock last cooled from a melt or was last recrystallized; these processes cause a blending of mineral material which "resets" the radioactive clock for purposes of making the necessary corrections mentioned above. *Sedimentary rocks* arising from the compression of sedimented particles of eroded mineral matter cannot be accurately dated using radioisotopic techniques. This is because the mineral matter in sedimentary rocks may be derived from a variety of sources of different ages. Consequently, radioisotope dating of a sedimentary rock would provide us only with an average value for the age of the rocks composing the sediments, and would not tell us anything about the time at which the sediments themselves were formed.

Most fossils are found in sedimentary rocks. This is not surprising, since any substantial alteration of the structure of a rock by heat or pressure would obliterate all traces of organic matter. It is an unfortunate circumstance from our standpoint, however, because there are no methods available for dating organic fossil matter older than about 40,000 years, and consequently the age of ancient fossils must be estimated by determining the age of the rocks in which they lie. The fact that most fossils are located in sedimentary rocks thus presents a problem in accurately determining age.

Several methods have been developed for estimating the age of sedimentary rocks. One method that gives a *minimum* age for unmetamorphosed sedimentary rocks is to date an intrusion of igneous rock; that is, if cracks in the sedimentary rock have been infiltrated (intruded) by molten rock that has later solidified, then we know

Table 2-2 GEOLOGICAL TIME SCALE*

ERA OR ERATHEM	SYSTEM OR PERIOD		SERIES OR EPOCH	ESTIMATED AGES OF TIME BOUNDARIES IN MILLIONS OF YEARS
Cenozoic	Quaternary		Holocene	
			Pleistocene	2–3
	Tertiary		Pliocene	12
			Miocene	26
			Oligocene	37–38
			Eocene	53–54
			Paleocene	65
Mesozoic	Cretaceous			136
	Jurassic			190–195
	Triassic			225
Paleozoic	Permian			280
	Carboniferous System	Pennsylvanian		
		Mississippian		345
	Devonian			395
	Silurian			430–440
	Ordovician			500
	Cambrian			570
Precambrian			Late	1700
			Middle	2500
			Early	

*In use by U.S. Geological Survey.

The Fossil Record

that the intrusion of molten rock must have occurred at a time after the formation of the sedimentary rock. Another method that has been used to provide a minimum age for sedimentary rock is to date glauconite, a greenish mica. Glauconite is often found in sedimentary rocks and is known to form in the rock in which it is found. These techniques, in combination with less direct ones, provide a reasonably good indication of the age of sedimentary rocks containing fossil material.

GEOLOGICAL TIME SCALE

Although a number of different systems have been used to describe geological time, most geologists now divide the history of the earth into four major eras. Beginning with the formation of the earth approximately 4.6 billion years ago, these are the Precambrian, Paleozoic (*paleo* = ancient; *zoic* = life), Mesozoic (*meso* = middle), and Cenozoic (*ceno* = modern). The time since the start of the Cambrian Period approximately 570 million years ago, which includes the Paleozoic, Mesozoic, and Cenozoic Eras, is sometimes referred to as the Phanerozoic (*phanero* = visible) Era. The Precambrian Period is arbitrarily divided into Early, Middle, and Late Precambrian (see Table 2-2).

PRECAMBRIAN FOSSILS

Numerous fossils have been found of organisms that have existed during the Phanerozoic Era, and they include representatives of all major phyla. Rocks from the Precambrian Period, which until recently was believed to possess a sparse fossil record, have yielded a number of important fossil discoveries. These discoveries suggest that primitive living systems may have evolved more than 3.2 billion years ago and demonstrate that living systems of gradually increasing diversity and complexity may have existed throughout the rest of the Precambrian Period.

The Fig-Tree and Onverwacht Geological Formations

The oldest fossils have been found in Precambrian rocks of the Fig-Tree and Onverwacht formations, which are located in the Transvaal of South Africa near the Swaziland border. These are sedimentary rocks that were formed over long periods of time by the gradual accumulation of mineral and other particles at the bottom of seas and lakes. Unlike most Precambrian rocks of sedimentary origin, the black cherts (carbon-containing rocks of microcrystalline quartz) and other stratified sediments of these Precambrian formations have not been substantially altered (metamorphosed) by heat and pressure since their formation.

The first Precambrian fossils, consisting of organic remains of what appear to be primitive microorganisms, were discovered in 1965 by E.S. Barghoorn and J.W. Schopf in samples of black chert from the Fig-Tree geological formation. Dated by the rubidium-strontium method, their age has been estimated at approximately 3.1 billion years.

Because these fossils are of minute size and are not visible to the naked eye, they were revealed only by the use of microscopic techniques. The smallest of the fossil microorganisms was detected by examining *surface replicas* of rock material in the electron microscope. In this technique, sections of the rock are finely polished and areas where fossil organic matter has been exposed at the surface by the polishing are etched with dilute hydrofluoric acid. This treatment dissolves inorganic rock material without destroying the organic matter of the fossil and thus further exposes the fossil organism on the rock surface. Since the rock material is opaque to an electron beam, the exposed organic matter cannot be viewed directly in the transmission electron microscope.* Instead, a replica of the rock surface is made by covering it with a thin film of electron-transparent material that becomes rigid when it solidifies. The film is removed intact from the surface, and its contours may then be "shadowed" with a heavy metal such as platinum. Since platinum metal is opaque to the electron beam, shadowing with platinum provides contrast for examining the surface replica under the electron microscope. The surface replica and shadowing techniques are illustrated in Figure 2–4. In some instances, Barghoorn and Schopf were able to remove the fossil organic matter from the rock surface and to examine it directly in the electron microscope without shadowing. In this case, the fossil appeared as an electron-dense body.

The shadowed surface replicas prepared by Barghoorn and Schopf from samples of Fig-Tree chert revealed rodshaped structures, demonstrated both in profile and in cross-section, which resemble modern rodshaped bacterial species (Fig. 2–5). These fossil structures are extremely small, varying in length from approximately 0.5 to 0.7 μm and in width from approximately 0.2 to 0.3 μm. What appears to be a cell wall, sometimes visible in cross-section, is seen to consist of two layers with a total thickness of 0.015 μm. This is similar in structure and dimension to cell walls of many present-day bacteria. The structures have been named *Eobacterium isolatum* (*eo* = dawn; *isolatum* = solitary, and describes its single-celled structure). Also visible by electron microscopy are threadlike organic filaments, up to 9 μm long, which are not similar to any known organism.

In other experiments, Barghoorn and Schopf cut very thin slices of Fig-Tree chert and examined them by light microscopy. These preparations revealed larger fossil structures, spheroid in shape

---

*The scanning electron microscope, which permits the examination of the contours of three-dimensional structures, was not well developed at the time Barghoorn and Schopf did their work (see Appendix I).

The Fossil Record

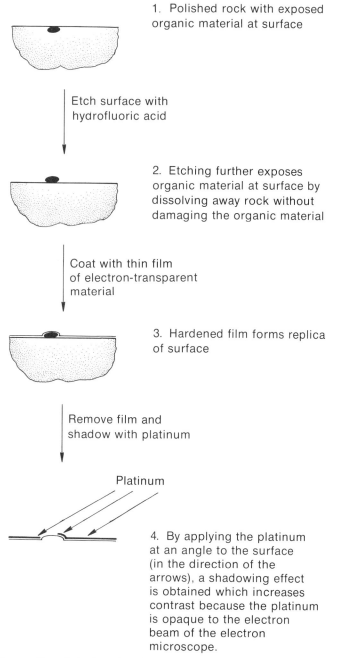

Preparation of shadowed surface replica for electron microscopy. **Figure 2–4**

and between 17 and 20 μm in diameter (Fig. 2–6). In general morphology they resemble some present-day groups of unicellular blue-green algae, as well as algae microfossils of younger Precambrian age. Barghoorn and Schopf have named them *Archaeo-*

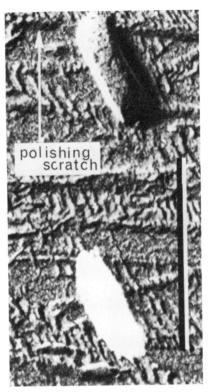

Figure 2-5 Electron micrographs showing *Eobacterium isolatum*. The top figure is a shadowed surface replica of a sample of Fig Tree chert; the bottom structure, which appears electron dense, is the preserved rod-shaped fossil which was dislodged from its original position at the rock surface. (*From* Barghoorn, E.S. and Schopf, J.W., *Science*, Vol. 152, p. 761, Figure 1, 1966. Copyright 1966 by the American Association for the Advancement of Science.)

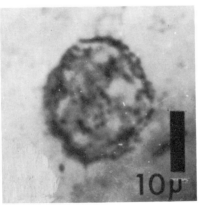 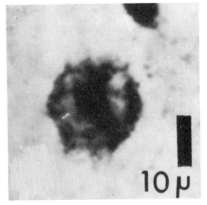

Figure 2-6 Barghoorn and Schopf examined thin sections of Fig-Tree chert in the light microscope and discovered these algae-like fossils, which are similar in morphology to some groups of modern unicellular blue-green algae. The fossil structures have been named *Archaeosphaeroides barbertonensis*. (*From* E.S. Barghoorn [1971] *Scientific American*, Vol. 224:33. May, 1971, p. 33. Copyright 1971 by Scientific American.)

The Fossil Record

sphaeroides barbertonensis (archaeo = primitive; *sphaeroides* = sphere; *barbertonensis* describes the Barberton region in which the Fig-Tree geological formation is located).

In addition to the morphological evidence, chemical analyses of Fig-Tree sediments have demonstrated the presence of several classes of complex organic compounds including *n*-alkanes and isoprenoids. This finding is of particular interest because two isoprenoid compounds detected in the sediments are pristane and phytane; these are usually thought to be degradation products of part of the chlorophyll molecule. Because chlorophyll is present in all known photosynthetic organisms and is considered essential for the process of photosynthesis, many workers have suggested that the presence of pristane and phytane in Fig-Tree sediments shows that photosynthesis had evolved by this time. Indeed, this finding has been interpreted by some scientists as biochemical evidence that *A. barbertonensis,* which appears to resemble some modern blue-green algae in morphology, may well have been a photosynthetic autotroph.

The presence of organic compounds in Precambrian sediments must, however, be interpreted with great caution. Recent evidence shows that in some instances organic compounds may have been introduced recently by ground water flowing through the formations. In addition, chlorophyll is not the only possible biological source of pristane and phytane. Pristane and phytane may also be derived from structurally similar compounds that are synthesized by many prokaryotic microbes, some of which may have been present at early stages of biological evolution.

Beneath the Fig-Tree formation lies the older Onverwacht formation, much of which consists of unmetamorphosed sedimentary rocks. The Onverwacht cherts, estimated by radioactive dating methods to be between 3.2 and 3.4 billion years old and by geological evidence to be possibly as old as 3.7 billion years, were originally examined for fossil material by A.E.J. Engel and his coworkers in 1968. Engel reported the discovery of a large number of "cup-shaped" and spherical microstructures which varied in size from 6 to 193 μm. Although their resistance to destruction by hydrofluoric acid is indirect evidence of their organic nature, the enormous size variation among these structures has made it difficult to interpret their relationship to modern organisms.

More recently, J. Brooks and M. Muir have examined samples of Onverwacht chert and have discovered spheroidal and filamentous fossil microstructures (Fig. 2–7). The spheroids are the more abundant, and unlike the structures described by Engel, tend to be uniform in size, although spheroids from older formations tend to be somewhat smaller (7 to 10 μm) than those from younger formations (15 to 20 μm). These structures appear to be the remains of living organisms and are generally assumed to be prokaryotes. This assumption is supported by the fact that the spheroids are very similar in morphology to unambiguously identified fossils of blue-green algae discovered by Schopf in the Late Precambrian Bitter Springs

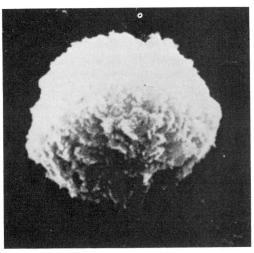

Figure 2-7 Specimens of etched Onverwacht chert examined directly in a scanning electron microscope which, because of its impressive depth of field, is particularly good for observing the surface contours of structures. *a.* A spheroid of organic matter is visible among the silica grains. *b.* Another spheroidal fossil at higher magnification exposed on the surface of an etched rock sample. (From J. Brook and G. Shaw [1973], *Origin and Development of Living Systems*, Academic Press, Inc., New York, pp. 285–286.)

formation in Northern Territory, Australia (approximately 1.0 billion years old).

In summary, Early Precambrian rocks from the Fig-Tree and Onverwacht formations contain what appear to be fossil remains of microorganisms that resemble present-day bacteria and blue-green algae and that have been classified as primitive, unicellular

The Fossil Record

prokaryotes. Although possessing little morphological complexity, some of these may well have been capable of photosynthesis.

Younger Precambrian Formations

Other fossil discoveries from Precambrian formations younger than the Onverwacht and Fig-Tree formations have yielded considerable information about the types of organisms present on earth during the remainder of the Precambrian Era. These discoveries provide strong although admittedly fragmentary evidence for a gradual evolution of living systems beginning with the simple unicellular prokaryotes of the Early and Middle Precambrian Period.

Limestone deposits discovered near the city of Bulawayo in southern Rhodesia and estimated to be approximately 2.7 billion years old, are generally considered to represent fossil colonies of living organisms (Fig. 2–8). The deposits closely resemble modern blue-green algal stromatolites; stromatolites are laminated sediments resulting from the accretion of limestone (calcium carbonate) on successive layers of sedimented communities of microorganisms. If the Bulawayan stromatolites were indeed produced by blue-green algae, as has been postulated by many researchers, then

Figure 2–8

These stromatolites were discovered in Bulawayo, Rhodesia, and are estimated to be approximately 2.7 billion years old. They are believed to have resulted from the accretion of limestone on successive layers of sedimented blue-green algae. (*From* Schopf, J.W. et al. [1971]. *Journal of Paleontology 45*, p. 477.)

oxygen-producing photosynthetic organisms were clearly established by the end of the Early Precambrian Period. This conclusion is supported by the discovery of single and clustered spheroidal structures resembling bacteria and blue-green algae in the Soudan Iron formation in Minnesota, which is about the same age as the Bulawayan limestones.

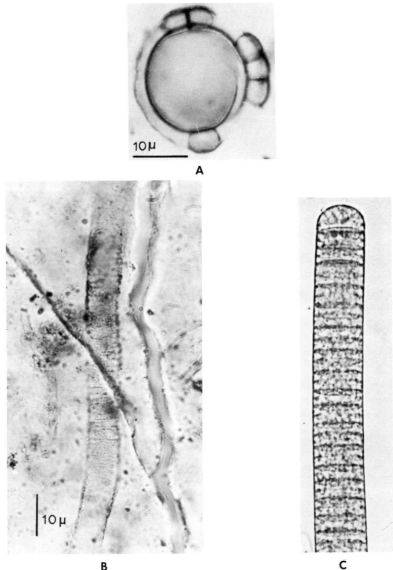

Figure 2-9 Fossil organisms from the 2 billion year old Gunflint Iron formation are shown in a and b. a. A spheroidal structure of uncertain relationship to existing living forms. The structure appears to consist of a thickwalled inner sphere surrounded by a thinwalled outer sphere, partially ruptured in this specimen. Smaller spheroids seem to be attached to the inner sphere b. A filamentous form closely resembling the modern blue-green alga, Oscillatoria, which is shown in c for comparison. (a and b from Barghoorn, E.S. and Tyler, S.A. Science, Vol. 147, pp. 571 and 566, Figs. 8 and 3, 1965. Copyright 1965 by the American Association for the Advancement of Science; c from Bold, H.C., The Plant Kingdom, 1970. Courtesy of Prentice Hall, Inc., Englewood Cliffs, New Jersey.)

The Fossil Record

The 1.9 billion year old Gunflint Iron formation near the Minnesota border in Ontario, Canada, is the youngest formation that has provided unequivocal evidence of the existence of living organisms in the Precambrian Period. The Gunflint chert has yielded a variety of unicellular spheroids and filamentous fossil structures (Fig. 2–9). These include fossils that resemble chemosynthetic iron bacteria and other diverse types of bacteria and blue-green algae, as well as fossils that do not resemble any present-day organisms. The formation also includes stromatolites that were almost certainly formed by blue-green algae. In addition to fossil structures, the Gunflint rocks contain a number of organic compounds including pristane and phytane.

The earliest examples of fossil organisms that may well represent eukaryotic green algae were discovered by P.E. Cloud and coworkers in the Beck Springs dolomite (compact limestone composed of calcium magnesium carbonate) in Southern California. Some of these fossils, estimated at between 1.2 and 1.4 billion years old, are shown in Figure 2–10. They appear to be cells containing dark inclusions which may be nuclei or other cytoplasmic organelles.

By far the best preserved and most varied fossil structures were discovered by Barghoorn and Schopf in the Late Precambrian Bitter Springs formation in central Australia (Fig. 2–11). The fossils, which are believed to be about one billion years old, include 20 certain or probable species of blue-green algae, 14 of which have been assigned to modern groups. It is thus apparent that blue-green algae

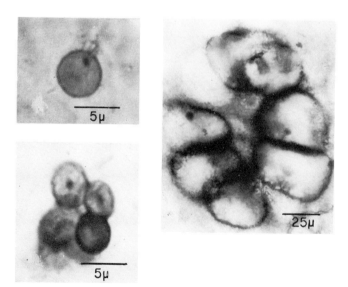

Figure 2-10 Examples of fossil structures discovered in the 1.2 to 1.4 billion year old Beck Springs dolomite. These fossils closely resemble extant eukaryotic green algae, and hence represent the oldest well-documented examples of eukaryotic cells. The internal dark spots and dark areas suggest the preservation of eukaryotic cell structures. (*From* Cloud, P.E., et al., [1969], *Proc. Nat. Acad. Sci., 62*:628.)

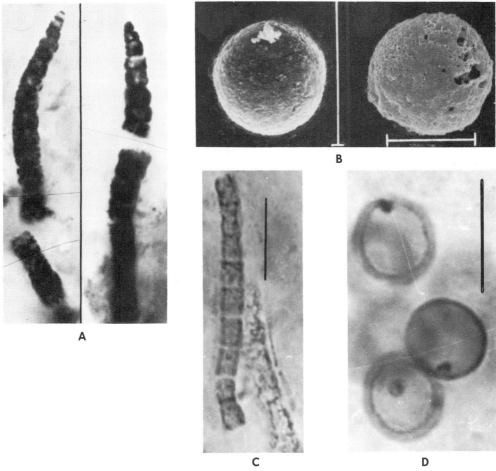

Figure 2-11 Fossil organisms discovered in the late Precambrian Bitter Springs formation (~ 1 billion years old), representing a variety of forms. a. Small, tapering filaments similar in morphology to modern filamentous blue-green algae such as *Oscillatoria* and *Nostoc.* b. A spheroidal blue-green alga strikingly similar in morphology and dimensions to the spheroidal structures discovered in the Early Precambrian Onverwacht formation (see Figure 2-7). c. Large filamentous form highly suggestive of modern filamentous blue-green algae. d. Spheroidal structures resembling extant small spherical green algae such as *Chlamydomonas.* (a from J.W. Schopf and J.M. Blacic [1971]. Journal of Paleontology 45:925–960; b from J.W. Schopf [1970], Journal of Paleontology 44:1–6; c and d from E.S. Barghoorn and J.W. Schopf, *Science*, Vol. 150, p. 338, Figs. 2 and 8, 1965. Copyright 1965 by the American Association for the Advancement of Science.)

were highly diversified by this time. In addition to blue-green algae, fossils resembling three species of bacteria, two genera of green algae (eukaryotes), and two possible species of fungi were reported.

The fossil record encompassing the last 400 million years of the Precambrian Period (1 billion to 600 million years ago) shows traces of early multicellular organisms. These include sponges, polychaete worms, sarcodines, and others. Multicellular organisms were therefore clearly established by the beginning of the Cambrian Period nearly 600 million years ago, and from then on living organisms began a period of relatively rapid evolution well documented by the Cambrian fossil record.

ORIGIN AND EARLY DEVELOPMENT OF BIOLOGICAL SYSTEMS

Evidence provided by the Precambrian fossil record, considered in relation to what we know about the structural and functional design of present-day organisms, suggests the following scenario concerning the origin and early development of biological systems. Although the scheme is highly speculative, its general outline is accepted as a reasonable reconstruction by the great majority of biologists.

Living systems may have arisen prior to 3.2 billion years ago and possibly earlier than 3.7 billion years ago. The materialistic theory of life's origin contends that the first cells were formed as a result of a long process of chemical evolution whereby simple organic compounds, synthesized abiotically on the primitive earth, obtained ever-increasing levels of complexity. In some manner, discrete morphological entities *capable of reproduction* were eventually formed. Precisely how this occurred is currently the subject of a great deal of speculation; we will consider various theories in later parts of this book. In any case, the first cells must have possessed at least the minimum metabolic capabilities guaranteeing successful reproduction, for the process of biological evolution requires not only reliable reproduction, but the production of *heritable* variation on which the pressures of natural selection may act.

Since the atmosphere of the earth probably lacked free oxygen, it seems reasonable to assume that the first cells were *anaerobic heterotrophs:* they existed in the absence of oxygen and depended for survival on organic compounds synthesized by abiotic means. Gradually, they evolved an increased ability to synthesize organic molecules from precursors available in the environment. As long metabolic pathways evolved, cells became less dependent on the environment for supplies of nutrient compounds. The evolution of these pathways was favored by natural selection, since a cell that evolved the capacity to synthesize any necessary compound could then survive without obtaining that compound as a nutrient from the environment and hence possessed a competitive advantage over metabolically less versatile cells. One of these metabolic innovations that eventually evolved was the ability to degrade glucose and to use the energy released in the process to form ATP.

A particularly significant metabolic breakthrough was the appearance of *anaerobic autotrophs:* some species of chemosynthetic bacteria and a wide range of photosynthetic prokaryotes. Photosynthesis endowed organisms with the ability to use the energy of solar radiation (light) to produce "energy-rich" ATP. Some of these photosynthetic organisms produced molecular oxygen as a by-product of the photosynthetic process. Oxygen-producing photosynthetic prokaryotes, some of which resemble modern forms of blue-green algae, were probably firmly established as early as 2.7 billion years ago, shortly before the beginning of the Middle Precambrian. The establishment of oxygen-producing photosynthesis had far-reaching effects. Eventually, the atmosphere of the earth became drastically changed as a result of the release of free oxygen into the air.

As oxygen accumulated in the atmosphere as a byproduct of photosynthesis, anaerobic prokaryotes that were not capable of surviving in the presence of oxygen were eliminated or were restricted to local environments that lacked oxygen. *Facultative anaerobes,* anaerobic organisms capable of living in either the absence or presence of free oxygen, probably became widely distributed. When the oxygen concentration in the air reached high levels, *aerobic heterotrophs* and *aerobic autotrophs* became the dominant types; these organisms required oxygen for survival.

This scheme is supported by the wide range of metabolic types found among present-day prokaryotes. These include heterotrophs; autotrophs, some of which produce oxygen as a byproduct of photosynthesis and some of which do not; strict anaerobes (killed by oxygen); facultative anaerobes; and strict aerobes (require oxygen).

About 1.5 billion years ago, eurkaryotic cells arose in some unknown way, probably from prokaryotic ones. By the time eukaryotic cells appeared, it is probable that all the basic metabolic patterns now found in present-day cells had already evolved. These eukaryotic cells, aerobic and hence adapted to survival in an oxygen–containing environment, became the structural unit of all autotrophic and heterotrophic multicellular forms.

REFERENCES

Barghoorn, E.S. (1971). The oldest fossils. *Scientific American, 224*:30 (May, 1971).
Barghoorn, E.S., and J.W. Schopf (1966). Microorganisms three billion years old from the Precambrian of South Africa. *Science, 152*:758.
Brooks, J., and G. Shaw (1973). *Origin and Development of Living Systems.* Academic Press, Inc., New York.
Calvin, M. (1969). *Chemical Evolution.* Oxford University Press, New York.
Levy, J., J.J.R. Campbell, and T.H. Blackburn (1973). *Introductory Microbiology.* John Wiley and Sons, New York.
Margulis, L. (1970). *Origin of Eukaryotic Cells.* Yale University Press, New Haven, Connecticut.
Wilson, E.O., *et al.* (1973). *Life on Earth.* Sinauer Associates, Inc., Stamford, Connecticut.

CHAPTER 3

Life and the Fitness of Chemical Elements

In our survey of the universal characteristics of living systems we observed, underlying the complexity and diversity of various cell types, a remarkable uniformity in chemical composition and metabolic design. Indeed, it is precisely this uniformity that provides strong evidence for the common evolutionary origin of living forms. We noted that all cells are surrounded by a plasma membrane and contain four main classes of organic molecules (proteins, nucleic acids, carbohydrates, and lipids), a number of smaller organic compounds, and a few inorganic substances.

Of course, it is a consequence of cellular specialization that the *specific* chemical compounds present in a cell will vary with each particular cell type, depending upon its structure and function. Thus, although all cells contain proteins, different proteins are found in different cell types. A liver cell, for example, contains specific proteins, nucleic acids, and other molecules that are not present in a red blood cell, a plant epidermal cell, a bacterial cell, or any other cell type. This is because liver cells do things that other cells do not. For instance, liver cells produce bile used in digestion, and thus they contain enzymes and other substances required for this specific task. But apart from such functional specializations, each cell possesses the basic chemical composition and metabolic organization characteristic of all other cell types.

CHEMICAL CONSTITUENTS OF CELLS

The cell's chemistry is primarily the chemistry of organic compounds and water. Water is the principal constituent of living organisms. It serves as the medium in which most cellular reactions occur and as such is essential to living systems. Anywhere from about 60 per cent to 95 per cent of the weight of active living tissue consists of water, although metabolically less active tissues such as bone contain considerably less water than this. Even seeds and spores, which in their dormant state possess an extremely low water content, require the uptake of water for germination. The increase in water triggers an enhancement of the metabolic rate, which precedes the synthesis of new cellular material and the resumption of cell division associated with germination. Because the metabolic reactions of the cell occur in an aqueous medium, the metabolism of the cell is influenced to a large extent by the specific chemical properties of water. In the next chapter, we will consider in greater detail the unique role of water in living systems.

By far the greater proportion of the *dry weight* of a cell consists of organic compounds. However, the relative concentration of the molecular constituents of different cell types is so variable that values for an "average" cell must be viewed as extremely approximate and should be accepted with considerable reservation (Tab. 3–1). For example, hemoglobin, the iron-containing protein that transports oxygen in the blood, represents about 95 per cent of the dry weight of mature human erythrocytes, whereas sperm cells and fat cells contain, respectively, very high proportions of nucleic acid (DNA) and lipids. Many plant cells contain very high levels of carbohydrates relative to other constituents.

Even within a particular cell type, the composition may vary over an enormous range depending upon the cell's physiological state or developmental stage. Thus the liver cells of rats, which contain approximately 17 per cent glycogen (dry weight) in the "normal" adult animal, may be depleted of glycogen under conditions of severe physiological stress. Values of more than 30 per cent glycogen (dry weight) are not uncommon in liver cells that are actively engaged in storing carbohydrates. During embryonic or regenerative growth of the liver, there is a relative increase in RNA, a decrease in glycogen content, and sometimes an increase in lipid levels.

ELEMENTAL COMPOSITION OF CELLS

If we examine the elemental composition of the chemical constituents of cells, we find that all cells contain primarily four ele-

Table 3–1 MOLECULAR COMPOSITION OF AN "AVERAGE" CELL

	APPROXIMATE PER CENT DRY WEIGHT
Protein	71
Lipid	12
Nucleic acid	7
Carbohydrates	5
Inorganic substances and other materials	5

Chemical Constituents of Cells

ments—hydrogen (H), carbon (C), nitrogen (N), and oxygen (O). Together these four elements constitute more than 99 per cent of the atoms present in living systems. Seven other elements are found in living tissue in smaller amounts, ranging from about 0.25 to 0.01 atoms per cent,* and are sometimes classified as "secondary" elements. These seven, in order of their abundance in mammals, are calcium (Ca), phosphorus (P), sulfur (S), sodium (Na), potassium (K), chlorine (Cl), and magnesium (Mg).

Fourteen other elements are known to be present in either plants or animals, but are needed only in trace amounts (less than 0.001 atoms per cent). Some of these "trace elements," such as manganese (Mn), iron (Fe), cobalt (Co), copper (Cu), zinc (Zn), and molybdenum (Mo), have been shown to be present in a wide variety of plants and animals, and are probably universally required for life. Others, including boron (B), silicon (Si), vanadium (V), chromium (Cr), fluorine (F), selenium (Se), tin (Sn), and iodine (I), have been proved to be essential in at least one species of plant or animal, but whether they have wider importance has not yet been resolved.

It is quite probable that other trace elements necessary for life will be discovered in the near future. For although it is in theory simple to test an organism's requirement for a trace element, in practice it is exceedingly difficult to isolate the organism effectively from a source of the element. Often an organism needs such minute amounts of a trace element that it may obtain sufficient quantities as contaminants on the walls of the vessel in which it is contained, in the nutrients and water it is being given, or even in airborne dust particles. Only recently have fluorine, silicon, tin, and vanadium been shown to be necessary trace elements in some organisms, and this was achieved only by raising rats and other small animals in rigidly controlled, contaminant-free environments.

The Primary and Secondary Elements

It is possible to make some useful generalizations concerning the functions, relative distributions, and importance of the various elements present in living organisms. All biologically important organic molecules contain hydrogen, carbon, and oxygen as their principal elements. In addition to these three elements, nitrogen and phosphorus are also present in nucleotides, the basic subunits of nucleic acids. All amino acids, the molecular building blocks of protein, also contain nitrogen, and a few possess sulfur. Phosphorus, nitrogen, or sulfur, or combinations of these, are occasionally encountered as additional components of carbohydrates and lipids. These latter three elements are also important constituents of many vitamins and other relatively small molecules such as coenzymes,

---

Atoms per cent is defined as the number of atoms of a particular element per 100 total atoms.

non-proteinaceous substances that are necessary for the activity of certain enzymes.

Many of the elements in living systems are derived from inorganic substances that exist dissolved in the aqueous matrix of the cell in the form of charged atoms or groups of atoms termed *ions*. The ions of major importance in the cell are H^+ (hydrogen ion), Na^+ (sodium ion), K^+ (potassium ion), Ca^{++} (calcium ion), Mg^{++} (magnesium ion), Cl^- (chloride ion), $SO_4^=$ (sulfate ion), $PO_4^=$ (phosphate ion), OH^- (hydroxide ion) and HCO_3^- (bicarbonate ion).

In any particular cell type, the ionic composition and ionic balance (relative proportions of different ions) are normally maintained at extremely constant values, although these values may vary somewhat in different cell types, even within the same organism. Even minor disturbances in ionic balance may result in cell damage or death.

The Trace Elements

Although the trace elements are required only in minute amounts, they are nevertheless absolutely essential to living systems. Most of the trace elements appear to be required for the activity of enzymes, or as components of metabolically important compounds such as hemoglobin (iron-containing), vitamin B_{12} (cobalt-containing), and thyroid hormones (iodine-containing), as well as others. A wide variety of plant and animal diseases are associated with deficiencies of certain trace elements; many such human diseases are prevalent even today. Lack of sufficient iron in the diet is a frequent cause of anemia, whereas iodine deficiency is associated with thyroid malfunction and consequent developmental and metabolic abnormalities. Until the recent widespread use of iodized table salt, thyroid malfunction was not uncommon in inland areas such as Switzerland and the American Midwest, where seafood, a rich source of iodine, is not generally available.

Chemical Selectivity of Living Organisms

In all, only some 25 elements have been shown to be used by living organisms, and of these the most frequently used are among the lighter elements. If we compare the relative amount of each element in the soft tissues of living organisms with the abundance of each element in the universe, in the earth's crust, and in sea water (Tab. 3-2), it becomes apparent that living organisms do not merely use those elements that are most available in the environment. Rather, during the course of biological evolution certain elements were selected for and others were rejected.

Why are these particular elements, and only these elements, present in the building blocks of living matter? The theory of evolution suggests to us that those elements which in some way endowed

Chemical Constituents of Cells

Table 3-2 RELATIVE ABUNDANCE OF ELEMENTS (ATOMS PER 100 ATOMS)

ELEMENT	UNIVERSE*	EARTH'S CRUST†	SEAWATER†	SOFT TISSUES OF LIVING ORGANISMS
Hydrogen	90.79	2.91	66.27	62.61
Helium	9.08	trace	trace	absent
Carbon	0.021	0.035	0.001	10.60
Nitrogen	0.042	0.003	trace	1.14
Oxygen	0.057	60.62	33.13	24.92
Sodium	trace[a]	2.50	0.27	0.027
Magnesium	0.002	1.79	0.033	0.019
Aluminum	trace	6.25	trace	absent
Silicon	0.003	20.62	trace	trace
Phosphorus	trace	0.071	trace	0.23
Sulfur	trace	0.017	0.017	0.28
Chlorine	trace	0.008	0.32	.028
Potassium	trace	1.38	0.006	.002
Calcium	trace	1.90	0.006	0.39
Iron	0.005	1.85	trace	trace

*Data recalculated from Edsall, J.T., and J. Wyman (1958). *Biophysical Chemistry,* Vol. I. Academic Press, Inc., New York.
†Data recalculated from Mason, B. (1966). *Principles of Geochemistry.* John Wiley and Sons, New York.
[a]Trace is <0.001 per cent.

living organisms with increased reproductive fitness would have been established in living systems through the process of natural selection. The reason for the selection of certain elements must somehow be related to the chemical properties of the individual atoms and of the compounds that these elements can form, for these elements do not exist in living systems as neutral atoms but rather as ions or as components of molecules. The chemical properties of particular atoms are, in turn, largely dependent upon their electronic structure. Thus, in order to understand the basis for the selectivity of living systems, we need to examine atomic structure in more detail than we did in Chapter 2, and we need to understand the relation between atomic and molecular structure.

ATOMIC AND MOLECULAR STRUCTURE

ATOMIC ORBITALS

We have mentioned previously that neutral atoms of each element possess a characteristic number of electrons which exist as negatively charged particles moving at extremely high velocities around the positively charged nucleus. In recent years, the development of the Quantum Theory has modified our view of atomic structure and has permitted us to describe the energies and positions of electrons in an atom.

The Quantum Theory tells us that electrons are most likely to be found in regions in space around the nucleus called *atomic orbitals;* each atomic orbital can accommodate two electrons. Within the region defined by an atomic orbital, there is *not* an equal probability of finding an electron in any one position at a given time. This means that if we observe the position of an electron in a particular orbital at a number of different times, we will find the electron more frequently in certain positions than in others. Consequently, atomic orbitals are often represented by shaded diagrams that indicate *electron density* in the orbital; the greater the probability of finding an electron in a particular position within the region defined by the orbital, the greater the intensity of shading in that position (Fig. 3–1). Note, however, that atomic orbitals and their electron-density diagrams indicate only the most *probable* location for electrons and *not* precise locations.

The shaded electron-density diagrams also represent the geometric shape of atomic orbitals, which are designated according to their shape as *s, p, d,* or *f.* For example, *s* orbitals are spherical, *p* orbitals possess a distorted dumbbell shape, and so on (Fig. 3–2).

Thus, according to the Quantum Theory, the atoms of each element are thought of as possessing a characteristic number of atomic orbitals in which electrons are permitted to exist. Taken together, these orbitals describe the most probable locations of electrons surrounding the nucleus of an atom. It is important to remember that atomic orbitals are a property of atoms and exist even if no electrons are present in them. The analogy to flight pat-

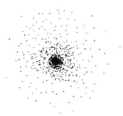

Figure 3–1 Electron density diagram of an atomic orbital. The density of the shading is proportional to the probability of finding an electron in that area of the orbital. The black dot represents the nucleus.

Atomic and Molecular Structure

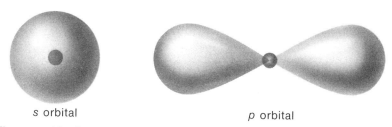

s orbital   p orbital

Figure 3-2

The geometric shape of s orbitals and p orbitals. The central dark region represents the nucleus.

terns around an airport is a good one for describing atomic orbitals: the allowed flight patterns exist even if no airplanes are flying within them.

ENERGIES OF ELECTRONS

From the preceding discussion, it should be apparent that an atomic orbital may be defined not only in terms of its geometric shape but also in terms of its distance from the nucleus. Hence a particular orbital may be thought of as describing the most probable distance, or the average distance, that an electron in this orbital is located from the atomic nucleus. This is an important consideration, since electrons in orbitals whose areas of greatest electron density are at a large distance from the nucleus will contain, on the average, greater potential energy relative to the nucleus than electrons in orbitals with electron-dense areas closer to the nucleus. This may be explained in terms of the attraction of a negatively charged electron to the positively charged nucleus. The further an electron is from the nucleus, the greater the energy it would release if it were to fall to the innermost orbital. This situation is analogous to the potential energy of a weight which is raised to discrete levels above the surface of the earth. The further the weight is raised, the greater its potential energy relative to the surface of the earth. The average distance of an electron from the nucleus of an atom is described by its *principal quantum number,* which can have positive integral values of 1, 2, 3, 4, and so forth. The principal quantum number is thus an important indication of the energy of an electron.

Atomic orbitals, then, may be described in terms of both their shape (s, p, d) and a principal quantum number (1, 2, 3). In considering orbitals *of the same shape,* the lower the principal quantum number, the lower the energy of an electron in that orbital. For example, electrons in 1s orbitals are *on the average* closer to the nucleus than those in 2s orbitals, and therefore possess lower energies. It is also true that for orbitals of any one principal quantum number, electrons in s orbitals have lower energies than those in p orbitals, which have lower energies than those in d orbitals, and so on.

The energy levels of atomic orbitals in atoms possessing more than one electron are shown in Figure 3-3. Each black dot in the diagram represents an orbital which can hold two electrons; thus there is one 1s orbital, one 2s orbital, three 2p orbitals, and so on. Each of the three 2p orbitals is shown to occupy an equivalent energy level. This is possible because p orbitals, unlike s orbitals, which

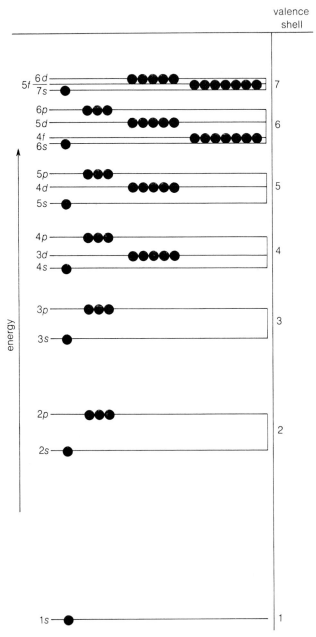

Figure 3-3 Orbital energy levels in multielectron atoms. Each atomic orbital is represented by a black dot and is capable of holding two electrons. (*From* R.G. Gymer [1973], *Chemistry: An Ecological Approach*, Harper and Row, Publishers, New York.)

Atomic and Molecular Structure

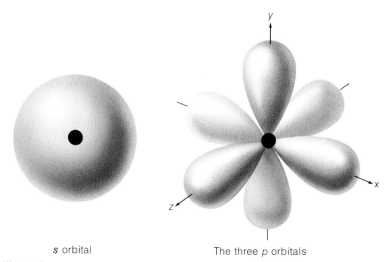

s orbital   The three p orbitals

An illustration of the differences in the three-dimensional character of s and p orbitals. The s orbitals are spherically symmetrical, whereas the three p orbitals of any one energy level are located at right angles to one another relative to the nucleus.

Figure 3-4

possess spherical symmetry, display directional properties relative to the nucleus. Thus, the three p orbitals lie at right angles to one another along imaginary x, y, and z axes through the nucleus (Fig. 3-4). The d and f orbitals also show directional properties, and there are, respectively, five and seven equivalent orbitals for each of these shapes. Also notice in Figure 3-3 that some of the energy levels, such as those for 2s and 2p, those for 3s and 3p, and those for 4s, 3d, and 4p, form groups with very similar energies, and that these groups are separated from each other by considerably larger energy differences. Groups of orbitals that possess similar energy levels are known as *valence shells* of an atom. We will see shortly that it is the electrons in the *outermost* valence shell of an atom that are particularly important in chemical bonding.

Under normal conditions, an atom, like any physical system, will exist in the state of lowest possible energy, also known as its *ground state,* simply because it is most stable in this form. Hence the arrangement of electrons in an atom in its ground state will be that of minimal energy. Consequently, in multielectron atoms in their ground state, *energy levels are, in general, filled with electrons in such a way that the lowest energy level is filled to capacity with electrons before the next higher level is used.* Thus, for example, nitrogen, which possesses seven electrons in its ground state, has a completed 1s orbital (two electrons), a completed 2s orbital (two electrons), and one electron in each of its three 2p orbitals.

Where energy levels are being filled that possess almost identical energies, as in the case of the 4s and 3d orbitals, it is not always possible to predict the electron configuration with accuracy. This is because electrons in the various energy levels exert forces on one another, and when we are considering energy levels with very similar energies, these forces become extremely important in determining the electron configuration of lowest energy.

CHEMICAL BONDING

Under certain conditions, atoms can combine with one another in relatively stable combinations to form *molecules;* molecules containing atoms of two or more different elements are termed *compounds*. The attractive force that effectively holds two atoms in a molecule together is known as a *chemical bond*. Since it is a general rule that physical systems tend toward a state of minimal possible energy, the essence of chemical bond formation is the decrease in energy that occurs when two atoms unite. In practical terms, then, the energy of a stable combination of atoms must be less than the total energies of the separate atoms.

As we mentioned earlier, the tendency of an atom to combine with another atom by the formation of a chemical bond between them, also known as the *chemical reactivity* of the atoms, is dependent almost exclusively on the arrangement of electrons around the nucleus of the interacting atoms. This is because the interaction between two atoms involves rearrangements of the electron clouds of each atom in response to the other. Indeed, such rearrangements involve primarily the electrons in the outermost valence shell of atoms. These outermost electrons are less tightly bound to the atom than are inner electrons, simply because of their increased distance from the nucleus. For this reason, the electrons in the valence shell are influenced to a greater degree than inner electrons by the attractive forces of another atomic nucleus in close proximity, and therefore these electrons are the ones that are primarily involved in the electron cloud rearrangements associated with bonding. Bonding, however, does not necessarily occur between any two atoms, but occurs only if the electron configuration of two atoms in close proximity is such that the two electron clouds can rearrange so as to minimize the energy of the system.

The rearrangement of electrons in atoms near each other occurs, then, because the electrons of each atom may be attracted to the nucleus of the other atom as well as to its own nucleus. In some extreme cases, electrons of one atom may be so strongly attracted to another atom as to be detached completely. For discussion purposes, it is convenient to separate the electron interactions involved in chemical bond formation into two main types: (1) those in which electrons are shared between two atoms (*covalent bonding*), and (2) those in which electrons are transferred from one atom to another (*ionic bonding*). As you will see from the following discussion, however, this division is in many respects artificial, since the great majority of electron interactions in chemical bonding involve neither complete sharing nor complete transfer of electrons, but rather some intermediate condition.

Atomic and Molecular Structure

Covalent Bonds

The formation of covalent bonds between atoms will concern us most, since it is these bonds that are most frequently encountered in the molecules present in living systems. Hydrogen, carbon, oxygen, nitrogen, phosphorus, and sulfur—the main atomic constituents of the organic molecules of cells—are involved primarily in covalent bonding.

The mechanism of covalent bond formation is well illustrated by the formation of a molecule of hydrogen gas from the reaction between two neutral hydrogen atoms. Each hydrogen atom possesses only one electron located in the 1s orbital. As two hydrogen atoms approach one another, the electron of each atom is attracted to the nucleus of the other. The result of this mutual attraction is a reduction in potential energy as the atoms come closer together, and an eventual overlap between the electron clouds of the two atoms. The atoms will continue their approach to one another, with increased overlap of electron clouds, until the repulsive forces between the two nuclei predominate. At the position of maximum approach, the energy of the combined atoms is less than that of the two isolated atoms, and they will thus tend to remain in this position. In this configuration, the two 1s orbitals may be thought of as merging into one larger orbital, termed a *molecular orbital,* in which the two electrons are strongly attracted to *both* nuclei. In the molecular orbital, the region between the two nuclei formed by the overlap of the two electron clouds contains a total electron density that is higher than in other regions of the orbital. Therefore, both electrons are, in a sense, shared between the two atoms, and this sharing of electrons is called a covalent bond (Fig. 3–5).

A single covalent bond is formed between two atoms by the mutual sharing of one pair of electrons. Once a covalent bond is formed between two atoms, it may be broken only by supplying an amount of energy, often called the *dissociation energy,* that is equal to the difference in the potential energies of the combined and uncombined atoms.

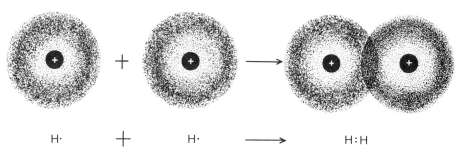

Figure 3–5

In covalent bonding between two hydrogen atoms, the overlap of two atomic orbitals forms a molecular orbital in which there is a concentration of electron density in the region between the two nuclei.

The shared pair of electrons that constitutes a covalent bond between two atoms is sometimes represented by two dots placed between the chemical symbols of the two atoms. Thus, two covalently bonded hydrogen atoms may be represented in shorthand as H:H. Often, however, a line is used in place of dots to indicate a shared electron pair, and any unshared pairs are not shown. The hydrogen molecule is thus represented as H—H. When the type of bonding is understood, the notation is often further simplified to H_2.

The formation of covalent bonds in molecules that are composed of more than two atoms is similar in principle to their formation in a simple diatomic molecule such as molecular hydrogen. Each covalent bond in the molecule requires a pair of shared electrons and two overlapping atomic orbitals, one from each atom.

Coordinate Covalent Bonds

We have represented the formation of a covalent bond by the overlap of two atomic orbitals from different atoms, and we have indicated that each of these orbitals contributes one electron to the bonding electron pair. Although in this example an electron is contributed by each atom, this need not necessarily be the case. Since atomic orbitals exist about an atom even if there are no electrons in them, a covalent bond may alternatively be formed by the overlap of an empty atomic orbital from one atom with an orbital from another atom which is filled with two electrons. In theory, bonding of this type may occur between any atoms that possess an unshared pair of electrons in their valence shell and atoms that have an empty orbital in their valence shells, but it most commonly occurs only with certain elements. Although the source of the bonding electron pair is only one of the atoms, the filled molecular orbital is identical in both cases, and the covalent bonds are indistinguishable regardless of the source of the electrons. However, in certain instances we shall encounter shortly, it is useful to indicate how a particular bond was formed. Consequently, when both electrons in a bonding pair come from one atom, the bond is called a *coordinate covalent bond.*

Multiple Bonding

Two bonded atoms may sometimes share more than one pair of electrons. This so-called *multiple bonding* may involve the sharing between two atoms of either two electron pairs *(double bonding)* or three electron pairs *(triple bonding)*. Multiple covalent bonds form in the same way as single covalent bonds; that is, they require available orbitals and electrons. We can think of a double bond, for example, as requiring two overlapping orbitals from each atom and two pairs of electrons. Oxygen, nitrogen, carbon, phosphorus, and sulfur, five of the six most common elements found in living systems

Atomic and Molecular Structure

(the other is hydrogen), are the main elements that participate in multiple bonding. We will see shortly that this ability to form multiple bonds is one property that helps to explain the unique fitness of these particular elements for living systems.

An example of double bonding is in the carbon dioxide (CO_2) molecule, where each of the two oxygen atoms is attached to the single carbon atom by a double bond. Each double bond requires the overlap of two oxygen orbitals with two orbitals from the carbon atom, and each of the overlapping orbitals accommodates two electrons. In chemical notation, a double bond may be represented by four dots between the two bonded atoms. Other electrons in the valence shell that do not participate in bonding are indicated by additional dots surrounding each atom. Thus carbon dioxide may be written

$$:\ddot{O}::C::\ddot{O}:$$

or, as in single covalent bonding, may be represented more simply by replacing each bonding pair of electrons with a line between the bonded atoms.

$$O=C=O$$

Again notice that in this type of notation, unshared electrons are not usually indicated.

The diatomic nitrogen molecule is an example of a molecule that is held together by a triple bond. This bond requires three orbitals from each atom that overlap with one another and three pairs of electrons. The triple bond in N_2 may be shown as

$$:N:::N: \quad \text{or} \quad N\equiv N$$

Bond Energies

Earlier, we defined a quantity called dissociation energy, which is equal to the energy needed to separate two bonded atoms in a diatomic molecule, such as H_2 or N_2. From the opposite point of view, we can see that the dissociation energy is a measure of the strength of the chemical bond between two atoms. In molecules of more than two atoms (*polyatomic* molecules), the energy required to break a bond between two particular types of atoms varies somewhat with the overall chemical structure of the molecule. Consequently, we speak of *bond energies* in polyatomic molecules rather than dissociation energy; bond energies represent approximate (or average) quantities of energy necessary to break a chemical bond between two particular atoms. Examples of bond energies for bonds joining certain atoms are given in Table 3–3. Notice that multiple bonds between the same atoms are stronger than single

Table 3-3 BOND ENERGIES OF SELECTED CARBON BONDS

BOND	ENERGY (kcal/mole[a])
C—C	83
C=C	148
C≡C	194
C—H	99
C—N	66
C—O	82

[a] In this case, 1 mole refers to 6.02×10^{23} bonds.

bonds, but that a double bond is not twice as strong, nor is a triple bond three times as strong, as a single bond.

Geometry of Molecules

Because molecular orbitals are formed by the overlap of atomic orbitals and because atomic orbitals are defined by their positions in space, a predictable geometric relationship must exist between atoms in a polyatomic molecule. Consider, for example, the water molecule (H_2O), in which bonding occurs by the overlap of the $1s$ orbitals of each of the hydrogen atoms with a separate $2p$ orbital of

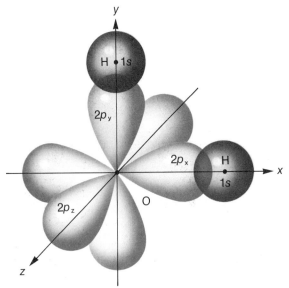

Figure 3-6 The overlap of orbitals in the water molecule. The $1s$ and $2s$ orbitals of oxygen are not shown. The $2p$ orbital is filled with a pair of electrons and is consequently not available for bonding.

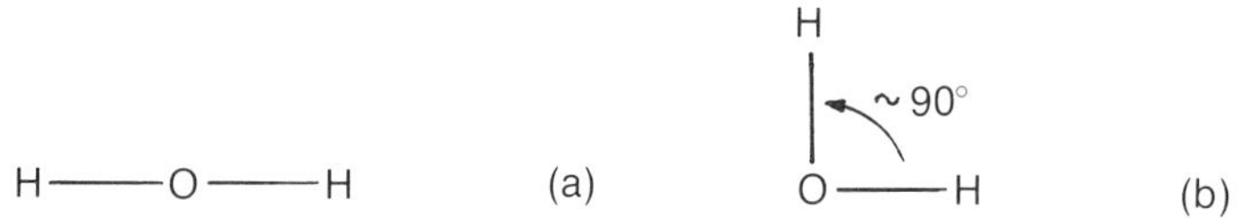

Atomic and Molecular Structure

H——O——H (a)

H
| ~90°
O——H (b)

Figure 3-7

a. A hypothetical linear configuration of water which does not occur because of the geometric constraints of orbital overlap. *b.* An H—O—H bond angle of about 90° would be predicted for the water molecule on the basis of orbital overlap. In fact, because of interactions with other electrons in the molecule, the bond angle is actually 104.5°.

oxygen. We have noted before that *s* orbitals are spherical, but that *p* orbitals are dumbbell-shaped and are located in space at right angles to one another. Thus, to obtain maximum overlap, the hydrogen atoms must approach the oxygen atom on the perpendicular axis of the orbitals (Fig. 3–6). We would therefore expect that the water molecule would not have a linear configuration (as in Fig. 3–7*a*), but that the bonds connecting the hydrogen atoms to the oxygen atom would form an angle (*bond angle*) of approximately 90° (Fig. 3–7*b*).

Actually, the bond angle in water is 104.5°, rather than the expected 90°, because of interactions between the bonding electrons and other electrons in the molecule; we will discuss these interactions in Chapter 4 when we consider the structure of water in greater detail. But it is nevertheless clear from this example that the shape of a molecule depends upon the electronic configurations of its constituent atoms.

The geometry of molecules is important in understanding the metabolic activities of the cell, since chemical reactivity *between* molecules depends partly upon the shape of the molecules. Thus, for example, a very large enzyme molecule in a living cell may completely lose its ability to catalyze a particular metabolic reaction simply because its geometry has been altered by the rearrangement of one or two atoms.

Although chemical structures are usually represented on paper as two-dimensional figures, molecules composed of more than three atoms are not necessarily planar. For instance, in the chemical compound methane (CH_4), normally written as

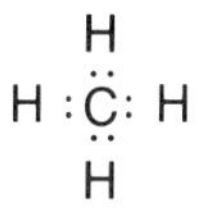

$$H:\overset{..}{\underset{..}{C}}:H \quad \text{with } H \text{ above and below}$$

or

$$H-\underset{H}{\overset{H}{\underset{|}{\overset{|}{C}}}}-H$$

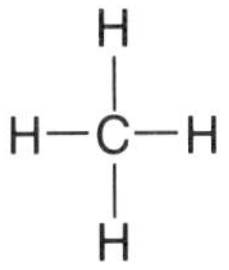

the hydrogen atoms are located about the central carbon atom at the corners of a regular tetrahedron, with H—C—H bond angles of

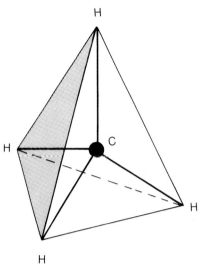

Figure 3-8 The three-dimensional configuration of methane (CH₄). The carbon atom is located at the geometrical center of a regular tetrahedron.

109° (Fig. 3–8). Methane, like many other organic compounds, thus possesses a defined non-planar structure.

Polar Molecules

We have pictured covalent bond formation in the diatomic hydrogen molecule as resulting from equal sharing of a pair of electrons between the two atoms. In addition, the entire molecule is electrically neutral, since it contains two protons and two electrons. In other molecules, however, two atoms involved in a covalent bond may not share the bonding electrons equally between them. That is, one of the atoms may tend to draw the electron pair away from the other atom, so that the shared electrons are, *on the average,* closer to one atom than to the other. Thus, although the entire molecule is still electrically neutral, there is a separation of charge *within* the molecule because the atom that tends to attract the bonding electrons will possess a slight negative charge relative to the other atom in the bonding pair. This separation of charge between two bonded atoms is called a *bond dipole moment.*

The tendency of an atom to attract the bonding electrons in a covalent bond is termed *electronegativity;* the greater the electronegativity of an atom, the greater is this attracting power. Oxygen, nitrogen, and chlorine are the most electronegative elements commonly found in living systems. Electronegativity values for some elements important to living systems are given in Table 3–4.

As a consequence of the geometric shape of molecules, electronegativity differences between atoms in covalently bonded compounds may, *but need not,* result in molecules that possess a negative and a positive end. Such molecules are called *polar molecules;*

Atomic and Molecular Structure

Table 3-4 ELECTRONEGATIVITIES OF ELEMENTS IMPORTANT TO LIVING SYSTEMS

ELEMENT	ELECTRONEGATIVITY
K	0.8
Na	0.9
Ca	1.0
Mg	1.2
P	2.1
H	2.1
C	2.5
S	2.5
N	3.0
Cl	3.0
O	3.5

non-polar molecules are those that do not possess a negative and a positive end.

Water is an excellent example of a polar compound. The large electronegativity of oxygen causes the electrons that are shared between oxygen and hydrogen to spend most of their time closer to the oxygen atom than to the hydrogen atom. Because of the geometric shape of the water molecule, this charge distribution results in the formation of a polar molecule, with the oxygen end of the molecule possessing a net negative charge and the hydrogen end possessing a positive charge (Fig. 3-9). Carbon tetrachloride (CCl_4), on the other hand, is non-polar. Like methane, carbon tetrachloride is a tetrahedrally shaped molecule (Fig. 3-10). Because of the difference in electronegativity between carbon and chlorine, each of the C—Cl bonds has a dipole moment. However, the geometric shape of carbon tetrachloride is such that all the bond moments cancel and consequently the molecule *as a whole* is non-polar.

The distribution of charge in the water molecule. As a result of its geometric shape and the electronegativity differences between hydrogen and oxygen, water is a polar molecule.

Figure 3-9

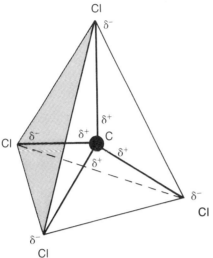

Figure 3-10 The carbon tetrachloride molecule is non-polar because it possesses geometric symmetry. Bond dipole moments resulting from electronegativity differences between carbon and chlorine will *not* result in a molecule with positive and negative ends.

Ionic Bonds

As the difference in electronegativity between two bonded atoms increases, the bonding pair is distorted more and more toward the more electronegative atom, and the character of the bond departs increasingly from pure covalency (or equal sharing of electrons). In the very extreme case, electrons in one atom are so strongly attracted by another atom that one or more electrons will actually be transferred to the other atom, creating a positively charged ion (*cation*) and a negatively charged ion (*anion*). The attraction between the atoms is then purely electrostatic in nature, and the bonding that results from this type of attraction is termed an *ionic bond*.

Solid sodium chloride (NaCl) is a good example of a compound held together by ionic bonds. The neutral sodium atom, which possesses only one electron in its outer electron shell, tends to lose this electron, forming a positively charged sodium ion

$$\cdot Na \rightarrow Na^+ + 1e$$

The chlorine ion, on the other hand, which in its neutral state possesses seven electrons in its valence shell, has a high tendency to accept an electron

$$\cdot \ddot{C}\ddot{l}\colon + 1e^- \rightarrow Cl^-$$

The positive sodium ion and the negative chloride ion may thus be held together by electrostatic attraction. It should be noted that although we have represented ion formation as a two-step process, the electron that is removed from sodium is best thought of as being drawn to the chlorine atom in a one-step process.

Ionic molecules such as sodium chloride may be represented in chemical shorthand in a number of ways. One of these depicts the

Atomic and Molecular Structure

configuration of electrons in the outer valence shell of each atom; for example

$$\text{Na}:\ddot{\text{Cl}}:$$

Sometimes only the charge on each atom is noted, as in

$$\text{Na}^+ \text{Cl}^-$$

More often, it is represented simply as NaCl when the type of bonding between the atoms is clearly understood.

Ionic Character

We have seen that covalent bonds may depart in varying degrees from pure covalency, or equal sharing of bonding electrons. Sometimes the term "amount of ionic character" is used to express the degree of deviation of a covalent bond from pure covalency.

The amount of ionic character of molecular bonds is an important factor in determining the behavior of a molecule when it is dissolved in water. This is an important consideration since, as we have repeatedly stressed, metabolic reactions in the cell occur in an aqueous medium. Molecules that are held together by purely ionic bonds separate almost completely into ions when dissolved in water; molecules that are held together by purely covalent bonds show virtually no tendency to ionize in solution.* In general, the more ionic the bond, the more readily ionization will occur. Sodium chloride thus ionizes almost completely in aqueous solution; the covalent bonds of methane show virtually no ionic character, and consequently methane displays virtually no ionization in aqueous solution. Ionization in water occurs because the negative ions in an ionic compound (or negative regions of a covalently bonded polar molecule) are attracted to the positive ends of polar water molecules, whereas the positive ions (or positive regions of a polar molecule) are attracted to the negative ends of water molecules. This process is referred to as *hydration,* and it occurs because it releases sufficient energy to break the chemical bonds holding the molecule together. The net result of hydration is that ions in water are surrounded by a layer of water molecules; that is, the ions are *hydrated.*

Weak (Secondary) Interactions Between Atoms in Living Cells

The covalent bonds that hold organic molecules together are *strong chemical bonds,* having bond energies ranging from approxi-

---

*Strictly speaking, ionic molecules are said to *dissociate* when dissolved in water since ions already exist in the solid aggregate; covalent molecules are said to *ionize.* In practice, the two words are often used interchangeably, and we will use the word "ionization" to denote both processes.

mately 40 to 200 kcal/mole.* In addition to these, there are a number of *weak interactions* between atoms, sometimes referred to as *secondary forces,* which are particularly important to biological systems. The secondary forces may occur between atoms in the same molecule or between atoms in different molecules. Thus not only are these weak interactions primarily responsible for stabilizing the specific three-dimensional shape of large molecules but they also influence the distribution of molecules in the cell. Although the cell is a highly complex entity, there is nothing magical about the structure it assumes. The arrangement of the various molecular components of the cell—what we call the structure of the cell—is determined mainly by secondary forces between atoms in separate molecules.

Ionic Bonds in Aqueous Solutions. One important weak interaction that occurs within cells is the ionic bond. Although the ionic bonds that hold an inorganic crystal together are exceedingly strong bonds, ionic bonds tend to be weak *in aqueous solution.* This is because the water molecules forming the hydration layer around ions partially shield oppositely charged ions from one another, and consequently the strength of the bond between the ions is greatly diminished. Ionic bonds in aqueous solution only have average bond energies of approximately 5 kcal/mole, and are hence much weaker than covalent bonds.

Hydrogen Bonding. Another type of weak interaction that is particularly significant within living cells is the *hydrogen bond,* which results from polar attraction between molecules. In this bond, a single hydrogen atom that is covalently bonded to a highly electronegative atom forms a weak chemical bond to another highly electronegative atom on the same or a different molecule (Fig. 3–11a). For the formation of a hydrogen bond, the highly electronegative atom normally must be nitrogen or oxygen. Since this bond results from polar attraction between atoms, it is not a strong bond, requiring on the order of 3 to 7 kcal per mole for its disruption.

Hydrogen bonding is responsible for much of the specific three-dimensional configuration of nucleic acids, proteins, and many other molecules of biological importance, and is thus of general significance in the metabolic reactions of the cell (Fig. 3–11b and c). Water forms hydrogen bonds very readily with other water molecules as well as with many other compounds. We will see in the next chapter that this ability to form hydrogen bonds endows water with many physical and chemical properties that make it uniquely fit as the solvent of living systems.

Ion-Dipole Interactions. An additional type of weak interaction frequently encountered within cells is the attraction between ions and polar molecules, as illustrated by the hydration of ions, discussed

---

*One kcal (kilocalorie) is equal to 1000 calories; a calorie is the amount of heat energy required to raise one gram of water from 14.5 to 15.5°C.

One *mole* refers here to 6.02×10^{23} bonds; see Chapter 4 for a more detailed definition.

Atomic and Molecular Structure

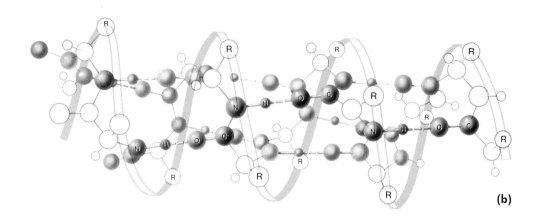

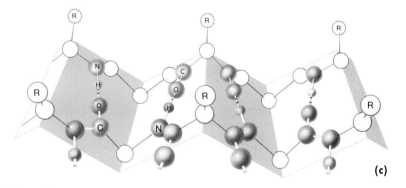

a. A general representation of hydrogen bond formation. X and Y, which may be either on the same or a different molecule, must be highly electronegative atoms, usually N or O. The hydrogen bond is represented by the dotted line. b. Hydrogen bonding between atoms in a single molecule may be responsible for specific configurations. This drawing represents a portion of a protein molecule which possesses a helical configuration (α-helix) due to hydrogen bonding. The ribbon emphasizes the turns of the helix; R represents a side group. (Adapted from P. Doty [1957], *Scientific American* 197:176, September, 1957.) c. Hydrogen bonds between separate protein molecules can bind the molecules together; the so-called β-configuration found in silk fibers is represented here. (Adapted from P. Doty [1957], *Scientific American* 197:176, September, 1957.)

Figure 3-11

above. As we shall see later, the solubility of ionic compounds in water is a result of these forces.

Hydrophobic Bonds

Hydrophobic bonding refers to the tendency of non-polar molecules, or non-polar regions of molecules, to associate (aggregate) in an aqueous environment so that their contact with water molecules is minimized. The name hydrophobic (hydro = water; phobic = hating) derives from the behavior of non-polar compounds in water. *Hydrophobic bonds,* although they are weak bonds with energies of less than one to about 3 kcal/mole, *are believed to be the most important forces stabilizing the specific configuration of large molecules.* In Chapter 7, we will discuss the role of hydrophobic bonding in determining the configuration of protein molecules.

CHEMICAL NOTATION: SOME PRACTICAL RULES

Appendix II discusses some practical rules for writing chemical structures of organic and inorganic compounds. Students not familiar with bonding or chemical notation should refer to these rules.

THE SUITABILITY OF H, C, N, O, P, AND S FOR THE MOLECULES OF LIVING SYSTEMS

The preceding discussion of atomic structure and chemical reactivity provides us with a basis for understanding why living systems are composed primarily of hydrogen (H), carbon (C), nitrogen (N), oxygen (O), phosphorus (P), and sulfur (S), and why the chemistry of life is based upon the chemistry of carbon-containing compounds. Since these elements are not simply those most available in the environment, we believe that they were selected during the course of evolution because of their particular suitability for the roles they play in living systems.

Living cells are characterized by a high level of complexity and organization. So that this ordered condition is not easily destroyed, it is essential that the molecular components of living systems be particularly stable. On the other hand, these molecules must not be so stable as to be inert, since living systems are dynamic entities and must possess the flexibility to respond to environmental challenges.

STRUCTURAL ROLES

Hydrogen, carbon, nitrogen, and oxygen, the most abundant elements in living systems, possess chemical characteristics that impart stability to the molecules they form. This stability is due to the fact that these four elements are among the smallest of all ele-

Molecules of Living Systems

ments. Small atoms, because of their electronic configurations and number of electron shells, generally form stronger chemical bonds than do heavier atoms. We know that hydrogen, for example, normally possesses only one electron; C, N, and O each possess only two electrons (a completed $1s$ shell) beneath their valence shells. In large atoms, the presence of many non-bonding paired electrons beneath the valence shell creates repulsive forces which weaken any bond between atoms. Small atoms, on the other hand, because of the small number of electrons beneath their valence shells, have minimum repulsive interference of this type. Bonding atomic orbitals can therefore overlap to a greater degree and thereby form a stronger bond.

Another factor that may increase bonding stability is the fact that O, N, and C, along with P and S, are the primary elements that show a tendency to form multiple bonds. Multiple bonds are stronger than single bonds and thus impart great stability to molecules in which they occur.

Aside from its ability to form strong bonds, carbon possesses other characteristics that make it uniquely fit as a "backbone" for biological molecules. Since C—C single bonds have bond energies that are about equal to single C—O, C—N, and C—H bonds (see Table 3–4), there is very little tendency for C—C bonds to break down preferentially, and for the carbon to form stronger and more stable bonds with oxygen, nitrogen, or hydrogen. Consequently, carbon atoms will react with other carbon atoms to form long-chained or branched molecules of great stability. The stability of carbon-containing compounds is further enhanced by the fact that carbon reacts with other atoms in such a way as to possess no unshared electron pairs in its valence shell. Hence carbon compounds cannot form coordinate covalent bonds, and are therefore relatively inert to reaction with other substances.

METABOLIC ROLES

Just as H, C, N, and O possess properties that make them particularly suited for structural compounds, P and S possess different properties, which allow them to play important roles in energy storage and utilization within the cell. Because of electrons underlying the valence shell, phosphorus (atomic number = 15) and sulfur (atomic number = 16) form weaker bonds than do hydrogen, carbon, nitrogen, or oxygen. Consequently, these bonds can be more easily broken in the metabolic reactions of the cell than bonds involving the latter elements. By reactions of this type, part of the potential energy present in these compounds can be released and utilized by the cell without destroying the structural molecules that make up the organism.

The instability of compounds containing phosphorus and sulfur relative to those composed of H, C, N, and O is due not only to the larger size of the phosphorus and sulfur atoms but also to the fact that under certain circumstances electrons in valence shell orbitals of phosphorus and sulfur can be rearranged so that additional empty orbitals (3d orbitals) become available for bonding. These empty orbitals permit the formation of additional covalent bonds (coordinate covalent bonds) by the donation of unshared electron pairs by other compounds, such as water. The formation of these additional bonds results in instability of the phosphorus- or sulfur-containing compounds, and they consequently break down with the release of chemical energy which may be used by the cell.

An understanding of the process of energy transfer, which is based upon the storage of energy in chemical bonds and the release of that energy by breaking chemical bonds, is central to an understanding of cellular processes. It will be discussed in great detail later. Here we wish only to point out that phosphorus and sulfur, as a consequence of their chemical characteristics, have been selected for a primary role in energy transfer, just as all elements present in living organisms have been selected throughout evolution for particular roles that they are uniquely fitted to play.

REFERENCES

Ahrens, L. H. (1965). *Distribution of the Elements on Our Planet.* McGraw-Hill Book Co., New York.

Baker, J. J. W., and G. E. Allen (1965). *Matter, Energy and Life.* Addison-Wesley Publishing Co., Palo Alto, California.

Bowen, H. J. M. (1966). *Trace Elements in Biochemistry,* Academic Press Inc., New York.

Bronk, J. R. (1973). *Chemical Biology.* Macmillan Inc., New York.

Companion, A. L. (1964). *Chemical Bonding,* McGraw-Hill Book Co., New York.

Deevey, E. S. (1970). Mineral cycles. *Scientific American, 223*:148.

DeRobertis, E. D. P., W. W. Nowinski, and F. A. Saez (1970). *Cell Biology.* W. B. Saunders Co., Philadelphia.

Doljanski, F. (1960). The growth of the liver with special reference to mammals. *International Review of Cytology, 10*:217.

Edsall, J. T., and T. J. Wyman (1958). *Biophysical Chemistry: Thermodynamics, Electrostatics, and the Biological Significance of the Properties of Matter,* Vol. I. Academic Press, Inc., New York.

Frieden, E. (1972). The chemical elements of life. *Scientific American, 227*:52.

Gymer, R. G. (1973). *Chemistry: An Ecological Approach.* Harper and Row, New York.

Mason, B. (1966). *Principles of Geochemistry.* John Wiley and Sons, New York.

Ryshkewitsch, G. E. (1963). *Chemical Bonding.* Reinhold Publishing Co., New York.

CHAPTER **4**

The Special Roles of Water and Carbon Dioxide

In 1913 Lawrence J. Henderson, a noted professor of biological chemistry at Harvard University, published a little book entitled *The Fitness of the Environment: An Inquiry into the Biological Significance of the Fitness of Matter.* In it, he set out to demonstrate that "a fit organism inhabits a fit environment." By this he meant that if we were to examine the physical environment of the earth, we could point to a number of factors, such as the availability of water, which make the earth a particularly suitable place for living organisms to develop on and to inhabit.

In support of his thesis, Henderson examined the physical and chemical properties of water, carbon dioxide, and the carbon compounds, and was able to show that the properties of these substances endowed them with characteristics which made them uniquely fit for specific functions of great importance to biological systems. We have already examined the fitness of carbon compounds for living systems, and we now turn to a discussion of water and carbon dioxide.

THE BIOLOGICAL ROLE OF WATER

Water is the principal constituent of all living, active cells, and is indispensable to life as we know it on this planet. It serves not only as the essential medium in which all metabolic activities of the cell occur but also as the matrix for the structural components of the cell. Water has the ability to dissolve such a large number of chemical substances that it is often referred to as the "universal solvent." This, of course, is inaccurate; there are many substances which will not dissolve in water, but it is certainly true that water is unequalled in this respect. Biological systems take advantage of the dissolving properties of water by using water to transport nutrients, waste products, and other substances into and out of cells. The ability of water to ionize many of the substances it dissolves (Chapter 3) is crucial to a wide range of metabolic processes, and water itself is directly involved in many of the metabolic reactions of the cell. By helping to damp temperature fluctuations on the earth, water also serves to make the environment more hospitable to living organisms.

Many of the exceptional properties of water are a consequence of the ability of water molecules to form strong hydrogen bonds with other water molecules. We have considered hydrogen bonding briefly before, and have pointed out that a hydrogen bond may form between a highly electronegative atom and a hydrogen atom which is covalently bonded to another highly electronegative atom. In order to understand fully both the molecular basis for hydrogen bonding in water and its significance in influencing the properties of water, we need to examine the molecular structure of water and the electronic charge distribution in the water molecule.

MOLECULAR STRUCTURE OF WATER

You will remember that the water molecule has the shape of an isosceles triangle, with an O—H bond length of 0.1 nm and an H—O—H angle of 104.5°. The strong electronegativity of the oxygen atom in water tends to pull the bonding electron pair away from each hydrogen atom, with the consequence that the regions around the hydrogen atoms are left with a net positive charge relative to the oxygen atom, which possesses a negative charge. In addition, because of repulsion between the four pairs of electrons surrounding the oxygen atom, these four electron pairs (the two unshared electron pairs in the valence shell of oxygen, as well as the two pairs

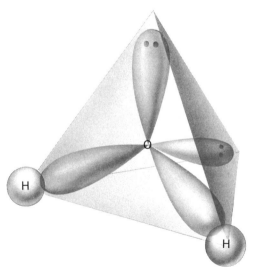

Figure 4–1 The electron charge clouds in the water molecule have directional properties and are oriented around the oxygen atom at the corners of a regular tetrahedron. Only one lobe of each oxygen orbital is shown. The two dots signify a pair of unshared electrons.

The Biological Role of Water

of bonding electrons) tend to establish their charge clouds geometrically as far from each other as possible. Thus, the four electron pairs have directional properties, and the electron clouds will tend to be located in space about the oxygen atom at the corners of a regular tetrahedron (Fig. 4–1). It is because of the nonlinear arrangement of its atoms and the resulting charge distribution that the water molecule is strongly polar and is capable of forming hydrogen bonds between molecules. In addition, the directional properties of its charge clouds specify a precise geometric arrangement of hydrogen bonds.

THE STRUCTURE OF ICE

The crystalline structure of ice, or solid water, is a result of the intermolecular (inter = between) forces (hydrogen bonds) between water molecules. Each water molecule is hydrogen-bonded to four others in a regular way. Each of the two hydrogen atoms in a single water molecule forms a single hydrogen bond with the oxygen atom in a separate water molecule; the oxygen atom, which possesses two regions of electron density corresponding to the unshared electron

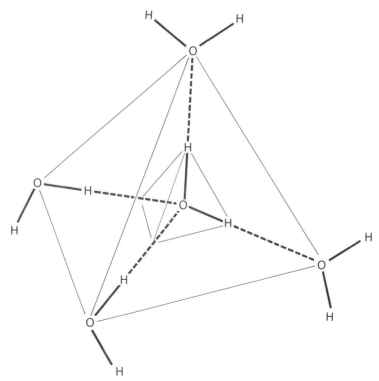

The three-dimensional arrangement of water molecules in ice. Hydrogen bonds are designated by dashed lines. (Adapted from R.G. Gymer [1973], *Chemistry: An Ecological Approach,* Harper and Row, Publishers, New York, p.237.)

Figure 4–2

pairs, will form two hydrogen bonds, each to a hydrogen atom in a third and fourth water molecule. Since the geometry of hydrogen bonding is specified by the directional properties of the charge clouds, the net result is that in ice each oxygen atom is located at the center of a regular tetrahedron of oxygen atoms, as shown in Figure 4-2. The distance between a hydrogen atom on one molecule and an oxygen atom on another is 0.18 nm.

The energy of the hydrogen bond in ice (and water) is variable, but is approximately 4.5 kcal/mole. This compares with an average value of 110 kcal/mole for the intramolecular (intra — within) covalent bond, the O—H bond. Because the hydrogen bonds in water are very much weaker than the covalent bonds, it follows that the hydrogen bonds are easier to break.

When ice melts at 0° C, approximately 15 per cent of the hydrogen bonds are broken, and the regular crystalline structure is destroyed. Liquid water at 0° C still contains many hydrogen-bonded

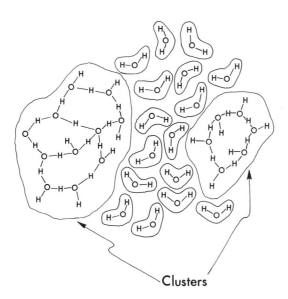

Figure 4-3 A proposed model for the structure of liquid water. Short-lived, hydrogen bonded clusters of water molecules are interspersed with non-hydrogen bonded molecules. (Adapted from G. Nemethy and H.A. Scheraga [1962], *Journal of Chem. Phys.*, 36:3382.)

The Biological Role of Water

molecules and is thought to possess a structure in which clusters of hydrogen-bonded molecules are interspersed with molecules which are not hydrogen-bonded to others. The structure is considered to be a dynamic one, however, with hydrogen bonds continually breaking and reforming, in such a manner that different water molecules are involved in hydrogen bonding at any given time (Fig. 4–3).

As heat is added and the temperature of the water increases, more hydrogen bonds are broken, and there are consequently fewer molecules involved in hydrogen-bonding clusters and more non-hydrogen-bonded molecules than at lower temperatures. However, even at the boiling point (100°C), liquid water contains a large number of hydrogen-bonded regions. Only when water enters the gaseous state as steam, or water vapor, are almost all of the hydrogen bonds broken.

With this background in the structure of water, we are in a good position to understand the molecular basis for a number of physical and chemical properties which endow water with a special fitness for biological systems.

PHYSICAL AND CHEMICAL PROPERTIES OF WATER

Density

It is a curious fact, and an extremely important one for living organisms, that ice floats in liquid water. In this respect, ice is unlike most solids which are denser than their corresponding liquid states. The density of ice is a result of its hydrogen-bonded structure. When ice melts, hydrogen bonds are broken, and the water molecules become more closely packed than they were in the crystalline structure. Hence the density of water at 0°C is greater than that of ice, and ice floats.

As the temperature of water is raised above 0°C, additional hydrogen bonds are broken, allowing more molecules to pack closely, but at the same time increased thermal expansion caused by increased molecular agitation tends to decrease density. The close packing of the molecules increases the density of the water more than thermal expansion decreases it until a point of maximum density is reached at approximately 4°C. At temperatures higher than this, thermal expansion becomes a more important influence on density than is the packing of the molecules, and the density of water begins to decrease. However, even at high temperatures, the density of ice is still considerably less than that of liquid water.

The unusual density characteristics of water are crucial to life on large portions of the earth's surface. Since water possesses a maximum density at about 4°C and since ice floats, bodies of water do not freeze readily with the onset of winter. As the temperature of the surface water falls, it will increase in density and will sink through the underlying waters until it reaches water with a temperature and density similar to its own. This sinking will, of course, displace warmer waters to the top. When the temperature of the surface waters reaches 4°C, the water will have reached its maximum density and will sink until it reaches water also at 4°C. This process will continue until the water temperature is uniformly about 4°C; only then will the surface waters cool below 4°C and eventually freeze. The formation of ice on the surface then further retards freezing by insulating the water beneath it.

Consider what the earth's environment would be like if ice did not float. As winter approached and ice froze at the surface of lakes and oceans, it would sink to the bottom. As the process continued each winter, bodies of water would fill up with ice and would eventually become solidly frozen. Because of the poor thermal conductivity of water, they would melt only a little at the top during the summer. The unavailability of liquid water in large areas would restrict life to tropical regions of the earth's surface.

Specific Heat

The specific heat of a substance is defined as the amount of heat necessary to raise one gram (g) of that substance 1°C in temperature. Relative to most other substances, liquid water has an extremely high specific heat (1.0 cal/g);* indeed, only liquid ammonia has a specific heat which is higher (1.23 cal/g).

It follows from our definition of specific heat that a substance which possesses a high specific heat relative to other substances is able to absorb a larger amount of heat without a change in temperature. The high specific heat of water is related to its hydrogen-bonded structure, but this relationship is not easily explained except on a qualitative level. It is due to the fact that molecules can absorb energy as vibrational energy, which brings about stretching and compression of chemical bonds within the molecule. Since water contains many hydrogen-bonded molecules, it can absorb during a 1°C rise in temperature more energy as vibrational energy associated with the hydrogen bonds than can molecules which are not involved in hydrogen bonding.

Why is the high specific heat of water important to biological systems? We know from our everyday experience that we feel hot when we exercise vigorously. This is because the metabolic activities of cells are associated with the production of large amounts of heat. It is important to the cell that this heat be dissipated quickly

---

*Water is the standard on which the definition of the calorie is based; one calorie (cal) is the amount of heat necessary to raise one g of water from 14.5 to 15.5°C. Thus the specific heat of water is exactly 1.0 cal/g only between these two temperatures. In actuality, the specific heat of water does not vary greatly with temperature, and consequently for our purposes we may consider liquid water to have a constant specific heat over the entire temperature range at which it is a liquid (0 to 100°C).

The Biological Role of Water

and efficiently so that it does not have an adverse effect on cellular structure and activity. For example, most of the cell's enzymes are active only within a narrow temperature range. They are often irreversibly damaged by a rise in temperature, which may disrupt hydrogen bonds and other forces that determine the specific three-dimensional configuration necessary for enzyme action. Since water is able to absorb relatively large amounts of heat without a significant temperature rise, the effects of heat produced by metabolic activity are minimized.

The high specific heat of water is also important in regulating the temperature of the environment by damping wide fluctuations in temperature. In order to understand this effect, we must remember that thermal (heat-related) properties of a substance, such as specific heat, possess two opposite aspects; that is, not only does water require a large amount of heat to raise its temperature, but it must give up a large amount of heat in order to drop in temperature.

The environmental effects of this property of water are particularly noticeable with regard to very large bodies of water, which undergo remarkably little temperature change throughout the year. Even smaller lakes and ponds change temperature at a slower rate and to a lesser degree than they would if water did not have such a high specific heat. The maintenance of a reasonably constant aquatic environment is, of course, extremely important to aquatic organisms, most of which do not regulate their body temperature and hence cannot tolerate rapid or extreme changes in water temperature. In addition to this effect, the large amounts of water on the surface of the earth create a more even distribution of heat around the earth, since water, because of its high specific heat, can move large quantities of heat in water currents.

Heat of Vaporization

The vaporization of a liquid, that is, the process by which a liquid is changed into the vapor state, requires heat. This is readily understood if we consider that a molecule leaving the liquid state and entering the vapor state needs energy in the form of heat to overcome the attractive forces between it and other molecules in the liquid. Simply, energy is needed to pull the molecules apart. In the case of water, a relatively large amount of heat is required to vaporize a molecule because of the strong hydrogen bonds that exist between it and the liquid. The heat of vaporization of water at 40°C is 574 cal/g, which is a considerably higher value than for any other substance that is liquid at room temperature. This means that in order to evaporate one gram of water at 40°C, 574 calories of heat are required.

Water's high heat of vaporization is important to living organisms in the dissipation of heat released by metabolic activity. The evaporation of sweat in animals has a cooling effect on the body since substantial amounts of heat are absorbed by the water in sweat when it evaporates. Loss of excess water through the leaves of plants by evaporation has a similar cooling effect.

As with the high specific heat of water, its high heat of vaporization has a moderating effect on temperature extremes in the environment. The evaporation of water in the tropics removes large quantities of heat; conversely, the condensation of water vapor (the reverse of vaporization) in colder regions releases large quantities of heat.

Heat of Fusion

Another property of water which is important in regulating environmental temperature is its high heat of fusion (80 cal/g). Only ammonia and a number of inorganic molecules held together by strong ionic bonds possess higher values.

The heat of fusion is defined as the amount of heat absorbed when a solid melts. (This value is, of course, equal to the amount of heat released when the liquid freezes). Water's high heat of fusion is due to the fact that relatively large amounts of heat are required to break down its regular hydrogen-bonded structure.

As the temperature of the air falls with the onset of winter and water begins to freeze, heat is released, warming the environment. In the spring, the melting of ice absorbs heat from the environment. Both of these effects help to moderate the temperature fluctuations associated with the change of the seasons.

Surface Tension

Surface tension is characterized by the tendency of the surface of a liquid to act as if it were covered by an elastic membrane; consequently the surface tends to assume the smallest possible surface area, approaching a spherical configuration. The surface tension arises because molecules at the surface of a liquid are attracted to molecules of the underlying liquid more than they are to molecules in the gaseous phase above them. The greater the intermolecular forces within the liquid, the greater the surface tension. Water, with its highly hydrogen-bonded structure, has the highest surface tension of all common liquids except mercury.

Because of its high surface tension, water will tend to rise within any capillary space it can adhere to. This is important to biological systems, since it helps bring water to the surface of the soil by promoting its movement through very fine spaces in the soil. In addition, it leads to the rise of water in the capillary structures of plants and thus assists the transport of water to great heights.

The Biological Role of Water

Solutions

Before we consider other properties of water relating to its ability to dissolve a wide variety of substances, we must digress for a moment to discuss some properties of solutions in general and of aqueous solutions in particular.

A *solution* is usually defined as a homogeneous mixture of two or more substances. The substance present in excess, or the substance which retains its physical state in solution, is called the *solvent*; the dissolved substance, or substances, are known as *solutes*. By "homogeneous" we mean that the mixture forms only one phase (either solid, liquid, or gas), and is entirely uniform in appearance even under the light microscope. Thus in a solution, molecules or ions are distributed uniformly throughout the solvent, and under normal conditions, the solute molecules will not settle out. For example, a solution is formed when sodium chloride (NaCl) is dissolved in water. The resulting mixture is entirely homogeneous and consists of one liquid phase. A solution may be differentiated from a *suspension*, which is a non-uniform mixture of two or more substances, such as sand and water, or oil particles and water (Fig. 4–4). Cell biologists are generally concerned with solutions in which water is the solvent and in which the solute is a solid.

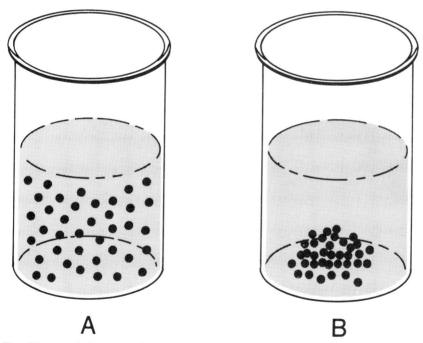

Figure 4–4

The difference between a solution and a suspension. *a.* In a solution, the solute particles (black dots) are uniformly distributed throughout the solvent and under normal conditions will not settle out. *b.* A suspension, in which the particles have settled out of solution.

CONCENTRATION OF SOLUTIONS. It is often necessary to know the number of solute particles per unit volume of solution, also known as the *concentration* of the solution. This is most conveniently expressed in terms of *molarity*, which is defined according to the following equation:

$$\text{Molarity} = \frac{\text{Number of moles of solute}}{\text{Liter of solution}}$$

where one *mole* is equal to 6.023×10^{23} particles. Thus a solution containing 6.023×10^{23} molecules of glucose per liter of water has a concentration of one molar (1 M).

FORMATION OF SOLUTIONS. The uniform dispersion of solute molecules throughout the solvent involves the disruption of attractive forces between solute particles and those between solvent particles. These attractive forces are known as *cohesive forces.* The hydrogen bonds between water molecules and the ionic attraction between sodium and chloride ions are both examples of cohesive forces present in a solution of sodium chloride. Other attractive forces occur in solutions; these are the *adhesive forces* between solute and solvent particles. Although factors other than energy considerations are involved in the formation of solutions, it is a general rule that a solute will tend to dissolve in a solvent if the adhesive forces are equal to or stronger than the cohesive forces. That is, the energy released in the formation of adhesive forces must be equal to or more than the energy required to break the cohesive forces. Because of this energy requirement, non-polar substances which cannot form strong adhesive forces with water are usually only sparingly soluble in water.

Although strong adhesive forces can sometimes result from chemical reaction between the solvent and solute particles, leading to the formation of entirely new substances, the most general type of adhesive forces arising in aqueous solutions in the cell is a result of a physical interaction between solute and solvent particles. This interaction is called *solvation* of the solute. When the solvent is water, the interaction is referred to as *hydration.* As we mentioned in Chapter 3, hydration arises from the attraction between charged solute particles (ions) and the polar water molecule. For example, when magnesium chloride ($MgCl_2$) is dissolved in water, strong adhesive forces are formed between water molecules and both the Mg^{++} and Cl^- ions as a result of hydration of the ions (Fig. 4–5).

It is important to remember that *all ions in water are hydrated,* and although we do not usually indicate this fact when writing chemical formulae of ions in aqueous solution, it should be clearly understood.

Dielectric Constant

The ability of water to dissolve ionic compounds is enhanced because of another property of water, its high dielectric constant.

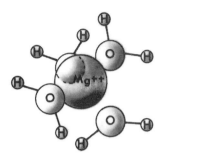 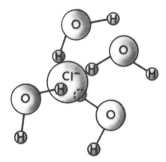

A schematic diagram illustrating the hydration of magnesium and chloride ions. Figure 4-5

The dielectric constant (D) of a liquid is defined by the equation

$$F = \frac{e_1 e_2}{Dr^2}$$

where F is the electrostatic force of attraction between two oppositely charged ions, e_1 and e_2, located in solution at a distance r from each other. This means that the force of attraction between two ions of opposite charge dissolved in a solvent is inversely proportional to the dielectric constant of the solvent and to the square of the distance between them. Thus, the higher the dielectric constant of a solvent, the less the attractive force between two ions which are located a given distance apart. Liquids with a high dielectric constant are able to dissolve a larger number of ionic compounds than those with low dielectric constants simply because the oppositely charged ions do not tend to attract one another and form insoluble arrays. At 20° C, water possesses a dielectric constant of 80, which is higher than that of any pure liquid except hydrogen cyanide. On the other hand, most organic solvents, such as ethyl alcohol and chloroform, have dielectric constants considerably lower than that of water. This is one of the reasons why sodium chloride, for example, is only sparingly soluble in these solvents.

In addition to its ability to dissolve a wide range of ionic compounds, water is able to dissolve polar covalent compounds which are able to form hydrogen bonds with it. Thus, for example, ammonia is very soluble in water (Fig. 4–6). Polar compounds which do not form hydrogen bonds with water generally are not soluble in water because the strength of the attractive forces between water

$$H\overset{\delta^+}{-}\underset{\underset{H}{|}}{\overset{\overset{H}{|}}{N}}\overset{\delta^-}{\cdots}H\overset{\delta^+}{-}\overset{\delta^-}{O}\diagdown H$$

The formation of hydrogen bonds between ammonia molecules and water molecules accounts for the high solubility of ammonia in water. Figure 4-6

and the polar compound (adhesive forces) is not great enough to overcome the cohesive forces in water due to hydrogen bonding.

THE FITNESS OF WATER

This brief survey of some of the physical and chemical properties of water gives us an appreciation of the importance of water to biological systems. We have seen that water is uniquely fitted not only to serve a central metabolic role within the organism but also to make this planet more hospitable to life as we know it. Henderson has expressed it more elegantly:

> Water, of its very nature, as it occurs automatically in the process of cosmic evolution, is fit, with a fitness no less marvelous and varied than that fitness of the organism which has been won by the process of adaptation in the course of organic evolution.

THE BIOLOGICAL ROLE OF CARBON DIOXIDE

Carbon dioxide (CO_2) is a linear, non-polar, highly symmetrical compound with the following chemical structure:

$$O=C=O$$

It exists as a free gas at ordinary temperatures and, like most carbon-containing compounds, is relatively inert to reaction with other substances. It is extremely soluble in water and tends to distribute itself between air and water in such a way that its concentration (amount per unit volume) in both substances is about equal. Because of this property, carbon dioxide is continually exchanging between the waters of the earth and the atmosphere. Thus, for example, if the concentration of carbon dioxide should rise in any aqueous medium relative to its concentration in the surrounding gaseous phase, then carbon dioxide will tend to leave the water and enter the gaseous phase until the concentrations of carbon dioxide in the two phases have been equalized.

Carbon dioxide is the ultimate source of carbon for all organic compounds present in living organisms. Autotrophic organisms convert carbon dioxide into simple organic compounds which may be used by plants and animals as nutrients. These may serve either as a source of energy or as substances from which other cellular constituents may be synthesized. The extraction of energy from these organic compounds results ultimately in the production of carbon dioxide as an end product. Thus carbon dioxide is continually "cycling through" living organisms. It is apparent that both the volatility and solubility of carbon dioxide, which promote its ready distribution in the environment, make it particularly suited for this key metabolic function (Fig. 4–7).

The Biological Role of Carbon Dioxide

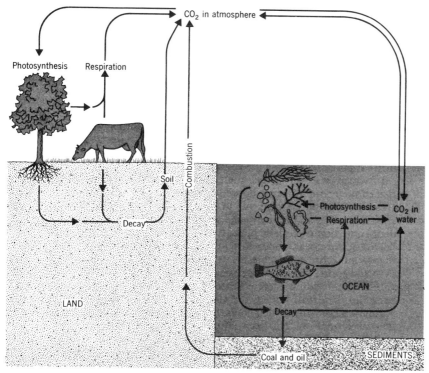

Figure 4-7

The carbon cycle. Carbon dioxide in water or in the atmosphere is assimilated by photosynthetic plants and some bacteria to produce organic compounds. These organic compounds may, in turn, be consumed as nutrients by non-photosynthetic organisms. The respiration of living organisms or their eventual decay returns carbon dioxide to the environment. (*From* G.C. Stephens and B.B. North [1974], *Biology*, © 1974, John Wiley & Sons, Inc., New York. Reprinted by permission.)

In addition to functioning as the source of carbon for living organisms, carbon dioxide is involved in the regulation of the hydrogen ion concentration within living organisms and, probably to a very minor extent, in marine environments. This function of carbon dioxide is extremely important to living organisms because they cannot tolerate large fluctuations in hydrogen ion concentration. Organisms normally maintain the hydrogen ion concentration of their cellular fluids within a remarkably narrow range. The hydrogen ion is critical to many metabolic reactions. A number of enzymes, for example, react with hydrogen ions and thus operate most efficiently at a specific hydrogen ion concentration. Changes from this optimum concentration may cause an alteration in the charge distribution on the enzyme molecule. This, in turn, may result in loss of enzymatic activity by disrupting the intramolecular forces which give the enzyme molecule its specific three-dimensional configuration. In addition to their intracellular effects, changes in the hydrogen ion concentration in oceans and lakes can be directly harmful to aquatic animals. In order to understand how carbon dioxide helps to regulate the concentration of hydrogen ions in an aqueous medium, we must first review some properties of acids and bases.

ACIDS AND BASES

Hydrogen ions (or protons, since they are hydrogen atoms without electrons) are often transferred from one chemical substance to another. If a substance donates protons in a chemical reaction, that substance is defined as an *acid*; if it accepts protons, it is defined as a *base*.

Thus in the chemical reaction

$$\text{HCl} + \text{H}_2\text{O} \longrightarrow \text{Cl}^- + \text{H}_3\text{O}^+ \quad (1)$$
(hydrogen chloride) (water) (chloride ion) (hydronium ion)

hydrogen chloride acts as an acid by donating a proton to water, which acts as a base.

It is important to understand that a substance *functions* as an acid only if it is involved in a reaction with a base that can accept a proton from it, simply because a proton cannot be lost from an acid unless it has somewhere to go. The same reasoning, of course, holds true for bases; for a substance to act as a base, it must accept a proton from an acid in a chemical reaction.

Some substances may act as acids under certain conditions and as bases under others. For example, in the reaction

$$\text{H}_2\text{O} + \text{NH}_3 \longrightarrow \text{OH}^- + \text{NH}_4^+ \quad (2)$$
(ammonia) (ammonium ion)

water functions *as an acid* by donating a proton to ammonia (acting as a base). However, when hydrogen chloride is placed in water as in our previous example (Equation 1), water functions *as a base* by accepting a proton from HCl (acting as an acid).

Notice that when an acidic substance is placed in water (Equation 1) hydronium ions (H_3O^+) are produced. These result from the release of hydrogen ions from the acid and their transfer to the solvent (water). Also notice that a base in water produces hydroxide ions; however, as in Equation 2 above, these ions do not necessarily originate from the base itself.

The chemical properties of acids and bases *in water* are due to the presence of hydronium ions (H_3O^+) and hydroxide ions (OH^-), respectively. The greater the concentration of hydronium ions (or hydroxide ions) in aqueous solution, the stronger are the acidic (or basic) properties of the substance. The concentration of hydronium ions in aqueous solution is, in turn, a function of the degree to which an acid donates hydrogen ions to water, and the concentration of hydroxide ions in solution is a function of the degree to which a base accepts hydrogen ions from water. Another way of saying the same thing is that the concentration of hydronium ions

The Biological Role of Carbon Dioxide

(or hydroxide ions) in solution depends upon the degree to which an acid (or base) ionizes in water.

STRENGTHS OF ACIDS AND BASES

The relationships discussed above allow us to formulate some simple definitions. *Strong acids,* those with strongly acidic properties, and strong bases, those with strongly basic properties, ionize almost completely in aqueous solution; *weak acids* and *weak bases,* those with weakly acidic and weakly basic properties, respectively, ionize only partially in water.

The difference between a strong acid and a weak acid (or a strong base and a weak base) is not, then, a qualitative one; the difference is simply one of the degree to which each ionizes in water. Thus, for example, if we were to add a small amount of the strong acid, hydrogen chloride (HCl), to water, within a very short time most of the acid would be converted (ionized) to hydronium ions and chloride ions, and only an extremely small amount of non-ionized HCl will still remain in the solution.

If, however, we added a weak acid, such as carbonic acid, to water, the following reaction would occur.

$$H_2CO_3 + H_2O \rightleftharpoons HCO_3^- + H_3O^+$$
(carbonic acid)   (bicarbonate ion)

But in this case only a small proportion of the acid will be ionized to HCO_3^- and H_3O^+, and the remainder will exist in solution in the non-ionized form (H_2CO_3). The reaction is written with double arrows to indicate that substances on both sides of the equation exist in solution.

CHEMICAL EQUILIBRIUM

The ionization reactions of acids and bases illustrate an important chemical principle. At a given temperature and pressure, any particular ionization reaction will eventually reach a state where the proportions of reactants (substances on the left side of the arrows) to products (substances on the right side of the arrows) always remain the same, and there is no *net* change in the system. Such reactions are said to have reached a state of *chemical equilibrium. Indeed, it is a characteristic of all chemical reactions that they will eventually reach a point of chemical equilibrium.*

---

*It is important for future discussions to point out that chemical reactions never reach equilibrium in living systems. Reaching equilibrium is a characteristic of the non-living state.

Chemical equilibrium is not a static condition but rather a dynamic one. Thus, even after equilibrium has been attained, reactants are continually being converted to products, and products are continually being converted back to reactants. However, at equilibrium the *rate* of conversion of reactants to products is exactly balanced by the *rate* of conversion of products to reactants. Hence in dilute solutions, the proportion of products to reactants in a particular reaction *at equilibrium* is fixed at a given temperature and pressure, and is equal to a constant, K, called the *equilibrium constant* of the reaction. For any reaction, K is defined by the following equation.

$$K = \frac{\text{concentration of each of the products multiplied together}^*}{\text{concentration of each of the reactants multiplied together}}$$

Thus, for the ionization of carbonic acid in water

$$K = \frac{\text{concentration of } HCO_3^- \times \text{concentration of } H_3O^+}{\text{concentration of } H_2O \times \text{concentration of } H_2CO_3}$$

According to chemical convention, concentration may be denoted by brackets, and therefore the equation above may be rewritten

$$K = \frac{[HCO_3^-] \times [H_3O^+]}{[H_2O] \times [H_2CO_3]}$$

In a dilute aqueous solution, the concentration of water does not change significantly during a chemical reaction, and consequently the ionization equation is generally rearranged so that the concentration of water is incorporated into K, such that

$$K \times [H_2O] = \frac{[HCO_3^-] \times [H_3O^+]}{[H_2CO_3]} = K_a$$

The new constant, K_a, is termed the ionization constant of the acid, and since it specifies the concentration of hydronium ions in a dilute aqueous solution of the acid, it is a measure of the acidity of the acid.

Because K_a is a constant, a change in the concentration of any one component in the reaction will result in changes in the concentration of other components. This is a consequence of Le Chatelier's principle, which states that *if any system at equilibrium is disturbed, the system will tend to change in such a way as to minimize or overcome the effect of the disturbance.* For instance, in our previous

---

*Attractive forces between oppositely charged ions in solution reduce the effect of the charge on the individual ions so that their concentrations seem lower than they actually are. The effective concentration of an ion in solution is called its *activity*, and the equilibrium constant, K, strictly should be defined in terms of the activities, rather than concentrations, of the products and reactants at equilibrium. However in dilute solutions, which we usually encounter in the living cell, activities approximate concentrations expressed in molarities.

example, if we were to disturb the equilibrium of the system by adding additional H_2CO_3 to the solution, a certain proportion of the newly added H_2CO_3 would ionize until (if possible) the original equilibrium between reactants and products was re-established. On the other hand, an increase in the concentration of HCO_3^- or H_3O^+, or both, would drive the reaction in the direction of increased H_2CO_3 production.

It should be apparent from the discussion above that if the ionization constant of an acid, or the equilibrium constant of any reaction, is very large, then at equilibrium most of the reactants are converted to products, and we say that the reaction tends to *go to completion*. The ionization of hydrochloric acid in water is a reaction of this type. The degree to which a reaction goes to completion is extremely important in metabolic systems and is determined by energy considerations which we will discuss in the next chapter.

pH

Since the hydronium ion is so important to living systems, a quantitative scale, the pH scale, is often used to indicate the acidity of aqueous solutions. The pH of an aqueous solution is defined as the negative logarithm of the hydronium ion concentration

$$pH = -\log_{10} [H_3O^+]$$

or

$$[H_3O^+] = 10^{-pH}$$

Thus, for example, if the concentration of hydronium ions in a solution is equal to 10^{-5} M, then the pH of the solution is 5.

The pH scale runs from zero to 14.* The midpoint on this scale, a pH of 7, is taken as the pH of a neutral solution. This is because the pH scale is based upon the hydronium ion concentration in pure water, in which the concentrations of hydronium and hydroxide ions are the same and are both equal to 10^{-7} M at 25°C (see Appendix III). Acidic solutions have pH values less than 7 and basic solutions have pH values greater than 7. The more acidic the solution, the lower the pH; the more basic the solution, the higher the pH (Fig. 4–8).

HYDRONIUM IONS AND HYDROGEN IONS

As we have seen, whenever free hydrogen ions (protons) are removed from an acid in aqueous solution, they are hydrated to

---

*In fact, the pH scale runs below zero and above 14; for example, a 10 M solution of a strong acid has a pH of −1.0. However, since such strongly acidic or basic solutions do not normally occur within the cell, the pH scale is usually thought of as extending only between zero and 14.

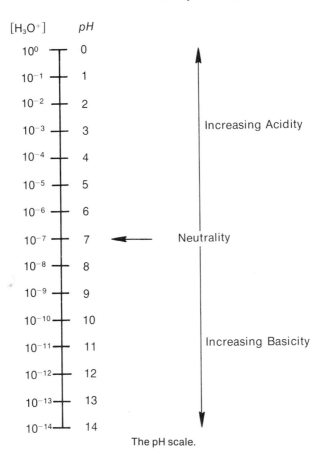

Figure 4-8 The pH scale.

form hydronium ions. For simplicity, however, it has been customary *to assume* the presence of hydronium ions in aqueous solution and to write all chemical formulae in terms of hydrogen ions. We will adopt this convention because it is so widely used, and we will henceforth refer only to hydrogen ions, *with the clear understanding that free hydrogen ions exist as hydronium ions in aqueous solution.*

BUFFERS

Some solutions, called buffers, tend to resist large changes in pH when hydrogen ions or hydroxide ions are added to them. Indeed, living systems are able to maintain a relatively constant pH in their internal environment because the cellular fluids act as buffers.

Solutions of amino acids and proteins are excellent buffers. This is because they can act as acids under some conditions and as bases under others, and thus by releasing or accepting hydrogen ions, they are able to counteract the effect of the addition of acidic or basic substances to the solution. We will discuss the molecular

The Biological Role of Carbon Dioxide

basis for the buffering capacity of amino acids and proteins in Chapter 7.

Proteins and amino acids play a significant role in pH maintenance within the cell. The proteins in blood plasma and the hemoglobin in red blood cells are particularly important in this respect.

Other types of buffers are solutions containing a weak acid (or a weak base) and the anion of the acid (or base). We can illustrate how this type of buffer regulates pH by considering as an example the carbonic acid (H_2CO_3)–bicarbonate (HCO_3^-) buffer system, which is of central importance to living organisms. The effective operation of this buffer system depends upon the unique physical and chemical properties of carbon dioxide.

THE CARBONIC ACID–BICARBONATE BUFFER SYSTEM

Cellular metabolism is characterized by the accumulation of acidic endproducts such as sulfuric acid, phosphoric acid, and a variety of organic acids whose effects must be buffered in order to prevent a drastic and harmful fall in intracellular pH. The carbonic acid–bicarbonate buffer system plays a particularly significant role in this respect.

When CO_2 is produced as a byproduct of metabolic activity, it is readily hydrated to form the weak acid, H_2CO_3. This, in turn, partially ionizes to H^+ and HCO_3^-, according to the equilibrium we have discussed above. These reactions may be summarized by the equation

$$H_2O + CO_2 \rightleftharpoons H_2CO_3 \rightleftharpoons H^+ + HCO_3^- \tag{3}$$

The buffering capacity of this system depends upon the fact that under normal conditions, bicarbonate ions are concentrated within the cell and in the extracellular tissue fluids of multicellular animals. Consequently, as hydrogen ions are produced as a result of cellular metabolism, they are eliminated by combination with bicarbonate ions to form carbonic acid, and the pH of the cellular fluids is thus maintained reasonably constant. The formation of carbonic acid, is, of course, a consequence of Le Chatelier's principle: The production of hydrogen ions disturbs the ionization equilibrium and drives the reaction in the direction of increased carbonic acid production.

The increased carbonic acid concentration, in turn, drives the equilibrium reaction (Equation 3) to the left, resulting in the dehydration of carbonic acid to CO_2. As the CO_2 is eliminated into the surrounding environment, either directly or through the respiratory systems of animals, the resulting decrease in CO_2 concentration favors additional dehydration of carbonic acid.

It should be emphasized that the entire process of pH maintenance within the cell is far more complicated than this, and involves a variety of other buffers and metabolic reactions. The phosphate buffer system ($H_2PO_4^- - HPO_4^{-2}$) as well as a number of protein buffers are, for example, particularly important in maintaining intracellular pH, but it is probably fair to say that the carbonic acid–bicarbonate buffer system is of central and universal importance to living systems.

Not only is the $H_2CO_3 - HCO_3^-$ buffer system important within living organisms but it also may be one of many buffers that help maintain the pH of the ocean at a reasonably constant value (pH of ~ 8.1). Although it was formerly believed that the pH of the ocean is regulated primarily by the $H_2CO_3 - HCO_3^-$ buffer system, it is now generally held that the clay sediments, compounds of aluminum, silicon, and oxygen, act as the major buffers in the ocean by absorbing and releasing hydrogen ions.

THE FITNESS OF CARBON DIOXIDE

The foregoing discussion demonstrates the importance of carbon dioxide to living systems, both as the ultimate source of organic compounds and in pH maintenance. We have attempted to demonstrate that it is the physical and chemical properties of carbon dioxide—its volatility, its ability to form the weak acid (carbonic acid) by hydration, and its tendency to distribute equally between the aqueous and vapor phases—that make carbon dioxide particularly suitable for these key metabolic roles.

REFERENCES

Dowben, R. M. (1971). *Cell Biology.* Harper and Row, New York.
Edsall, J. T., and T. Wyman (1958). *Biophysical Chemistry,* Vol. 1. Academic Press, Inc., New York.
Gymer, R. G. (1973). *Chemistry: An Ecological Approach.* Harper and Row, New York.
Henderson, L. J. (1913). *The Fitness of the Environment.* The Macmillan Co., New York.
Tuttle, W. W., and B. A. Schottelius (1969). *Textbook of Physiology.* The C. V. Mosby Co., St. Louis, Missouri.
White, E. H. (1964). *Chemical Background for the Biological Sciences.* Prentice-Hall Inc., Englewood Cliffs, New Jersey.

CHAPTER 5

Cellular Energetics

We have mentioned previously that cells continuously expend energy, which they require for a number of essential metabolic processes. These include the synthesis of cellular constituents and other cellular products from nutrient substances, the transport of certain chemical substances into or out of the cell, and cell motility, as well as a number of specialized processes such as movement of chromosomes during mitosis in eukaryotic cells, contraction of muscle cells, transmission of electrical impulses by nerve cells, and so forth. These processes include not only those involved with such activities as growth and reproduction but also those concerned with simply maintaining the *status quo,* the everyday "housekeeping" functions of maintenance and repair. In other words, a cell expends energy merely to stay alive — to support its highly ordered, complex structure. A cell deprived of an energy source will become chemically and structurally disorganized and will die very quickly. This requirement for a continuous supply of energy is a fundamental characteristic of all living systems.

Beyond the obvious necessity, what is the fundamental reason for this energy requirement and, in practice, how do cells obtain, store, and utilize energy? In order to answer this question with a reasonable degree of sophistication, we need first to understand what energy is and then to consider certain energy relationships which are known as the Laws of Thermodynamics.

CHEMICAL THERMODYNAMICS

ENERGY AND THE FIRST LAW OF THERMODYNAMICS

Energy may be defined simply as the capacity to do work. Work may take a variety of forms such as lifting a weight, synthesizing a complex molecule, heating a house, reading this textbook, and so on. Energy may be quantified, and it is obvious that some work requires more energy than does other work.

Energy exists in a number of forms—matter,* light, heat, electricity, chemical energy, motion, and others—and may be converted from one form into another, but in the process energy is neither created nor destroyed. This principle is known as the First Law of Thermodynamics (or the Law of Conservation of Energy) and is often stated in another way: *The energy of the universe is constant.*

In talking about any process involving energy transfer, it is necessary to distinguish between the *system,* which may be defined as the specific portion of matter under study, and its *surroundings,* or environment (Fig. 5–1). The surroundings include the rest of the universe; that is, everything except the system under study. A system may be a chemical reaction, a machine, an organism, or any other thing that we wish to focus our attention upon. When a system undergoes change, it will either absorb or release energy. It follows that if the energy of the universe is constant (the First Law of Thermodynamics), then any energy change that results when a system goes from its initial to its final state must be exactly equaled by a change *in the opposite direction* in the energy content of the surroundings. For convenience we sometimes refer to a *closed* (or *isolated*) system, one in which there is no energy in any form exchanged between the system and its surroundings. The energy in any closed system is, by definition, constant.

Energy transformations in physical and chemical processes are usually accompanied by a transfer of heat to or from the surroundings. Processes that occur with a release of heat are termed *exothermic*; an example of an exothermic process is the burning of wood or other fuel. *Endothermic* processes, such as the melting of ice, are those in which heat is absorbed from the surroundings. In respect to the utilization of energy in the form of heat, biological processes differ basically from many physical ones in which heat energy is capable of performing work, as, for example, in the inter-

---

*Matter is convertible into energy, as described by Einstein's equation $E = mc^2$, where E = energy, m = mass and c = the velocity of light.

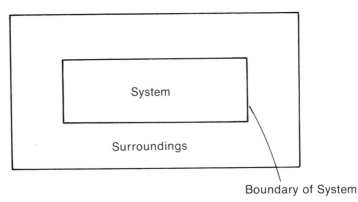

Figure 5–1 Energy transfers within a system may be considered as separate from those that occur in the surroundings.

Chemical Thermodynamics

nal combustion engine, which converts heat energy into mechanical energy by causing heat to flow from a hot to a cooler body. All processes within an organism occur at nearly constant temperature; indeed, we have mentioned previously that even a rise in temperature of a few degrees Centigrade may have adverse effects on the structural components and enzymatic systems of living cells. Thus living organisms have no mechanism for utilizing heat energy to perform work. Consequently, heat is generally useless as an energy source for biological systems except, of course, for maintaining body temperature in warm-blooded animals.

Living systems use primarily chemical energy, or more precisely, they use the energy of chemical bonds to drive metabolic processes. It is true that photosynthetic cells are able to capture radiant (light) energy from the sun, but this is quickly converted into the chemical bond energy of the ATP molecule, where it is stored for use in energy-requiring cellular processes. It is for this reason that in our consideration of energy utilization by the cell, we will restrict our discussion primarily to energy relationships associated with chemical reactions.

THE SECOND LAW OF THERMODYNAMICS

The First Law of Thermodynamics tells us that energy is fully interconvertible in any given process, but it does not tell us whether it is energetically possible for a particular process to occur spontaneously (without the input of energy). Common experience tells us that certain processes do not occur spontaneously; such processes are referred to as *unnatural processes.* For example, we know that we need not be particularly concerned that the episode illustrated in Figure 5–2 will come to pass: A rock can roll downhill spontaneously, but it cannot roll uphill unless we expend energy to make it do so.

In order to illustrate the principle of thermodynamic spontaneity with an example more appropriate to the chemical processes that occur in living systems, let us consider the following exothermic reaction

$$C_6H_{12}O_6 + 6O_2 \longrightarrow 6CO_2 + 6H_2O + \text{Energy released}$$

(glucose)   (oxygen)   (carbon dioxide)   (water)

This reaction, the production of 6 molecules of carbon dioxide and 6 molecules of water from 1 molecule of glucose and 6 molecules of oxygen is a *natural process*; that is, it could occur spontaneously in a closed system, without any input of energy from the surroundings. Like all spontaneous reactions, it is an energy-yielding reaction. Energy is released (made available) as a consequence of the reaction, because the energy of the initial state (the potential energy in glucose and oxygen) is greater than the energy of the final

Figure 5-2 An example of an unnatural physical process. (© 1963 Saul Steinberg. From THE NEW WORLD [Harper]. Originally in The New Yorker.)

state (the potential energy in carbon dioxide and water). The reverse reaction, the formation of 1 molecule of glucose and 6 molecules of oxygen from 6 molecules of carbon dioxide and 6 molecules of water is an unnatural process that requires energy, and hence cannot occur spontaneously.

From what we know about the First Law of Thermodynamics we

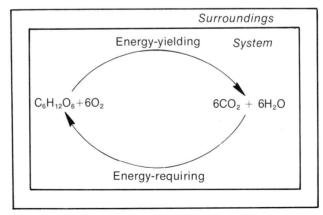

Figure 5-3 Completely reversible processes like the one illustrated here cannot occur.

Chemical Thermodynamics

would expect that all energy-yielding reactions would be completely reversible. If we were able to capture all the energy released in an energy-yielding reaction, it seems as though it should be possible to use this energy to drive the reaction in the opposite (energy-requiring) direction. Indeed, it should be possible to run the reaction back and forth endlessly: In one direction, energy would be released, and this energy could then be used to drive the reaction in the other direction (Fig. 5–3). Although such an energy transfer would in no way violate the First Law, we find that the process cannot occur as we have described it. This leads us to the conclusion that the energy released by an energy-yielding reaction is not sufficient to drive the same reaction in the opposite direction.

This example illustrates a fundamental property of all natural processes: their irreversibility.* A natural process such as a spontaneous chemical reaction is irreversible because not all the energy released in the initial reaction is available for running the reaction in the opposite direction. To account for this situation, a quantity called *entropy* (from the Greek word meaning transformation) has been defined; it is a measure of the amount of energy that becomes unavailable for work as a consequence of any natural process. *The universe, or any system plus its surroundings, tends spontaneously toward a state of increasing entropy.* This statement, which emphasizes the irreversibility of all natural processes, is known as the Second Law of Thermodynamics. It can be shown by statistical physics that entropy is analogous to randomness or lack of order. So the Second Law of Thermodynamics may be restated: *The universe, or any system plus its surroundings, tends spontaneously toward a state of increasing randomness or disorder.* In fact, the ultimate fate of the universe is to reach a state of maximum entropy or complete randomness. This state, which has been referred to as *entropic doom,* is associated with the unavailability of any energy to do useful work. The Second Law, therefore, points to the *direction* of processes in time—always in a direction toward increasing entropy—and has prompted the British physicist Sir Arthur Eddington to refer to the Second Law as "time's arrow."

Although the concept of entropy appears somewhat complicated and esoteric, it is really a simple one which we encounter in our everyday experience. The tendency to disorder in the natural world is inescapable. Gardens become overgrown with weeds, brick walls crumble, our living places become sloppy, automobiles fall into disrepair—unless we do work (expend energy) to reverse the

---

*The irreversibility of natural processes is illustrated most dramatically by examples involving heat transfer. We know that a man cannot warm his hands by holding a snowball, thus transferring heat from the snowball to his hands and in the process making the snowball colder and his hands warmer. The First Law does not prohibit an energy transfer of this type, but heat simply does not flow spontaneously from a cold body to a warmer one. It can be made to do so, but only with the input of energy; for example, by using a device such as a refrigerator, which transfers heat from its cooling coils to its external surroundings.

process. *Thus disorder is more probable in nature than order.* What this means is that highly ordered states are not expected to occur very often. For example, from a molecular standpoint, we might find at any given moment all the molecules of gas in the top part of a bottle, or a mixture of several gases spontaneously separating into pure substances, but such ordered conditions would be highly improbable. In fact, it is possible to demonstrate a simple relationship between entropy and probability: The entropy of a system will increase if the system goes from a less probable (relatively ordered) to a more probable (relatively less ordered) state.

In summary, then, all natural processes are spontaneous and irreversible and proceed with an increase in entropy. Some processes that, in a closed system, are not spontaneous and proceed with a decrease in entropy may be forced to occur by providing the system with additional energy in an appropriate form.* The important lesson of the Second Law is that a process can never result in a net decrease in the entropy of the universe. Consequently, if the entropy of a system decreases, then there must be a larger increase in the entropy of the surroundings, such that the net result of the process is an increase in entropy.

There are a number of ways by which entropy increases (Fig. 5–4). It should be apparent from the previous discussion that highly ordered, complex entities possess less entropy than more disorganized ones. A brick house possesses less entropy than a pile of bricks; a living cell possesses less entropy than a cell homogenate. At the molecular and atomic levels, there is generally an increase in entropy in going from a solid to a liquid state, and a further increase in going from the liquid to the gaseous state. When a solid is dissolved in solution, there is also an increase in entropy. Complex molecules possess less entropy than simpler ones; for instance, glucose has less entropy than water or carbon dioxide, and a protein—composed of hundreds of linked amino acids—has less entropy than a single amino acid. Entropy is also increased by a rise in temperature, since the average thermal motion of atoms and molecules increases. In fact, most energy transformations are not 100 per cent efficient and result in a loss of heat to the surroundings; the entropy of the surroundings is thereby increased.

In our previous example, the spontaneous reaction

$$C_6H_{12}O_6 + 6O_2 \longrightarrow 6CO_2 + 6H_2O + \text{Energy released}$$

results in an increase in entropy both in the system, since the molecular products are simpler in structure than the reactants, and in the surroundings, since heat is released as a result of the reaction. If the reaction is forced in the opposite direction by supplying energy, there is a *decrease* in the entropy of the system. We find, however, that heat

---

*There are other processes which are said to be forbidden because they may not occur under any circumstances.

Chemical Thermodynamics

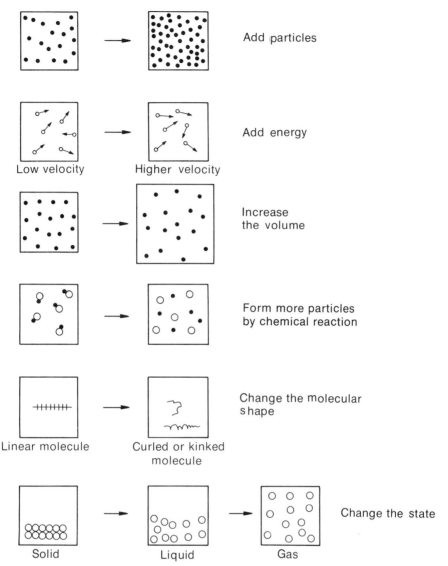

Figure 5-4

A diagram showing some of the ways entropy may increase. (Adapted from *Energetics, Kinetics and Life: An Ecological Approach* by G. Tyler Miller Jr. © 1971 by Wadsworth Publishing Company, Inc., Belmont, California 94002. Reprinted by permission of the publisher.)

lost to the surroundings during the process increases the entropy in the surroundings to a greater extent than the entropy decreases in the system (Fig. 5–5). Thus when the system and its surroundings are considered together, the net result is an increase in entropy.

What we are really saying is that the Second Law of Thermodynamics allows us to predict whether or not a particular chemical reaction, or any process, can occur spontaneously, and that an increase in entropy is a necessary condition for thermodynamic spontaneity. We can thus think of increasing entropy as the "force" that drives a chemical reaction.

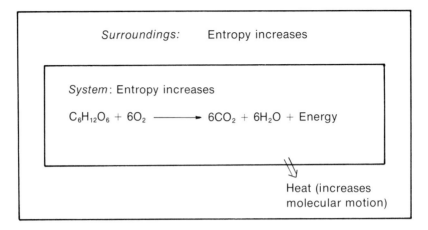

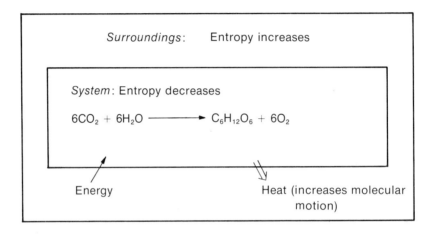

Figure 5-5 All processes occur with an increase in entropy when the system and its surroundings are considered together.

FREE ENERGY

In practice, it is not always easy to calculate the net entropy change that occurs in the system and its surroundings as a result of a particular chemical reaction, and it is therefore often difficult to tell whether or not a chemical reaction is thermodynamically feasible (spontaneous). This difficulty was overcome by J. Willard Gibbs, an American physicist and mathematician who is responsible for much of the theory of chemical thermodynamics. Gibbs was able to derive a mathematical expression by which energy changes *in the system alone* could be used to predict thermodynamic feasibility, or the overall tendency for a reaction to proceed. According to the relationship derived by Gibbs, the net energy changes that occur

Chemical Thermodynamics

during a chemical reaction are described by the change in free energy (ΔG)* of the reaction. The free energy change takes into account the total energy change when a reaction occurs and the energy that becomes unavailable for work (entropy factor). Thus, *at constant temperature and pressure*

Free energy change = Total energy change − unavailable energy†
(or reaction tendency)

A reaction occurs spontaneously if it results in a *decline* in free energy; that is, if the total free energy of the products is less than the total free energy of the reactants, then free energy is released during the process and the free energy of the reaction is said to decrease. The free energy released during the process becomes available for work, and for this reason free energy is often referred to as "available energy." All this leads to the conclusion that *any spontaneous reaction proceeds toward a state of lowest possible free energy.* By convention, ΔG is given a minus value if a reaction results in a decrease in free energy (i.e., if it is spontaneous); a positive value of ΔG signifies that the reaction will not proceed spontaneously and requires energy to make it go. Reactions that occur with a decrease in free energy are called *exergonic,* and those that proceed with an increase in free energy, and thus may occur only if energy is provided, are referred to as *endergonic* reactions.

Whereas the sign of ΔG determines whether or not a reaction is spontaneous, the magnitude of ΔG represents the amount of energy that *potentially* may be released or absorbed during the course of the reaction. Thus a negative ΔG value indicates the *maximum* amount of work, or useful energy, available from the reaction *if it goes to completion* and a positive ΔG indicates the *minimum* amount of work, or energy, theoretically necessary to force the reaction to completion.

STANDARD FREE ENERGY

We have indicated above that the ΔG of a reaction must be defined with reference to a specific temperature and pressure. This is, of course, merely another way of saying that the value of ΔG varies with the temperature and pressure at which the reaction takes place. If the reaction occurs in aqueous solution, as it does within the cell, then we must specify temperature and *concentration,* rather than pressure. It is apparent that if we want to compare the relative

---

*The Greek letter Δ, or delta, signifies an incremental change in a particular quantity; G refers to free energy, in honor of Gibbs.

†This expression is a summary of the formal equation $\Delta G = \Delta H - T\Delta S$ where ΔG = change in free energy, ΔH = average change in chemical energy or heat content (enthalpy) at constant temperature and pressure, T = temperature in degrees Kelvin (°K) and ΔS = change in entropy. By convention, ΔH is given a minus value if the chemical energy or heat content of the products is less than that of the reactants.

thermodynamic spontaneity as well as the amount of energy absorbed or released for a number of different chemical reactions, we must calculate all ΔG values at the same temperature and concentration. For relative comparisons, a set of standard reaction conditions is used. It is most often defined as a temperature of 25° C and a concentration *of all reactants and products* of one molar (one mole per liter of solution). In order to indicate that a change in free energy was determined for a particular reaction under these so-called *standard state conditions,* the symbol $\Delta G°$, the standard free energy change, is used.

It should be obvious that reactions in the living cell rarely occur at standard state conditions of temperature and concentration, and that $\Delta G°$ values provide only a relative comparison between different reactions. In this regard, it is important to remember that a reaction that is spontaneous under standard state conditions is not necessarily spontaneous under other conditions of temperature and concentration, as for example those conditions that exist within the cell.

FREE ENERGY AND CHEMICAL EQUILIBRIUM

It is possible to relate the standard free energy change of a reaction to the equilibrium constant, K, of the reaction. Since any spontaneous reaction proceeds towards a state of lowest possible free energy, it follows that in any spontaneous chemical reaction, if a certain mixture of reactants and products has a lower free energy than either the reactants alone or the products alone, then the reaction will proceed until this equilibrium mixture is attained. The free energy change of a reaction, then, may be thought of as a measure of the tendency for the reaction to change to a state of chemical equilibrium; that is, to a state of minimum free energy. Thus a reaction with a negative $\Delta G°$ will proceed *to the right under standard state reaction conditions* (a temperature of 25° C and an initial concentration of all reactants and products of one molar) to reach a state of chemical equilibrium, and hence we say that the reaction, as we have written it, occurs spontaneously to yield the indicated products. On the other hand, a reaction with a positive $\Delta G°$ will proceed *under standard state conditions to the left* to reach chemical equilibrium, and we say that the reaction, as we have written it, does not occur spontaneously to yield the indicated products on the right side of the equation. This leads us to the conclusion that a reaction which is spontaneous in one direction is not spontaneous in the other direction. This is an important point because it emphasizes the fact that we call a reaction spontaneous if it proceeds to the right *as we have written the equation.*

For a reaction at equilibrium it can be shown that

$$\Delta G° = -4.6T \log K \quad (1)$$

Chemical Thermodynamics

where T = temperature in degrees Kelvin and K = the equilibrium constant of the reaction.

This mathematical relationship between $\Delta G°$ and K is best understood if we consider the following equilibrium reaction:

$$A + B \rightleftharpoons C + D$$

and thus,

$$K = \frac{[C] \times [D]}{[A] \times [B]}$$

If K is equal to 1 at 25°C, we know that the concentration of the products (C and D) is equal to the concentration of the reactants (A and B) at equilibrium. By substituting this value of K into Equation 1 on page 92, we find that $\Delta G° = 0$ (since the logarithm of 1 is equal to 0). If K is greater than one, then *at equilibrium* the concentration of products exceeds the concentration of reactants, and $\Delta G°$ for the reaction is a negative number (since the logarithms of all numbers greater than 1 are positive). In this case we say that the equilibrium position favors the products, and thus the reaction, occurring under standard state conditions, is spontaneous. If K is less than 1, the concentration of products at equilibrium is less than the concentration of reactants, and $\Delta G°$ for the reaction is a positive number (numbers less than 1 have negative logarithms). The equilibrium position thus favors the reactants and the reaction is not spontaneous under standard state conditions.

FREE ENERGY COUPLING

The equilibrium position of a reaction may be shifted by changing the temperature at which the reaction occurs, or by changing the concentration of reactants or products. In this way, it is often possible to cause an otherwise nonspontaneous reaction to proceed.

In addition to altering the temperature of a reaction or the concentration of reactants or products, it is possible to use the energy produced by a spontaneous reaction (exergonic or "downhill" reaction) to drive a nonspontaneous one (endergonic or "uphill" reaction). A *coupled reaction* of this type is possible if the total free energy of the process decreases. For example, consider the following two reactions:

$$A \longrightarrow B \qquad \Delta G = +1.2 \text{ kcal/mole}$$
$$C \longrightarrow D \qquad \Delta G = -5.0 \text{ kcal/mole}$$

The reaction A $\longrightarrow$ B will not occur spontaneously because it is associated with an increase in free energy. C $\longrightarrow$ D, on the other hand, proceeds with a free energy decrease and hence occurs spontaneously. By coupling the two reactions, the endergonic one

(A ⟶ B) can be made to proceed since the overall free energy change for both reactions considered together is negative.

Individual Reactions:
$$A \longrightarrow B \qquad \Delta G = +1.2 \text{ kcal/mole}$$
$$C \longrightarrow D \qquad \Delta G = -5.0 \text{ kcal/mole}$$

Coupled Reaction:
$$A + C \longrightarrow B + D \qquad \Delta G = -3.8 \text{ kcal/mole}$$

Energy coupling is no abstract concept confined to chemical processes. Consider, for example, the operation of a wind-up clock. As the mainspring on the clock unwinds (a spontaneous process), the free energy released in the process is used to move the hands of the clock (a nonspontaneous reaction). Indeed, energy coupling is the mechanism by which all nonspontaneous processes are made to proceed.

The ADP-ATP System in Energy Transfer

Cells use the coupled reaction principle to drive endergonic metabolic processes. In fact, much of the metabolic machinery of the cell is designed merely to couple energy-yielding reactions such as the breakdown of nutrient molecules with energy-requiring ones such as the synthesis of complex cellular components. The energy-yielding metabolic reactions are not, however, usually *directly* coupled to the energy-requiring reactions. This would be entirely too inflexible, as it would always require the energy-yielding reaction to occur at the same time and in the same place in the cell as the energy-requiring reaction. Actually, we know from our studies of cell structure that catabolic reactions, which involve the release of

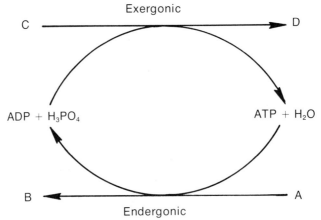

Figure 5-6 The ATP-ADP system. Energy released in exergonic reactions (e.g. C → D) is stored in the ATP molecule and may be used subsequently to drive endergonic reactions (e.g. A → B).

Chemical Thermodynamics

energy, are often spatially separated in the cell from the energy-requiring ones. For example, the primary function of mitochondria in eukaryotic cells is to carry out the breakdown of certain nutrient molecules—exergonic processes which provide most of the energy used in other parts of the cell.

Instead of direct coupling, the cell uses an intermediary cyclic system which serves to store the energy released by exergonic reactions and shuttle it to the site of endergonic ones. The operation of the system is illustrated in Figure 5-6. Free energy released in exergonic reactions of the cell is used to drive the endergonic synthesis of ATP (adenosine triphosphate) from ADP (adenosine diphosphate) and phosphoric acid, according to the following equation:

$$ADP + H_3PO_4 \rightleftharpoons ATP + H_2O \qquad \Delta G° = +7.3 \text{ kcal/mole}$$
(phosphoric acid)

When needed for an energy-requiring process, the energy which is stored in the ATP molecule may be released in the reverse of the reaction above:

$$ATP + H_2O \rightleftharpoons ADP + H_3PO_4 \qquad \Delta G° = -7.3 \text{ kcal/mole}$$

This reaction, termed the hydrolysis (hydro = water; lysis = breakdown) of ATP because it involves the breakdown of ATP by water, is coupled in the cell to energy-requiring reactions, which are thereby able to proceed.

The ATP–ADP system is the primary energy-transferring mechanism in all cells. The use of ATP as an energy-carrier molecule is an extremely flexible way of controlling the flow of energy in a cell. If the amount of ATP in the cell declines, then energy-yielding processes are speeded up or energy-requiring ones are slowed down, or both, until the deficit is made up. The reverse may occur if levels of ATP are plentiful. Because of its central importance in providing energy for numerous metabolic processes, ATP is often called the "energy currency" of the cell. It is important to remember, however, that there is nothing magic or mysterious about the ATP molecule. Many introductory textbooks in cell biology refer to ATP as possessing "high energy bonds," and this terminology unfortunately implies not only that there is something unique about the chemical bonds in the ATP molecule but also that energy is released when these chemical bonds are broken. This is misleading, since we have seen in our discussion of chemical bonding that the disruption of chemical bonds requires relatively large amounts of energy. ATP molecules possess high energy bonds only in the sense that their *hydrolysis* proceeds spontaneously with a relatively high negative ΔG. That is, the hydrolysis of ATP to ADP and phosphoric acid is brought about by the breakdown of a certain bond in the ATP molecule and the formation of other bonds to make the products; the bond that is split is referred to as a high energy bond because the products of the overall hydrolysis reaction have considerably less free energy than the reactants. Indeed, the large amount of free energy that is released upon hydrolysis is related to the chemical structure of the molecule as a whole, and is not dependent solely on the particular bond that is broken by hydrolysis.

Because the term high energy bond is widely used, we will adopt the same terminology to be consistent, but we must keep in mind its inaccuracy. High energy bonds are generally classified as those whose hydrolysis is accompanied by a decrease in free energy exceeding approximately 5 kcal/mole.

The chemical structure of ATP, and its hydrolysis to form ADP and phosphoric acid, are shown in Figure 5–7. Notice that ATP is composed of three parts: a nitrogen-containing organic compound called adenine; ribose, a five-carbon sugar; and three phosphate groups which are linked together by phosphoric acid anhydride bonds (P—O—P). Also notice that the terminal phosphate group is removed from ATP during hydrolysis to ADP. The high energy bond that is split during this reaction (labeled A) is shown by convention as a squiggle (~) to denote that its hydrolysis results in a relatively large negative ΔG of reaction. There is another high energy bond (B) in ATP that may also be hydrolyzed under certain conditions.

Although the hydrolysis of each of the high energy bonds in ATP possesses a $\Delta G°$ of -7.3 kcal/mole, under conditions *inside the cell* each of these bonds shows a free energy of hydrolysis between approximately -8.5 and -14.5 kcal/mole. The exact value depends on the intracellular pH, the concentration of phosphoric acid, the concentration of Mg^{++} ions, and the concentration of ATP relative to that of ADP, among other factors.

In general, any phosphate-containing compound has a high energy of hydrolysis if it possesses the following configuration, in which a phosphate group is bonded to another atom bearing a double bond.

$$-\overset{\overset{X}{\|}}{Y}-O\sim\overset{\overset{O}{\|}}{\underset{\underset{OH}{|}}{P}}-OH$$

In addition, nitrogen- and sulfur-containing compounds with the general configurations

$$-\overset{\overset{O}{\|}}{C}\sim S \quad \text{or} \quad -\overset{\overset{H}{|}}{N}\sim\overset{\overset{O}{\|}}{\underset{\underset{OH}{|}}{P}}-OH$$

are also high energy compounds. As we shall see, many of these compounds play an important role in cellular metabolism. None,

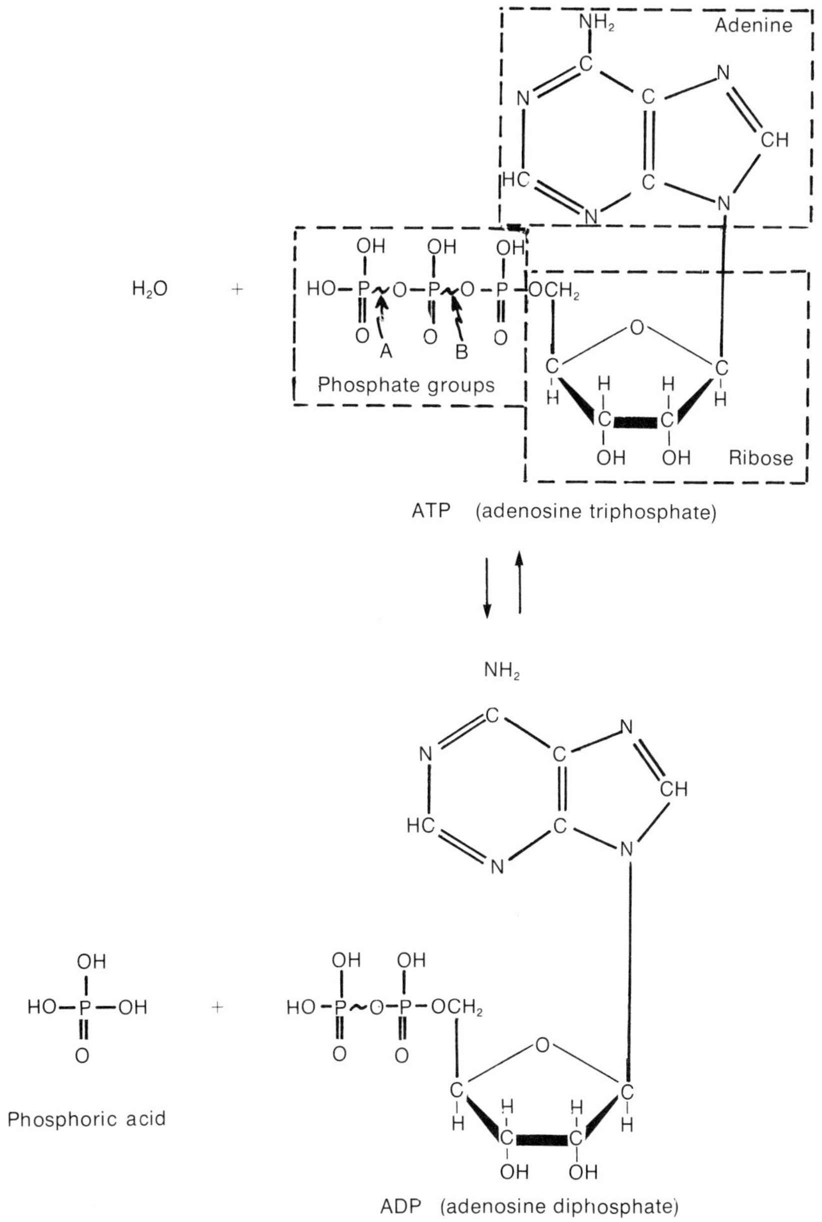

Chemical Thermodynamics

Figure 5-7. The hydrolysis of ATP to ADP and phosphoric acid. To make the reaction easier to follow, the molecules are shown in their un-ionized forms, although they would be ionized within the cell at pH 7.0.

however, is as universally important as ATP, which serves as the energy source for most energy-requiring processes in the cell.

Mechanism of Free Energy Coupling in Cells

In practice, two reactions cannot be coupled simply by allowing the energy-yielding reaction (ATP hydrolysis) to proceed in the vicinity of the energy-requiring reaction. Rather, the energy must be channeled in some way, so that the energy obtained from the exergonic reaction is used to drive the endergonic one, and is not merely dissipated as heat. In living cells, when an endergonic reaction is coupled to the hydrolysis of ATP, two completely separate reactions do not actually occur. Instead, the coupling is usually achieved by one or more successive "group transfer" reactions, in which the high energy bonds of ATP are transferred to another molecule. Assume, for example, that the endergonic reaction of C + D to form E is coupled to the hydrolysis of ATP. The two separate reactions and the net reaction may be written

Endergonic reaction: $C + D \rightleftharpoons E$ (2)

Exergonic reaction: $ATP + H_2O \rightleftharpoons ADP + HO-\underset{\underset{OH}{|}}{\overset{\overset{O}{\|}}{P}}-OH$ (3)

Overall reaction: $C + D + ATP + H_2O \rightleftharpoons E + ADP + HO-\underset{\underset{OH}{|}}{\overset{\overset{O}{\|}}{P}}-OH$ (4)

In the cell, the reaction may actually proceed in the following way

Endergonic reaction: $ATP + C \rightleftharpoons C \sim \underset{\underset{OH}{|}}{\overset{\overset{O}{\|}}{P}}-OH + ADP$ (5)

Exergonic reaction: $C \sim \underset{\underset{OH}{|}}{\overset{\overset{O}{\|}}{P}}-OH + D + H_2O \rightleftharpoons E + HO-\underset{\underset{OH}{|}}{\overset{\overset{O}{\|}}{P}}-OH$ (6)

Overall reaction: $C + D + ATP + H_2O \rightleftharpoons E + ADP + HO-\underset{\underset{OH}{|}}{\overset{\overset{O}{\|}}{P}}-OH$ (7)

In Equation 5, a high-energy phosphate group of ATP is transferred to C, which thus becomes a high-energy compound. The reaction is an endergonic one (its equilibrium lies to the left), but it proceeds because it is coupled by virtue of the common intermediate compound C~P(=O)(OH)—OH to Reaction 6, which is highly exergonic (its equilibrium lies *far* to the right).

One way of looking at the process is that Reaction 6, by removing C~P(=O)(OH)—OH, pulls Reaction 5 to the right. When we say that the exergonic reaction drives the endergonic one, or that the overall reaction proceeds with a negative ΔG, we mean that the equilibrium of the exergonic reaction lies further to the right than the equilibrium of the endergonic reaction lies to the left. We say the reactions are "coupled" because a product of one reaction is a reactant of the other. Thus the only condition under which the energy from an exergonic reaction can be channeled so that it drives an endergonic reaction occurs when the two reactions *share a common intermediate compound*.

The coupling of two reactions need not necessarily occur by the same mechanism as the one shown here. Two reactions can be coupled as long as they share a common intermediate compound and as long as the overall reaction proceeds with a negative ΔG. In our later discussions of cellular metabolism, we will see that metabolic reactions are coupled in a number of different ways.

CHEMICAL KINETICS: REACTION RATE AND PATH

The Second Law of Thermodynamics tells us whether a given reaction (or process) is energetically feasible, not whether it will occur. Neither does the Second Law say anything about the *rate* of the reaction or the reaction pathway. It is concerned only with the energetic feasibility of going from the initial state (reactants) to the final state (products), and says nothing about how the process actually occurs. Consider, for example, the reaction

$$C_6H_{12}O_6 + 6O_2 \rightleftharpoons 6CO_2 + 6H_2O$$

$$\Delta G° = -686 \text{ kcal/mole}$$

$$K = 10^{500}$$

We know from the values for $\Delta G°$ and K that this reaction will proceed spontaneously to completion when it occurs, and that it will release a large amount of free energy. However, if we were to place solid glucose in a sealed container with oxygen gas at a pressure of 1 atmosphere and leave the container at 25°C, the reaction would proceed so slowly as to be negligible. In other words, although the reaction is energetically feasible, it does not occur at an appreciable rate under these conditions. The overall reaction also does not specify whether carbon dioxide and water are produced in a one-step reaction (as in the combustion of glucose), or as a result of a complex metabolic pathway involving many intermediate compounds (as in the breakdown of glucose in the living cell); that is, it does not tell us the path by which the reaction occurs. In our attempts to understand cellular metabolism we are, of course, interested in both of these factors: the path, because it tells us how a specific chemical process occurs in the living cell, and the rate, because metabolic reactions of the cell must occur rapidly in order to be useful. The branch of chemistry that involves the study of the paths and rates of chemical reactions is known as *chemical kinetics*.

REACTION RATES

The rate of a reaction may be defined as the number of molecules (or atoms) reacting per unit time. Reaction rate has been explained by a combination of two theoretical models, the so-called *collision theory* and the *transition state theory*.

Collision Theory

The collision theory provides a qualitative description of the factors that affect reaction rate and proposes that reactions occur as a result of collisions between the reacting species. We know, however, that the rate of most organic reactions is exceedingly slow, and often negligible, at temperatures normally encountered in living organisms, even if the reactants are present at high concentrations and the reaction possesses a large negative free energy. Since the frequency of collisions is large at high concentrations of reactants, we may conclude that most collisions between reactants do not result in a reaction. To account for this observation, the collision theory proposes that in order to cause a reaction, a collision must occur not only with the proper orientation for the reacting groups to come together but also with sufficient energy to bring about the necessary electronic and atomic rearrangements in the reacting atoms or molecules (Fig. 5–8). Thus, according to the collision theory, the rate of reaction depends upon the number of "effective" collisions that occur in a given time; that is, the number of collisions possessing the orientation and sufficient energy to cause a reaction.

The minimum amount of energy necessary to bring about a reaction is called the activation energy (free energy of activation); its value is characteristic of a given reaction occurring under defined conditions. Consequently, if the activation energy of a particular re-

Chemical Kinetics: Reaction Rate and Path

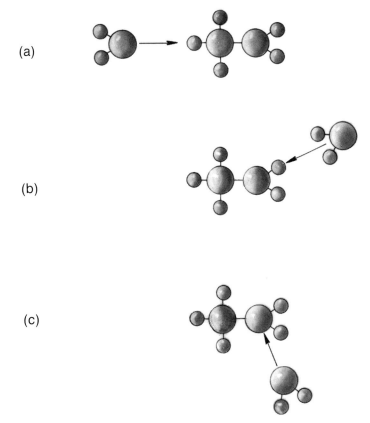

Figure 5-8
Two molecules must collide in a specific orientation (as depicted, for example, in a) in order to react; collisions in other orientations (such as those shown in b and c) do not bring about a reaction.

action is large, then most molecules will not collide with enough energy to react.*

It is useful to think of the activation energy as an energy barrier that prevents the reaction from occurring: The reaction can proceed only when the barrier is surmounted, even if the overall reaction possesses a large negative ΔG. The energy requirement of a reaction is often illustrated by an energy diagram, shown in Figure 5-9.

Transition State Theory

The transition state theory expands upon the collision theory by proposing that an unstable *transition state,* or *activated complex,* is formed when reactants collide with the proper orientation and with

---

*With respect to energies of molecules, we know that all molecules in a population do not have the same energy of motion but that the largest proportion will have some average energy, while some few have energies considerably larger, and some have energies smaller, than the average.

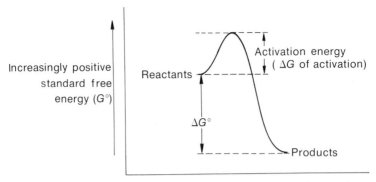

Figure 5-9 Free energy diagram of an exothermic reaction. Although the reaction possesses a negative ΔG value, it will not occur unless an amount of energy at least equal to the activation energy is supplied.

a collision energy at least equal to the activation energy. This unstable complex is assumed to be in equilibrium with the reacting molecules, and can dissociate either to give the new products of the reaction or to produce the original reactants. The overall reaction rate is determined by the concentration of the activated complex at equilibrium; the greater the concentration of activated complex, the greater will be the frequency of its dissociation into products and hence the greater the reaction rate.

The formation of an activated complex is often represented by the following general equation, which describes the reaction of A_2 and B_2 molecules to yield AB molecules.

$$A_2 + B_2 \rightleftharpoons [A_2B_2] \rightarrow 2AB$$
$$\text{(reactants)} \quad \text{(activated complex)} \quad \text{(products)}$$

It is perhaps easier to visualize this process if we illustrate it diagrammatically, as in the free energy diagram shown in Figure 5–10.

Factors Influencing Reaction Rate

A number of factors have been shown to alter the rate of a reaction. For example, raising the temperature of the reacting system will increase both collision frequency and the energy of collision by increasing the average motion of the molecules. The absorption of light energy by certain molecules has a similar effect; indeed, such *photochemical reactions* are believed to have been one of the primary means by which simple organic compounds were produced in the atmosphere of the primitive earth. Increasing the concentration of the reactants will increase the probability of collisions and hence the frequency of effective collisions. Collision frequency is also increased by breaking the reactants into smaller particles.

Another way to increase the rate of a reaction is by the use of substances called *catalysts*. When mixed with the reactants, the

Chemical Kinetics: Reaction Rate and Path

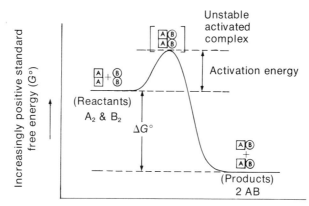

Free energy diagram illustrating the formation of the unstable activated complex. (Adapted from ENERGETICS, KINETICS AND LIFE: AN ECOLOGICAL APPROACH by G. Tyler Miller Jr. © 1971 by Wadsworth Publishing Company, Inc. Belmont, California 94002. Reprinted by permission of the publisher.)

Figure 5-10

catalyst apparently increases the reaction rate by lowering the activation energy for the reaction. Hence in the presence of the catalyst a greater proportion of reactant molecules will possess the energy necessary to form the activated complex than in the absence of the catalyst (Fig. 5–11).

Catalysts lower the activation energy by altering the path of the reaction. They may, for example, combine with reactant molecules

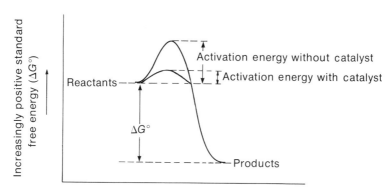

Free energy diagram illustrating the effect of a catalyst on the activation energy of a reaction. By decreasing the activation energy of a reaction, a catalyst increases the rate of the reaction.

Figure 5-11

to form an activated complex that requires less energy for its formation than did the original activated complex.

It is a characteristic of catalysts that they are not used up or permanently altered during a chemical reaction. Because of this, a catalyst can be used repeatedly, and consequently only small amounts are required to catalyze a particular chemical reaction. It is important to understand that catalysts do not operate by altering the equilibrium constant (K) or the free energy change (ΔG) of a reaction. Hence a catalyst is not able to make a reaction occur that is thermodynamically unfeasible. Instead, by increasing the rate of the forward as well as the reverse reaction, a catalyst is merely able to speed up the approach to chemical equilibrium.

Enzymes: The Biological Catalysts

In Chapter 3, we noted that living systems are composed of organic compounds which, because of their chemical structure, are quite stable and relatively unreactive. Indeed, almost all the chemical reactions that normally occur in the cell require a large activation energy and would not proceed, or would proceed at exceedingly slow rates, at temperatures found in living organisms were it not for the presence of the biological catalysts called *enzymes*. By determining the metabolic capabilities of the cell, enzymes play a critical role in every living system.

Nomenclature of Enzymes. By convention, an enzyme is named by adding the suffix *-ase* to a term which describes the reaction the enzyme catalyzes. For example, the enzyme that catalyzes the formation of fumaric acid by the removal of hydrogen atoms from succinic acid is called succinic dehydrogenase. Of course, succinic dehydrogenase also catalyzes the reverse reaction, the addition of hydrogen atoms to fumaric acid to form succinic acid. Therefore, in the case of freely reversible reactions, it is somewhat arbitrary how the enzyme is named, and for this reason succinic dehydrogenase might well have been called "fumaric hydrogenase." However, enzymes are usually named in a way that describes the dominant direction of the reaction as it occurs in the cell. A few enzymes were given names before any nomenclature convention was adopted, and these names generally end with *-in*.

Chemical Structure of Enzymes. We have mentioned before that enzymes are proteins, and that the instructions for their synthesis are contained in the DNA code. Each different enzyme has a characteristic amino acid sequence. Because of chemical affinities between certain amino acids and repulsive forces between others, the amino acid sequence imparts to the enzyme molecule a specific three-dimensional configuration stabilized by intramolecular forces. These may include strong forces such as covalent interactions as well as weaker forces such as the electrostatic attraction between oppositely charged portions of amino acid molecules (ionic forces), hydrophobic bonds, and hydrogen bonds and other polar forces.

In some enzymes, the protein is normally complexed with other non-proteinaceous substances called *cofactors,* which are required for the catalytic activity of the enzyme. It should be emphasized that in enzymes of this type, *neither the protein portion nor the cofactor has any catalytic activity by itself.*

Cofactors may include metal ions or certain organic molecules. If the cofactor constitutes an integral part of the structure of the enzyme molecule and is not readily dissociable under normal intracellular conditions, it is referred to as a *prosthetic group.* In some cases, the cofactor is an organic molecule which may reversibly dissociate from the protein portion of the enzyme and which, therefore, is not an integral part of the enzyme molecule. This type of cofactor is referred to as a *coenzyme.* The coenzyme molecule is not usually attached to the protein, and associates with the protein only during catalysis. In a sense, then, a coenzyme is one of the substrates of the reaction. Many prosthetic groups and coenzymes are manufactured from *vitamins,* and it is for this reason that vitamins are essential to normal metabolism. Since many organisms are unable to synthesize certain vitamins, these compounds must be obtained in the diet.

Specificity of Enzymes. Because each enzyme has a unique molecular structure, and because the catalytic activity of an enzyme is dependent upon its molecular structure, enzymes are *specific* in their action. Different enzymes, however, possess different degrees of specificity. Some, such as succinic dehydrogenase, are highly specific, and catalyze only a single reaction (in both directions). Others have wider specificity, and catalyze the same general reaction involving different molecules of the same type. Papain, for example, is an enzyme that is able to catalyze the hydrolysis of covalent bonds between different amino acids in proteins.

Effects of Temperature and pH on Enzyme Activity. The rate of any chemical reaction is increased by an increase in temperature, and enzyme-catalyzed reactions are no exception *as long as the increase in temperature does not adversely affect the structure of the enzyme molecule.* The temperature dependence of enzyme-catalyzed reactions explains why insects and other animals which do not regulate their body temperature are lethargic when the environmental temperature is low and show increased activity as the temperature, and hence their metabolic rate, increases. Indeed, the chirping frequency of crickets depends upon metabolic rate and has been reported to be a very reliable indication of environmental temperature.

Because enzymes are proteins, they are subject to thermal inactivation; that is, the weak intramolecular bonds which contribute to the specific three-dimensional shape of an enzyme molecule are disrupted at high temperatures. This process, known as *denaturation* when the characteristic configuration of a molecule is destroyed by whatever means, results in the loss of catalytic activity. The net effect of an increase in temperature on the rate of an enzyme-catalyzed reaction may be summarized as follows: At temperatures below the denaturation point of the enzyme, the rate is increased by a rise in temperature; at temperatures beyond this point, the rate decreases as the effects of denaturation predominate.

The denaturation temperature of different enzymes varies. Many enzymes are rapidly inactivated at temperatures exceeding 40°C. Others, such as those present in organisms adapted to growth in hot springs, are active at temperatures as high as 70°C, or sometimes even higher.

The rate of an enzyme-catalyzed reaction is also affected by the pH of the medium in which the reaction takes place; that is, each enzyme shows optimal activity at a certain pH and lessened activity at pH values below and above this point. Although most enzymes show a pH optimum around neutrality, some enzymes may have optimal activity at very low, and others at very high, pH values. For example, pepsin, which catalyzes the hydrolysis of proteins in the acid environment of the stomach, shows a pH optimum of about 2; the enzyme arginase, which catalyzes the breakdown of the amino acid arginine, has optimal activity at about pH 9.5.

The pH dependence of enzyme catalysis may be explained on the basis of charge effects. Since the pH may influence the charge on both the enzyme and substrate molecules, it plays an important role in determining the catalytic activity of the enzyme. Certain charged groups may be essential to catalytic activity, and these may exist in a particular charged state only within a certain pH range. At extremes of pH, alterations in the charge on the enzyme molecule may lead to the disruption of intramolecular forces and consequent denaturation.

It should be mentioned at this point that temperature and pH effects on individual enzymes are often determined under artificial situations outside the living cell, and it is possible, indeed probable, that enzymes may show quite different characteristics under conditions encountered within the cell, such as when they are bound to cellular membranes. We must keep in mind that eukaryotic cells are highly compartmentalized, and that particular enzymes are generally isolated within specific structural organelles or in specific regions. This is also true, but in a more limited sense, with prokaryotic cells, where many metabolic reactions take place on cellular membranes. We know very little about pH and temperature differentials in localized regions within an individual cell, but because of the pronounced effect of temperature and pH on enzyme activity, it is tempting to speculate that localized changes in temperature and pH within a cell may serve to regulate metabolic activity.

Efficiency of Enzymes. Most enzymes work extremely fast, and it has been reported that some enzymes are able to catalyze as many as one million reactions per minute. Indeed, enzymes are considerably more efficient than other types of catalysts.

Mechanisms of Enzyme Action. It is not clearly understood how an enzyme lowers the activation energy of a specific reaction, although it is apparent that the specific amino acid sequence of the enzyme molecule and its three-dimensional shape are crucial in this respect. Indeed, in addition to adverse effects on enzyme activity caused by alteration in the shape of an enzyme molecule, a mutation in the DNA of the cell resulting in a change of a single amino acid in an otherwise normal amino acid sequence will sometimes completely destroy the activity of the enzyme.

The importance of enzyme structure to the action of enzymes has been explained on the basis of the so-called "lock and key" hypothesis. According to this hypothesis, the first step in an enzyme-catalyzed reaction is the attachment of the reactants (also referred to as *substrates*) to a particular acceptor area, or "active site," on the surface of the enzyme molecule to form an unstable enzyme-substrate complex. The formation of the enzyme-substrate complex brings the reactants into close proximity to one another and in some way lowers the activation energy of the reaction, possibly by increasing the probability of interaction between reactant molecules, and thus facilitates the conversion of reactants to products. The products are then released from the active site of the enzyme, and the enzyme is able to pick up more reactant molecules and repeat the process (Figs. 5-12 and 5-13).

According to the lock and key hypothesis, the specificity of an enzyme depends upon the fact that only certain substrates, because of their precise molecular conformation and charge distribution, can fit onto the active site of the enzyme, much as a key fits into a lock. The precise geometrical relationship that must exist between the enzyme and substrate also provides an explanation for the importance of enzyme structure to the catalytic process: Alteration of the amino acid sequence or three-dimensional shape of the enzyme molecule may change the active site in such a way that substrate molecules can no longer attach.

More recently, the lock and key hypothesis has been modified by D.E. Koshland and his coworkers, who have suggested that the active site of an enzyme, instead of being rigid, is flexible, and that binding to the correct substrates induces specific conformational changes in the enzyme molecule which result in catalysis (Fig. 5-14). This "induced fit" hypothesis is supported by physical data which show that certain enzymes change in conformation during binding with their substrates. In an attempt to provide a more rigorous explanation for enzyme catalysis, Koshland has also proposed an "orbital steering" hypothesis. This suggests that enzymes are somehow able to align the outermost electrons of substrate molecules and thus promote the overlap of electrons which, as described in Chapter 3, is necessary for chemical bond formation.

Inhibition of Enzyme Activity. The catalytic activity of enzymes can be destroyed not only by denaturation but also by certain chemical substances called *inhibitors,* which produce their effects by binding to the enzyme molecule. Chemical inhibition of this type can be *reversible* or *irreversible.*

REVERSIBLE INHIBITION. Reversible inhibition may occur by several mechanisms.

(1) If the inhibitor substance is structurally similar to the sub-

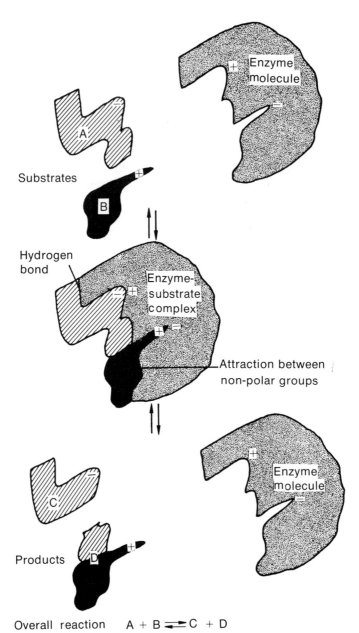

Figure 5–12 Diagrammatic representation of the lock and key hypothesis of enzyme action. Notice that the same enzyme catalyzes the reaction in either direction, hence our designation of molecules as substrates and products is arbitrary. (Drawing modified from J.D. Watson [1970], *Molecular Biology of the Gene,* W.A. Benjamin, Inc., New York.)

strate molecule, the inhibitor may attach loosely to the active site of the enzyme and thus prevent the binding of the substrate. Because this type of inhibition can be reversed by increasing the concentration of substrate, it is called *competitive inhibition,* since the substrate and inhibitor appear to compete for the active site. The degree of inhibition not only depends on the relative concentrations

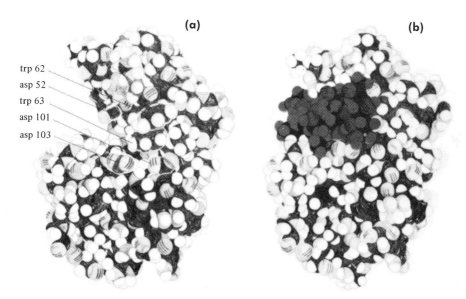

Figure 5-13

"Space-filling" molecular models of (a) the enzyme lysozyme without bound substrate and (b) the enzyme-substrate complex. A space-filling model gives a good representation of the three-dimensional structure of a molecule, because it indicates the effective volume taken up by the nucleus and electrons of each atom in the molecule. Each sphere or partial sphere in the model represents a different atom. (*From* S.L. Miller and L.E. Orgel [1974], *The Origins of Life on the Earth*, Prentice-Hall, Inc., Englewood Cliffs, New Jersey, p.67.)

of substrate and inhibitor but also on the affinity of the substrate and the inhibitor for the enzyme molecule.

Competitive inhibition is the basis for the antibacterial action of sulfanilamide and related sulfa drugs. These compounds compete with para-aminobenzoic acid for the active site of a bacterial enzyme which catalyzes the synthesis of an essential coenzyme from para-aminobenzoic acid and other precursors. Since humans are not able to synthesize this particular coenzyme and must obtain it from their diet, their metabolism is not affected by the sulfa compounds.

(2) A second type of reversible inhibition is not affected by increasing the substrate concentration, and is therefore termed *non-competitive inhibition*. This type of inhibition results when the inhibitor substance binds to some site on the enzyme molecule other than to the active site. This binding is thought to induce a conformational change in the enzyme molecule which does not interfere with the attachment of the substrate to the active site, but which greatly reduces the catalytic activity of the enzyme.

(3) Many reversible inhibitors show properties characteristic of both competitive and non-competitive inhibition.

IRREVERSIBLE INHIBITION. The other type of chemical inhibition, irreversible inhibition, occurs when an inhibitor attaches so tightly to the enzyme as to form an extremely stable complex. This may occur by binding of the inhibitor to the active site of the enzyme or to another site on the enzyme where it affects catalytic activity.

① *Catalysis occurs*

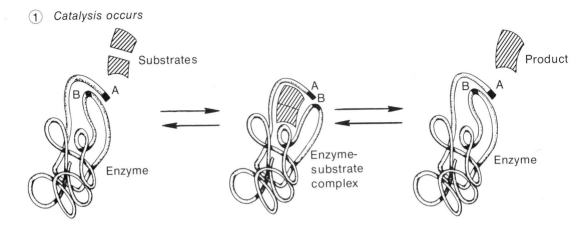

② *Catalysis does not occur*

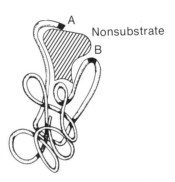

Figure 5–14 The "induced fit" hypothesis of enzyme action, as proposed by D.E. Koshland. As visualized in the diagrams drawn here, catalysis will occur only if chemical groups A and B of the enzyme align properly following attachment of substrate; the wrong substrate molecule will prevent the proper alignment of A and B, and catalysis will not take place. As in Figure 5–12, note that the same enzyme catalyzes the reaction in both directions.

REACTION PATH

So far, we have confined our discussion of chemical kinetics to reaction rates, and have not mentioned the other important aspect of kinetics, the path or *mechanism* by which reactions occur. In future chapters, however, we will consider a number of metabolic reaction pathways in great detail, and will demonstrate the importance of the path of a reaction to the efficiency of cellular metabolism.

LIFE, THERMODYNAMICS, AND KINETICS

How does the Second Law apply to energy transfer and utilization in living systems, and what does the Second Law tell us about energy requirements during the origin and development of living systems? We know that cell structure is extremely ordered and thus

Life, Thermodynamics, and Kinetics

highly improbable. Energy is therefore required to maintain this condition, since the breakdown of cellular organization is associated with a large increase in entropy and thus is favored thermodynamically.

Cells preserve their low entropy by increasing the entropy of their surroundings. This increase in entropy occurs in two ways. Cells obtain complex nutrient substances from the environment and, in the process of extracting energy from them, break them down into simpler substances which they return to the environment. In addition, heat that is lost in these and other chemical transformations in the cell increases the entropy of the surroundings. The end result is a net increase in entropy when both the organism and its surroundings are considered as a whole.

The origin of living systems initially involved the formation of increasingly complex organic molecules from the simple gases present in the atmosphere of the primitive earth. The production of these compounds was associated with a decrease in entropy and hence their synthesis required a supply of energy, which was available from the surroundings in the form of ultraviolet radiation, lightning, heat, radioactivity and other sources. Eventually primitive, self-replicating morphological entities (anaerobic heterotrophs) were formed that were capable of maintaining their organized structure (low entropy) by extracting energy from abiotically synthesized molecules present in the environment. ATP, used as an energy donor for energy-requiring reactions, was probably first obtained from the environment, but eventually selection pressure favored the establishment of metabolic pathways which were capable of coupling energy-yielding reactions with ATP production. Crude at first, these pathways became increasingly efficient at extracting free energy from available organic compounds, particularly as enzymes evolved which permitted kinetically difficult reactions to occur.

The evolution of photosynthetic autotrophs further decreased the dependence of living systems on the environment for obtaining necessary organic compounds. Capable of utilizing the energy of the sun's rays to form ATP, photosynthetic organisms were provided with an abundant source of chemical energy which permitted the synthesis of complex organic molecules from carbon dioxide and water—a process which proceeds with a decrease in entropy and, without an external energy supply, would not occur. The organic materials synthesized by photosynthetic organisms could then be ingested by heterotrophic organisms and broken down to supply energy for essential, energy-requiring metabolic processes.

It is perhaps useful to think of energy as flowing in one direction through biological systems: Energy is neither created nor destroyed, but in the process of life the energy of the sun's rays, capable of performing useful work, is eventually transformed into heat energy that results in the random motion of molecules and that, in this form, is useless to living organisms. Entropy increases in the process, useful energy is lost and so, also, the Second Law is obeyed. It is thus apparent that biological processes, like all others, are governed by the laws of thermodynamics. Indeed, the struggle for survival is, in a very basic sense, a struggle to maintain low entropy.

REFERENCES

Bent, H. A. (1965). *The Second Law.* Oxford University Press, New York.
Blum, H. F. (1955). *Time's Arrow and Evolution.* Princeton University Press, Princeton, New Jersey.
Crawford, F. H. (1963). *Heat, Thermodynamics and Statistical Physics.* Harcourt, Brace and World, Inc., New York.
Howland, J. L. (1968). *Introduction to Cell Physiology: Information and Control.* Macmillan, Inc., New York.
Koshland, D. E., Jr., and M. E. Kirtley (1966). Protein structure in relation to cell dynamics and differentiation. In *Major Problems in Developmental Biology,* 25th Symposium, p. 217. Academic Press, Inc., New York.
Lehninger, A. L. (1965). *Bioenergetics.* W. A. Benjamin, Inc., New York.
Miller, G.T. (1971). *Energetics, Kinetics and Life: An Ecological Approach.* Wadsworth Publishing Co., Inc., Belmont, California.
Rosenberg, E. (1971). *Cell and Molecular Biology—An Appreciation.* Holt, Rinehart and Winston, Inc., New York.
Villee, C. A. (1972). *Biology.* W. B. Saunders Co., Philadelphia.

CHAPTER 6

The Primitive Earth and Its Environment

The discoveries of what appear to be fossil organisms in the Precambrian Fig-Tree and Onverwacht geological formations tell us that living systems were probably present on the earth at least as early as 3.2 billion years ago. If this in fact was the case, then the first living organisms must have appeared sometime during the 1.4 billion years following the earth's formation approximately 4.6 billion years ago. The actual time when living systems appeared cannot be estimated with greater precision than this. Some biologists believe that the formation of living organisms occurred within as little as one million years after the formation of the earth, and others, stressing the improbability of such an event, think that hundreds of millions of years would be necessary. In fact, all arguments to the contrary, there is no way of estimating how much time is required: The development of life on the primitive earth is a historical process of the kind that cannot be repeated, and in actuality it may have taken a very short time or very long time.

If the first organisms were, as we believe, similar in basic design to contemporary organisms, then our studies of modern organisms can tell us a great deal about the general structural and functional characteristics of the earliest living systems. Although their metabolism and structure were undoubtedly far simpler than that of the simplest contemporary unicellular prokaryotes, the earliest organisms must have used many of the same organic compounds found in their modern descendants.

It is probably fair to say that the vast majority of modern biologists accept some version of the materialistic theory of life's origin. Indisputably, there is considerable disagreement concerning the precise details of how life evolved, but at the same time there is remarkably little controversy about the overall scheme—that living systems were formed as a result of a gradual and continuing evolution of organic material. The wide acceptance of this theory is based in large part on the ease with which organic compounds characteristic of contemporary living systems can be produced in laboratory experiments, using starting materials and experimental conditions which simulate those thought to have been present during the early history of the earth. Indeed, these so-called *simulation experiments,* performed by a large number of investigators, have produced not only many of the simple organic building blocks of modern organisms but also larger organic molecules and polymers similar or identical to the complex molecules found in biological systems. In addition, it has even been possible, again recreating primitive earth conditions, to form molecular aggregates which display a number of the physical and chemical characteristics common to living cells.

If we are to believe that simulation experiments represent a plausible model for the origin of living systems on the primitive earth, then our first task is to demonstrate that the experimental conditions that have been chosen for the various simulation experiments are neither far-fetched nor unreasonable. To do this, we must begin with a discussion of what we know, or more precisely, what we have been able to deduce, about the environment on the primitive earth.

THE FORMATION OF THE EARTH

THE ORIGINAL KANT—LAPLACE THEORY

In the second half of the eighteenth century, the German philosopher Immanuel Kant and Pierre de LaPlace, a French mathematician and astronomer, independently proposed that the sun and planets were formed by local aggregations of solid particles and gas within a vast, rotating cosmic dust cloud. The *Kant–LaPlace theory* for the origin of the solar system was soon abandoned, however, because the theory seemed unable to account for the fact that the

Figure 6–1 The solar system. Note that the planets revolve around the sun in a single plane. (Adapted from H.E. Newell [1967], NASA Publication EP–47.)

The Formation of the Earth

sun possesses much less of the rotational motion, or angular momentum, of the solar system than do the planets and their satellites (Fig. 6-1).

THE COLLISION THEORY

To explain the increased rotational motion of the planets relative to the sun, the formation of the solar system was subsequently described in terms of the so-called *collision theory,* which was widely accepted by astronomers until the early 1940s. According to the collision theory, the earth and other planets were derived from the sun as a result of a near collision between it and another star. It was envisioned that the large gravitational field of the approaching star would have torn incandescent material from the sun; this material, thrown into orbit around the sun, would have thus acquired most of the rotational motion of the solar system, and would have given rise to the planets upon cooling.

For a number of reasons, the collision theory is no longer acknowledged as a reasonable explanation for the formation of the solar system. In the first place, it is considered extremely unlikely, given the vast distances which normally separate stars in space, that two stars could have passed one another closely enough to produce a gravitational force of sufficient magnitude to tear away bodies as large as our planets. This conclusion is supported by evidence which suggests that there are a number of planetary systems similar to our own, and consequently it seems extremely improbable that all these systems would have been formed by near-collisions. Indeed, even if our sun had been involved in a near-collision with another star, it can be calculated that the material torn from the sun would have possessed so much kinetic energy that it would have disintegrated.

THE COLD ACCRETION THEORY

For the past several decades, geologists and astronomers have generally accepted a modern version of the Kant-LaPlace theory known as the *cold accretion theory.* This theory overcomes the original objections to the Kant-LaPlace theory by explaining how a planetary system with rotational properties similar to our own could have been formed from a rotating, turbulent mass of cosmic dust and gas (Fig. 6-2).

It is possible to show that the aggregation of dust particles and gas in a cosmic cloud would eventually have caused the cloud to collapse into a disc, with particles rotating in the plane of the disc about its center. The cold accretion theory contends that initially the sun was formed in the center of the disc by the accretion of gas and solid particles. The physical law of the conservation of angular momentum tells us that the rotational motion of the particles in the cloud must have been conserved during the

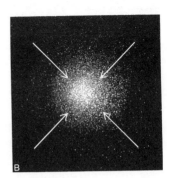
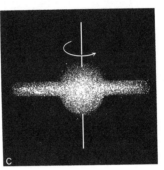
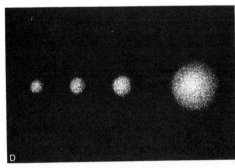

Figure 6-2 The formation of the solar system from a turbulent cloud of cosmic dust and gas, as envisioned by the cold accretion theory. *a.* Dust particles and gas molecules form a turbulent cloud. *b.* and *c.* The cloud collapses into a disc. *d* and *e.* Successive stages of further aggregation, resulting in the formation of the sun and the planets. (From Pasachoff, J.M. [1977] *Contemporary Astronomy: An Introduction.* Philadelphia, W. B. Saunders Company.)

The Formation of the Earth

aggregation process. Consequently, it is assumed that most of the angular momentum of the original dust cloud was transferred to particles traveling in orbit about the newly formed sun; these particles eventually condensed in concentric bands around the sun, forming the planets. Thus, once the dust cloud had formed into a disc, all rotation would have taken place within the plane of the disc. The cold accretion theory is supported by the observation that all the planets in our solar system revolve around the sun in orbits which form a single plane, and that this plane is the same as the plane of rotation of the sun.

According to the cold accretion theory, the dust cloud that formed the solar system was initially at a relatively low temperature. However, as the earth increased in size and mass as a result of the accretion process, the center of the earth would have been compressed by gravitational force. The heat generated by this pressure as well as by the decay of radioactive elements, would have greatly increased the temperature of the earth's interior, causing it to melt.

There is considerable disagreement concerning the extent to which the *surface* of the earth was heated during this process. At one extreme are those who contend that the accumulation of material occurred so slowly that the heat energy released in the process was radiated away without heating the surface of the earth to more than several hundred degrees Centigrade. Scientists at the other extreme believe that the accretion process was rapid and that a great deal of heat, produced in a short time, could not have been dissipated faster than it accumulated; eventually this heat would have melted the entire earth including the surface. The majority of scientists now favor the view that the earth was once entirely molten, although a significant minority disagree with this conclusion. If indeed the surface was molten, it probably cooled down substantially within several hundred million years of the earth's formation, perhaps even to temperatures similar to those on the earth today. This estimate is based upon the fact that the oldest known rocks in the earth's crust have been dated by radioisotopic methods at about 3.9 billion years. Since the age of a rock refers to the time of its most recent solidification, we know that significant crustal cooling must have occurred at least by this time and probably before.

THE AGE OF THE EARTH

How long ago was the earth formed? Estimates of the earth's age have been obtained from several independent sources. By analyzing crustal rocks of different ages for their lead isotope content and by making a number of complicated assumptions about the geological history of the earth, it is possible to infer the age of the earth. By this method, an age of 4.6 billion years is obtained.

Other independent values for the age of the earth have been obtained by dating meteorites and rocks from the moon. Meteorites are objects which fall to the earth from space; they are generally believed to come from the asteroid belt which is made up of small bodies occupying orbits mostly between Mars and Jupiter. Most of the currently accepted theories assume that the moon and the meteorites were formed at about the same time and from the same cosmic dust cloud as the earth. Meteorites, dated by the potassium–argon, rubidium–strontium and uranium–lead methods, show ages of 4.6 billion years. Lunar rocks from Apollo missions Twelve, Fourteen, and Fifteen have been dated at 4.5 billion years, and lunar soils show ages of 4.6 billion years.

Since the results of these three independent methods of estimating the age of the earth agree so well, it seems reasonable to assume that the earth was formed approximately 4.6 billion years ago. Although this estimate may prove to be inaccurate by several hundred million years, a potential error of this magnitude will not affect our arguments concerning the origins of life.

THE ATMOSPHERE OF THE PRIMITIVE EARTH

Although there are not sufficient geochemical data to provide us with a precise and conclusive description of conditions prevailing on the surface of the primitive earth, it is possible to construct some reasonable models by drawing on a large amount of available circumstantial evidence. It should be mentioned from the outset, however, that indirect evidence is often subject to a wide variety of interpretations, and that there is a great deal of controversy particularly with regard to the composition of the primitive atmosphere. To some degree, therefore, the conclusions of the following discussion must represent the predilection of this writer.

ESCAPE OF THE ORIGINAL ATMOSPHERE

One type of evidence which tells us something about the processes occurring during the formation of the earth's atmosphere is obtained by comparing the average relative abundances of the chemical elements in the whole earth and in the universe (Tab. 6–1). These data show that the earth has exceedingly small amounts of the inert noble gases (helium, neon, argon, krypton, and xenon) compared to the amounts of these gases in the universe as a whole. Since only helium and hydrogen gas (H_2) can escape from the earth's gravitational field today, we are led to the conclusion that the heavier noble gases must have escaped from the original small, cool dust particles before the earth was fully formed and thus before any significant gravitational field had developed.

If the noble gases were lost, then it is also reasonable to assume that small volatile molecules with masses less than that of xenon (the heaviest of the noble gases) would have been lost to a large degree at the same time. These would include molecules such as methane (CH_4), ammonia (NH_3), water (H_2O), nitrogen gas (N_2), oxygen gas

The Atmosphere of the Primitive Earth

Table 6-1 RELATIVE ABUNDANCES OF SOME ELEMENTS IN THE UNIVERSE AND ON EARTH (ATOMS/10,000 ATOMS OF SILICON)*

ELEMENT	ATOMIC WEIGHT	UNIVERSE†	WHOLE EARTH‡
Elements forming volatile substances			
Hydrogen	4.0	4.0×10^8	84
Carbon	12.0	35,000	70
Nitrogen	14.0	66,000	0.2
Oxygen	16.0	215,000	35,000
Fluorine	19.0	16	3
Sulfur	32.1	3,750	1,000
Chlorine	35.5	90	32
Rock-forming elements			
Sodium	23.0	440	460
Magnesium	24.3	9,100	8,900
Aluminum	27.0	950	940
Silicon	28.1	10,000	10,000
Phosphorus	31.0	100	100
Potassium	39.1	30	40
Calcium	40.1	490	330
Titanium	47.9	20	20
Manganese	54.9	70	40
Iron	55.8	6,000	13,500
Nickel	58.7	270	1,000
Inert gases			
Helium	4.0	3.1×10^7	3.5×10^{-7}
Neon	20.2	86,000	1.2×10^{-6}
Argon	39.9	1,500	5.9×10^{-4}
Krypton	83.8	0.51	6.0×10^{-8}
Xenon	131.3	0.04	5.0×10^{-9}

*Table after R.G. Gymer (1973), *Chemistry: An Ecological Approach.* New York, Harper and Row, p. 432.

†From H.E. Seuss and H.C. Urey (1956), *Reviews of Modern Physics,* 28:53.

‡From B. Mason (1966), *Principles of Geochemistry,* 3rd Edition. New York, John Wiley and Sons, pp. 52, 210.

(O_2), carbon monoxide (CO), carbon dioxide (CO_2), hydrogen sulfide (H_2S), and other simple gases formed from the light elements hydrogen, carbon, nitrogen, oxygen, and sulfur (Tab. 6-2). Indeed, an examination of the relative abundance of these elements in the whole earth as compared to the universe shows that they too were lost from the material forming the earth, but to a lesser degree than the noble gases.

To account for the amounts of hydrogen, carbon, nitrogen, oxygen, and sulfur remaining on the earth, we must assume either that volatile molecules of these elements were adsorbed strongly to the original dust particles or that these elements were held to the

Table 6-2 A COMPARISON OF THE APPROXIMATE ATOMIC OR MOLECULAR MASSES OF VARIOUS VOLATILE SUBSTANCES

SUBSTANCE	CHEMICAL FORMULA	APPROXIMATE MASS
Hydrogen molecule	H_2	2
Helium	He	4
Methane	CH_4	16
Ammonia	NH_3	17
Water	H_2O	18
Neon	Ne	20
Nitrogen molecule	N_2	28
Carbon monoxide	CO	28
Oxygen molecule	O_2	32
Hydrogen sulfide	H_2S	34
Argon	Ar	40
Carbon dioxide	CO_2	44
Krypton	Kr	84
Xenon	Xe	131

dust particles in a chemically combined, non-volatile form. Hydrogen and oxygen, for example, may have been held as water in hydrated minerals or in a wide range of other non-volatile, inorganic compounds as they are today; nitrogen may have been present in ammonium salts; carbon may have been held in the form of graphite or carbonates. Highly polar compounds will be strongly bound to quartz dust particles by processes involving covalent and ionic bond formation (Fig. 6-3). Non-polar gases, particularly the unreactive noble gases that possess filled valence shells, will be held only weakly. In addition, a number of simple organic compounds, including formaldehyde ($H_2C=O$), hydrogen cyanide ($HC \equiv N$), and cyanoacetylene ($HC \equiv C-C \equiv N$) among others, have been identified in cosmic dust clouds. It is therefore likely that organic compounds were synthesized in the dust cloud from which the earth was formed, and that these organic compounds contributed to the amounts of hydrogen, carbon, nitrogen, and oxygen that were retained. This idea is supported by the discovery of a variety of organic compounds in carbonaceous meteorites; it has been shown that these compounds were most probably produced by abiotic means (see Chapter 8).

During the cold accretion process, then, the light elements were bound physically or chemically to dust particles in the cosmic cloud. As the earth formed, its temperature rose, and substances composed of these elements would have been physically released from the dust particles or decomposed into their volatile compounds. As the temperature increased, however, the earth grew in size

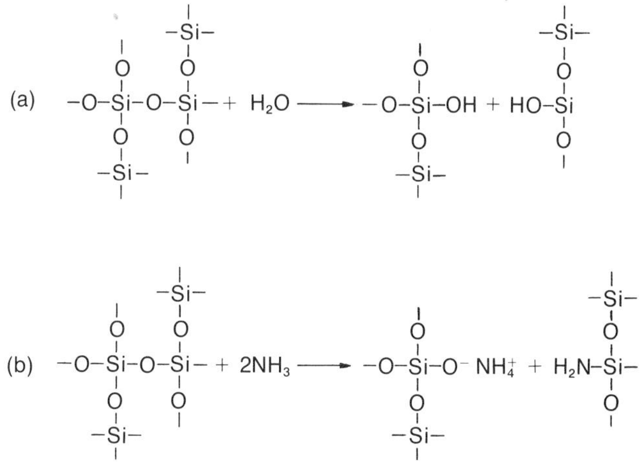

The Atmosphere of the Primitive Earth

(a) $-O-\underset{\underset{-Si-}{\overset{O}{|}}}{\overset{\overset{-Si-}{|}}{Si}}-O-\underset{\underset{-Si-}{\overset{O}{|}}}{\overset{\overset{O}{|}}{Si}}- + H_2O \longrightarrow -O-\underset{\underset{-Si-}{\overset{O}{|}}}{\overset{\overset{O}{|}}{Si}}-OH + HO-\underset{\overset{O}{|}}{\overset{\overset{-Si-}{|}}{Si}}-$

(b) $-O-\underset{\underset{-Si-}{\overset{O}{|}}}{\overset{\overset{-Si-}{|}}{Si}}-O-\underset{\underset{-Si-}{\overset{O}{|}}}{\overset{\overset{O}{|}}{Si}}- + 2NH_3 \longrightarrow -O-\underset{\underset{-Si-}{\overset{O}{|}}}{\overset{\overset{-Si-}{|}}{Si}}-O^- \ NH_4^+ + H_2N-\underset{\overset{O}{|}}{\overset{\overset{-Si-}{|}}{Si}}-$

Figure 6-3 Chemical bond formation between quartz and (a) water and (b) ammonia. Only a portion of the quartz O—Si—O lattice is shown.

and mass, and the very gravitational forces that were elevating the temperature of the earth would have led to the retention in the earth's atmosphere of volatile compounds released from the interior of the earth. This outgassing most probably took place at temperatures between 300°C and 1500°C, and would have resulted in the release of H_2, N_2, NH_3, H_2O, CH_4, CO_2, CO, and H_2S. Much of the outgassing probably occurred by volcanic activity, but we do not know how long the process took (Fig. 6-4). As the surface of the earth cooled (if it was

Figure 6-4 This photograph of an erupting volcano gives an indication of the vast amounts of gas involved. (From W.E. Stokes [1966], *Essentials of Earth History*, Prentice-Hall, Inc., Englewood Cliffs, New Jersey.)

ever very hot), water vapor began to condense and to form oceans and lakes.

In summary, the scarcity of the noble gases on the earth as compared to their abundance in the universe suggests that the earth, during its formation, lost its original atmosphere, and that a secondary atmosphere was generated by an outgassing process caused by heating of the earth's interior.

CHEMICAL PROPERTIES OF THE PRIMITIVE ATMOSPHERE

So far, we have described the processes that we believe produced the atmosphere of the primitive earth, but although we have indicated that a number of different gases were released by outgassing of the interior, we have said nothing about the *relative* amounts of different gases in the primitive atmosphere. As we will see, this is of crucial importance in assessing the plausibility of modern theories on the origin of life, for the synthesis of organic compounds from simple gases will take place only when the composition of gases is such that it forms a *reducing* mixture. In order to understand what we mean by the word "reducing," we must first digress to explain the processes of chemical oxidation and reduction. The concept of oxidation and reduction will be particularly important throughout the remainder of this book, and therefore you should take particular care to understand it thoroughly.

Oxidation-Reduction Reactions

The term *oxidation* was originally used to describe a chemical reaction involving the combination of oxygen with another substance, and *reduction* was used to describe the opposite process; that is, the removal of combined oxygen from a substance. For example, magnesium (Mg) is *oxidized* in the reaction

$$2\ Mg + O_2 \longrightarrow 2\ MgO \tag{1}$$

and carbon dioxide is *reduced* in the reaction

$$CO_2 + 4H_2 \longrightarrow CH_4 + 2H_2O \tag{2}$$

More recently, we have come to think of oxidation and reduction in more general terms, and oxidation is now defined as the *loss of electrons in a chemical reaction* and reduction as the *gain of electrons in a chemical reaction*. In applying these definitions, we find that the processes of oxidation and reduction are most obvious in reactions involving ions. For example, in the reaction

$$Cu^+ + Fe^{+++} \longrightarrow Cu^{++} + Fe^{++} \tag{3}$$

The Atmosphere of the Primitive Earth

Cu^+ gains an additional positive charge by losing an electron, and is therefore said to be oxidized to Cu^{++}; at the same time, Fe^{+++} gains an electron, becoming less positively charged, and is said to be reduced to Fe^{++}. Note that oxygen is *not* involved in this reaction.

Another important feature of Equation 3 (above) is that electrons lost by one substance are gained by another. This is always true, because electrons do not simply float around loose. In this respect, such a reaction is analogous to an acid-base reaction: Just as a substance cannot act as an acid unless its protons are taken up by a base, a substance cannot be oxidized unless its electrons are accepted by another substance. To emphasize the reciprocal nature of the processes of oxidation and reduction, such reactions are referred to as *oxidation-reduction reactions,* or *redox reactions* for short.

Since in an oxidation-reduction reaction, the oxidized substance donates electrons that reduce another substance, the oxidized species is often referred to as a *reducing agent*; conversely, the substance which is reduced (gains electrons) in a chemical reaction is called the *oxidizing agent*. In Equation 3, Cu^+ is the reducing agent and Fe^{+++} is the oxidizing agent. A strong oxidizing agent is, then, one that has a high affinity for electrons in a chemical reaction; strong reducing agents have a strong tendency to give up electrons. This, again, is analogous to the situation in acid-base reactions. Whether a substance will function as a reducing or an oxidizing agent depends on the other substance with which it reacts, just as water, for example, can act as an acid or as a base depending upon the substance with which it is reacting.

Oxidation-Reduction Involving Covalent Compounds

The processes of oxidation and reduction are obvious with reactions involving ions or compounds held together with ionic bonds, where it is simple to trace the loss or gain of electrons. Oxidation and reduction are not so readily apparent in reactions involving covalent compounds. Consider, for example, the reaction

$$H_3C-\underset{\underset{H}{|}}{\overset{\overset{H}{|}}{C}}-OH + \tfrac{1}{2}O_2 \longrightarrow H_3C-\overset{\overset{H}{|}}{C}=O + H_2O \qquad (4)$$
$$\text{(ethyl alcohol)} \hspace{4em} \text{(acetaldehyde)}$$

in which ethyl alcohol is oxidized to acetaldehyde (and molecular oxygen is reduced to water). In this reaction, a pair of electrons as well as a pair of hydrogen ions are removed from ethyl alcohol during its oxidation to acetaldehyde by molecular oxygen. The net effect of this oxidation is the removal of two hydrogen atoms (two protons plus two electrons) from ethyl alcohol. *In fact, most oxidations of organic compounds in living systems occur in this way, and are called dehydrogenations; most biological reductions are hydrogenations.*

In order to view this process in terms of the loss and gain of electrons, we may regard the oxidation as proceeding in two steps—first, the removal of electrons, followed by the dissociation of the acid

$$H_3C-\underset{H}{\overset{H}{C}}-OH \longrightarrow H_3C-\underset{H}{\overset{H^+}{C}}-OH^+ + 2e^-$$

$$\downarrow$$

$$H_3C-\overset{H}{C}=O + 2H^+$$

Of course, this is only a convenient way of looking at the process. The electrons and protons need not be transferred independently, and thus Equation 4 can be thought of as representing a transfer of two hydrogen atoms from ethyl alcohol to molecular oxygen. All this leads to the conclusion that in redox reactions involving covalent compounds, it is often but not always helpful to think of oxidation as the addition of oxygen or the removal of hydrogen, and to think of reduction as the removal of oxygen or the addition of hydrogen. We must, of course, keep in mind that electron transfer is the fundamental aspect of the oxidation-reduction process.

To complete our discussion of redox reactions, let us reconsider the reactions illustrated in Equations 1 and 2 above, in terms of electron transfer. In Equation 1, Mg is oxidized by molecular oxygen and, of course, molecular oxygen is reduced at the same time. The transfer of electrons in this reaction can best be illustrated by Lewis electron dot structures.

$$\cdot Mg\cdot \; + \; :\!\ddot{O}\!::\!\ddot{O}\!: \longrightarrow 2\;Mg\!:\!\ddot{O}\!:$$

Thus, electrons of the Mg atom may be thought of as being transferred to the more electronegative oxygen atom, and Mg is consequently oxidized. In the same sense, molecular oxygen is reduced in this reaction because it gains electrons from Mg.

Equation 2 may be depicted as

$$:\!\ddot{O}\!::\!C\!::\!\ddot{O}\!: \; + \; 4\;H\!:\!H \longrightarrow H\!:\!\underset{H}{\overset{\ddot{H}}{C}}\!:\!H \; + \; H\!:\!\ddot{O}\!:\!H$$

In this case, the electrons in CO_2 may be thought of as being associated with the more electronegative oxygen atoms; consequently, carbon dioxide is reduced to methane, since the carbon in methane may be thought of as accepting the electrons of hydrogen. H_2 is oxidized in this reaction, since its electrons may be thought of as being lost to the more electronegative oxygen atom.

The Atmosphere of the Primitive Earth

Thus, although we can rationalize the redox process in both Equations 1 and 2 in terms of electron transfer, when possible it is probably easier in reactions of this type to view oxidation and reduction in terms of the addition or removal of oxygen or hydrogen, as we did for the reaction in Equation 4.

Oxidation States

Most elements can exist in a number of different *oxidation states*. For example, carbon can be in a fully oxidized form, in a fully reduced form, or in some intermediate state of oxidation. Carbon dioxide, the most highly oxidized form of carbon, may be reduced in successive stages to methane, the most fully reduced one-carbon compound (Fig. 6–5).

In the organic compounds of biological systems, carbon is more reduced than it is in carbon dioxide. It has been observed in simulation experiments that organic compounds (reduced carbon compounds) are not synthesized, except perhaps in trace amounts, in an atmosphere containing molecular oxygen (free oxygen) such as the present atmosphere of the earth (Tab. 6–3). Indeed, if an organic compound is placed in contact with an atmosphere containing an excess of free oxygen, the organic compound will eventually be converted into carbon dioxide.

The Reducing Atmosphere

How does all this apply to our speculations about the composition of the primitive atmosphere? We can see from the previous discussion that molecular oxygen is a good oxidizing agent (because of its tendency to accept electrons from another atom), and hydrogen gas is a good reducing agent (because of its tendency to donate electrons to another atom). Thus an atmosphere containing free oxygen gas would have oxidizing properties; that is, it would tend to convert substances into more highly oxidized forms. Con-

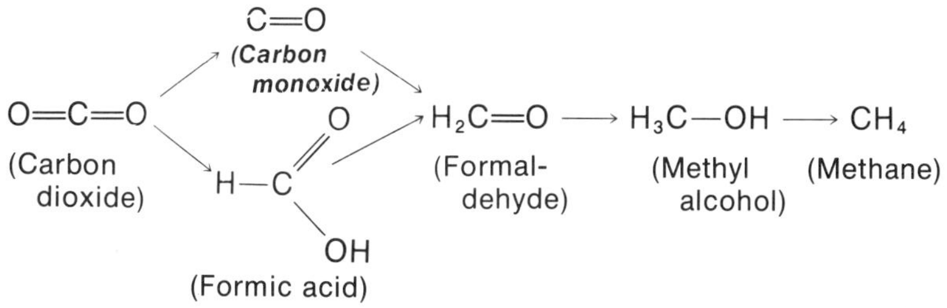

Figure 6–5 The oxidation states of carbon are illustrated by the reduction of carbon dioxide in successive steps to methane. Carbon monoxide and formic acid are in the same oxidation state because their structures differ by one water molecule.

Table 6-3 COMPOSITION OF CLEAN, DRY AIR IN THE PRESENT ATMOSPHERE OF THE EARTH*

GAS	PER CENT BY VOLUME
Nitrogen (N_2)	78.09
Oxygen (O_2)	20.95
Argon (Ar)	0.93
Carbon dioxide (CO_2)	0.03
Neon (Ne)	0.002
Helium (He)	each less than 0.001
Methane (CH_4)	
Krypton (Kr)	
Nitrous oxide (N_2O)	
Hydrogen (H_2)	
Ozone (O_3)	
Xenon (Xe)	

*From D.H. Kenyon and G. Steinman (1969), *Biochemical Predestination.* New York, McGraw-Hill Book Company, Inc.

versely, an atmosphere containing free hydrogen gas would have reducing properties, and substances would tend to be converted into more highly reduced forms. We will use the term "reducing atmosphere" to include any equilibrium mixture of gases that does not contain molecular oxygen. Such an atmosphere excludes direct, extensive oxidation of organic compounds. It is, of course, obvious that some gaseous mixtures will be relatively more reducing (or less oxidizing) than others. Thus, for example, an atmosphere containing an equilibrium mixture of CH_4, NH_3, H_2, and H_2O is strongly reducing, whereas one containing CO, N_2, and H_2O is considerably less so.

Because organic compounds are not synthesized in an oxidizing atmosphere and are unstable in the presence of free oxygen, it has been generally assumed that there was no significant amount of free oxygen in the primitive atmosphere. Thus, although there is considerable disagreement concerning the exact composition of the original atmosphere and the strength of its reducing character (i.e., strongly reducing or only moderately so), it is commonly believed that the primitive atmosphere certainly was not oxidizing.

Although this is a convenient and essential conclusion for proponents of modern theories of life's origin, what evidence is there that this was in fact the case? In the first place, many geologists believe that free oxygen was not among the gases extruded from the interior of the primitive earth during the outgassing process. Their reasoning is based upon the assumption that volcanic gases on the primitive earth would have been formed in equilibrium with molten rock (magma) in the earth's mantle containing reduced metallic iron. Because the metallic iron has subsequently concentrated in the earth's core, the gases from early volcanoes should have been considerably more reduced than contemporary volcanic gases, which presumably are formed in equilibrium with magmas not containing reduced metallic iron. Since free oxygen is not detectable in gases from modern volcanoes or hot springs, and is not found trapped in igneous rocks or meteorites, these geologists conclude that the primitive atmosphere did not contain free oxygen.

Other indirect evidence supports the same conclusion. At the present time, free oxygen is added to the atmosphere in two ways. By far the largest proportion of O_2 is formed as a byproduct of the biological process of photosynthesis. Photosynthesis can be safely ruled out as a mechanism of oxygen production on the primitive earth, because we do not believe that photosynthetic organisms, or any organisms for that matter, were present during the earliest stages of the earth's history. Indeed if we did, there would be little point in this discussion.

Free oxygen is also formed in relatively small amounts by the ultraviolet light-induced breakdown (photolysis) of water in the upper atmosphere, a reaction that proceeds as follows.

$$2\ H_2O \longrightarrow 2\ H_2 + O_2$$

The lighter hydrogen gas produced as a result of this reaction escapes from the earth's gravitational field, leaving free oxygen in the atmosphere. During the early history of the earth, however, free oxygen would have been removed from the primitive atmosphere by combining with reduced mineral compounds on the surface of the earth's crust, and hence it is argued that oxygen would not have accumulated to any significant degree until a surface layer of minerals had been oxidized.

The composition of Precambrian mineral deposits has been taken by some geologists as further evidence supporting the view that free oxygen was lacking in the primitive atmosphere. For example, gold-uranium deposits of the Dominion Reef and Witwatersrand geological system in South Africa, more than 2 billion years old, contain considerable amounts of uraninite (UO_2) and sulfides of iron, lead, and zinc (FeS, PbS, and ZnS). Even at extremely low oxygen concentrations, uraninite is oxidized to UO_3, and the metal sulfides are oxidized to their corresponding sulfates. Lead sulfide, for instance, is oxidized to lead sulfate ($PbSO_4$) at a partial pressure of O_2 corresponding to 10^{-63} atmospheres at 25°C.

Although the presence of uraninite and metal sulfides, as well as reduced minerals in other Precambrian sediments, points to the virtual absence of oxygen in the atmosphere approximately 2 billion years ago, evidence of this type is by no means conclusive. It is always possible that these minerals were laid down under local reducing conditions that were not representative of the rest of the atmosphere. Indeed, ferrous and mercurous sulfides (FeS and HgS)

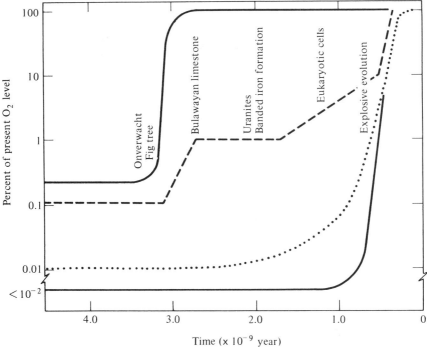

Figure 6–6 Four different models estimating the time course of oxygen appearance in the atmosphere. The upper and lower solid lines represent, respectively, high and low estimates by Miller and Orgel (1974). A model by Rutten (1970) is represented by the dashed line, and a model by Berkner and Marshall (1965) is shown by the dotted line. (From S.D. Miller and L.E. Orgel [1974], *The Origins of Life on the Earth,* Prentice-Hall, Inc., Englewood Cliffs, New Jersey, p. 52.)

have been found in recent sediments that we know were deposited after the atmosphere had become highly oxidizing.

The overall balance of evidence seems strongly in favor of a model predicting that the primitive atmosphere contained only negligible amounts of free oxygen gas. Apart from this conclusion, there is considerable disagreement on the time course of oxygen appearance in the atmosphere. Four possible models are shown in Figure 6–6, each obtained by making different assumptions. All these models are highly speculative, and it is clear that other models would be consistent with the available data.

COMPOSITION OF THE PRIMITIVE ATMOSPHERE

In addition to the conclusion that the primitive atmosphere in all probability was not oxidizing, we can deduce some further information about its composition. One extreme view, originally suggested by Harold Urey, is that the primitive atmosphere was strongly reducing, consisting primarily of molecular hydrogen, methane, ammonia, and water. Urey has argued that although H_2 was probably escaping into space while the earth was condensing, hydrogen is so abundant in the universe (more than 90 per cent of all atoms are hydrogen) that excess H_2 would have been present in the early atmosphere either as a residue from the original cosmic

The Atmosphere of the Primitive Earth

dust cloud, or as a consequence of outgassing, or from both sources. In the presence of excess hydrogen at moderate temperatures, carbon, nitrogen, and oxygen would have existed as methane, ammonia, and water. Because ammonia is so soluble in water, Urey has calculated that most of the volatile nitrogen would have existed as ammonium ion (NH_4^+) in the primitive ocean, providing that the buffer capacity of the ocean was sufficient to maintain its pH around 8.

The concept of a highly reduced atmosphere on the primitive earth is indirectly supported by spectroscopic data showing the presence of small amounts of NH_3 and CH_4 as well as large amounts of H_2 in the gaseous atmosphere of Jupiter. (Because of Jupiter's low temperature, water is presumably frozen lower in the atmosphere and hence is not detected in the upper gaseous atmosphere). It has been suggested that the cold temperature and large mass of Jupiter, as compared to the relatively high temperature and small mass of Earth, would have retarded the loss of Jupiter's atmosphere, and that consequently the present atmosphere of Jupiter is similar to that in the original cosmic dust cloud. Similar data, obtained from spectroscopic analysis of the tails of comets, suggest that comets are composed of frozen water, ammonia, and methane—compounds presumably present in the dust cloud from which the solar system (which includes comets) was derived.

There are scientists who reject Urey's model. They contend that although the primitive atmosphere did not contain free oxygen, it also was *not* strongly reducing and consisted primarily of a mixture of CO_2, N_2, and H_2O. This view is based upon two main assumptions. The first assumption is that any residual hydrogen gas remaining on the earth after its condensation would have escaped at the latest within 300,000 years, an exceedingly short time in the history of the earth and much faster than estimated by Urey and others. Secondly, it is assumed that CO_2 and N_2 would have been expelled in large amounts from the interior of the earth, and in the absence of excess free hydrogen, would not have been reduced further. According to this view, volcanic gases reaching the surface of the primitive earth had the same composition as contemporary volcanic gases, which consist primarily of CO_2 and N_2 and virtually lack CH_4 and NH_3. As we might imagine, the latter assumption has been questioned by those who believe that because primitive volcanic gases were presumably in equilibrium with magmas containing an excess of reduced iron, these gases would have been considerably more reduced than those from modern volcanoes.

A third model, which incorporates features of both the models discussed above, has gained considerable support. Proposed by H. D. Holland and his collaborators, this model contends that during the early history of the earth (approximately the first 0.5 billion years) the atmosphere was strongly reducing, and that this period was followed by a second stage, lasting about 2 billion years, in which the character of the atmosphere was less reducing. During a third stage (approximately 2 billion years ago to the present), the atmosphere would have become increasingly oxidizing as free oxygen

Table 6-4 COMPOSITION OF THE ATMOSPHERE DURING VARIOUS STAGES IN THE EARTH'S HISTORY, AS PROPOSED BY HOLLAND (1962)

	STAGE 1[a]	STAGE 2[a]	STAGE 3[a]
Major components (Partial pressure greater than 10^{-2} atm)	CH_4 $H_2(?)$	N_2	N_2 O_2
Minor components (Partial pressure between 10^{-2} and 10^{-4} atm)	$H_2(?)$ H_2O N_2 H_2S NH_3 Ar	H_2O CO_2 Ar	Ar H_2O CO_2
Trace components (Partial pressure between 10^{-4} and 10^{-6} atm)	He	Ne He CH_4 $NH_3(?)$ $SO_2(?)$ $H_2S(?)$	Ne He CH_4 Kr

[a]Stage 1 = 4.5 to 4.0 billion years ago; Stage 2 = 4.0 to 2.0 billion years age; Stage 3 = 2.0 billion years to the present. (From H.D. Holland [1962], Model for the evolution of the earth's atmosphere. In A.E.J. Engel, H.L. James, and B.F. Leonard [Eds.], *Petrologic Studies,* Boulder, Colorado, Geological Society of America.)

accumulated primarily as a result of photosynthesis. Details of this model are summarized in Table 6–4.

At the present time, it is impossible to tell which, if any, of these three models is correct, or whether the primitive atmosphere actually possessed a reducing character and a gaseous composition somewhat different than has been supposed.

As we will see shortly, the exact composition of the primitive atmosphere is relatively unimportant in simulation experiments designed to test the validity of modern theories on the origin of life. As long as a mixture of gases contains carbon, nitrogen, oxygen, and hydrogen in a form more highly reduced than a mixture of CO_2, N_2, and H_2O,* organic compounds of biological importance are readily synthesized. For example, results obtained in simulation experi-

---

*Some simple organic compounds, including formaldehyde, are produced in simulation experiments from a mixture of CO_2, N_2, and H_2O, but no organic compounds characteristic of biological systems have been identified.

ments using a strongly reduced CH_4, NH_3, H_2, and H_2O mixture are similar to those obtained when a less reduced $CO-N_2-H_2-H_2O$ mixture is used.

FREE ENERGY SOURCES ON THE PRIMITIVE EARTH

A mixture of CH_4, NH_3, H_2, and H_2O, or any other mixture of gases likely to have been present in the primitive atmosphere, will not produce significant quantities of organic compounds if placed in a dark container at room temperature. Even if the formation of a particular organic compound from simple gases is thermodynamically feasible, we know that this does not necessarily mean that the reaction will occur at a detectable rate. This is because the gases need to overcome activation barriers in order to react with one another. Of course, most reactions forming organic compounds from simple gases are, in addition, thermodynamically unfeasible, and hence proceed with a positive ΔG. Thus free energy is required to force the reaction to proceed. In many cases, the synthesis of complex organic compounds proceeds through the endergonic formation of highly reactive intermediate compounds that can combine to form new compounds with little or no energy input. Consequently, the production of organic compounds in simulation experiments requires not only a reducing mixture of simple gases but also a supply of free energy capable of promoting the reaction.

Table 6-5 ESTIMATES OF AVAILABLE ENERGY SOURCES ON THE PRIMITIVE EARTH*

SOURCE	ENERGY (calories/cm²/year)
Ultraviolet light from the sun	
300 to 250 nm	2837
250 to 200 nm	522
200 to 150 nm	39.3
Below 150 nm	1.7
Electric discharges	4.
Shock waves	1.1
Radioactivity	0.8
Heat from volcanoes	0.13
Cosmic rays	0.0015

*From S.L. Miller and L.E. Orgel (1974). *The Origins of Life on the Earth.* Englewood Cliffs, New Jersey, Prentice-Hall, Inc.

The various sources of energy on the primitive earth and the amounts of each potentially available for prebiotic chemical synthesis are shown in Table 6–5. These include ultraviolet light from the sun, electric discharge, heat from volcanoes, cosmic rays, heat and high energy particles produced from the disintegration of radioactive isotopes, and shock waves produced from collisions of comets or meteorites with the atmosphere of the earth.

It is not easy to assess the relative importance of each of these energy sources for prebiotic synthesis. The fact that a source of energy is potentially available does not mean that it is used efficiently for the production of organic compounds.* Electric discharge and ultraviolet light are estimated to have been the most important energy sources. Although electric discharge appears to be more efficient than ultraviolet light for producing organic compounds in simulation experiments, energy in the form of ultraviolet light was available during prebiotic times in a considerably larger quantity than energy as electric discharge, assuming that the amount of thunderstorm activity was approximately the same then as it is now. In addition, simulation experiments using shock waves as an energy source are reported to produce extremely high yields of organic compounds, and hence shock waves may also have been an important source of energy for prebiotic synthesis.

Not every wavelength of ultraviolet light is equally effective in promoting the synthesis of organic compounds from a mixture of simple gases. This is because only light that is absorbed by a reacting molecule (or atom) is effective in producing a chemical change, and a particular molecule is only able to absorb light of certain wavelengths. In order to understand why this last statement is true, let us examine briefly the process of light absorption.

ABSORPTION OF LIGHT IN CHEMICAL SYNTHESIS

When light is absorbed by a molecule (or atom), the energy of the light is transformed into the kinetic energy of an electron, and the electron is transferred from its original orbital to an orbital of higher energy farther from the nucleus. When this occurs, the molecule (or atom) is said to be in an *excited state* because it has additional energy.

In such a process involving the interaction of light with matter, light behaves as if it were composed of discrete packets of energy called *quanta* or *photons*. The amount of energy in a photon varies with the wavelength of light; light of a short wavelength possesses more energy per photon than light of a longer wavelength. In order to undergo the electronic transition required to reach an excited state, a molecule must absorb *one photon* that contains exactly the amount of energy needed to transfer the electron to a higher

---

*By "efficiency" we mean the weight of organic compounds produced per standard unit of energy using a given mixture of reactants.

energy level. The molecule may not absorb a fraction of a photon, but must absorb one complete photon. Consequently, the molecule may only absorb light of a wavelength that corresponds to a photon of precisely the right amount of energy. Different molecules absorb characteristic wavelengths of light because each molecule requires a different amount of energy to undergo a transition to an excited state. In certain cases, a molecule may absorb more than one wavelength of light if it is capable of undergoing electronic transitions involving different electrons, each transition requiring a characteristic amount of energy.

An excited molecule may behave in a number of different ways. (1) It may lose its excitation energy in the form of heat (increased molecular motion), and thus return to its original, low-energy state (*ground state*). (2) It may lose its excitation energy by the emission of light (and some heat). This process, called fluorescence, will be discussed further in Chapter 12. (3) It may transfer its excitation energy to another molecule by collision. (4) It may enter into a chemical reaction. In this case, the excited electron may be ejected from the molecule and the positively charged molecule may then react with another substance. Alternatively, the molecule may dissociate into fragments, which are usually highly reactive and may take part in subsequent reactions to form other molecules.

The majority of photochemical reactions occur by the formation of reactive fragments. The most common reactive fragments are *free radicals*; these owe their reactivity to the fact that they possess unpaired valence electrons. For example, a methyl radical and a hydrogen atom may be formed by the photochemical dissociation of methane, as follows

$$\text{H–CH}_3 \longrightarrow \cdot\text{CH}_3 + \text{H}\cdot$$

(methane) → (methyl radical) + (hydrogen atom)

Other high energy sources such as electric discharge and particles emitted as a result of radioactive disintegration also bring about the formation of reactive fragments.

Gases likely to have been present in the primitive atmosphere, such as H_2, CH_4, H_2O, CO_2, CO, NH_3, and H_2S, all absorb ultraviolet light of wavelengths below about 200 nm, although NH_3 and H_2S can also absorb small amounts of ultraviolet light of somewhat higher wavelengths (Fig. 6–7). Consequently, most of the ultraviolet light of wavelengths above 200 nm was not available for early prebiotic synthesis. However, many of the simple organic compounds which would have been formed initially, such as formaldehyde ($H_2C=O$) and acetaldehyde ($H_3C-\overset{\overset{\text{H}}{|}}{C}=O$), absorb at higher wavelengths, and

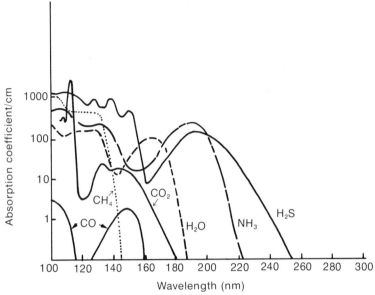

Figure 6–7 Absorption of NH_3, H_2S, CO_2, CO, H_2O, and CH_4 at various wavelengths. Molecular hydrogen (H_2) only absorbs below 100 nm. The absorption coefficient is a measure of the amount of light absorbed by a given concentration of substance. (Redrawn from S.L. Miller and L.E. Orgel [1974], *The Origins of Life on the Earth,* Prentice-Hall, Inc., Englewood Cliffs, New Jersey, pp. 57–58.)

therefore much larger amounts of ultraviolet light energy would have been available for further synthesis involving these compounds.

The present oxidizing atmosphere of the earth is not transparent to wavelengths of light below about 300 nm. This is partially a result of the presence in the atmosphere of molecular oxygen, which absorbs ultraviolet light of wavelengths below about 220 nm. In addition, ultraviolet light between 220 and 300 nm is absorbed by ozone (O_3) present in the upper atmosphere. Ozone is produced in a photochemically initiated process by the irradiation of oxygen molecules with ultraviolet (UV) light of short wavelengths (approximately 175 to 210 nm), according to the following reactions

$$O_2 + \text{short wavelength UV light} \longrightarrow O + O$$

$$O + O_2 \longrightarrow O_3$$

Most of the ultraviolet light absorbed by ozone is dissipated as heat in the upper atmosphere.

Since the available evidence suggests that free oxygen was not present in the primitive atmosphere, ozone would not have been produced, and consequently short wavelength ultraviolet light would have penetrated the early atmosphere. However, as significant amounts of free oxygen accumulated in the atmosphere as a result of the ultraviolet photolysis of water, or of photosynthesis, or of both processes, an ozone screen would have developed in the upper atmosphere. This screen prevented short wavelength ultraviolet light from penetrating the lower atmosphere. Thus an important energy source for the prebiotic synthesis of organic compounds was excluded. However, because the ozone screen shielded the surface of the earth from high intensity ultraviolet radiation, it permitted evolving organisms to inhabit shallow waters and, eventually, terrestrial habitats.

THE "PRIMITIVE BROTH"

We have neglected to discuss one very important point with regard to the stability of organic compounds once they had been synthesized on the primitive earth. From what we know about chemical reactions, it should be obvious that high energy sources that form reactive fragments by the dissociation of reactant molecules will also eventually bring about the breakdown of the products of the reaction. Indeed, if the products are thermodynamically unstable, the reactants will be favored in any such process.

If complex organic molecules were to accumulate at all on the primitive earth, they had to be removed from the high energy sources. This could have occurred in a number of ways. Compounds formed in the atmosphere by the action of ultraviolet light and electric discharge might, for example, have been removed from the at-

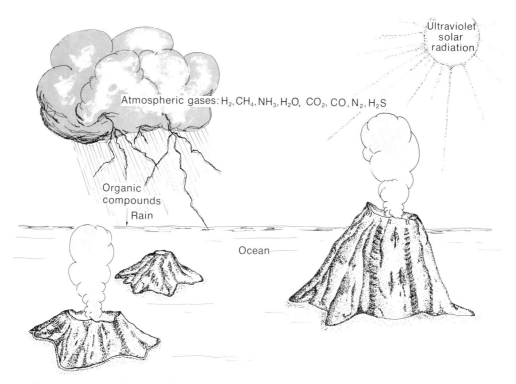

Organic compounds synthesized in the atmosphere accumulated in the oceans.   Figure 6–8

mosphere by gravitational settling, and may have accumulated in the oceans. Gravitational settling would be enhanced by the aggregation of product molecules and by the action of rain. Once dissolved in the oceans, these compounds would have been protected from destruction by electric discharges and by ultraviolet light in the atmosphere. It is estimated that ultraviolet light could only have penetrated the surface regions of the ocean because of the accumulation in the ocean of ultraviolet-absorbing compounds. This idea is supported by the observation that brownish tars, which absorb strongly in the ultraviolet, are produced in simulation experiments by the action of electric discharge on a reducing mixture of CH_4, NH_3, H_2, and H_2O. Compounds produced with the aid of terrestrial sources of energy, such as heat from volcanoes and high energy particles produced by the disintegration of radioactive isotopes, might also have been dissolved in the ocean and thus removed from the further effects of heat and ionizing radiation. It is because of the theoretical accumulation of organic compounds in the primitive ocean that the ocean at this stage is often referred to as the "primitive broth" or "prebiotic soup" (Fig. 6–8).

THE ENVIRONMENT ON THE PRIMITIVE EARTH: A SUMMARY

In summary, then, we can construct a probable model describing the early history of the earth. The earth was formed from a turbulent cloud of cosmic dust and gases. During the initial stages of the aggregation of material forming the earth, volatile compounds were probably lost from the dust particles. Carbon, nitrogen, hydrogen, oxygen, and sulfur must have been retained by the dust particles in non-volatile form. In later stages of aggregation, the compression of material as well as the decay of radioactive elements resulted in the production of large amounts of heat, which eventually melted the interior of the earth. Whether or not the surface also melted is a matter of intense debate; but it has been generally assumed that, in any case, the surface cooled to contemporary temperatures within several hundred million years of the earth's formation. When the temperature at the surface fell below 100°C, water condensed into vast oceans.

The formation of the primitive atmosphere was a result of outgassing from the earth's interior, largely due to volcanic activity. As the temperature of the earth's interior rose, the non-volatile compounds of carbon, nitrogen, hydrogen, oxygen, and sulfur were pyrolyzed to a mixture of NH_3, N_2, H_2, CH_4, H_2O, CO, CO_2, and H_2S. Although we do not know the exact composition of the primitive atmosphere, it undoubtedly had reducing properties. Whether or not it was strongly reducing, or only moderately so, is undecided, but everyone seems to agree that, at the most, only trace amounts of molecular oxygen were present.

Simple organic compounds were synthesized from the mixture

of atmospheric gases with the aid of energy sources on the primitive earth. These included ultraviolet radiation, electric discharges, heat from volcanoes, radioactivity, cosmic rays, and shock waves. Organic compounds accumulated in the ocean, which has become referred to as the "primitive broth." It is thus generally assumed that life originated in the oceans from these organic compounds.

REFERENCES

Brooks, J., and G. Shaw (1973). *Origin and Development of Living Systems.* Academic Press Inc., New York.
Calvin, M. (1969). *Chemical Evolution.* Oxford University Press, New York.
Gymer, R.G. (1973). *Chemistry: An Ecological Approach.* Harper and Row, New York.
Kenyon, D.H., and G. Steinman (1969). *Biochemical Predestination.* McGraw-Hill Book Co., New York.
Miller, S.L., and L.E. Orgel (1974). *The Origins of Life on the Earth.* Prentice-Hall, Inc., Englewood Cliffs, New Jersey.
Orgel, L.E. (1973). *The Origins of Life: Molecules and Natural Selection.* John Wiley and Sons, New York.

CHAPTER 7

The Chemical Nature of Biological Systems

The current widespread interest in the origin of living systems is a direct outgrowth of recent advances in our understanding of cellular structure and function. Modern biology has led us progressively to the conclusion that living organisms are complex physical-chemical systems whose structure and function can be explained in terms of evolutionary theory and described by the interaction of their molecular components according to the well-established laws of chemistry and physics. The description of living systems in these terms, although often resulting in oversimplification, has provided a scientific basis for serious consideration of the materialistic theory of the origin of life: the view that life arose as one step in a continuous process of chemical evolution by the chance formation and interaction of a variety of organic molecules.

In the previous chapter, we mentioned that experimental evidence supporting the materialistic theory is derived primarily from the accumulated results of a large number of simulation experiments. These experiments, designed to mimic conditions that might reasonably have prevailed on the primitive earth at the start of the process of chemical evolution, have yielded molecules similar to those of which modern organisms are composed. It is, of course, an implicit assumption in the design of simulation experiments that the first organisms were similar chemically to those existing today—an assumption which the vast majority of biologists accept, and which we have attempted to support by circumstantial evidence derived from the fossil record and from the implications of evolutionary theory.

Before we can judge for ourselves the significance of the results obtained in simulation experiments, we must be familiar with the chemical characteristics of living systems, so that we can interpret the experimental results with some degree of sophistication. This task is not as difficult as it might appear at first, for in spite of the complexity of even the simplest prokaryotic cell, cellular structure and function are concerned primarily with the four classes of biochemical substances: proteins, nucleic acids, carbohydrates, and lipids. Consequently, we will now turn to a brief consideration of the chemical structures and properties of these substances.

PROTEINS AND THEIR SUBUNITS, THE AMINO ACIDS

Proteins are quantitatively the most numerous organic compounds in the living cell, and are important in practically all aspects of cell structure and function. As we have already seen, many proteins are enzymes. The specific enzymes in a particular cell determine its metabolic capabilities, and hence variations in enzyme content account for the metabolic differences between cell types. Proteins also form much of the structural material of the cell, and are thus directly responsible for the structural variability between different cells and different organisms.

In addition to their general enzymatic and structural roles, proteins are responsible for a number of specialized physiological functions within the organism. Many of the hormones are proteins; these serve to regulate and coordinate various physiological activities. Antibodies, which are important in protecting animals from disease, are also composed of protein. Proteins are responsible for muscular contraction and for oxygen transport from the lungs to the tissues. They are involved in electron transport during respiration and in photosynthesis—a process crucial to ATP production.

Although proteins have a variety of other functions as well, these examples serve to illustrate the enormous functional diversity of the proteins. This diversity of proteins is a result of their chemical structure. Proteins are polymers of 20 common amino acids and a few additional amino acids that are formed by the chemical modification of several of these 20 amino acids after they have been incorporated into proteins. Each protein has a unique amino acid sequence which endows it with a specific shape and function; these characteristics are, of course, directly attributable to the specific properties of the amino acids of which the protein is composed. Even for an average-sized protein of about 250 amino acid units, the number of possible sequences of amino acids is practically limitless. Consequently, a very large number of different proteins, with different chemical properties, can exist.

AMINO ACIDS

Amino acids, as their name suggests, possess both an acidic region (a *carboxyl group*, $-\overset{\overset{\displaystyle O}{\|}}{C}-OH$) and a basic region (an *amino group*, $-NH_2$), as well as a third region referred to as a *side chain.* This third region is usually denoted in chemical notation by the letter *R,* indicating that its chemical composition varies from one amino acid to the next. We can thus write the general structure of the naturally occurring amino acids as follows

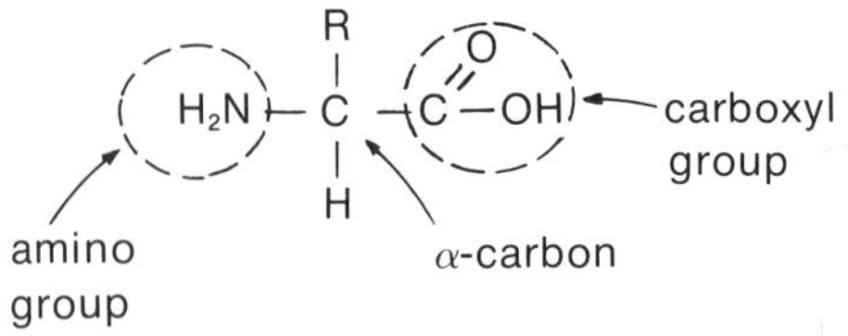

$$\underset{\text{amino group}}{H_2N}-\underset{\underset{H}{|}}{\overset{\overset{R}{|}}{C}}-\underset{\text{carboxyl group}}{\overset{\overset{O}{\parallel}}{C}-OH}$$

α-carbon

The carbon immediately adjacent to the carboxyl group is called the *α–carbon* (alpha-carbon). With one exception, amino acids occurring in proteins are all α–amino acids; that is, the carboxyl and amino groups are both attached to the α–carbon.*

In aqueous solution within the physiological pH range (in the region around pH 7.0), the carboxyl and amino groups attached to the α–carbon will tend to exist in ionized forms:

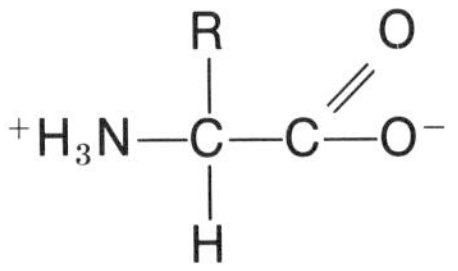

$$^+H_3N-\underset{\underset{H}{|}}{\overset{\overset{R}{|}}{C}}-\overset{\overset{O}{\parallel}}{C}-O^-$$

Only at a low pH, well below that normally encountered within healthy, living cells will protons combine with the acidic groups ($-\overset{\overset{O}{\parallel}}{C}-O^-$), thus producing a positively charged molecule, according to the equation

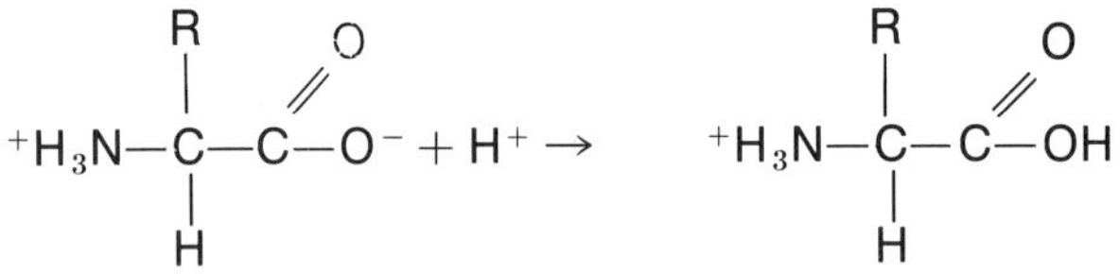

$$^+H_3N-\underset{\underset{H}{|}}{\overset{\overset{R}{|}}{C}}-\overset{\overset{O}{\parallel}}{C}-O^- + H^+ \rightarrow {}^+H_3N-\underset{\underset{H}{|}}{\overset{\overset{R}{|}}{C}}-\overset{\overset{O}{\parallel}}{C}-OH$$

At very high pH values, again outside the normal physiological pH range, the basic groups ($-NH_3^+$) will give up a proton, and the mole-

---

*In writing the chemical structures of organic compounds, it is unimportant in most cases how we orient chemical groups relative to one another. Thus

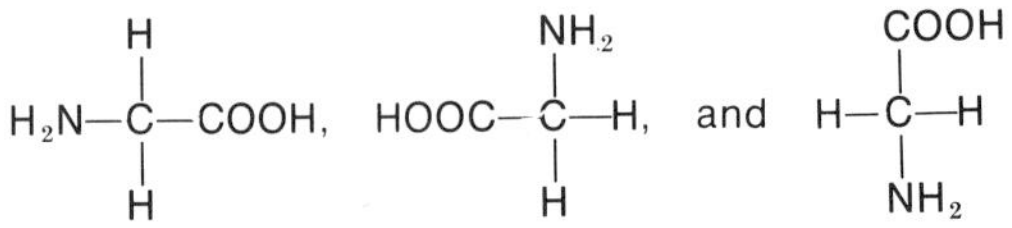

$$H_2N-\underset{\underset{H}{|}}{\overset{\overset{H}{|}}{C}}-COOH, \quad HOOC-\underset{\underset{H}{|}}{\overset{\overset{NH_2}{|}}{C}}-H, \quad \text{and} \quad H-\underset{\underset{NH_2}{|}}{\overset{\overset{COOH}{|}}{C}}-H$$

all represent the same compound.

Proteins and Their Subunits, the Amino Acids

cule will become negatively charged:

$$^+H_3N-\underset{H}{\underset{|}{\overset{R}{\overset{|}{C}}}}-\overset{O}{\overset{\|}{C}}-O^- \rightarrow \quad H_2N-\underset{H}{\underset{|}{\overset{R}{\overset{|}{C}}}}-\overset{O}{\overset{\|}{C}}-O^- + H^+$$

Since the structures of different amino acids vary only with respect to their *R*-groups (side chains), it follows that the specific chemical properties that distinguish one amino acid from another must be dependent on the chemical nature of the particular side chain of each amino acid. Indeed, it is possible to categorize amino acids according to the general characteristics of their *R*-groups.

Neutral Amino Acids

The neutral amino acids, shown in Figure 7–1, are so named because they have uncharged *R*-groups at neutral pH (and within the physiological pH range). Also in Figure 7–1, the neutral amino acids have been further subdivided on the basis of their affinity for an aqueous medium, that is, their hydrophilic (water-loving) or hydrophobic (water-hating) nature. Valine, leucine, isoleucine, methionine, and phenylalanine are strongly hydrophobic, and because of this their side chains gain stability when removed from an aqueous environment. Consequently, when these amino acids are present in a protein, they tend to fold into the interior of the protein molecule away from the surrounding aqueous phase.

The remainder of the neutral amino acids are either weakly hydrophobic or weakly hydrophilic; these can be found either in the interior of a protein molecule or on the surface in contact with the aqueous medium. The *R*-groups of glycine and alanine are nonpolar but only weakly hydrophobic due to their small size. Proline, also weakly hydrophobic, differs from the other naturally occurring amino acids because the nitrogen of its amino group forms a ring structure with the side chain. This changes the amino to an *imino* group ($-\overset{|}{N}H$), and thus technically proline is an *imino acid*. As we shall see, the rigid ring structure of proline has very important effects on the three-dimensional configuration of proteins.

Serine, threonine, tyrosine, tryptophan, and cysteine (ciś tē ēń) are weakly hydrophilic because their side chains are somewhat polar. These side chains do not, however, ionize at physiological pH values and hence remain uncharged. In some textbooks, cystine (ciśtēn)

$$HOOC-\underset{NH_2}{\underset{|}{\overset{H}{\overset{|}{C}}}}-CH_2-S-S-CH_2-\underset{NH_2}{\underset{|}{\overset{H}{\overset{|}{C}}}}-COOH$$

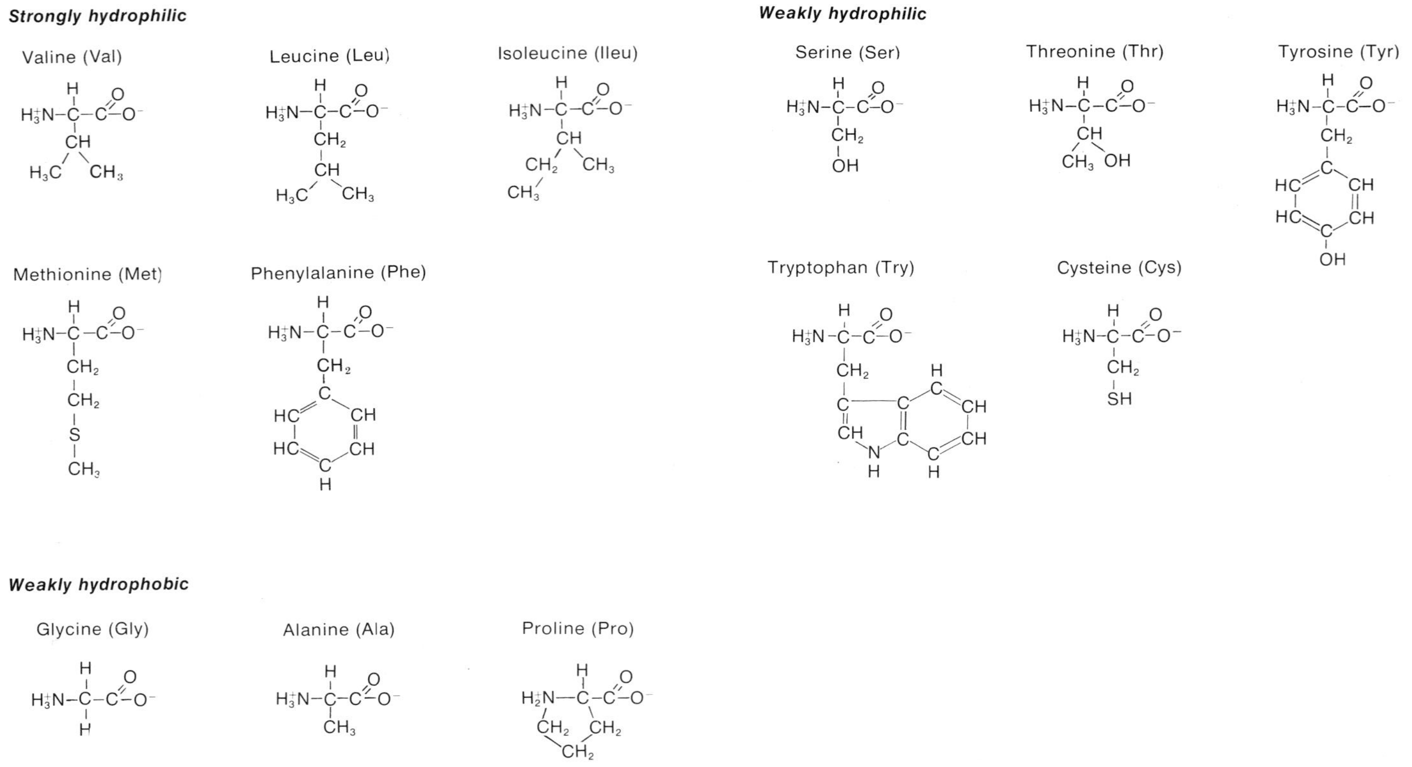

Figure 7-1 The chemical structures of the neutral amino acids as they would exist at physiological pH. The accepted abbreviation for each amino acid is given in parentheses.

Proteins and Their Subunits, the Amino Acids

is listed as a separate amino acid. Cystine is actually the oxidized form of cysteine; it is produced by the oxidation of the sulfhydryl groups (—SH) in two cysteine molecules, resulting in the formation of a covalent disulfide bond between them, as shown in the following equation

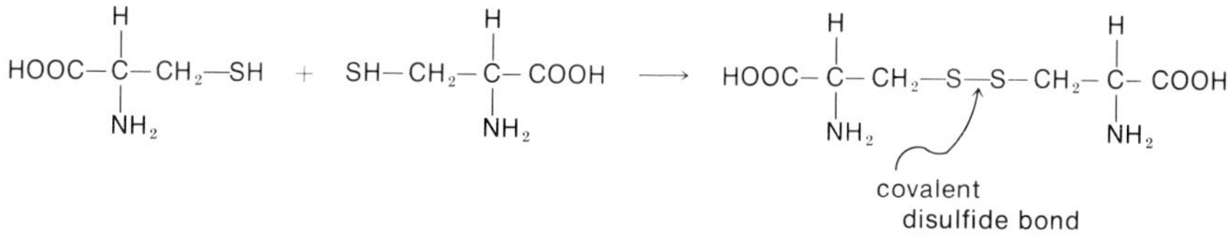

$$\text{HOOC}-\underset{\underset{NH_2}{|}}{\overset{\overset{H}{|}}{C}}-CH_2-SH \;+\; SH-CH_2-\underset{\underset{NH_2}{|}}{\overset{\overset{H}{|}}{C}}-COOH \;\longrightarrow\; \text{HOOC}-\underset{\underset{NH_2}{|}}{\overset{\overset{H}{|}}{C}}-CH_2-S-S-CH_2-\underset{\underset{NH_2}{|}}{\overset{\overset{H}{|}}{C}}-COOH$$

covalent disulfide bond

Acidic Amino Acids and Their Derivatives

The structures of the acidic amino acids, aspartic acid and glutamic acid, are shown in Figure 7–2. The side chain of each of these amino acids terminates in a carboxyl group, which gives up a proton (acts as an acid) in aqueous solution at neutral or basic pH. Hence the acidic amino acids possess an extra negative charge at physiological pH. Because of the negative charge on their side chains, aspartic acid and glutamic acid are strongly hydrophilic. When they are present in proteins, they tend to be oriented at the surface of the protein molecule with their side chains in contact with the surrounding aqueous medium.

Asparagine and glutamine, also shown in Figure 7–2, are

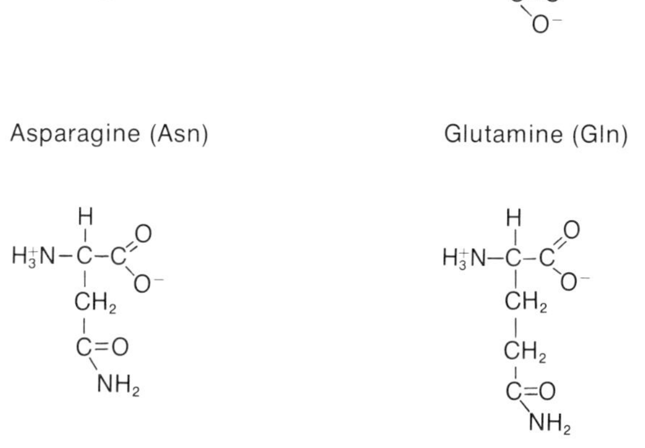

Aspartic acid (Asp)

Glutamic acid (Glu)

Asparagine (Asn)

Glutamine (Gln)

The acidic amino acids and their amides at physiological pH.

Figure 7–2

amides of aspartic acid and glutamic acid, respectively, in which the hydroxyl of the side chain carboxyl group is replaced by an $-NH_2$ group. The amide group ($-\overset{\overset{O}{\|}}{C}-NH_2$) in asparagine and glutamine is not ionized but is extremely polar, and these amino acids are therefore strongly hydrophilic.

Basic Amino Acids

Lysine, histidine, and arginine are classified as basic amino acids (Fig. 7–3). They are strongly hydrophilic because they possess side chain nitrogen-containing groups that accept protons (act as bases) in aqueous solution at neutral or acidic pH. Consequently, they usually carry a positive charge within the physiological pH range.

Since the side chains of the acidic amino acids act as weak acids (donating protons) at neutral and basic pH, and the side chains of the basic amino acids act as weak bases (accepting protons) at neutral and acidic pH, amino acids help to buffer the intracellular pH. As we shall see, the chemical characteristics of the

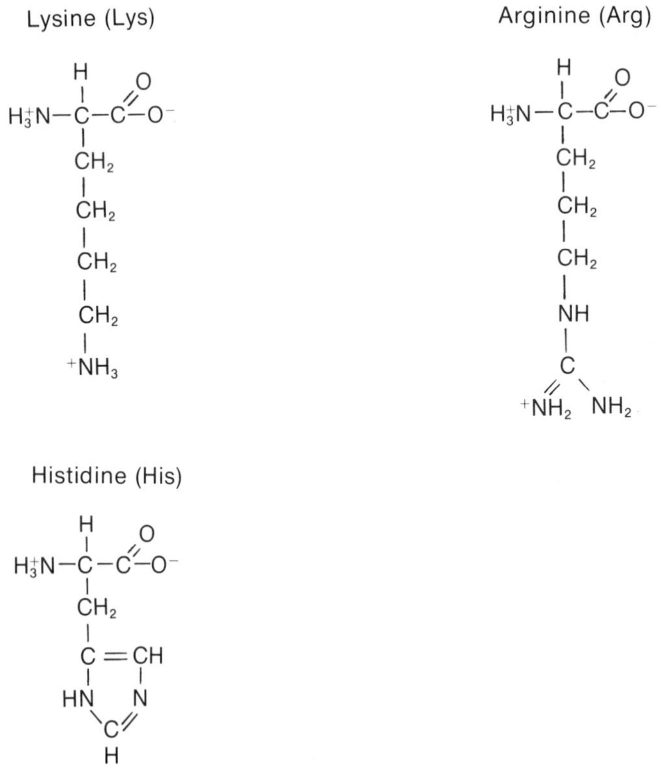

Figure 7–3 The basic amino acids at physiological pH. Note that histidine is represented as uncharged; its side-chain nitrogen will accept a proton around pH 6.0, slightly below physiological pH.

Proteins and Their Subunits, the Amino Acids

side chains are retained when amino acids are linked together to form proteins. Indeed, properties of a protein such as its solubility, charge, and buffering ability, as well as its overall configuration, are determined by the position and types of side chains in the protein.

Optical Activity

It is apparent from the general formula for an α-amino acid that the α-carbon has four different atoms or groups of atoms attached to it (with the exception of glycine; see Fig. 7–1). These four groups are situated about the α-carbon at the corners of a regular tetrahedron, in the same general configuration as the hydrogens in methane. There are two different ways of attaching these four constituents to the central carbon atom, and as shown in Figure 7–4, these two configurations are mirror images. Because they are mirror images, there is no way that the two forms can be superimposed on one another. The carbon atom to which the four different groups are attached is known as an *asymmetric carbon,* and the two possible configurations are called *optical isomers* of the compound. Optical isomers are designated as D or L to distinguish the two forms from each other. For example, the two optical isomers of alanine are D-alanine and L-alanine (Fig. 7–5).

Optical isomers can be distinguished on the basis of their ability to rotate the plane of polarized light. Polarized light is light whose waves are propagated (vibrate) in only one plane, rather than in all planes as in ordinary light (Fig. 7–6). If polarized light is shined through a solution of one optical isomer, the plane of the polarized

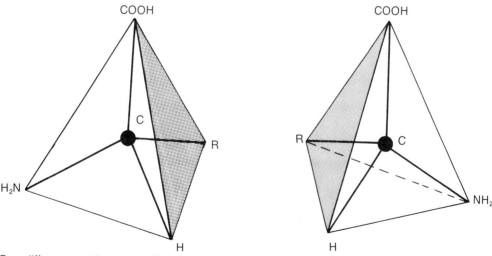

Four different constituents may be bonded to the central carbon atom in two configurations that are mirror images of each other.

Figure 7–4

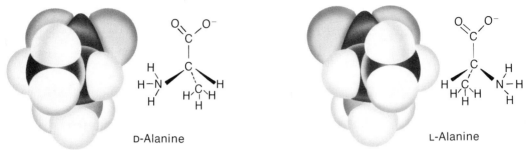

Figure 7–5 Space-filling molecular models of D- and L-alanine. Note that the two optical isomers are not superimposable. (After E.O. Wilson et al. [1973], *Life on Earth*, Sinnauer Associates, Inc., Stamford, Connecticut, p. 65.)

light will be rotated in one direction, whereas a solution of the other optical isomer (the mirror image of the first) will rotate the plane of light in the opposite direction. This is why compounds with asymmetric carbon atoms are said to possess *optical activity*. All amino acids are optically active (except for glycine, which has two identical groups attached to the α-carbon), and therefore only one configuration is possible.

The D- and L-forms of any optically active compound are defined by their structural relationship to arbitrarily designated D- and L-forms of glyceraldehyde, a three-carbon sugar with a single asymmetric carbon atom. However, the D and L designations do not refer to the direction in which polarized light is rotated, but rather to the structural relationship of an isomer to the configuration of D- and L-glyceraldehyde.

Solutions of organic material can be optically inactive either because all molecules in the solution are themselves optically inactive, as in an aqueous solution of glycine, or because the D- and L-isomers of a compound are present in equal amounts. In this latter case, the rotation of polarized light in one direction by the D-isomer

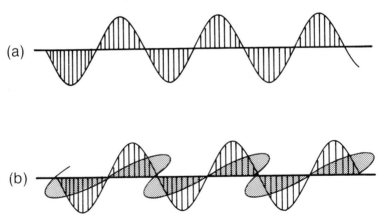

Figure 7–6 Polarized light consists of waves propagated in a single plane, for example, only in the plane of the paper as in *a*. Ordinary light consists of waves propagated in all planes; as an example, the propagation of waves in two planes is shown in *b*.

Proteins and Their Subunits, the Amino Acids

will exactly cancel the rotation in the opposite direction produced by the L-isomer. Solutions containing equal amounts of the D- and L-isomers of a molecule are called *racemic mixtures*.

Living organisms contain a large number of optically active constituents. For example, with a few rare exceptions, all of the amino acids in living organisms are of the L-form; nucleic acids are composed of D-nucleotides but not L-nucleotides; D-glucose is important in metabolism, but L-glucose is not used. The discovery of optically active compounds in living organisms was unexpected, and prompted Pasteur to write, "I am inclined to think that life, as it appears to us, must be a product of the dissymmetry of the universe...."

Why have living organisms evolved so that they use only L-amino acids, D-nucleotides, D-glucose, and other specific optical isomers? It is apparent that this question involves two separate aspects: (1) Why do living systems use only one form of optical isomer (for example, L-methionine, but not D-methionine), rather than using both? (2) If it is selectively advantageous for living organisms to use one form of optical isomer but not both, why do they use the particular isomeric form that they do?

These questions are particularly important with regard to discussions on the origin of life, and two pieces of evidence bear directly on the problem. The first is the observation that although optically active molecules may be produced in simulation experiments, the D- and L-isomers of a molecule are always formed in equal numbers. This leads to the conclusion that the "primitive broth" was probably optically inactive, consisting of racemic mixtures of optically active molecules. Some workers have challenged this conclusion. They contend that factors such as the direction of the earth's magnetic field and the production of polarized light from the sun might have been sources of asymmetry that would have favored the production of one isomeric form over the other. Thus, they argue, the availability of a particular isomer would have been an important factor in determining which of the two isomers was incorporated by the first living organism. Although these ideas may be correct, most workers agree that such factors could not have produced significant asymmetry to have had an important effect on the proportions of available D- and L-isomers in the primitive environment. Indeed, if optically active compounds were synthesized on the earth, there are good data indicating that they would not have been stable for long periods of time and would soon have formed racemic mixtures. We will assume, therefore, that the primitive ocean did not display optical activity.

The second point concerns the stability and chemical characteristics of D- and L-isomers of a molecule. It has been shown that *in an optically inactive environment,* the D- and L-isomers of a molecule have identical chemical properties and the same degree of stability. Thus an equilibrium mixture contains equal amounts of the two isomers.

With these points in mind, it is possible to propose a reasonable hypothesis to explain why living organisms selected one of the two isomeric forms of a molecule, but not both, from the racemic mixture of the "primitive broth." It appears that organisms could achieve a significant economy with such a strategy; that is, their metabolic machinery would have to be considerably more extensive in order to accommodate both the D- and L-isomers of molecules. For instance, we know that a specific fit is required between an enzyme and its substrate in order for catalysis to occur. Thus, an enzyme capable of catalyzing a reaction involving one isomer (for example, L-serine) would probably not be able to catalyze the reaction involving its mirror image (D-serine), simply because the mirror image form would not fit into the active site of the enzyme. Therefore, if an organism utilized both L-serine and D-serine, it would require two separate enzymes rather than one (Fig. 7–7).

If we accept the conclusion that it was selectively advantageous for living systems to use only one isomeric form of a particular molecule, we must still wonder why L-amino acids and D-nucleotides were selected rather than D-amino acids and L-nucleotides, or for that matter, any combination of isomeric forms (restricted, of course, to one isomeric form of each molecule). Most biologists contend that the initial choice between D- and L-isomers was entirely random, but that once the choice was made, it was retained throughout evolution. If life arose only once, we can assume that the first organism contained only L-amino acids. Alternatively, if life arose more than once, organisms possessing all L-amino acids, rather

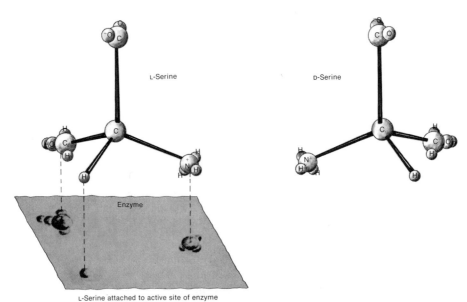

Figure 7–7   A schematic representation of the importance of molecular configuration to enzyme activity. L-serine is able to attach to the active site of the enzyme, but no amount of rotation of the D-serine molecule will permit it to fit into the active site. Consequently, an altogether different enzyme, the mirror image of the one depicted here, would be needed to catalyze a reaction involving D-serine.

Proteins and Their Subunits, the Amino Acids

than some L- and some D-, must have acquired a sufficient advantage to eliminate all competitors, for some reason or reasons we do not completely understand.

THE PEPTIDE BOND

Amino acids are joined together in chains by *peptide* bonds ($-\overset{\overset{O}{\|}}{C}-\overset{|}{\underset{H}{N}}-$; amide linkages) formed between the α-carboxyl group of one amino acid and the α-amino group of the adjacent amino acid. This polymerization reaction, shown in Figure 7-8, is sometimes referred to as a *dehydration condensation* because it involves the removal of the equivalent of one molecule of water.

Several amino acids linked together by peptide bonds are called *peptides;* long chains, consisting of more than eight or ten amino acids, are usually referred to as *polypeptides,* although the terminology is by no means exact. Amino acids in a peptide or polypeptide chain are called *amino acid residues.*

Notice in Figure 7-8 that peptides and polypeptides have a free amino group at one end and a free carboxyl group at the other end. The amino acid carrying the free amino group is called the *N-terminal residue;* the amino acid carrying the free carboxyl group is called the *C-terminal residue.* The terminal amino and carboxyl groups are, of course, ionized in the intracellular pH range. All other amino or carboxyl groups, except those present in side chains, are involved in peptide bonds.

Figure 7-9 shows a space-filling model of the backbone of part

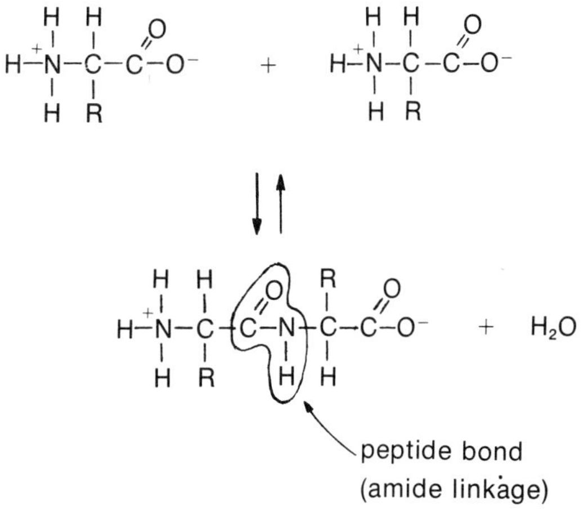

Formation of a peptide bond between two amino acids.

Figure 7-8

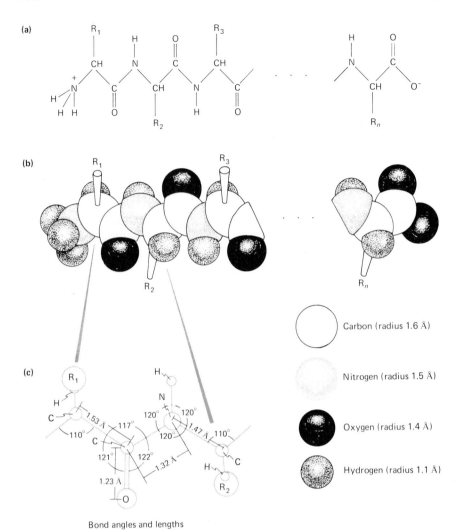

Figure 7-9 Chemical structure of the backbone of a polypeptide chain. *a.* Conventional structure of a polypeptide. *b.* A space-filling molecular model of the polypeptide chain, with R-groups omitted. *c.* Bond lengths and bond angles involved in the backbone structure. (From G.H. Haggis *et al.* [1964], *An Introduction to Molecular Biology,* John Wiley and Sons, New York.)

of a polypeptide chain, as well as the bond lengths and bond angles involved. The peptide bond is considerably shorter than the normal C—N single bond, and therefore has considerable double bond character. (The C—N linkage of the peptide bond has a length of 0.132 nm; the normal C—N single and double bond lengths are, respectively, 0.147 nm and 0.128 nm.) The partial double bond character of the peptide bond tends to hold the groups of six atoms, shown in Figure 7-10, in a rigid, planar configuration, and the only rotation allowed is the twisting of this unit about the bonds connecting it to the two α-carbons (Fig. 7-11). This restriction is particularly important in determining the possible configuration of a protein.

Proteins and Their Subunits, the Amino Acids

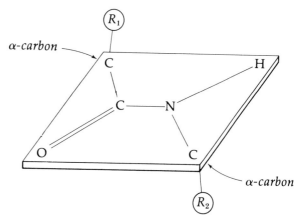

Planar structure of the atoms involved in a peptide bond. **Figure 7-10**

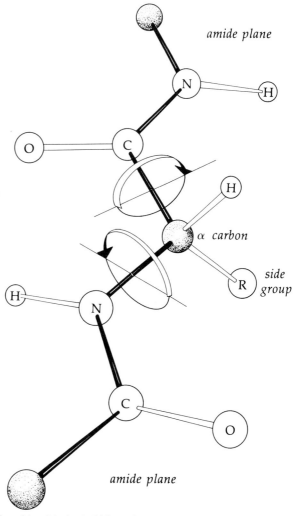

Because the atoms involved in a peptide bond tend to be held in a planar configuration, amino acids in a polypeptide chain can rotate only at the bonds adjacent to α-carbon atoms. In addition, not all such rotations are permitted because of spatial interference between bulky side chains. (From R.E. Dickerson, and I. Geis [1969], *The Structure and Action of Proteins*, Harper and Row, Publishers, New York.) **Figure 7-11**

Also as a consequence of the peptide bond structure, the electron distribution is such that both the carbonyl (—C=O) oxygen and the amide (—N—H) hydrogen are particularly good candidates for hydrogen bonding, since the oxygen carries a slight negative charge and the hydrogen carries a slight positive charge.

PROTEIN STRUCTURE

A protein may be defined as one or more polypeptide chains folded into a characteristic three-dimensional configuration. The specific configuration of a protein, which follows inevitably from the sequence of amino acids and their individual chemical properties, is required for its biological activity. Protein configuration is generally described in terms of four levels of organization: primary (I°), secondary (II°), tertiary (III°), and quaternary (IV°). The distinction between these various levels is not, however, always a sharp one.

Primary Structure

The primary structure of a protein refers to the order, or sequence, of amino acids in its polypeptide chain (or chains). Bovine insulin, a hormone involved in the regulation of glucose metabolism, was the first protein to be sequenced. As shown in Figure 7–12, it consists of two polypeptide chains held together by disulfide bonds. The two chains, denoted A and B, consist of 21 and 30 amino acid residues, respectively. Although bovine insulin is a small protein, the determination of its amino acid sequence was a difficult, time-consuming task, and was completed in 1953 after many years of work by F. Sanger and his colleagues; in 1958, Sanger was awarded the Nobel Prize in recognition of his accomplishment.

It is now clear that the sequence of amino acids is identical in all molecules of any one specific type of protein. A change in even one amino acid residue may adversely affect the biological activity of the protein. The relationship between primary structure and biological activity is well illustrated in studies of hemoglobin, the oxygen-carrying protein within red blood cells. Normal hemoglobin contains four polypeptide chains: two identical alpha (α) chains, each containing 141 amino acids, and two identical beta (β) chains, each containing 146 amino acids. There is an inherited disease in humans, called sickle cell anemia, that may be traced to the fact that in both β-chains a particular glutamic acid residue (the sixth amino acid residue from the N-terminal end) is replaced by a valine residue, as a result of a mutation in the DNA of affected individuals. The net effect is the replacement of the negative charge on the glutamic acid side chain by the uncharged side chain of valine. The α-chains and all other amino acid residues in the β-chains are identical to those in normal hemoglobin.

Proteins and Their Subunits, the Amino Acids

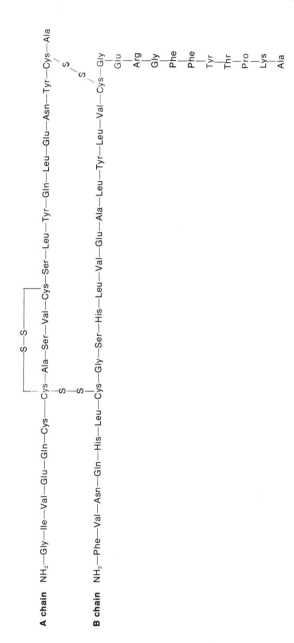

Figure 7-12 The primary structure of bovine insulin. (Data from F. Sanger [1952], Advances in Protein Chemistry 7:2.)

This seemingly minor structural change greatly influences the biological function of the hemoglobin molecule. It causes a decrease in the solubility of hemoglobin when it is deoxygenated, and consequently the hemoglobin forms crystals within the red blood cells. This, in turn, causes the red blood cells to change from the normal shape of a biconcave disc to a crescent or sickle shape (Fig. 7–13). The distorted cells tend to get caught in the capillaries and thus interfere with normal oxygen flow to the tissues. Because of their altered shape, sickled cells also have a markedly increased mechanical fragility and tend to be disrupted in the circulation, which leads to reduced numbers of oxygen-carrying red blood cells.

It should not be concluded from the above examples that all amino acid substitutions necessarily alter biological activity in such a profound way. Some substitutions appear to have little or no effect on biological activity, although there may be subtle effects that remain undetected. It is apparent, however, that certain amino acid residues in a particular protein are more important than others in determining the chemical properties and specific configuration of the protein molecule.

In summary, then, the primary structure of a protein tells us the number, kinds, and sequence of amino acids. It provides virtually no information, however, about the protein's three-dimensional configuration.

Secondary Structure

The secondary structure of a protein defines the degree to which the polypeptide chain (or chains) is stabilized by hydrogen bonds in the form of an α-*helix* or *pleated sheet*.

In the α-helix, the oxygen of each α-carbonyl group is hydrogen bonded to the amide nitrogen of the third amino acid from it along the polypeptide chain (Fig. 7–14). In this way, the first amino acid in a polypeptide chain is hydrogen bonded to the fourth amino acid in

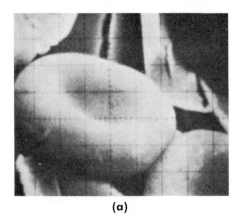 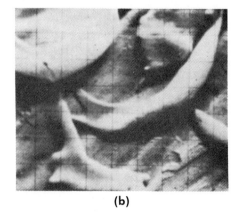

(a)   (b)

Figure 7–13 Scanning electron micrographs of (a) normal red blood cells and (b) red blood cells from individuals with sickle cell anemia. (From Brescia et al. [1974], *Chemistry: A Modern Introduction*, W. B. Saunders Co., Philadelphia.)

Proteins and Their Subunits, the Amino Acids

the chain; the second amino acid is hydrogen bonded to the fifth amino acid, and so on. Each peptide bond remains planar (see Figure 7–10), and the side chains of the amino acids project outward from the axis of the helix.

Pleated sheet structures are formed by regular hydrogen bonding either between extended, separate polypeptide chains, or between extended portions of the same chain folded back on itself. In this case, the amino acid side chains project above and below the hydrogen bonded sheets. Two conformations of pleated sheets are possible. In one, the so-called parallel pleated sheet, the hydrogen bonded polypeptide chains, or portions of a chain, run in the same direction relative to their N–terminal and C–terminal ends; in the antiparallel pleated sheet, the two hydrogen bonded chains run in opposite directions (Fig. 7–15).

Insoluble *fibrous proteins,* including the proteins of silk, wool, hair, nails, and others, have an uninterrupted secondary structure with more or less extended chains. For example, wool is composed of dead cells containing polypeptide chains of α-keratin, each of which is in the form of an α-helix. If the α-keratin is stretched, the hydrogen bonds stabilizing the α-helix are broken, and the chains attain the parallel pleated sheet conformation. However, when the tension is removed, the chains tend to re-form an α-helical configuration.

A number of factors may prevent the formation of long, uninterrupted regions of α-helix or pleated sheets, and consequently most proteins do not have the regular secondary structure characteristic of the fibrous proteins. For example, if a proline residue is present in the polypeptide chain, the α-helical configuration is interrupted. This occurs because the amide nitrogen of the proline residue, when involved in peptide bond formation, does not possess an attached hydrogen atom, and thus a hydrogen bond cannot be formed at this point in the chain (see Figures 7–1 and 7–8). In addition, hydrogen bonds in both the pleated sheets and α-helical structures may be disrupted by unfavorable side chain interactions, such as the presence of two bulky side chains next to each other or the electrostatic repulsion produced by two similarly charged side chains. When two cysteine residues on the *same* polypeptide chain form a disulfide bond between them, this will also distort the α-helical or pleated sheet configuration.

Tertiary Structure

The tertiary structure describes the way the regions of secondary structure are oriented with respect to one another and defines the specific three-dimensional configuration of a protein. For example, the α-helix does not completely describe the structure of fibrous proteins such as α-keratin. In hair, a number of chains in the form of an α-helix are twisted together into a multistranded cable.

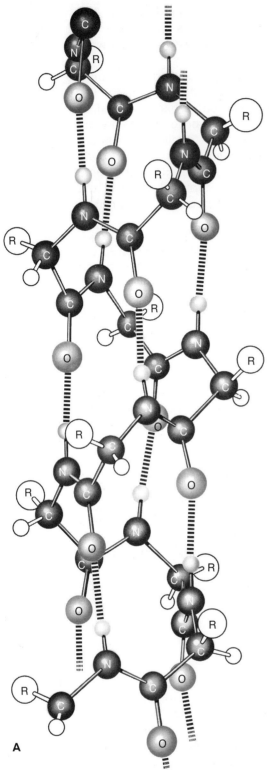

Figure 7–14 The α-helix. a. This "ball and stick" molecular model of the polypeptide backbone clearly demonstrates the hydrogen bonding pattern between —C=O and —NH groups.

Proteins and Their Subunits, the Amino Acids

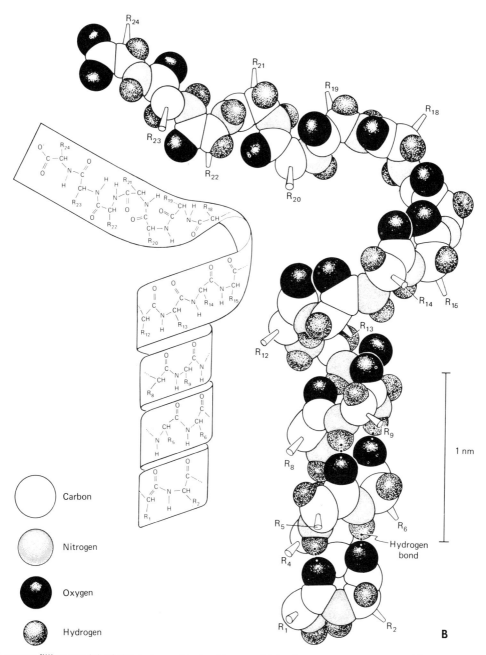

b. A space-filling model of the polypeptide backbone, which gives a more accurate representation of the configuration of an α-helix.

Figure 7–14
(Continued)

This cable, stabilized by numerous disulfide bonds between the individual chains, constitutes the tertiary structure of hair protein.

In our discussions of cellular metabolism, we are primarily interested in the water-soluble *globular proteins,* which include the enzymes, oxygen-carrying proteins, protein hormones, and others.

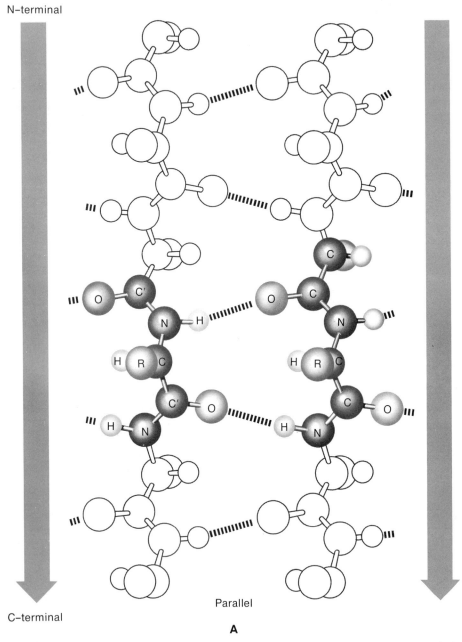

A

Figure 7–15 Hydrogen bonding in *(a)* parallel pleated sheet structure, in which all chains run in the same direction, and *(b, on opposite page)* anti-parallel pleated sheet, in which adjacent chains run in opposite directions. (From G.H. Haggis *et al.* [1964], *An Introduction to Molecular Biology*, John Wiley and Sons, Inc., New York.)

Proteins and Their Subunits, the Amino Acids

Globular proteins do not form long extended filaments like the molecules of fibrous proteins, but rather have a compact, somewhat spherical shape. A typical globular protein consists of sections of secondary structure interspersed with irregularly ordered regions. The entire polypeptide chain is folded in a highly specific manner to

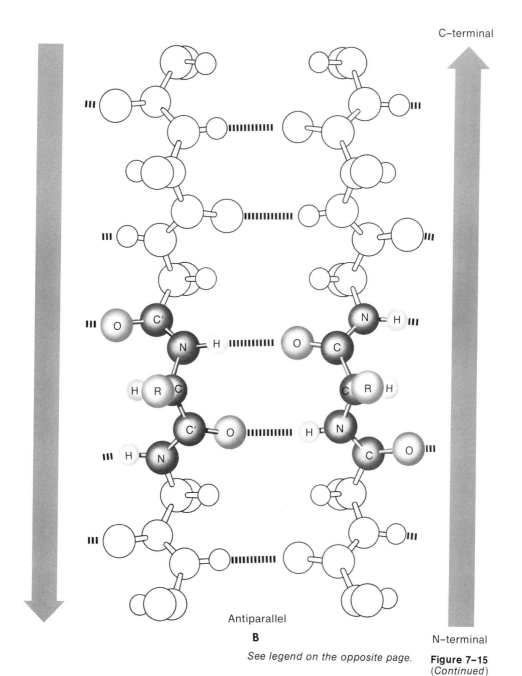

Antiparallel
B

See legend on the opposite page.

Figure 7–15 *(Continued)*

form the characteristic three-dimensional conformation of the protein.

Tertiary structure is stabilized by various weak molecular forces as well as by covalent linkages in the form of disulfide bonds or occasional *isopeptide bonds*—peptide linkages formed between a *side chain* amino group and a *side chain* carboxyl group. Some of the forces which determine tertiary structure are shown in Figure 7–16.

The three-dimensional configuration of a protein may be worked out by a complicated technique called *x-ray diffraction* or *x-ray crystallography.* In this technique, an x-ray beam is passed through a crystal of protein, and the patterns of reflections are used to determine the position of individual atoms in the crystal. X-ray diffraction analysis of protein structure is not easy, and consequently the tertiary structure of only a few proteins has been determined. One of these is lysozyme, an enzyme which fragments the polysaccharide backbone of the cell walls of certain bacteria. The tertiary structure of lysozyme is shown in Figure 7–17. It consists of α-helical regions, a region of antiparallel pleated sheet, and irregu-

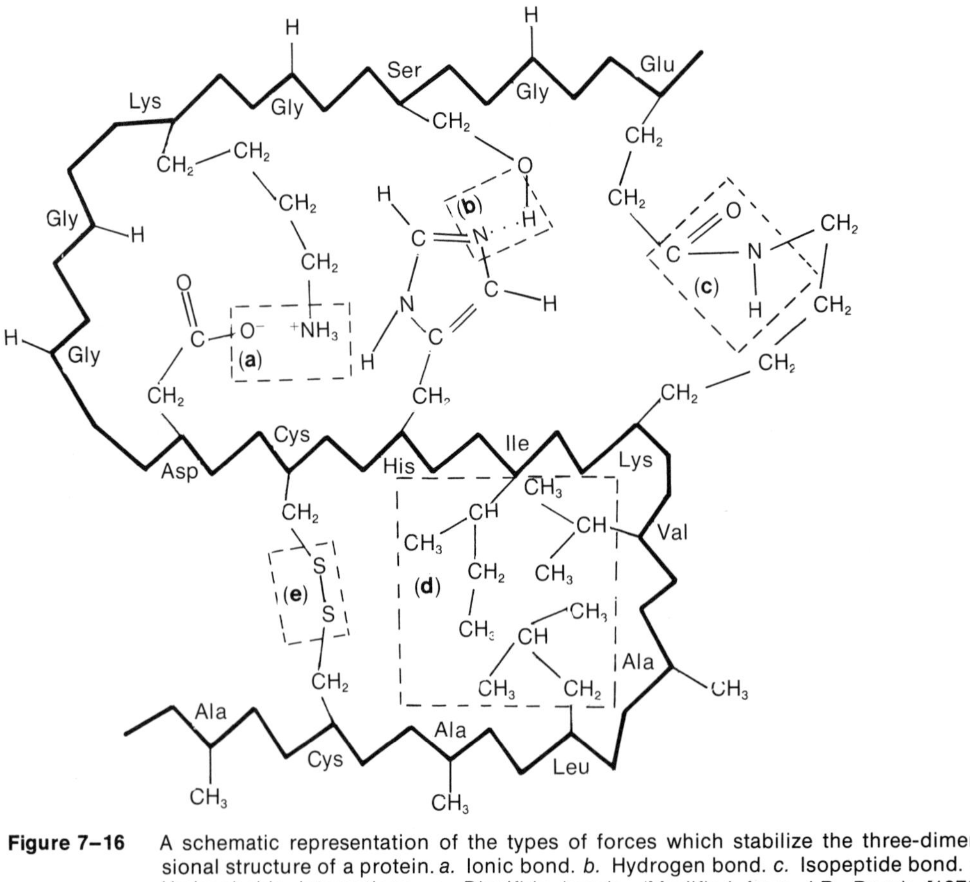

Figure 7–16 A schematic representation of the types of forces which stabilize the three-dimensional structure of a protein. a. Ionic bond. b. Hydrogen bond. c. Isopeptide bond. d. Hydrophobic interactions. e. Disulfide bonds. (Modified from J.R. Bronk [1973], *Chemical Biology,* Macmillan, Inc., New York, p. 104.)

Proteins and Their Subunits, the Amino Acids

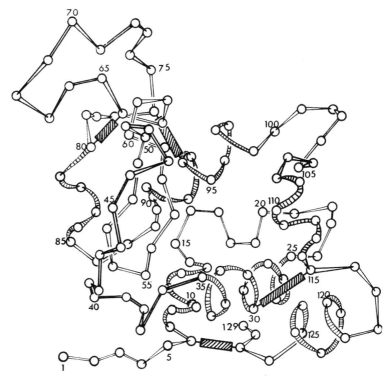

Figure 7–17
A model of the three-dimensional structure of the enzyme lysozyme, illustrating regions of secondary structure and the specific tertiary structure stabilized by many of the forces illustrated in Figure 7–16. The molecule consists of 129 amino acid residues in a single chain. α-helical regions exist between amino acid residues 5 to 15, 24 to 34, and 88 to 96. There is a short region of anti-parallel pleated sheet in the sharp turn formed by residues 42 to 54. The polysaccharide substrate of lysozyme attaches to the active site — a cleft in the molecule extending from upper left to lower right.

larly-ordered regions, all oriented in a characteristic three-dimensional configuration.

There does not appear to be any complicated "mechanism" which determines the precise folding of a protein. The specific three-dimensional configuration assumed by a particular protein is simply the most favored structure thermodynamically — the configuration of lowest energy, the one allowing the greatest total bond energy stabilizing the structure. Thus, given the proper sequence of amino acids, the protein will attain the proper configuration spontaneously. This conclusion is supported by the following observations.

1. If a protein is *denatured*; that is, if its tertiary structure is destroyed by high temperatures, extremes of pH, or other chemical means, biologically inactive polypeptide chains result. When the denatured polypeptide chains are removed from the denaturing conditions, the chains of some proteins will spontaneously assume their native configuration with full biological activity (*renaturation*).

2. Studies of the three-dimensional structure of proteins by x-ray crystallography have demonstrated that hydrophobic amino acid side chains tend to be buried in the interior of protein molecules removed from aqueous solvent, and the hydrophilic side chains tend to be located on the surface exposed to the surrounding

aqueous medium. This configuration appears reasonable from a thermodynamic standpoint, since it maximizes the number of chemical bonds. Such a structure has lower energy than one in which the side chains are oriented in such a way that they cannot form stabilizing chemical bonds with each other or with the solvent.

The net effect of the precise folding, then, is to determine a unique three-dimensional configuration for each protein. It should be stressed, however, that this configuration may change with alterations in pH, temperature, or solvent, or by binding of other substances such as inhibitors or metallic ions. In other words, the natural intracellular configuration (the *native* configuration) is not necessarily the most stable one under all environmental conditions.

Quaternary Structure

Not all proteins possess quaternary structure. This term refers to the spatial arrangement of polypeptide chains in proteins com-

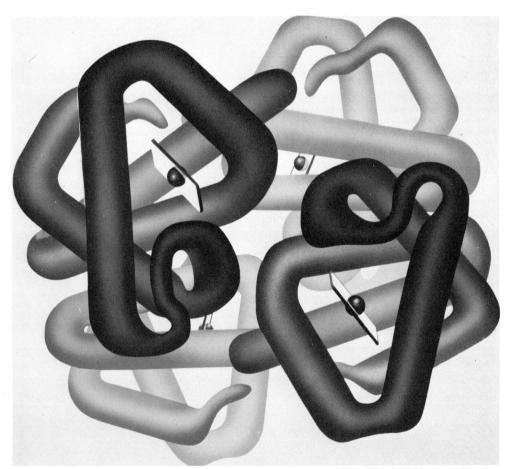

Figure 7–18   Quaternary structure of the hemoglobin molecule. (From R.E. Dickerson and I. Geis [1969], *The Structure and Action of Proteins,* Harper and Row, Publishers, New York.)

Proteins and Their Subunits, the Amino Acids

posed of two or more polypeptide chains which are not covalently bound together. In such a protein, each polypeptide chain is called a *subunit* and has its own tertiary structure. Quaternary structure is stabilized by weak interactions.

Hemoglobin was the first protein for which a complete quaternary structure was determined. The hemoglobin molecule consists of two α-chains and two β-chains, each possessing a characteristic tertiary structure. These four subunits are oriented in a specific configuration relative to one another to form the quaternary structure of the hemoglobin molecule (Fig. 7–18).

CONJUGATED PROTEINS

Not only does hemoglobin consist of four polypeptide chains arranged in precise quaternary structure, but each polypeptide chain has associated with it a *heme group*: an iron-containing organic ring structure belonging to a class of compounds called porphyrins (Fig. 7–19). Each heme group is capable of binding a molecule of oxygen (O_2), and consequently one hemoglobin molecule can carry four oxygen molecules. It is important to note, however, that an intact hemoglobin molecule consisting of polypeptide chains and attached heme groups, all in the proper configuration, is required for oxygen-carrying function.

Proteins like hemoglobin that have as part of their structure an organic or inorganic constituent that is *not* composed of amino acids

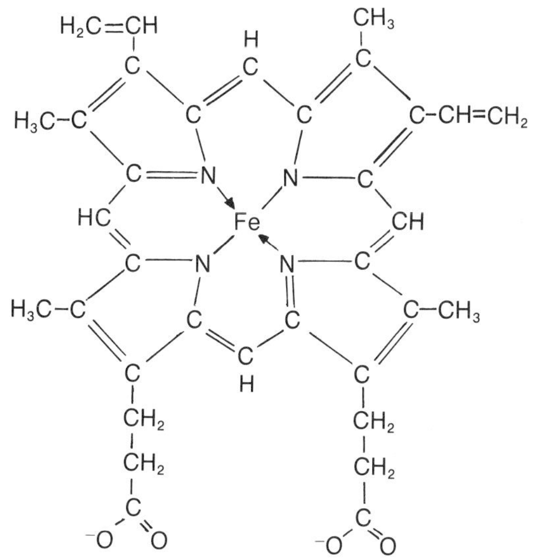

Figure 7–19 Chemical structure of heme. One heme group lies buried within a crevice of each polypeptide chain; the hydrophobic vinyl ($-CH=CH_2$) groups of the heme are surrounded by hydrophobic amino acid side chains, and the $-\overset{\overset{\displaystyle O}{\|}}{C}-O^-$ groups lie next to positively charged side chain nitrogens of lysine and arginine. The metal-containing ring structure is called a *porphyrin*; the attached side chains and sometimes the type of metal vary in different porphyrin compounds.

are called *conjugated proteins*; the non-amino acid part of the conjugated protein is called the *prosthetic group.* You will recall that some enzymes have prosthetic groups, and hence these belong to the general class of conjugated proteins.

Porphyrins are particularly important in the metabolism of living systems. Not only are porphyrins present in hemoglobin but they are found as prosthetic groups in a number of enzymes that play a crucial role in cellular metabolism. One class of enzymes, the cytochromes, are heme-containing proteins that are responsible for the transfer of electrons during the cellular oxidation of organic compounds and in the process of photosynthesis.

Not all porphyrins function as prosthetic groups. Indeed, the light-catalyzed reactions of photosynthesis depend on chlorophyll pigments, which are magnesium-containing porphyrin molecules.

CARBOHYDRATES

Carbohydrates serve a variety of functions in living systems. Not only are they an important energy source but they are essential constituents of the nucleotide subunits of nucleic acids. In addition, carbohydrates and their derivatives function as structural elements that help to provide rigidity in the cell walls of prokaryotes, fungi, and higher plants, and in the exoskeletons of invertebrate animals (arthropods).

The simplest carbohydrates are the *monosaccharides* (mono = one; sakchar = sugar), sometimes referred to as the *simple sugars.** These, like the amino acids of proteins, can combine to form short chain polymers called *oligosaccharides* (oligo = a few), consisting of two to ten monosaccharide residues. Longer chain polymers are termed *polysaccharides,* although it should be kept in mind that the distinction between oligosaccharides and polysaccharides is a purely arbitrary one. In contrast to proteins, the carbohydrate polymers important to biological systems are composed of only one or two different monosaccharide units, and consequently the polysaccharides do not possess the molecular and functional diversity of the proteins.

MONOSACCHARIDES

Monosaccharides have the general chemical formula $(CH_2O)_x$ and are further characterized by the fact that they do not give smaller carbohydrate units upon acid hydrolysis. Monosaccharides are often classified according to the number of carbon atoms they contain. Thus 3-, 4-, 5-, 6- and 7-carbon monosaccharides are termed, respectively, trioses, tetroses, pentoses, hexoses, and heptoses.

---

*Not to be confused with the common domestic sugar, sucrose, which is not a monosaccharide.

Carbohydrates

All natural monosaccharides are optically active, and most, but not all, monosaccharides occurring in living cells are in the D-configuration. As we mentioned earlier with respect to the optical activity of amino acids, D- and L-optical isomers are designated with reference to the arbitrarily defined configurations of glyceraldehyde, a triose which has only a single asymmetric carbon atom

```
        CHO                          CHO
         |                            |
    H—C—OH                       HO—C—H
         |                            |
       CH₂OH                        CH₂OH

    D-glyceraldehyde             L-glyceraldehyde
```

The D-isomer of glyceraldehyde is written with the —OH group on the right of the asymmetric carbon atom and the hydrogen on the left. The assignment of the D- or L-configuration for monosaccharides with more than one asymmetric carbon atom is based by convention solely on the orientation of the —OH and —H groups around the carbon atom immediately adjacent to the —CH₂OH group. If that particular —OH group is on the right, the D-isomer results; the L-isomer is the mirror image of this form. Thus, for example, the D- and L-isomers of glucose have the following configurations

```
        CHO                          CHO
         |                            |
    H—C—OH                       HO—C—H
         |                            |
    HO—C—H                        H—C—OH
         |                            |
    H—C—OH                       HO—C—H
         |                            |
    H—C—OH                       HO—C—H
         |                            |
       CH₂OH                        CH₂OH

      D-glucose                    L-glucose
```

The most important monosaccharides from a biological standpoint are the two hexoses, glucose and fructose, and the two pentoses, ribose and deoxyribose. Within the cell, these monosaccharides exist primarily as closed rings of atoms in equilibrium with only a small amount of the linear, open chain form. The formation of a ring structure from the linear form of D-glucose is illustrated in Figure 7–20.

In the ring structure of glucose, the ring is considered to lie perpendicular to the plane of the paper. This is indicated by shading the side of the molecule that projects out from the page toward the reader; the unshaded portions project behind the page away from the reader. The —H, —OH and —CH₂OH groups lie above or below the ring. In the D-forms of monosaccharides, the —CH₂OH

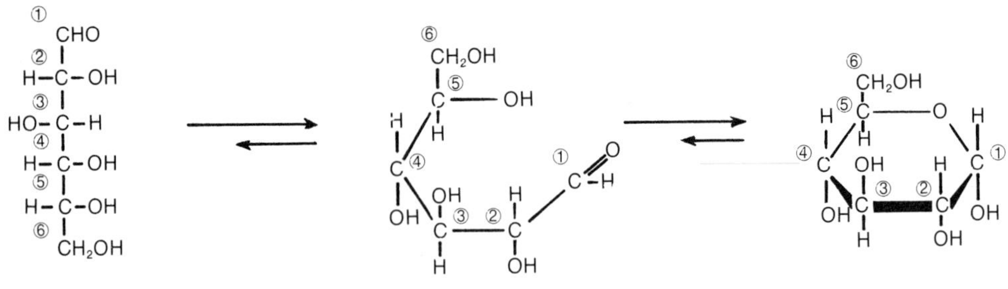

D-glucose (linear form)

D-glucose (ring structure)

Figure 7–20 Formation of the ring structure of D-glucose from the linear form.

group projects above the ring, when the ring is oriented as in Figure 7–20. By convention, the carbon atoms of the ring are numbered as shown; as an additional shorthand, the C's in the ring are omitted and understood to lie at the angles of the ring.

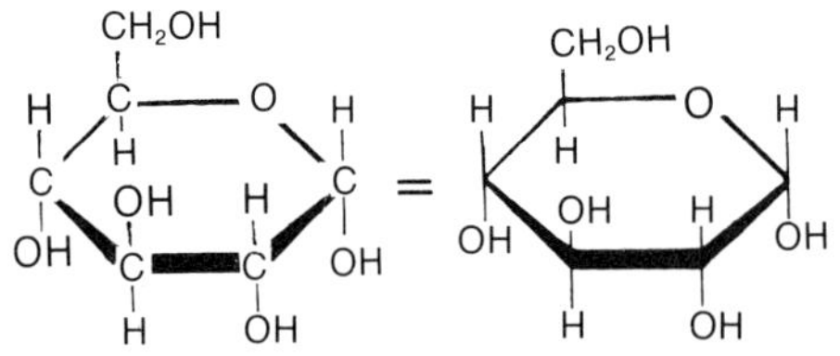

When D-glucose is in its linear form, the carbon atom at position 1 (C–1) is *not* asymmetric because it has only three different chemical groups attached to it (see Figure 7–20). However, when D-glucose is converted from its linear to its cyclical form, C–1 becomes asymmetric because it now has four different chemical groups surrounding it. Consequently, the cyclization of D-glucose results in the formation of two additional isomers, the α- and β-forms of D-glucose

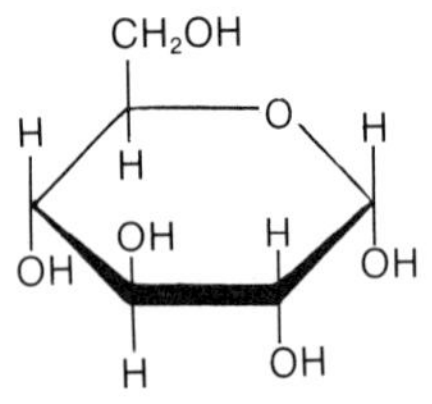

α-D-glucose

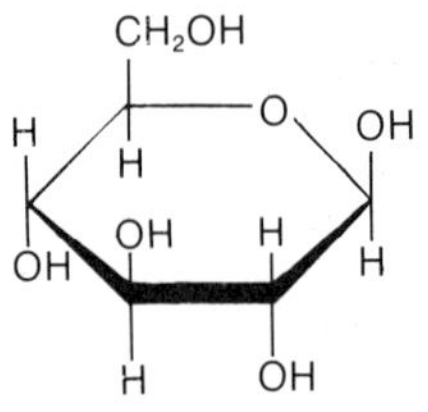

β-D-glucose

There are also, of course, α- and β- forms of L-glucose.

The α- and β- forms of a monosaccharide are called *anomers,* and the carbon atom which becomes asymmetric as a result of cyclization is called the *anomeric carbon.* If the hydroxyl group attached to the anomeric carbon is on the same side of the ring as the —CH_2OH group (C–5 or C–6, depending on the particular monosac-

Carbohydrates

charide), the β-configuration results; if it is on the opposite side, then the α-configuration results. In solution, there is an equilibrium between the α- and β- forms; about two thirds of D-glucose is in the β- form at equilibrium.

The distinction between the α- and β- forms of glucose (and of other monosaccharides which form a ring structure) is not merely an esoteric one, of interest only to carbohydrate chemists. Some carbohydrate polymers of the cell are composed of only the α- form of a monosaccharide, others are composed of only the β- form, and still others are composed of both. This is important because the chemical and physical properties of a polymer are influenced by the particular form (or forms) of monosaccharide of which the polymer is composed. For example, a polymer of α-D-glucose units (as in cellulose) has properties quite different than one formed from β-D-glucose units (as in starch).

Fructose, another common cellular hexose, has the following structure

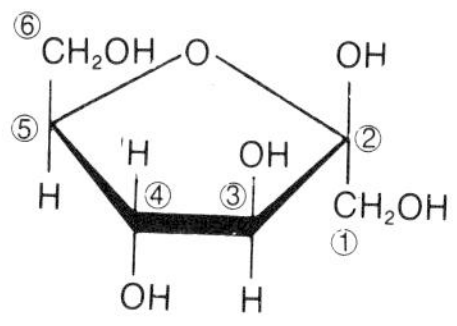

β-D-fructose

Unlike glucose, fructose forms a five-membered ring structure with the anomeric carbon at position 2 in the ring.

Glucose and fructose are of central importance in the processes of energy transfer within the cell and between living organisms. Both monosaccharides are produced as a result of photosynthesis. Plants and other autotrophs are able to convert fructose into glucose, and the glucose is then stored in the form of starch or cellulose, long chain polymers of glucose. Animals, fungi, and other heterotrophs, which are dependent on autotrophic organisms for survival, obtain glucose by enzymatically breaking down one or both of these polymers into their glucose subunits, and then extracting energy from the glucose molecule in a complex series of metabolic reactions.

The pentoses, ribose and deoxyribose, form a five-membered ring structure similar to that of fructose

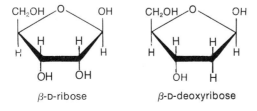

β-D-ribose   β-D-deoxyribose

Deoxyribose is identical to ribose, with the exception that the hydroxyl group attached to C–2 in ribose is replaced in deoxyribose by a hydrogen atom—hence the name deoxyribose. Like other monosaccharides, ribose and deoxyribose are optically active, and living organisms use only the D-isomers. Ribose is important as a constituent of RNA and of the ubiquitous energy-storage compound, ATP; deoxyribose is a constituent of DNA.

OLIGOSACCHARIDES

In living organisms, the most commonly encountered oligosaccharides are the *disaccharides,* formed by the elimination of water between two monosaccharide units. Two important representative disaccharides are maltose and sucrose. Maltose consists of two α-D-glucose residues linked between the C–1 of one glucose molecule and the C–4 of the other.

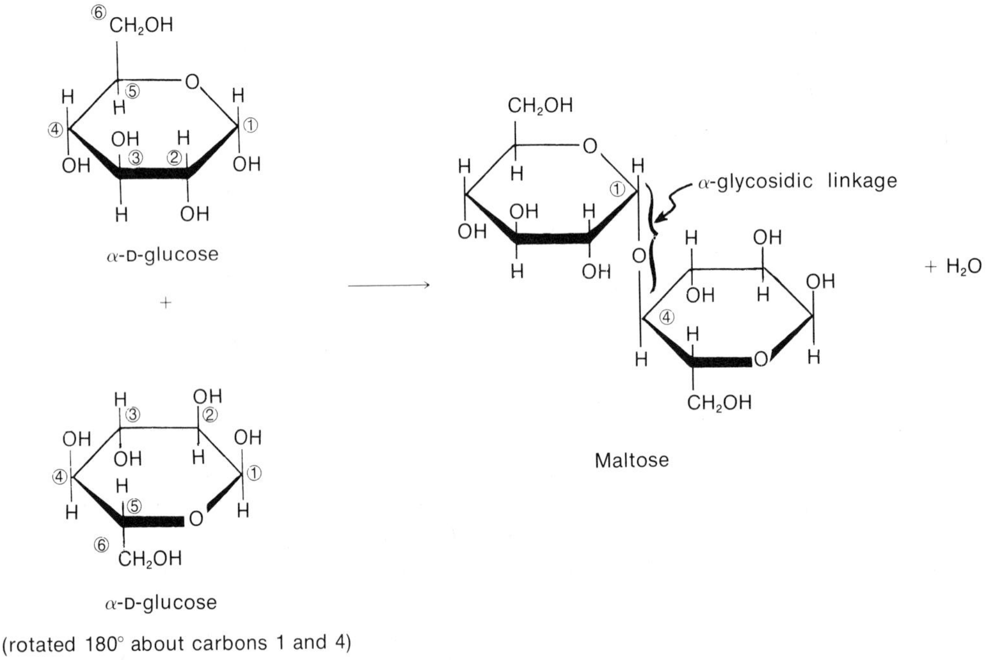

α-D-glucose

+

α-D-glucose
(rotated 180° about carbons 1 and 4)

Maltose

+ H_2O

α-glycosidic linkage

The C—O—C linkage between two monosaccharides is called a *glycosidic linkage,* and is formed when the hydroxyl group on the anomeric carbon of one monosaccharide (C–1 in the case of glucose) reacts with the —OH group of another monosaccharide. The glycosidic linkage is termed either α- or β- depending on the configuration of the —OH group on the anomeric carbon involved in the reaction. For example, in maltose, the anomeric carbon involved in the bond between the two glucose residues is in the α-configuration, so the glycosidic bond in maltose is termed an α-glycosidic linkage. In

Carbohydrates

referring to a specific glycosidic linkage, it is also customary to indicate which carbon atoms are joined together. Consequently, the glycosidic bond in maltose is designated as α-(1 → 4), meaning that the two monosaccharides are joined by an α-glycosidic bond between the C–1 of one monosaccharide and the C–4 of the other.

Sucrose, common cane and beet sugar, consists of α-D-glucose and β-D-fructose linked between the C–1 of glucose and the C–2 of fructose.

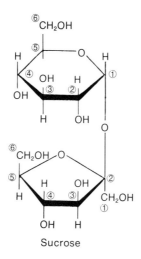

Sucrose

Here the glycosidic linkage is between the anomeric carbons of both glucose (C–1) and fructose (C–2), and technically the linkage is α with respect to glucose and β with respect to fructose. Hence, the linkage should actually be designated α, β-(1 → 2), although it is usually referred to as simply α-(1 → 2).

POLYSACCHARIDES

The bulk of carbohydrates are present in cells as polysaccharides, long chain polymers of monosaccharides linked together by glycosidic bonds. Three polysaccharides are widely distributed in living systems; these are *cellulose, starch,* and *glycogen,* all polymers of glucose.

Cellulose

Cellulose is the most abundant polysaccharide in nature and an important structural component of plants. Indeed, more than half the carbon in higher plants is in the form of cellulose, an insoluble, rigid, unbranched polymer varying in size from several hundred to more than 10,000 β-D-glucose units. The glucose residues are joined together in cellulose by β-(1 → 4) glycosidic linkages (Fig. 7–21a).

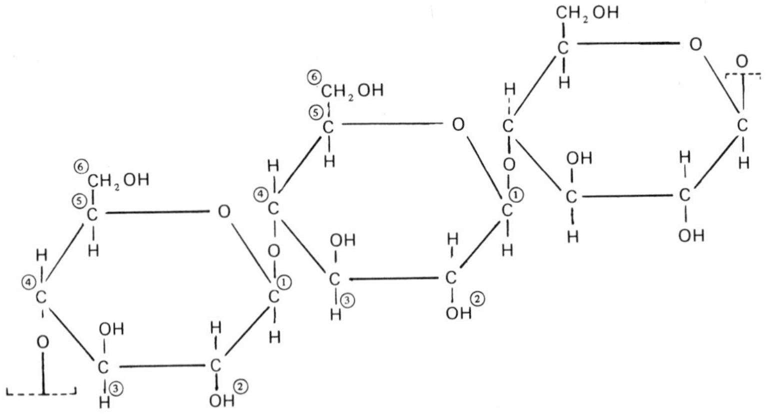

(a) Cellulose; β–(1 → 4) linkage

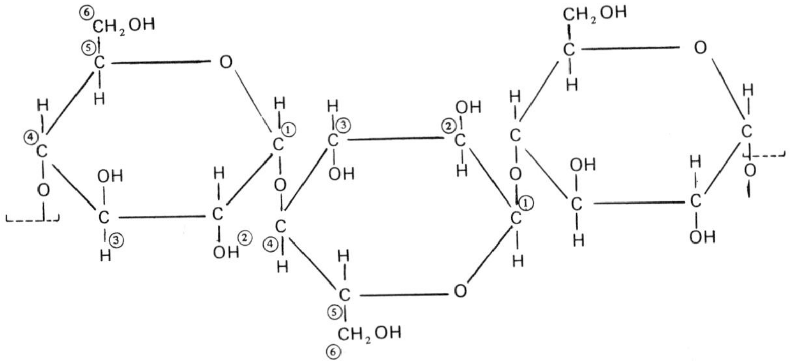

(b) Amylose; α–(1 → 4) linkage

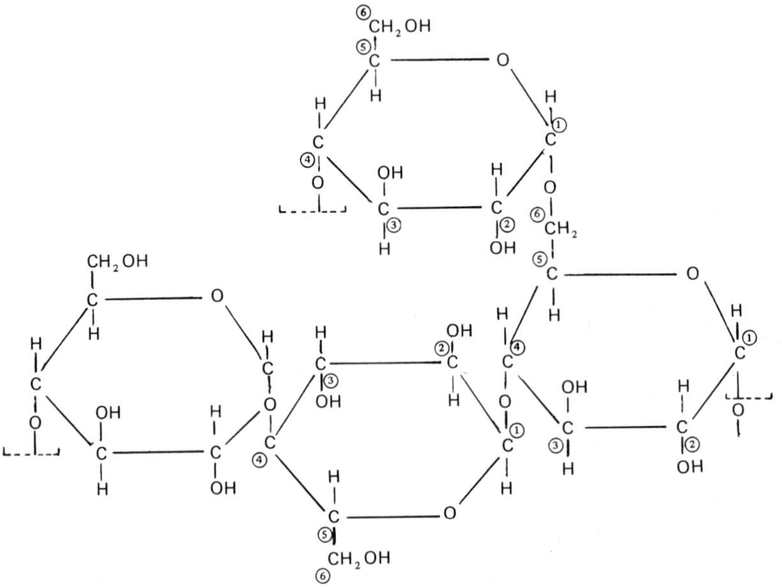

(c) Amylopectin; chains with α–(1 → 4) linkages crosslinked with α–(1 → 6) linkages

Figure 7–21 Portions of the cellulose, amylose, and amylopectin molecules. Both cellulose and amylose are unbranched chains which differ in structure only with respect to the way the glucose units are bonded together. Amylopectin is a branched molecule in which linear chains, similar in structure to those of amylose, are crosslinked by α- (1 → 6) glycosidic linkages.

Carbohydrates

It is important to point out that cellulose, like all polysaccharides, is not of one specific size, but rather may be a range of sizes, depending simply on the number of glucose residues linked together. A major source of cellulose is wood, which consists of empty cell walls composed primarily of cellulose and lignin, a complex ring polysaccharide.

Starch and Glycogen

Starch and glycogen are common storage forms of carbohydrates in plants and animals, respectively. Starch is the primary energy reserve in plants. In animals, glycogen is used for energy storage in liver and muscle and serves as a readily accessible source of glucose to stabilize blood sugar levels. However, the main energy reserves in animals are fats.

Starch is a mixture of two types of polymers found together in intracellular granules. One of these is *amylose,* an unbranched α-(1 $\rightarrow$ 4) polymer of glucose (Fig. 7–21*b*). Amylose thus differs from cellulose only in the type of glycosidic linkage between glucose units. The other polysaccharide polymer in starch is *amylopectin,* a branched molecule in which linear chains of α-(1 $\rightarrow$ 4) glucose polymers, like those composing amylose, are crosslinked by α-(1 $\rightarrow$ 6) linkages (Fig. 7–21*c*).

Glycogen is similar in structure to amylopectin, although it is more highly branched. Glycogen molecules tend to be much larger than those of amylopectin, and may be composed of up to 500,000 glucose units. Glycogen exists in animal cells in particles that are, however, much smaller than the starch granules in plant cells.

The individual molecules of α-(1 $\rightarrow$ 4) glucose polymers assume a helical configuration with about six glucose residues per turn (Fig. 7–22). The —OH groups of such helical polymers do not tend to form stable hydrogen bonds with adjacent helices because of the limited contact area available between two cylindrically shaped objects. Instead, the helices form hydrogen bonds with the surrounding water molecules. Hence glycogen and starch are nonfibrous and somewhat soluble.

In contrast, the β-(1 $\rightarrow$ 4) glucose polymers of cellulose are linear rather than helical. Because of this, many of the —OH groups on one chain may easily form hydrogen bonds with other —OH groups on an adjacent chain. Consequently, cellulose is found in closely packed fibers composed of 100 to 200 parallel chains held together by hydrogen bonds between the chains. The rigidity and solubility of cellulose are caused by these stabilizing crosslinks between the individual chains, as well as by the fact that, because of the crosslinking, few —OH groups in cellulose are available to interact with the aqueous solvent.

One additional polysaccharide of particular importance to a large number of living organisms is *chitin,* the structural substance of the exoskeleton of arthropods and of the cell walls of molds. Chitin is a linear, unbranched molecule consisting of N-acetyl glucosamine (a derivative of glucose) joined by β-(1 $\rightarrow$ 4) glycosidic linkages.

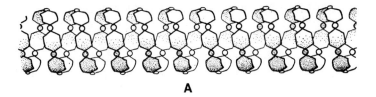

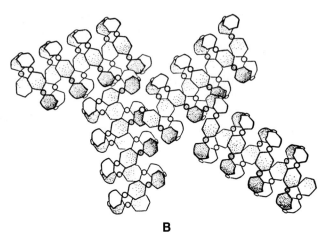

Figure 7-22 Three-dimensional configuration of polymer chains of (a) amylose and (b) amylopectin. (From Ambrose, E.J., and D.M. Easty [1970], *Cell Biology*, Addison-Wesley Publishing Co., Inc., Menlo Park, California.)

LIPIDS

Lipids comprise a diverse class of chemical substances that are characterized by their solubility in non-polar organic solvents (such as ether, chloroform, and benzene) and their low solubility in water. The solubility characteristics of lipids may be attributed to the fact that they possess significant non-polar regions and low dielectric constants, and are thus extensively hydrophobic, although they may also contain some hydrophilic groups.

Like proteins and carbohydrates, lipids have more than one function in biological systems. Certain types of lipids serve as energy storage compounds. Others are important as components of cellular membranes, as hormones, and as vitamins. We will consider four main categories of lipids: (1) fatty acids and fats, (2) phospholipids, (3) isoprenoid lipids, and (4) steroids. These categories include most but by no means all lipid compounds of importance to living organisms.

Lipids

FATTY ACIDS AND FATS

Fatty acids are carboxylic acids with the general formula $R-\overset{\overset{O}{\|}}{C}-OH$, where R is a long chain composed exclusively of carbon and hydrogen (*hydrocarbon chain*). The carboxylic acid group is strongly polar, and ionizes at intracellular pH by losing a proton. The long hydrocarbon chain is, however, non-polar and insoluble in water. Consequently, fatty acids have both hydrophilic and hydrophobic character.

When a solution of fatty acid in alcohol is layered on an aqueous surface, the alcohol dissolves in the water and the fatty acid molecules will orient themselves in a monomolecular layer at the surface with their polar, hydrophilic regions dissolved in the water and their hydrophobic tails in the air above the surface of the water. If fatty acid molecules are completely surrounded by water, they may form aggregates, called *micelles,* in which the hydrophobic regions of the fatty acid molecules are oriented toward the interior of the micelle away from the aqueous phase, and their hydrophilic groups are at the surface in contact with the surrounding water. Depending upon factors such as concentration and temperature, micelles may be in the form of small spheres or bimolecular layers (Fig. 7–23). In fact, the cleaning action of soaps, which are potassium or sodium salts of fatty acids, depends on their ability to trap lipid-soluble material in the hydrophobic regions of micelles.

Two fatty acids which are common in storage fats of animals are palmitic acid and stearic acid (Fig. 7–24 *a* and *b*). These two fatty acids are termed *saturated* fatty acids because their hydrocarbon portions contain only single bonds. Oleic acid and linoleic acid (Fig. 7–24 *c* and *d*) are common *unsaturated* fatty acids. Fatty acids with more than one double bond, such as linoleic acid, are sometimes called *polyunsaturated.*

Unsaturated fatty acids are more frequently encountered in living organisms than saturated ones. In addition, most of the fatty acids in animals contain an even number of carbon atoms. This is not surprising since, as we shall see, fatty acids are synthesized in animal cells by the polymerization of two-carbon units. Until recently, it was thought that fatty acids with odd numbers of carbon atoms were unimportant in living systems. It has now been shown, however, that odd-numbered fatty acids are present in all cells as a minor component of the total fatty acids, and that they are considerably more common in plants than in animals.

Within living cells, most fatty acids do not occur in an uncombined state, but rather are constituents of other lipid compounds such as fats, waxes, and phospholipids.

A fat is formed by the combination of fatty acid molecules with glycerol. One, two, or three fatty acid molecules may combine with glycerol forming, respectively, a *mono-, di-* or *triglyceride*. Triglycerides are, however, by far the most common form of fats in living organisms.

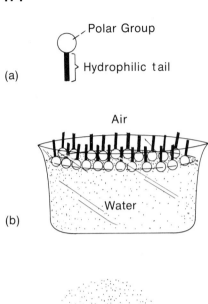

Figure 7–23 Behavior of fatty acid molecules in aqueous systems. *a.* Schematic representation of a fatty acid molecule. *b.* Formation of a surface film on water. *c.* Cross-section through a spherical micelle. *d.* Cross-section through a bimolecular layer.

The formation of a representative triglyceride, tripalmitin, from glycerol and three molecules of palmitic acid is shown in Figure 7–25. This reaction, in which an acid (in this case, a fatty acid) and an alcohol (glycerol) react with the elimination of water to form an ester

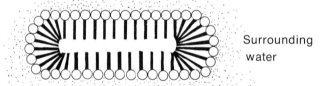

) is called an *esterification*.

The three fatty acid constituents in any particular molecule of triglyceride may be different, and need not even be all saturated or all unsaturated. It seems likely that the process by which fatty acids are incorporated into triglycerides is a random one, and although individual species have distinctive mixtures of fatty acids in their

Lipids

(a) $CH_3-CH_2-CH_2-CH_2-CH_2-CH_2-CH_2-CH_2-CH_2-CH_2-CH_2-CH_2-CH_2-CH_2-CH_2-COOH$

Palmitic acid (16 carbon atoms)

(b) $CH_3-CH_2-CH_2-CH_2-(CH_2)_{10}-CH_2-CH_2-CH_2-COOH$

Stearic acid (18 carbon atoms)

(c) $CH_3-CH_2-(CH_2)_5-CH_2-CH=CH-CH_2-(CH_2)_5-CH_2-COOH$

Oleic acid (18 carbon atoms)

(d) $CH_3-CH_2-CH_2-CH_2-CH_2-CH=CH-CH_2-CH=CH-CH_2-(CH_2)_5-CH_2-COOH$

Linoleic acid (18 carbon atoms)

Figure 7-24 Fatty acids commonly found in living systems. Palmitic and stearic acids are saturated fatty acids; oleic and linoleic acids are unsaturated.

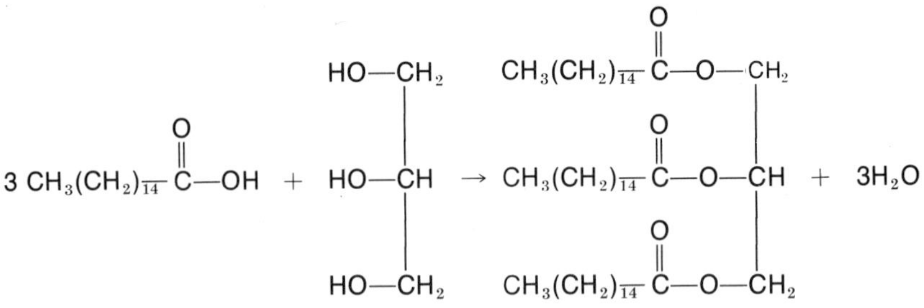

$$3\ CH_3(CH_2)_{14}\overset{O}{\underset{\|}{C}}-OH\ +\ \begin{array}{c}HO-CH_2\\ |\\ HO-CH\\ |\\ HO-CH_2\end{array}\ \rightarrow\ \begin{array}{c}CH_3(CH_2)_{14}\overset{O}{\underset{\|}{C}}-O-CH_2\\ |\\ CH_3(CH_2)_{14}\overset{O}{\underset{\|}{C}}-O-CH\\ |\\ CH_3(CH_2)_{14}\overset{O}{\underset{\|}{C}}-O-CH_2\end{array}\ +\ 3H_2O$$

Palmitic acid   Glycerol   Tripalmitin
(a fatty acid)   (an alcohol)   (triglyceride fat; an ester)

Figure 7–25   Fats are formed by the combination of fatty acids and glycerol. The formation of a triglyceride fat, tripalmitin, is shown here. Monoglycerides and diglycerides are formed by the reaction of glycerol with one and two fatty acid molecules, respectively.

triglycerides, these probably only reflect the types of fatty acids normally synthesized in organisms of a particular species.

Fats and fatty acids are the main energy storage molecules in animals. Although starch is the most common energy reserve in plants, this is not true without exception. Many seeds, for example, have large reserves of fats; these are the source of corn oil, peanut oil, and so forth. Fats and fatty acids yield over twice as much usable energy in the form of ATP as do carbohydrates. The use of fats rather than carbohydrates as energy reserves is appropriate to mobile organisms such as animals, since they are able to store in half the weight the same amount of energy as can plants that store starch as reserve material.

Fats cannot be used directly by cells as a source of energy, but first must be broken down into glycerol and fatty acids. This is accomplished with the aid of enzymes called lipases. Both the fatty acids and glycerol may then be degraded by cells in a series of energy-yielding reactions.

PHOSPHOLIPIDS

The simplest phospholipids are *phosphatidic acids,* in which the free —OH group of a diglyceride is esterified to the strong acid, phosphoric acid:

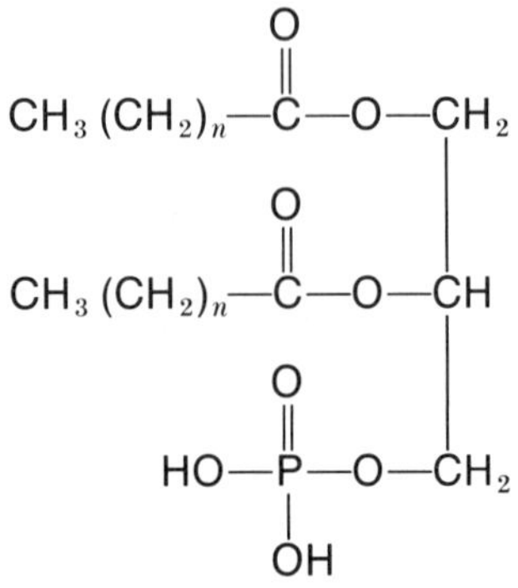

$$\begin{array}{c}CH_3(CH_2)_n-\overset{O}{\underset{\|}{C}}-O-CH_2\\ |\\ CH_3(CH_2)_n-\overset{O}{\underset{\|}{C}}-O-CH\\ |\\ HO-\overset{O}{\underset{\|}{P}}-O-CH_2\\ |\\ OH\end{array}$$

General structure of a phosphatidic acid

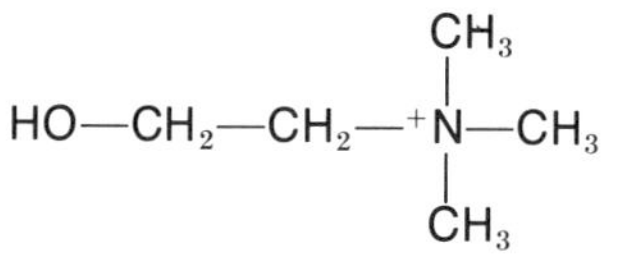

Lipids

$$\text{HO—CH}_2\text{—CH}_2\text{—}{}^+\text{N(CH}_3\text{)}_3$$

Choline

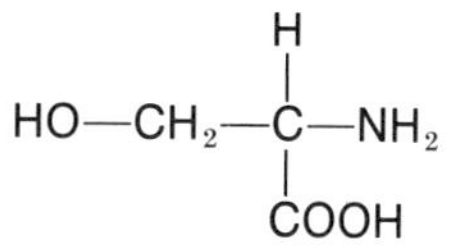

$$\text{HO—CH}_2\text{—}\underset{\underset{\text{COOH}}{|}}{\overset{\overset{\text{H}}{|}}{\text{C}}}\text{—NH}_2$$

L-serine

$$\text{HO—CH}_2\text{—CH}_2\text{—NH}_2$$

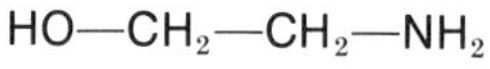

Ethanolamine

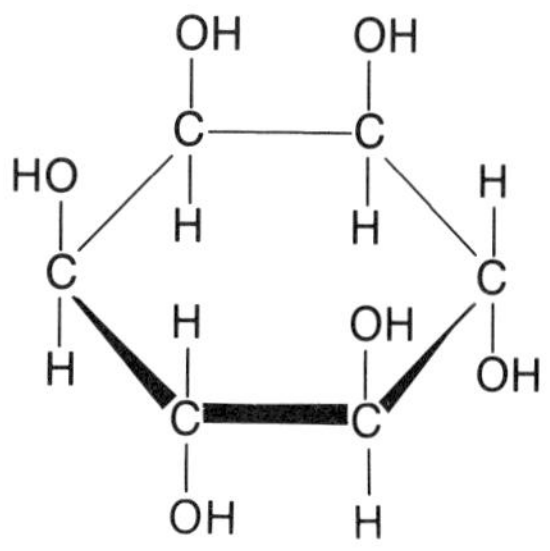

Inositol

Examples of some substances that may be covalently bonded to phosphatidic acids to form phospholipids. **Figure 7–26**

Because the composition of the two hydrocarbon chains may vary from one phosphatidic acid molecule to another, the term phosphatidic acid refers to a general class of compounds and not to one specific compound. As with triglycerides, the two hydrocarbon chains in phosphatidic acids are not necessarily similar.

The most important phospholipids in cells are derivatives of phosphatidic acids and have other substances, often nitrogen-containing bases, linked to one of the phosphate hydroxyl groups. Examples of substances that may be linked to phosphatidic acids include choline, ethanolamine, serine, inositol, and others (Fig. 7–26). One particularly important class of cellular phospholipids are the *lecithins* (phosphatidylcholines). Lecithins are constituents of cell membranes and are composed of choline bonded to a phosphatidic acid:

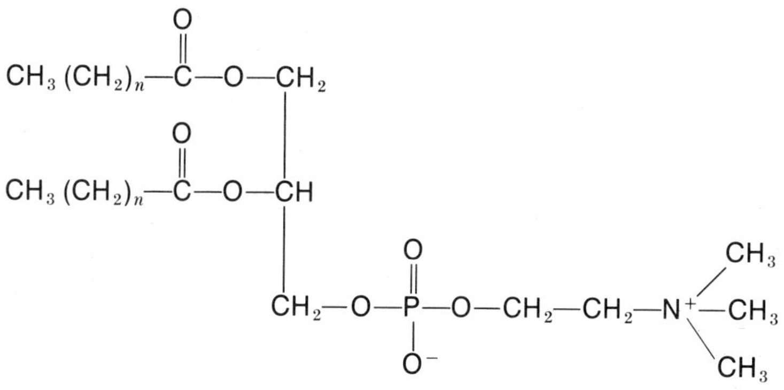

General structure of a lecithin

In aqueous solution around neutral pH (intracellular pH), both the nitrogen base and the acidic phosphate group are ionized, as shown above. Lecithins and other phospholipids thus possess a charged, polar end and a hydrophobic tail, and in this respect resemble fatty acid molecules.

The derivatives of phosphatidic acids exist in cells primarily in the L- rather than the D- configuration. Here again is another example of the chemical selectivity of living organisms.

It is important to remember that, like phosphatidic acids, lecithins and other phospholipids represent classes of compounds, since the composition of the hydrocarbon chains in individual molecules is variable. It has been reported, for example, that the lecithins in cell membranes of humans consist of more than 20 different molecules.

Since lecithins and other related phospholipids are (along with proteins and sometimes other lipids) principal constituents of cellular membranes, the behavior of phospholipids in aqueous systems has been studied by a number of investigators. When placed in water, phospholipids form monomolecular surface films, bilayers, and spherical micelles similar to those formed by fatty acids in water (see Figure 7–23). Indeed, it has been assumed that the tendency of

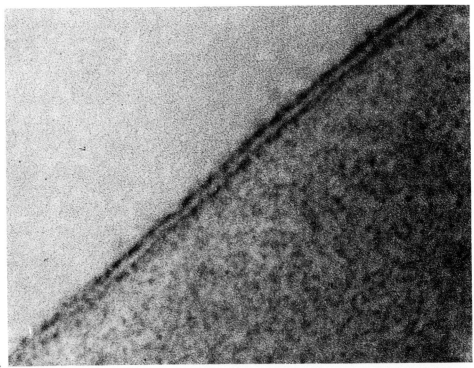

Figure 7–27   a. Electron micrograph of the plasma membrane of a human red blood cell (237,000 ×). (*From* Robertson, J.D. [1964]. In *Cellular Membranes in Development* [M. Locke, ed.], Academic Press, Inc., New York.)

Illustration continued on the opposite page.

Lipids

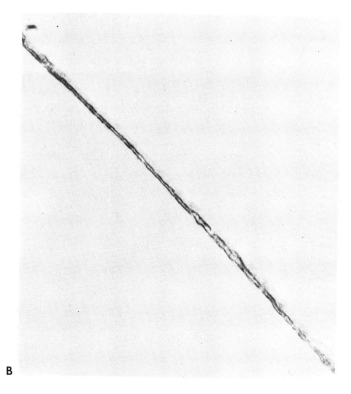

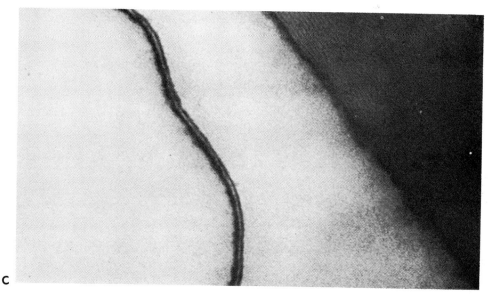

b. Electron micrograph of phospholipid bilayer 64,500×. (From Wolfe, S.L. [1972]. *Biology of the Cell,* Wadsworth Publishing Company, Belmont, California.) c. Electron micrograph of phospholipid bilayer with adsorbed protein. (From Ambrose, E.J. and D.M. Easty [1970]. *Cell Biology,* Addison-Wesley Publishing Co., Menlo Park, California.)

Figure 7-27 *(Continued)*

phospholipids to orient themselves spontaneously in this manner may have been an important factor in the origin of cell membranes.

This assumption is supported by several lines of evidence. It has been shown that phospholipid bilayers have dimensions similar to those of natural membranes, and under the electron microscope show patterns of electron density similar in appearance to those of cell membranes (Figure 7-27 a and b). In other experiments, W. Stoeckenius added a water-soluble protein to artificial phospholipid bilayers in aqueous medium, and demonstrated that the surface of the bilayers thickens—an indication of the binding of the protein to the lipid. In dimensions and appearance, these lipoprotein artificial membranes are even more similar to natural cell membranes than are simple phospholipid bilayers (Fig. 7-27c).

ISOPRENOID LIPIDS

Isoprenoid lipids are a group of lipids whose members are structurally related to isoprene:

$$CH_2=\underset{\underset{CH_3}{|}}{C}-CH=CH_2$$

Two isoprenoid lipids, pristane and phytane, are of particular interest to us because of the fossil evidence which demonstrates their presence in early Precambrian sediments (see Chapter 2). As you will remember, it was suggested that pristine and phytane may be breakdown products of the phytol tail of chlorophyll. The structures of pristane and phytane are shown in Figure 7-28 a and b; their similarity to phytol (Fig. 7-28c) lends support to this idea. Other scientists, however, have pointed out that the pristane and phytane in Precambrian sediments may well have been derived from isoprenoid compounds present in non-photosynthetic organisms.

Other important isoprenoid lipids are the *carotenoids,* a class of light-sensitive pigments synthesized only by prokaryotes and higher plants. One of these is β-carotene (Fig. 7-29a), which functions in many plants as a light-absorbing photosynthetic pigment. Vitamin A (Fig. 7-29b), necessary for vision in vertebrates, is synthesized by animals from β-carotene, which must be obtained in their diets.

Steroids

The steroids are a family of lipid compounds which have a common multiple-ring structure. Cholesterol, one important cellular steroid, is found in cell membranes of many organisms and is the cellular precursor to cortisone, testosterone, estrogen, progesterone, and other steroid hormones (Fig. 7-30). The D vitamins, close relatives of cholesterol, are important in the assimilation of calcium.

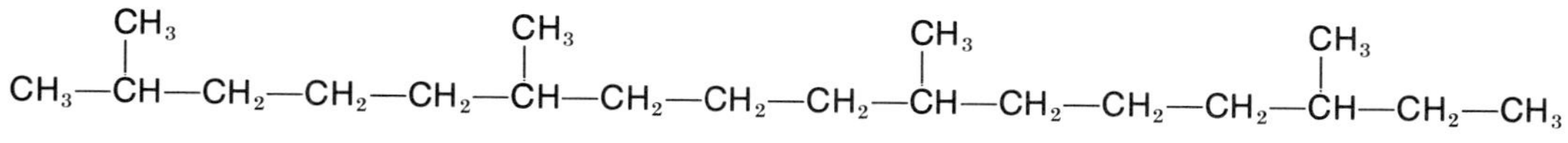

Lipids

$$CH_3-CH(CH_3)-CH_2-CH_2-CH_2-CH(CH_3)-CH_2-CH_2-CH_2-CH(CH_3)-CH_2-CH_2-CH_2-CH(CH_3)-CH_2-CH_2-CH_3$$

Phytane

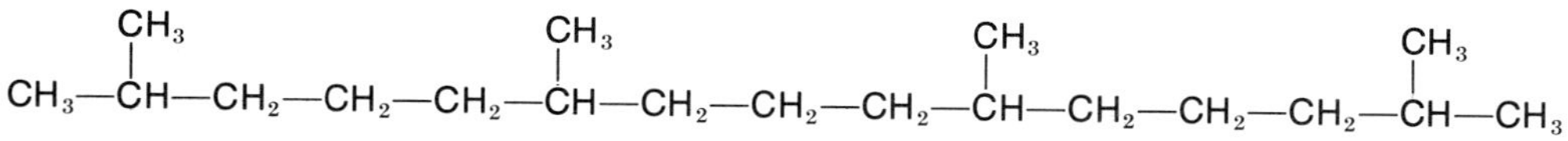

$$CH_3-CH(CH_3)-CH_2-CH_2-CH_2-CH(CH_3)-CH_2-CH_2-CH_2-CH(CH_3)-CH_2-CH_2-CH_2-CH(CH_3)-CH_3$$

Pristane

$$CH_3-CH(CH_3)-CH_2-CH_2-CH_2-CH(CH_3)-CH_2-CH_2-CH_2-CH(CH_3)-CH_2-CH_2-C(CH_3)=CH-CH_2OH$$

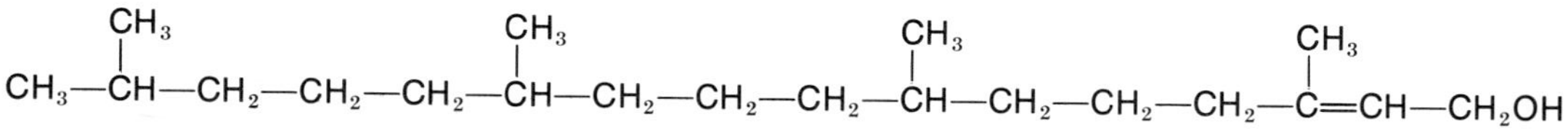

Phytol

Figure 7-28 Because of the close structural resemblance of the isoprenoid lipids pristane and phytane to the phytol tail of chlorophyll, many researchers have suggested that pristane and phytane are breakdown products of the chlorophyll molecule. Thus the presence of pristane and phytane in early Precambrian sediments has been interpreted as evidence for the existence of photosynthetic organisms more than 3 billion years ago (see Chapter 2).

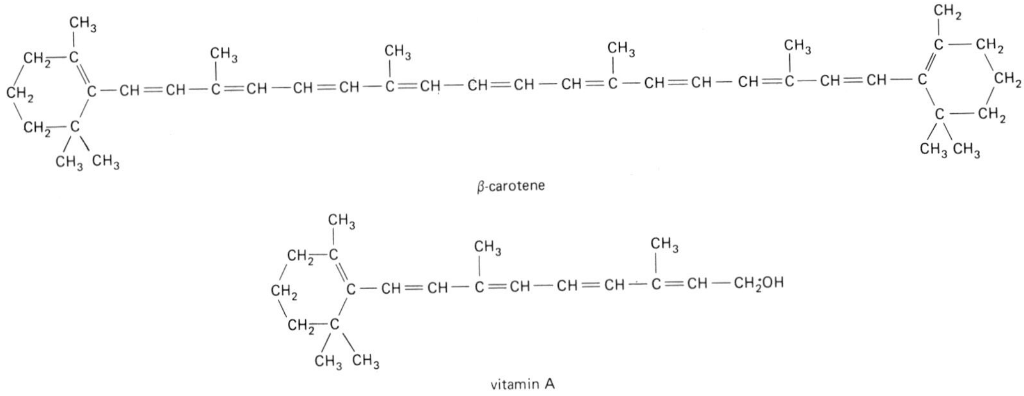

Figure 7–29 Chemical structures of β-carotene and its derivative, vitamin A. (From J.R. Bronk [1973], *Chemical Biology,* Macmillan, Inc., New York, p. 153.)

NUCLEIC ACIDS

There are two classes of nucleic acids: DNA (deoxyribonucleic acid) and RNA (ribonucleic acid). All the hereditary (genetic) information of the cell, that is, all the information necessary to reproduce and maintain a new organism, is stored in coded form in molecules of DNA. The DNA is replicated and distributed to daughter cells during cell division, and in this way all the hereditary information accumulated over billions of years of evolution is passed from cell to cell and from one generation of an organism to another. With the aid of RNA, this information is expressed as specific patterns of protein synthesis.

An understanding of the processes of DNA replication and protein synthesis is central to an understanding of cell function. Consequently, these processes will be discussed in detail later, and it is more appropriate to consider much of the molecular structure of nucleic acids in the context of those discussions. Therefore at this point we will present only a brief outline of the composition and structure of nucleic acids.

NUCLEOTIDES

Like proteins and polysaccharides, nucleic acids are polymers composed of subunits linked together by the elimination of water (dehydration condensation). The *nucleotide subunits* of nucleic acids are, however, structurally rather more complex than amino acids or monosaccharides. Indeed, nucleotides are themselves composed of three different molecules covalently bonded together, also by dehydration condensation; these are (1) a nitrogen-containing organic base, (2) a pentose sugar, and (3) phosphoric acid.

The chemical structures of molecules found as constituents of nucleotides are shown in Figure 7–31. The nitrogenous bases are

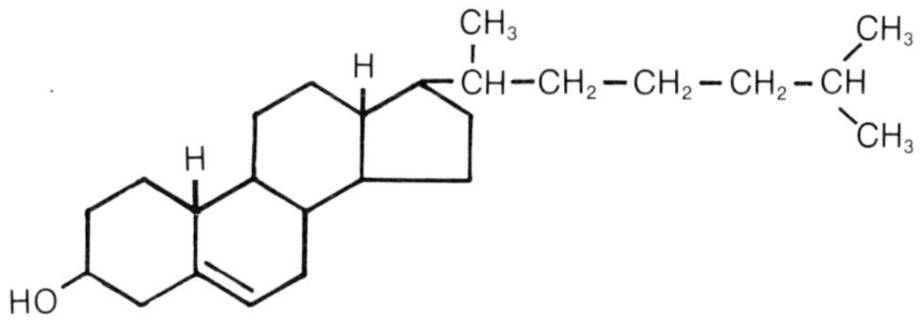

Nucleic Acids

Cholesterol

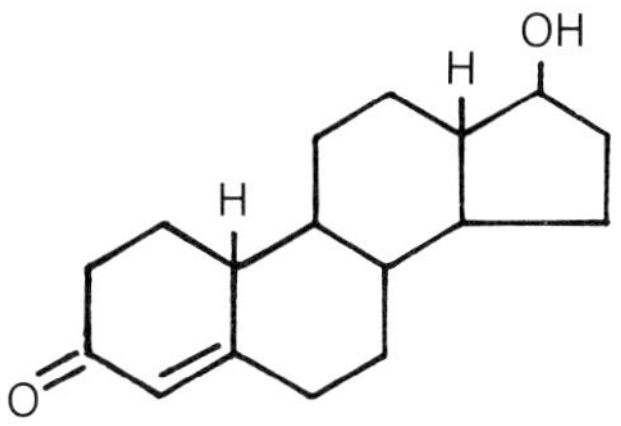

Testosterone

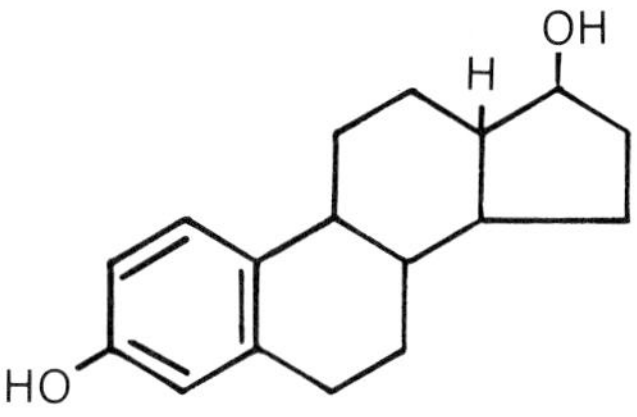

Estradiol-17β (An Estrogen)

Figure 7-30
The most abundant steroid in animal tissue is cholesterol, an important constituent of cell membranes of animals and the precursor of testosterone, estrogens, and other steroid hormones. The seemingly trivial differences in structure between testosterone and the estrogens have profound effects on biological activity.

derivatives of either *purine* or *pyrimidine,* nitrogen-containing ring compounds

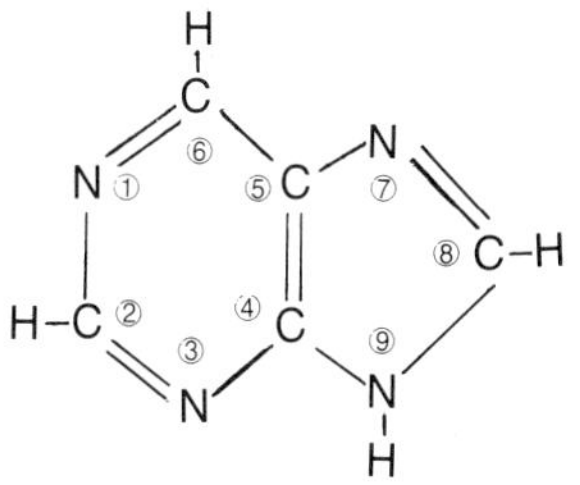

Purine

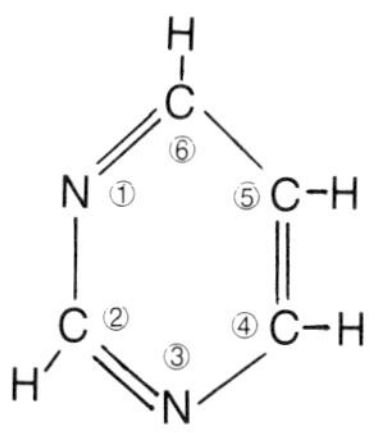

Pyrimidine

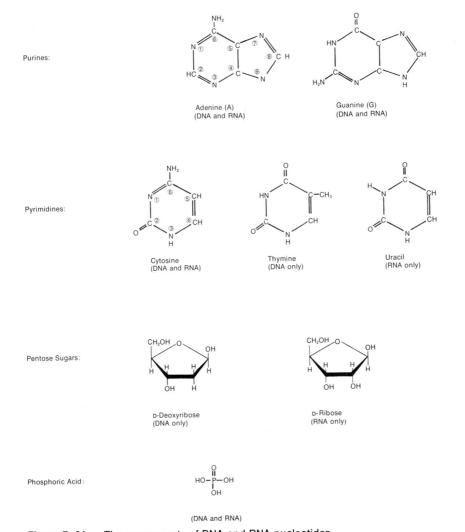

Figure 7–31 The components of DNA and RNA nucleotides.

Nucleotides of DNA and RNA differ somewhat in structure. A DNA nucleotide is formed from the pentose sugar deoxyribose, phosphoric acid, and any one of the four nitrogenous bases adenine (A), guanine (G), cytosine (C), or thymine (T). RNA nucleotides, on the other hand, are formed from the pentose sugar ribose, phosphoric acid, and either adenine, guanine, cytosine, or uracil (U). Thus the chemical composition of RNA nucleotides differs from that of DNA nucleotides in two respects: RNA nucleotides contain ribose rather than deoxyribose, and, when present, uracil rather than thymine. Figure 7–32 illustrates the way the individual components of a nucleotide are linked together. Also demonstrated by Figure 7–32 is the fact that nucleotides can be regarded as phosphorylated derivatives of *nucleosides;* nucleosides are combinations of a ni-

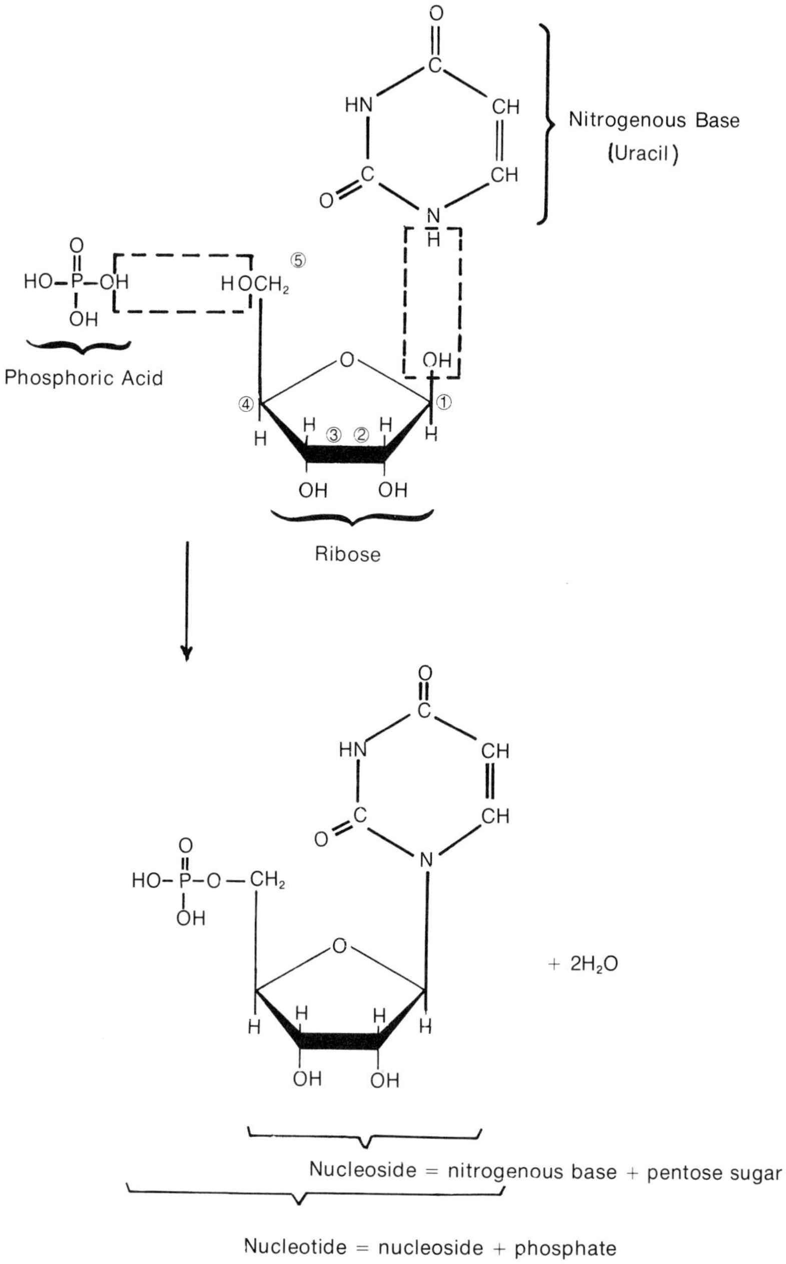

Nucleic Acids

Figure 7-32 Formation of a nucleotide from its component molecules by dehydration condensation.

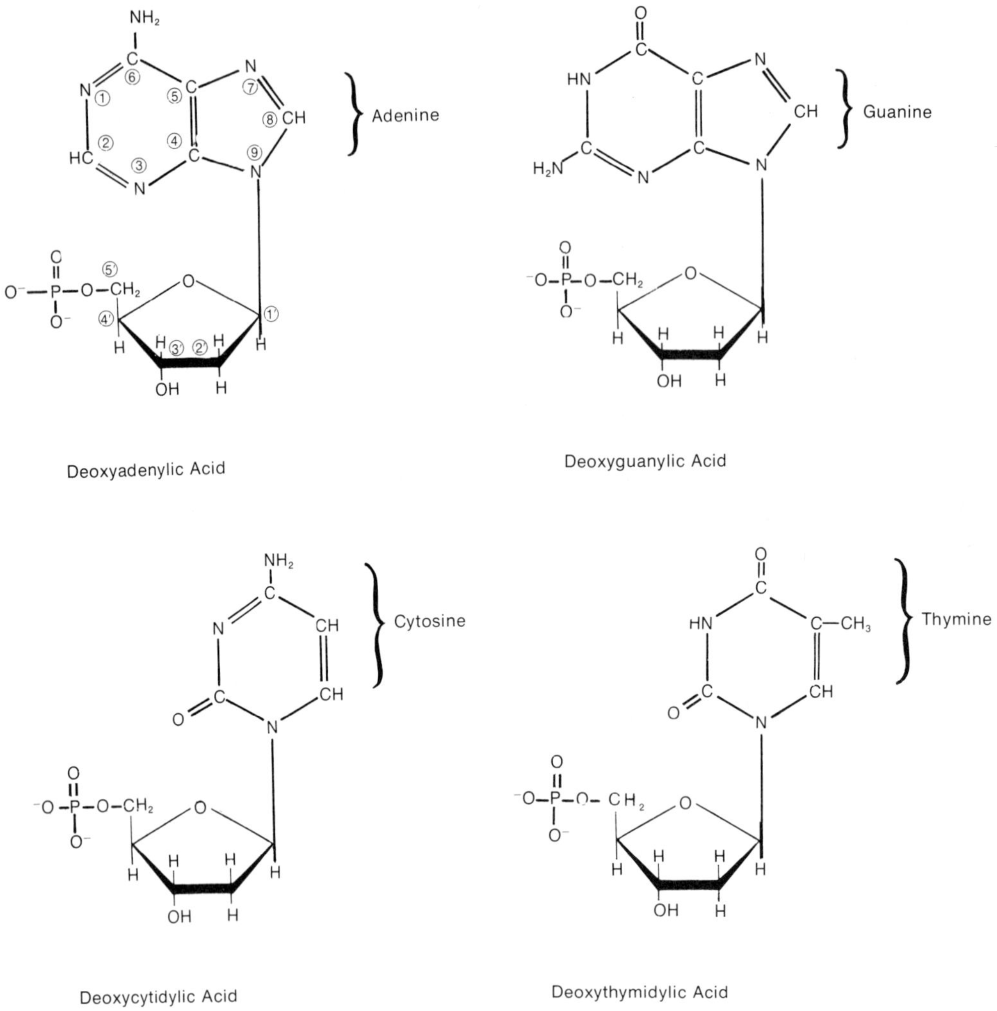

Figure 7-33 The four common nucleotides of DNA. The RNA nucleotides have the same structure, with the exception that deoxyribose is replaced by ribose, and thymine is replaced by uracil. Uridylic acid, the RNA nucleotide containing uracil, is shown in Figure 7-32.

trogenous base and a pentose sugar without an attached phosphate group.

Figure 7-33 shows the chemical structures of the four common nucleotides of DNA. It is apparent that a nucleotide contains two ring structures, one corresponding to the nitrogenous base and one to the pentose sugar. In order to differentiate one from the other, the numbers associated with the pentose ring are denoted 1′ (1 prime), 2′ (2 prime), etc., as indicated in the structure for deoxyadenylic acid (Fig. 7-33).

Each of the different nucleotides and nucleosides of RNA and DNA has a specific name. These are summarized in Table 7-1. It is

Nucleic Acids

Table 7-1 NOMENCLATURE OF THE NUCLEOSIDES AND NUCLEOTIDES

BASE	SUGAR	NUCLEOSIDE (BASE PLUS SUGAR)	NUCLEOTIDE (BASE PLUS SUGAR PLUS PHOSPHATE)
		DNA	
Adenine	Deoxyribose	Deoxyadenosine	Deoxyadenosine monophosphate (dAMP) Deoxyadenylic acid
Guanine	Deoxyribose	Deoxyguanosine	Deoxyguanosine monophosphate (dGMP) Deoxyguanylic acid
Cytosine	Deoxyribose	Deoxycytidine	Deoxycytidine monophosphate (dCMP) Deoxycytidylic acid
Thymine	Deoxyribose	Deoxythymidine	Deoxythymidine monophosphate (dTMP) Deoxythymidylic acid
		RNA	
Adenine	Ribose	Adenosine	Adenosine monophosphate (AMP) Adenylic acid
Guanine	Ribose	Guanosine	Guanosine monophosphate (GMP) Guanylic acid
Cytosine	Ribose	Cytidine	Cytidine monophosphate (CMP) Cytidylic acid
Uracil	Ribose	Uridine	Uridine monophosphate (UMP) Uridylic acid

important for future discussions that you become thoroughly familiar with this nomenclature system.

Nucleotides of DNA and RNA may occasionally contain nitrogenous bases other than those listed in Figure 7–31. For example, DNA of certain higher plants and animals contains trace amounts of 5-methylcytosine; in some bacterial viruses, 5-hydroxymethyluracil is present in place of thymine; in still other bacterial viruses, cytosine is replaced by 5-hydroxymethylcytosine. This latter base may also have one or two glucose residues attached by a glycosidic bond to the hydroxyl group at the C–5 position (Fig. 7–34). RNA also may contain smaller quantities of other bases besides A, G, C, and U. We will return to the role of these so-called "odd bases" when we discuss the mechanism of protein synthesis.

Nucleotides themselves have important functions in the cell. As we have already seen, nucleotides can add one or two additional phosphate groups to become di- and triphosphorylated. Two such compounds, ATP (adenosine triphosphate) and ADP (adenosine diphosphate), are important in energy storage and transfer. Nucleotides are also important constituents of many coenzymes. One

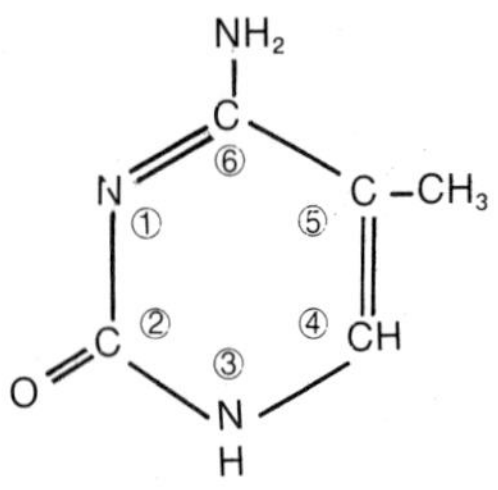

5-Methylcytosine

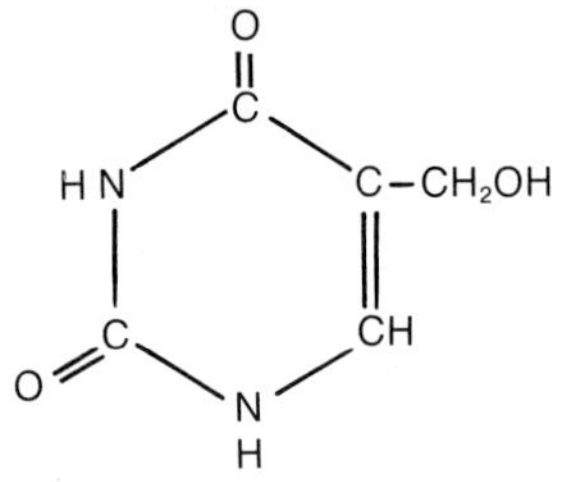

5-Hydroxymethyluracil

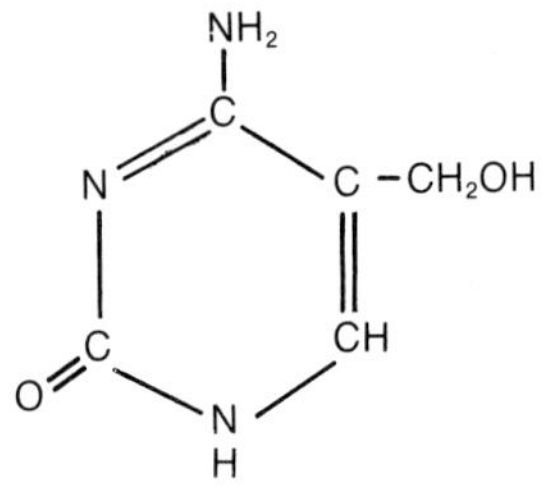

5-Hydroxymethylcytosine

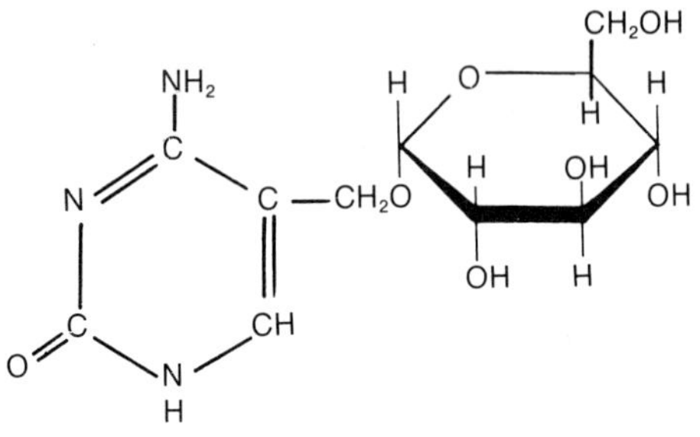

Glucosylated-5-hydroxymethylcytosine

Figure 7-34 Examples of uncommon bases found in certain kinds of DNA. Many such "odd bases" also occur in RNA.

nucleotide found in virtually all cells is 3′,5′-cyclic AMP (adenosine monophosphate)

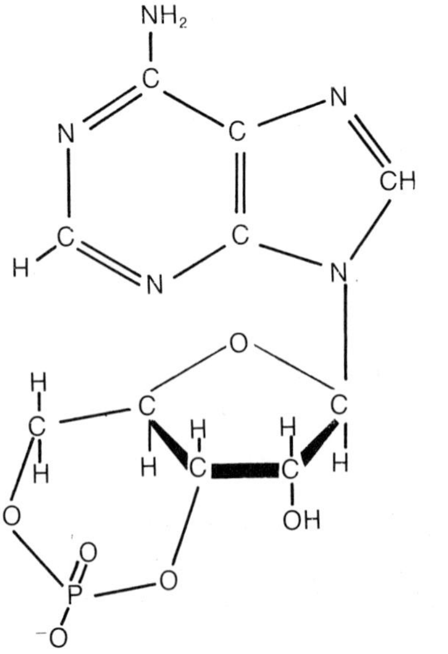

3′,5′-cyclic AMP

Nucleic Acids

Cyclic AMP serves as an activator of a number of enzymes and also acts as an intracellular mediator of the effects of many hormones.

POLYNUCLEOTIDES

Nucleotides are joined together to form a *polynucleotide chain* by a covalent linkage between the phosphoric acid residue of one nucleotide and the 3' carbon of the sugar on the next nucleotide. This linkage is often called a 3',5' *phosphodiester bond,* because the phosphate is esterified to two OH groups, one attached to the 3' carbon and one attached to the 5' carbon. The backbone of a polynucleotide chain thus consists of alternating sugar and phosphate units. This is illustrated in Figure 7–35, which shows segments of both DNA and RNA polynucleotide chains.

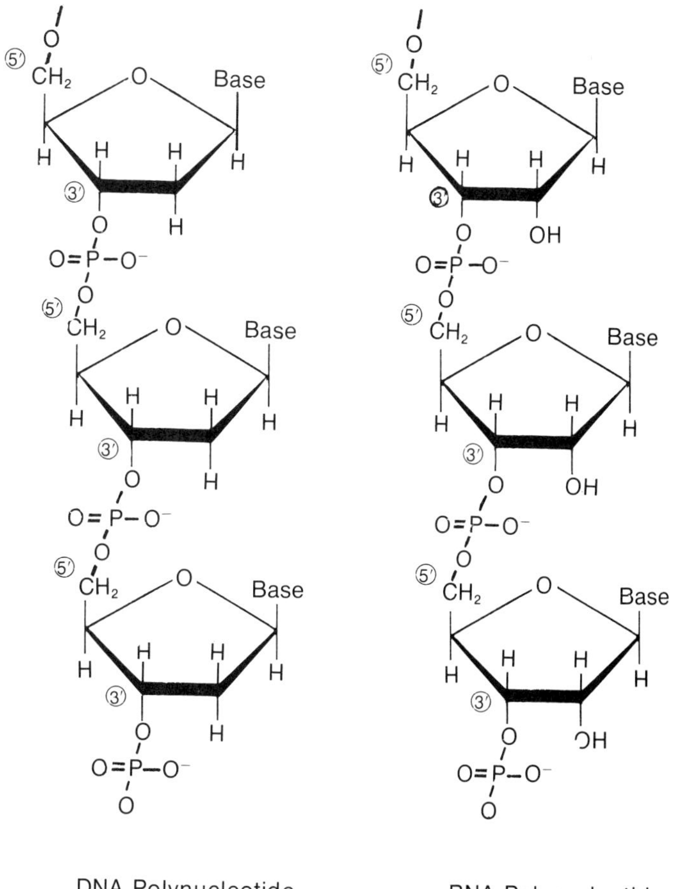

DNA Polynucleotide   RNA Polynucleotide

Figure 7–35 Portions of DNA and RNA nucleotide chains. In DNA, the attached bases may be adenine (A), guanine (G), cytosine (C), or thymine (T); in RNA, they may be adenine, guanine, cytosine, or uracil (U). Note the phosphodiester bond between the 3' carbon of one sugar residue and 5' carbon of another.

The sequence of nucleotides in DNA and RNA is the key to their genetic functions, just as the sequence of amino acids determines the biologic activity of a particular protein. Even though both DNA and RNA are usually composed of only four different nucleotides, the number of possible sequences of nucleotides is enormous in a large polymer.

As we shall see later, RNA usually exists as a single-stranded polynucleotide chain and can have an extremely specific three-dimensional structure. DNA usually consists of two nucleotide chains hydrogen bonded together in a specific manner in the form of a double helix.

REFERENCES: CHAPTER 7

Ambrose, E.J., and D.M. Easty (1970). *Cell Biology.* Addison-Wesley Publishing Co., Menlo Park, California.
Bronk, J.R. (1973). *Chemical Biology.* Macmillan Inc., New York.
Dickerson, R.E., and I. Geis (1969). *The Structure and Action of Proteins.* Harper and Row, Publishers, New York.
Dyson, R.D. (1974). *Cell Biology: A Molecular Approach.* Allyn and Bacon, Inc., Boston.
Edwards, N.A., and K.A. Hassall (1971). *Cellular Biochemistry and Physiology.* McGraw-Hill Book Co., Ltd., London.
Gymer, R.G. (1973). *Chemistry: An Ecological Approach.* Harper and Row, Publishers, New York.
Miller, S.L., and L.E. Orgel (1974). *The Origins of Life on the Earth.* Prentice-Hall, Inc., Englewood Cliffs, New Jersey.
White, A., P. Handler, and E.L. Smith (1973). *Principles of Biochemistry,* 5th Ed. McGraw-Hill Book Co., Inc., New York.

CHAPTER 8

Prebiotic Synthesis of Organic Compounds

Now that we have considered the composition and properties of the major chemical constituents of living systems, we are in a good position to review the results of simulation experiments designed to provide information concerning the course of chemical evolution on the prebiotic earth. It is reasonable to assume that this process, which resulted in the production of increasingly complex organic compounds and led to the eventual appearance of living systems, should be reproducible in the laboratory, at least to a limited extent. Indeed, both Oparin and Haldane, the originators of the materialistic theory of life's origin, suggested that their ideas could be tested by laboratory experimentation instead of being accepted or rejected on faith.

Of course, there are difficulties inherent in such an approach. Although the results of simulation experiments may support the Oparin-Haldane hypothesis, this does not constitute proof that events on the primitive earth necessarily followed the postulated course. But it is possible, from the information we can infer about the environment on the prebiotic earth, to derive reasonable conditions for carrying out simulation experiments. The results of these experiments can then indicate to us whether the transition from simple gases to complex molecules of biological significance is a plausible occurrence or an impossible one. Thus, although we can never hope to duplicate the exact mechanism by which living systems arose on earth (if, in fact, they did!), we can certainly gather experimental information that may suggest to us a crude outline of the processes that may have been important in the evolution of living systems. In fact, it has been so easy to obtain both simple and complex biochemical compounds in simulation experiments that a large number of biologists, including many of those originally skeptical, now support at least some version of the materialistic theory. Of course, this support is derived primarily from the assumption that the first self-replicating and self-sustaining living system, the ancestral living organism, was recognizably similar to organisms existing today, at least in terms of its chemical composition.

From our knowledge of present-day living organisms, what properties might we reasonably ascribe to the simplest ancestral organism? To restate the question, what are the minimal chemical and structural characteristics which would be necessary to sustain the first living cell as a self-replicating entity, if we assume that it was unambiguously related to modern organisms? Although we have no sure way of knowing the answer, a list of possible characteristics is shown in Figure 8–1. We have provided our hypothetical organism with enzymes needed to synthesize DNA, RNA, and protein from abiotically produced compounds such as amino acids, purines, pyrimidines, ribose, deoxyribose, and phosphoric acid. We have assumed that its genetic information is stored in molecules of DNA, and that it possesses a mechanism which utilizes RNA for decoding this information as specific patterns of protein synthesis. We have presumed that all the metabolic reactions of the cell occur in an aqueous matrix as they do now, and that the entire entity is surrounded by a crudely discriminating limiting membrane composed of lipoprotein. It also seems reasonable to endow our ancestral cell with the ability to synthesize its limiting membrane, again from chemical precursors available in the prebiotic soup. Finally, we have assumed that energy to sustain its replicative activities was derived from abiotically synthesized energy-rich compounds (such as ATP), and that all other metabolic capabilities and structural characteristics are later evolutionary developments.

We do not mean to imply that this hypothetical ancestral organism is necessarily the simplest possible self-replicating unit. Indeed, other "organisms" can be imagined based on far simpler mechanisms of information storage and retrieval, and possessing characteristics quite different from the ones we have described here. In fact, many biologists have proposed such models for the first ancestral cell. We have, however, intentionally endowed our model ancestral organism with chemical characteristics clearly related to those of present-day living organisms. The purpose of constructing such a "conservative" model is to define the minimal types of organic compounds necessary to form what we consider to be a simple living system, based on the design of extant living organisms. If simulation experiments produce these needed compounds, then we have accumulated some experimental evidence to support the materialistic theory. Of course, it is possible to contend that the first self-sustaining entity did not in any way resemble modern organisms, and that characteristics of present-day cells have entirely replaced earlier ones. Although this may well be true, it is not considered likely; furthermore, it is impossible to design valid experiments to test such a presumption.

It would perhaps be well at this point to clarify what we mean by "chemical evolution," for a thorough understanding of the term is crucial to later discussions. We have previously defined chemical evolution as that process which began with the formation of simple

Prebiotic Synthesis of Organic Compounds

Cell Components	Function	Chemical Components (Implied)
DNA	Replication	Deoxyribose Guanine, adenine, cytosine, thymine Phosphoric acid
RNA	Protein synthesis	Ribose Guanine, adenine, cytosine, uracil Phosphoric acid
Enzymes	DNA, RNA, and protein synthesis; membrane synthesis	Amino acids
Surface membrane	Maintenance of internal cellular environment	Lipids (fatty acids, glycerol); proteins
Cytoplasm	Solvent system	All of the above in water; inorganic ions
	Energy source	ATP, other energy-rich compounds

Figure 8-1 Hypothetical characteristics of the simplest self-replicating cellular entity. (Adapted from Margolis, L. [1970] Origin of Eukaryotic Cells, New Haven, Connecticut, Yale University Press.)

organic compounds from the gases in the primitive atmosphere. These organic compounds, attaining states of gradually increasing complexity, eventually gave rise to the first living cells. This process is summarized schematically in Figure 8-2. In using the term "chemical evolution," we do not mean to imply that selection pressures, *in the classic sense of evolutionary theory,* were the driving force for this process. Rather, we envision chemical evolution occurring as a result of natural chemical processes. The thermodynamic barriers inherent in proceeding from a state of high entropy to a state of lower entropy are overcome by the input of energy from the primitive environment. It is essential to remember that there is no premeditated "grand design" inherent in the process, no vital force, no desire on the part of simple gaseous substances to become living organisms. Thus chemical evolution proceeded in a random fashion, subject to the laws of chemistry and physics.

Simon Black has published a thought-provoking theoretical analysis of the course of chemical and biological evolution. Because the thermodynamic equilibrium involved in the spontaneous polymerization of simple organic compounds is exceedingly unfavorable, Black concludes that the formation of polymers in the prebiotic soup would be extremely slow. Consequently, he proposes that the process of chemical evolution, as we have described it in Figure

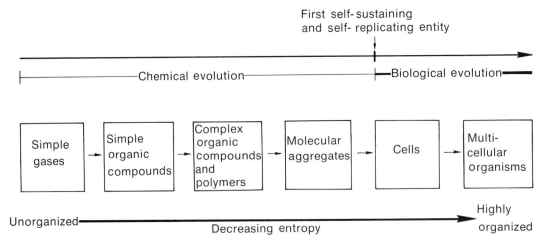

Figure 8-2 Schematic representation of the progression of chemical and biological evolution.

8-2, would not proceed to any significant extent without a driving force—something to provide the process with direction, to channel the energy input. Since hydrophobic interactions appear to be the dominant factor determining intermolecular associations in biochemical molecules, Black argues that the tendency of organic compounds in water to form hydrophobic associations represents the thermodynamic potential which is the driving force of both chemical and biological evolution. Thus Black contends that the complex catalytic pathways in living organisms evolved as a direct result of the thermodynamic tendency of organic compounds in water to attain a state of lower energy by the formation of hydrophobic bonds: Catalytic pathways represent a means by which the rate of hydrophobic associations between organic compounds is enhanced. As Black puts it:

> *Life arose, not in spite of these* [thermodynamic] *barriers, but because of them.* Evolution's driving force, confined as by a dam, carved elaborate channels through which organic compounds might flow in search of a lower level of free energy. The channels are catalytic pathways in living cells. . . . From the foregoing analysis it appears that life arises and persists as a vain search for a stable state. Because of the barriers to its direct achievement, a second search arises to reach this state [resulting in the formation of enzyme-catalyzed metabolic pathways]. But high stability and functional activity are incompatible. The triumph of the search for function is evident in the elaborate perfection of the cell; the failure to achieve stability is evident in death; the persistence of the search for a more stable state is evident in the survival of species.*

Black's view of the evolutionary process, although not accepted by all biologists, is worthy of serious consideration because it

---

*S. Black (1973), A theory on the origin of life. Advances in Enzymology *38*:193.

Prebiotic Synthesis of Simple Organic Compounds

provides one of the few reasonable mechanisms to explain the processes of chemical and biological evolution on a molecular level. In the next chapter, we will return to this point when we discuss mechanisms by which living entities may have evolved from complex organic compounds.

PREBIOTIC SYNTHESIS OF SIMPLE ORGANIC COMPOUNDS

AMINO ACIDS

Amino acids were the first organic compounds of biological interest to be identified as products of a laboratory experiment performed under simulated earth conditions. This experiment, conducted in 1953 by Stanley Miller, then a graduate student of Harold Urey at the University of Chicago, opened the new field of prebiotic chemistry. Because Miller's experiment is the prototype of other simulation experiments, we shall describe it in detail.

To perform his experiment, Miller constructed a glass *spark discharge apparatus,* shown in Figure 8–3. Water was added to the flask (A), and the entire apparatus was then evacuated of all air. Hydrogen, methane, and ammonia (1:2:2 by volume) were added through the stopcock (D) without allowing any air to enter the apparatus. The water in the small flask was then heated to boiling,

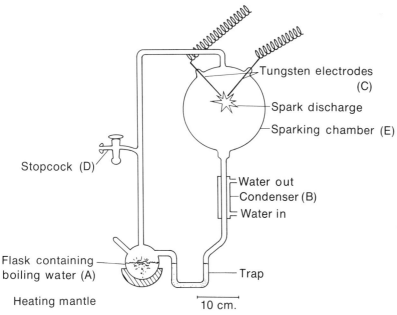

Figure 8-3

The spark discharge apparatus used by Stanley Miller in an experiment designed to simulate primitive earth conditions. (Redrawn from S. L. Miller [1955]. J. Am. Chem. Soc. 77:2352.)

producing steam which circulated clockwise through the apparatus. The spherical sparking chamber (E) contained two tungsten electrodes (C), separated by a gap of about 1 cm. These were attached to a high-frequency sparking coil (tesla coil), and a continuous spark discharge was produced between the electrodes. The convection currents, caused by the circulation of the steam through the apparatus, moved the gases past the sparking electrodes. Steam was condensed below the sparking chamber in the vicinity of the water-cooled condenser (B). In this way, the non-volatile products formed in the sparking chamber were washed through the trap and into the small flask, where they remained. Any volatile products would continue to recirculate, along with the steam and original gases, through the apparatus and past the spark.

Miller's experiment was designed to mimic as closely as possible conditions believed to have existed during early stages of chemical evolution. The spark served as a source of energy for chemical synthesis, and the mixture of reduced gases, containing no free oxygen, was designed to simulate the composition of the primitive atmosphere. The flask of boiling water, representing early stages in the history of the ocean, served as a means of producing steam and circulating the gases in the apparatus. In addition, the water served as a reservoir for removing non-volatile products from the high energy source; this is necessary, as you remember, to prevent decomposition of the products. With a great deal of imagination, the apparatus might be regarded as representing the synthesis of organic compounds by the action of lightning in the upper atmosphere, followed by washing of the products into the primitive ocean by rain.

The experiment was allowed to run for about a week with continuous sparking, after which the accumulated products in the small flask were subjected to a detailed chemical analysis. A number of organic compounds of biological interest were identified. These included amino acids and a number of other organic compounds, shown in Figure 8–4.

The following control experiments proved unequivocally that Miller's results were not due to contamination by microorganisms present in the spark discharge apparatus: (1) Negligible amounts of amino acids were produced when the experiment was performed as before, but without sparking. (2) Before the experiment was run, the apparatus was filled with water and gases and then sterilized for 18 hours at 130°C. The same results were obtained as in the original experiment. (3) When tested, none of the individual amino acids obtained in the experiment displayed optical activity; if the amino acids had been derived from living organisms, the L-form would be expected to predominate.

Miller himself acknowledged that the results of his experiment were unexpected in two respects. In the first place, relatively few chemical substances were obtained. This is surprising, since prior to Miller's experiment, most organic chemists would have thought that small amounts of a very large number of different compounds would

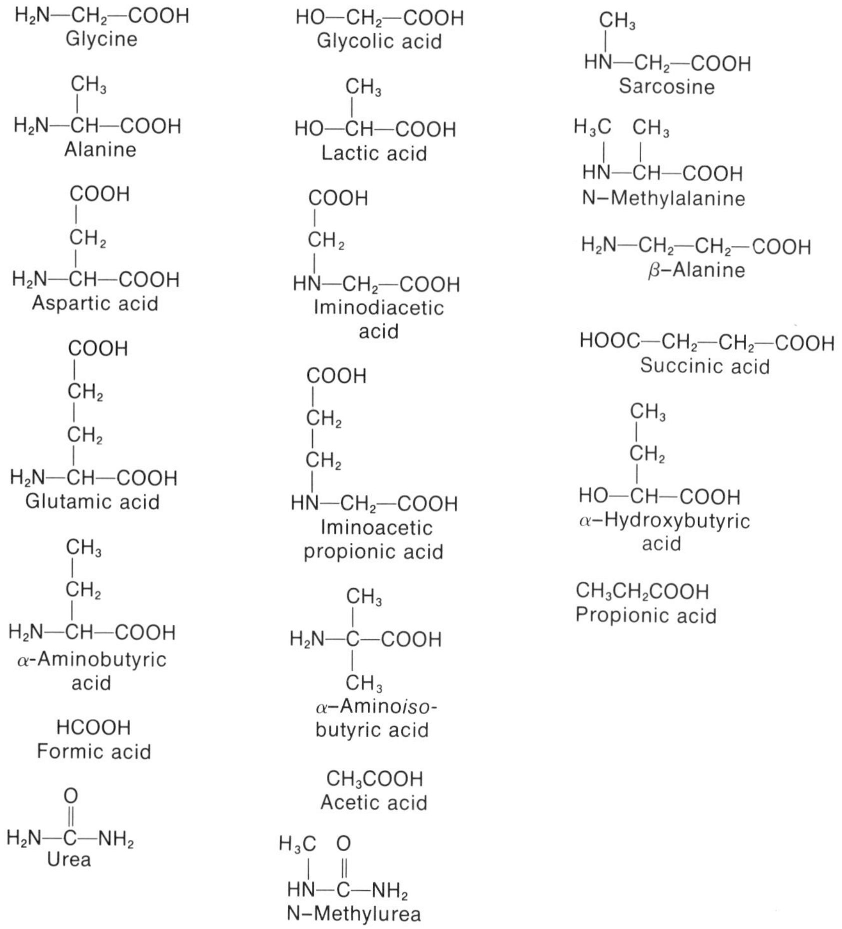

Prebiotic Synthesis of Simple Organic Compounds

H_2N-CH_2-COOH
Glycine

$$\begin{array}{c} CH_3 \\ | \\ H_2N-CH-COOH \end{array}$$
Alanine

$$\begin{array}{c} COOH \\ | \\ CH_2 \\ | \\ H_2N-CH-COOH \end{array}$$
Aspartic acid

$$\begin{array}{c} COOH \\ | \\ CH_2 \\ | \\ CH_2 \\ | \\ H_2N-CH-COOH \end{array}$$
Glutamic acid

$$\begin{array}{c} CH_3 \\ | \\ CH_2 \\ | \\ H_2N-CH-COOH \end{array}$$
α-Aminobutyric acid

HCOOH
Formic acid

$$\begin{array}{c} O \\ \| \\ H_2N-C-NH_2 \end{array}$$
Urea

$HO-CH_2-COOH$
Glycolic acid

$$\begin{array}{c} CH_3 \\ | \\ HO-CH-COOH \end{array}$$
Lactic acid

$$\begin{array}{c} COOH \\ | \\ CH_2 \\ | \\ HN-CH_2-COOH \end{array}$$
Iminodiacetic acid

$$\begin{array}{c} COOH \\ | \\ CH_2 \\ | \\ CH_2 \\ | \\ HN-CH_2-COOH \end{array}$$
Iminoacetic propionic acid

$$\begin{array}{c} CH_3 \\ | \\ H_2N-C-COOH \\ | \\ CH_3 \end{array}$$
α-Amino*iso*butyric acid

CH_3COOH
Acetic acid

$$\begin{array}{cc} H_3C & O \\ | & \| \\ HN-C-NH_2 \end{array}$$
N–Methylurea

$$\begin{array}{c} CH_3 \\ | \\ HN-CH_2-COOH \end{array}$$
Sarcosine

$$\begin{array}{cc} H_3C & CH_3 \\ | & | \\ HN-CH-COOH \end{array}$$
N–Methylalanine

$H_2N-CH_2-CH_2-COOH$
β-Alanine

$HOOC-CH_2-CH_2-COOH$
Succinic acid

$$\begin{array}{c} CH_3 \\ | \\ CH_2 \\ | \\ HO-CH-COOH \end{array}$$
α–Hydroxybutyric acid

CH_3CH_2COOH
Propionic acid

Figure 8–4

Organic compounds obtained by sparking a mixture of H_2, CH_4, NH_3, and H_2O. Four amino acids used in proteins as well as a number of other important metabolic compounds are among the products.

have resulted from sparking a mixture of CH_4, NH_3, H_2O and H_2. Second, many of the substances formed in the experiment are important compounds in all living systems.

Miller was interested in studying the kinetics of amino acid production in his experiment; that is, the rate and chemical pathway by which amino acids were synthesized. To do this, the spark discharge apparatus was modified so that samples could be withdrawn without interrupting the experiment, and the concentrations of NH_3, hydrogen cyanide (H—C≡N), aldehydes (R—$\overset{H}{\underset{}{C}}$=O), and amino acids were then determined at regular intervals during the course of a run. As shown in Figure 8–5, large amounts of hydrogen cyanide and aldehydes were synthesized within the first 30 hours. High concentrations of hydrogen cyanide, aldehydes, and ammonia were maintained until about 110 hours had elapsed, and then they declined rapidly. Total amino acids increased steadily for approximately

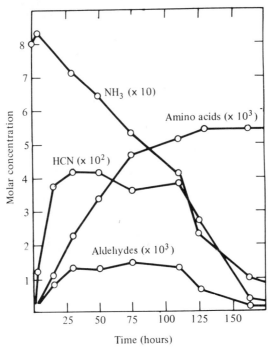

Figure 8-5 The concentrations of NH_3, HCN, aldehydes, and amino acids at various time intervals during the sparking of a mixture of H_2, NH_3, CH_4, and H_2O. (*From* S. L. Miller and L. E. Orgel [1974]. *The Origins of Life on the Earth.* Englewood Cliffs, New Jersey, Prentice-Hall, Inc. Used by permission.)

5 days, after which their concentrations leveled off. It thus appeared that HCN, NH_3, and aldehydes were intermediate products which disappeared from the spark discharge apparatus as they were converted into more complex products.

Of course, Miller did not choose to analyze for NH_3, HCN, and aldehydes purely on whim. These compounds are known to react in aqueous solution to produce amino acids as follows:

$$\underset{\text{(aldehyde)}}{R-\underset{|}{\overset{H}{C}}=O} + \underset{\text{(hydrogen cyanide)}}{H-C\equiv N} + NH_3 \rightleftarrows \underset{\text{(a nitrile compound)}}{R-\underset{\underset{NH_2}{|}}{\overset{\overset{H}{|}}{C}}-C\equiv N} + H_2O$$

$$R-\underset{\underset{NH_2}{|}}{\overset{\overset{H}{|}}{C}}-C\equiv N + 2H_2O \rightarrow \underset{\text{(amino acid)}}{R-\underset{\underset{NH_2}{|}}{\overset{\overset{H}{|}}{C}}-\overset{\overset{O}{\|}}{C}-OH} + NH_3$$

This series of reactions, termed the *Strecker synthesis,* is the oldest known synthesis of amino acids. With further experimentation,

Prebiotic Synthesis of Simple Organic Compounds

Miller was able to show that the amino acids produced in the spark discharge apparatus were probably formed by the Strecker synthesis as a result of the condensation of NH_3, HCN, and aldehydes in aqueous solution.

Miller's results, by demonstrating the feasibility of obtaining organic compounds of biological interest in a simulation experiment, served as a catalyst for further experimentation in prebiotic chemistry. Other researchers, using heat, ultraviolet light, high energy β-particles (K^{40} radioactive decay), and electric discharge as energy sources, as well as various reducing mixtures of simple gases, also obtained amino acids as products. As a result of these experiments, nearly all the naturally occurring amino acids have been identified as products of simulation experiments. Surprisingly, as long as there is a source of hydrogen, carbon, nitrogen, and oxygen, and as long as the gaseous mixture is a reducing one, the exact composition of the mixture appears to be unimportant to the production of amino acids. So far, however, it has not been possible to obtain amino acids from an atmosphere that contains molecular oxygen (O_2), or from a mixture as oxidized as CO_2, N_2, and H_2O.

In summary, then, amino acids are obtained from simple reducing mixtures of gases, using a variety of free energy sources. The important point is that conditions leading to amino acid production need not be carefully defined. We thus do not have to invoke some highly improbable set of circumstances in order to find experimental support for the materialistic theory.

In the two decades since the initial report of Miller's research, a large number of simulation experiments have been performed in an attempt to produce not only amino acids but also other organic compounds of importance to present-day living systems. Some of these experiments used simple gaseous reactants such as various combinations of N_2, NH_3, CO, CO_2, CH_4, H_2O, H_2, and H_2S. Other experimenters used as starting materials *reactive intermediate compounds* that had been shown to be formed in other syntheses. These compounds are characterized by their tendency to react at a fast rate among themselves or with other compounds, spontaneously or with minimal energy input. Examples of reactive compounds are the aldehydes and HCN produced in spark discharge experiments.

Nitriles (compounds such as HCN containing the chemical group—C≡N) and aldehydes apparently played an extremely important role in the process of chemical evolution. Not only are they involved as reactive intermediates in the production of amino acids but various nitriles and aldehydes also are important in the formation of many other compounds of biological importance, as we shall see. Nitriles and aldehydes owe their reactivity to the tendency to add other chemical groups into their structure because of the multiple bonding of the carbon atom in the —C≡N and $-\overset{\overset{\displaystyle H}{|}}{C}=O$ groups.

In addition to amino acids, many other compounds including sugars, nitrogenous bases, fatty acids, and porphyrin ring structures have been produced under more or less primitive earth conditions. Instead of attempting to give an exhaustive review of all the pertinent prebiotic experiments, we will consider only several examples illustrating the production of various classes of biologically important small organic compounds.

PURINE AND PYRIMIDINE BASES

Adenine was the first nitrogenous base to be synthesized under primitive earth conditions. This was done in 1960 by Juan Oro, who found that adenine as well as amino acids were formed in significant amounts by heating a concentrated aqueous solution of ammonium cyanide (NH_4CN) for 24 hours at 90°C. Ammonium cyanide is obtained by dissolving HCN in an aqueous solution of ammonium hydroxide; since ammonia was probably a constituent of the primitive atmosphere, a ready source of ammonium hydroxide (formed by NH_3 dissolved in water) would probably have been available for the production of ammonium cyanide. In the synthesis of adenine from ammonium cyanide it can be shown that adenine is formed as a result of the polymerization of five molecules of HCN. Although the chemical pathway, or mechanism, by which this condensation occurs is a fascinating topic for further study by those with a strong background in chemistry, it is beyond the scope of our discussion here. Because of the importance of adenine in biochemical systems as a component of both ATP and the nucleic acids, the ease with which adenine is obtained from HCN is considered by many biologists to lend impressive support to the materialistic theory.

One potential problem in the synthesis of adenine from HCN is that reasonably high concentrations of adenine (greater than 0.01 M) are required. HCN cannot be concentrated by evaporation because it is too volatile, and therefore if adenine were to be produced from HCN on the primitive earth, some other mechanism for concentrating HCN must have been available. One possible means was by freezing. We know that if dilute aqueous solutions of organic compounds are cooled to temperatures below 0°C, ice begins to form. Since organic compounds are not soluble in ice, they are left behind in the liquid portion. As the temperature continues to drop and ice is increasingly formed, the organic compound becomes more and more concentrated in the liquid solution. Eventually, at a temperature characteristic for each particular solute, even the solute will freeze out as solid crystals; this temperature is called the *eutectic point*. The eutectic point for mixtures of HCN and water is −21°C, and consequently at temperatures just above this point, very high concentrations of HCN in solution are obtained. Since good yields of adenine are formed from HCN at temperatures below 0°C, freezing may have been an important concentrating mechanism in the prebiotic production of adenine.

Purines also have been obtained in other experiments. In one of these, C. Ponnamperuma and his coworkers found that adenine and

Prebiotic Synthesis of Simple Organic Compounds

guanine could be produced by shining ultraviolet light on aqueous solutions of HCN in the absence of oxygen.

Other important prebiotic syntheses involve the production of pyrimidines. It was found that cytosine could be obtained in good yield by reacting cyanoacetylene (H—C≡C—C≡N) directly with aqueous cyanic acid (HN=C=O). Uracil may then be produced by the hydrolysis of cytosine.

Uracil and cytosine also have been synthesized by exposing cyanoacetylene, cyanic acid, or cyanogen (N≡C—C≡N) to electric discharges. These reactions are of special interest to us because of the recurrent role of nitriles in prebiotic chemistry. Cyanoacetylene is a reactive nitrile formed in all prebiotic reactions that produce HCN, cyanogen is obtained from HCN in an electric discharge, and cyanic acid is easily formed by the hydrolysis of cyanogen in neutral or weakly alkaline solutions. Thus all three compounds were very likely available for chemical syntheses on the primitive earth.

MONOSACCHARIDES

It has been known since the work of Butlerov in the middle of the nineteenth century that sugars are formed when formaldehyde (H—C(H)=O) is dissolved in a basic solution. Since formaldehyde is one of the aldehydes that are readily formed in simulation experiments, Butlerov without knowing it performed a plausible experiment in prebiotic chemistry. By the use of modern analytical techniques, it has since been determined that a wide range of sugars including trioses, tetroses, pentoses, and hexoses is produced in this reaction. Among these are glucose, fructose, ribose, xylose, and other sugars important in the metabolism of living systems.

In another experiment, Oro synthesized deoxyribose from aqueous solutions of formaldehyde and acetaldehyde (CH_3C(H)=O), or of glyceraldehyde and acetaldehyde, heated at 50°C in the presence of ammonium hydroxide (NH_4OH) or calcium oxide.

FATTY ACIDS AND GLYCEROL

Good yields of acetic acid (CH_3—C(=O)—OH) and propionic acid (CH_3—CH_2—C(=O)—OH) have been obtained by the action of electric discharge on a mixture of methane and water. In addition, highly branched, longer chain fatty acids are produced in very small yields. However, these latter fatty acids are not characteristic of those in living systems, which with rare exceptions consist of linear, unbranched chains.

Another chemical reaction, not carried out specifically as a prebiotic experiment, has bearing on the problem of fatty acid synthesis on the primitive earth. When carbon dioxide (CO_2) and ethylene ($CH_2{=}CH_2$) are irradiated with gamma (γ) rays at high pressure, long chain linear fatty acids of up to 40 carbon atoms result. Since the reactants and conditions of the experiment are simple ones, it is reasonable to believe that fatty acids could have been easily obtained under prebiotic conditions.

One substantial gap in prebiotic chemistry has been the failure to find glycerol in simulation experiments. Glycerol is, of course, needed to form fats from fatty acids.

ORGANIC COMPOUNDS IN METEORITES

In addition to the results of simulation experiments, there is another source of data that supports the view that organic compounds may have formed on earth under prebiotic conditions. This evidence is derived from the chemical analysis of *carbonaceous*

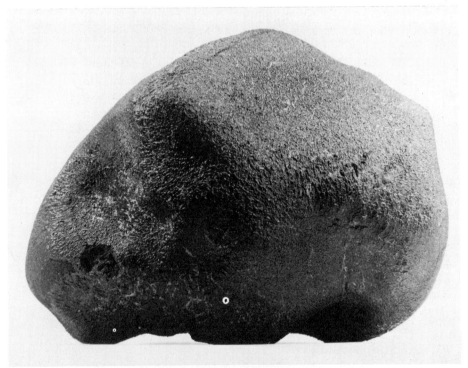

Figure 8-6 A fragment of the Murchison meteorite. Unlike the great majority of meteorites collected on the earth, the Murchison meteorite is soft and crumbly, and it contains carbon; it is therefore classified as a carbonaceous condrite. (Photograph by Jon Brenneis, Berkeley, CA. Reproduced with permission.)

Prebiotic Synthesis of Simple Organic Compounds

chondrites striking the earth. These are soft crumbly meteorites that possess a high carbon content.

From our point of view, the most important carbonaceous chondrite exploded into fragments over the town of Murchison, Australia, on September 28, 1969 (Fig. 8–6). Pieces were collected within a short time after impact, and these were subjected to detailed chemical analysis in laboratories around the world. One such analysis for amino acids is shown in Table 8–1. All the amino acids detected in the meteorite were found in a recent experiment of Miller's, in which he sparked a mixture of CH_4, N_2, and H_2O containing traces of NH_3, and analyzed the products with more sensitive techniques than were available at the time of his first experiment.

Table 8–1 AMINO ACIDS DETECTED IN THE MURCHISON METEORITE*

AMINO ACID	μg/g OF METEORITE
Glycine	6.0
Alanine	3.0
Glutamic acid	3.0
Valine	2.0
Aspartic acid	2.0
α-amino*iso*butyric acid	2.0
Norvaline	2.0
Proline	1.0
α-aminobutyric acid	1.0
Sarcosine	1.0
N-ethylglycine	1.0
N-methylalanine	1.0
β-alanine	0.5
*Iso*valine	0.3
Pipecolic acid	0.1
β-aminobutyric acid	0.1
β-amino*iso*butyric acid	0.1
γ-aminobutyric acid	0.1

*Data from S. L. Miller, and L. E. Orgel (1974), *Origins of Life on the Earth*. Englewood Cliffs, New Jersey, Prentice-Hall, Inc., p. 194.

Following the discovery of amino acids in fragments of the Murchison meteorite, a search was made for purines and pyrimidines. Although a number of pyrimidines were found, none of these is similar to pyrimidines occurring in biological systems.

It does not appear that the organic compounds present in the Murchison meteorite are contaminants from terrestrial organisms. In the first place, since samples were collected soon after impact there was little opportunity for extensive terrestrial contamination. Second, the amino acids with asymmetric carbon atoms are present as racemic mixtures of D- and L-forms. Third, a number of amino acids detected in the meteorite do not occur in terrestrial living organisms, nor do any of the pyrimidines.

The available evidence thus suggests that organic compounds in the Murchison meteorite come from extraterrestrial sources and that they are most probably of abiotic origin. Their presence tends to confirm our ideas, at least about early stages of chemical evolution. Indeed, the discovery of amino acids in the Murchison meteorite is not at all surprising in view of the detection (mentioned in Chapter 6) of HCN, formaldehyde, acetaldehyde, cyanoacetylene, and a number of other simple organic compounds in cosmic dust clouds.

A large number of fragments from other carbonaceous chondrites have been analyzed for organic compounds, and the results, extensively reported in the literature, have been the subject of considerable controversy because of alleged contamination by terrestrial organisms. These results have not been discussed here because the Murchison meteorite appears to be the only one for which terrestrial contamination has been effectively ruled out.

As an aside, other information derived from studies on the Murchison meteorite tends to support the authenticity of Precambrian fossils. No structural entities similar to those discovered in Precambrian sediments have been detected in the Murchison meteorite. This is a reassuring result, for if evidence had been found of morphological, cell-like entities associated with the meteorite, we would either have to question the significance of the Precambrian fossil evidence or assume that living organisms had evolved on meteorites.

PREBIOTIC SYNTHESIS OF COMPLEX ORGANIC COMPOUNDS

The experiments we have discussed, as well as others we have not mentioned, demonstrate the production of essentially all types of small organic compounds required for the evolution of simple living systems similar to those described in Figure 8–1. We will now turn to a consideration of the formation of more complex organic substances from these simple ones.

Prebiotic Synthesis of Complex Organic Compounds

DEHYDRATION CONDENSATION REACTIONS

As we discussed in Chapter 7, nucleosides, nucleotides, proteins, nucleic acids, polysaccharides, and fats are all composed of smaller units linked together by *dehydration condensation,* according to the general equation:

$$A\text{—}OH + H\text{—}B \rightleftharpoons A\text{—}B + H_2O$$

Such reactions are unfavorable from a thermodynamic standpoint, and thus require the input of energy in order to occur. There are several ways in which the energy may be effectively applied.

The formation of a dehydration condensation product would be favored if it were possible by some means to remove from the reacting system the water produced as a result of the reaction. As we noted in Chapters 4 and 5, this would have the effect of driving the reaction above to the right, toward increased formation of the polymerized product (A—B).

Although it is possible to carry out condensation reactions in this way, many researchers have argued that the necessary anhydrous conditions would have existed only in very localized environments on the primitive earth; for example, in regions near volcanoes where the temperature was far enough above the boiling point of water to force the reaction to occur but not so high as to cause destruction of the organic compounds. Consequently, these workers have concluded that such anhydrous syntheses would have had limited applicability on the prebiotic earth. Furthermore, since most prebiotic models suggest that simple organic compounds accumulated in the primitive ocean, it seems reasonable to believe that an alternate mechanism must have been available for carrying out dehydration syntheses in the presence of water. This complicates the problem considerably, for under aqueous conditions water must be eliminated from the reacting molecules while the system is dissolved in a great excess of water. Such a mechanism requires a large input of energy, since the more water there is available, the less the reaction is favored from a thermodynamic standpoint.

Within the aqueous environment of the living cell, energy-requiring condensation reactions are coupled to the energy-generating hydrolysis of high energy phosphoric anhydride linkages in ATP (Chapter 5). However, because ATP itself must have been formed by the dehydration condensation of adenine, ribose, and phosphoric acid, probably in aqueous solution, researchers have looked for other mechanisms that might have preceded phosphoric anhydride linkages for bringing about dehydration condensations in aqueous solutions.

Substances that may have functioned as primary condensing agents during early stages of chemical evolution are HCN and some related nitrile compounds. The multiple carbon-nitrogen bonds in certain nitriles are capable of absorbing a water molecule by preferentially reacting with the organic substances to be linked together, instead of reacting directly with the water molecules of the solvent. The removal of H_2O between two reacting molecules can occur by this method because the negative ΔG of the reaction of the nitrile with water is large enough to overcome the positive ΔG of the condensation reaction, such that the entire process proceeds with a negative ΔG (Fig. 8–7).

In addition to HCN, cyanamide ($NH_2-C\equiv N$) and its dimer (dicyandiamide) have been studied extensively as condensing agents. It is probable that both these compounds were available on the primitive earth, for they have been produced by electron beam irradiation of CH_4-NH_3-H_2O mixtures and by the action of ultraviolet light on aqueous solutions of ammonium cyanide.

NUCLEOSIDES

The formation of nucleosides under reasonable simulations of prebiotic conditions has been surprisingly difficult. Ponnamperuma and his coworkers have reported the synthesis of deoxyadenosine and adenosine. Adenosine was formed by the action of ultraviolet irradiation on an aqueous mixture of adenine, ribose, and phosphoric acid. Interestingly enough, adenosine was not formed unless phosphoric acid was present, even though there is no phosphate in adenosine. In their synthesis of deoxyadenosine, the Ponnamperuma group found that phosphoric acid could be replaced by HCN. It has been postulated that HCN may serve as a condensing agent, promoting the linkage of deoxyadenosine to adenine by the mechanism discussed above.

Gerhard Schramm and his colleagues also have reported synthesizing adenosine and deoxyadenosine by incubating adenine and ribose (or deoxyribose) with polymetaphosphate ethyl ester (PMPE) in non-aqueous solution. PMPE (Fig. 8–8) is obtained by heating phosphorus pentoxide (P_2O_5) in chloroform and diethyl ether, and its use in prebiotic synthesis was suggested by the biological role of phosphoric acid anhydrides such as ATP. Although PMPE has been shown to bring about a variety of dehydration condensations, its use

Condensation reaction	A—OH + H—B	⟶ A—B + H_2O	ΔG is positive
Reaction of nitrile with water	C + H_2O (Nitrile)	⟶ C · H_2O	ΔG is negative
Net reaction	A—OH + H—B + C	⟶ A—B + C · H_2O	ΔG is negative

Figure 8–7 One way to bring about a dehydration condensation is to couple the removal of water from the substances to be united (an endergonic reaction) with an endergonic reaction in which another compound (C) combines with the water.

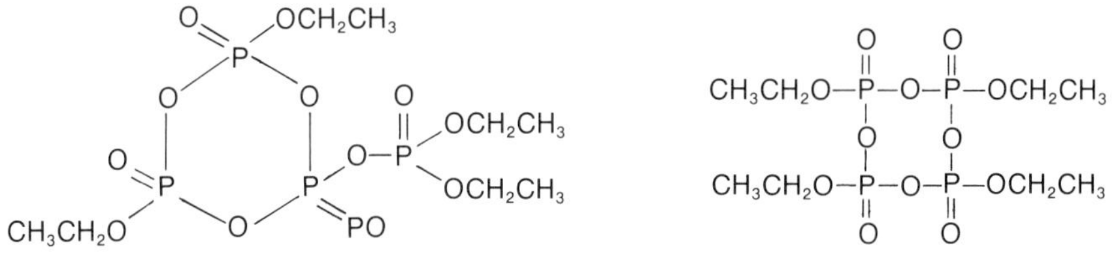

Prebiotic Synthesis of Complex Organic Compounds

Figure 8-8
Two possible chemical structures of polymetaphosphate ethyl ester (PMPE).

in prebiotic experiments has been severely criticized. PMPE is highly unstable in water, and hence must be used under anhydrous conditions. Moreover, neither phosphorus pentoxide nor PMPE has been found to occur naturally on the earth. Schramm, however, contends that PMPE may have been formed on the primitive earth in a two-step process in which H_3PO_4 polymerized at high temperatures (greater than 300°C) in water-free regions to yield polymetaphosphate; this was then esterified by reaction with appropriate organic compounds, again in the absence of water.

NUCLEOTIDES

Ponnamperuma has produced nucleotides by the phosphorylation of nucleosides. One method which gave good yields was heating nucleosides for 2 to 4 hours at 130°C in the presence of inorganic acid phosphates such as NaH_2PO_4 or $Ca(H_2PO_4)_2$; more moderate conditions, such as heating for longer lengths of time at reduced temperatures, were also effective in producing nucleotides.

In experiments by other workers, the nitrile compounds cyanamide and dicyandiamide have been used in aqueous solution to bring about the phosphorylation of nucleosides in the presence of phosphoric acid. Products include mono-, di- and triphosphorylated nucleotides (e.g., AMP, ADP, and ATP).

Schramm, again using PMPE, has produced nucleotides by the phosphorylation of nucleosides in the presence of excess PMPE without any solvent (PMPE is itself a liquid).

PEPTIDES AND POLYPEPTIDES

Thermal Condensation of Amino Acids

In 1958, Sidney Fox and Kaoru Harada reported that amino acid polymers could be synthesized by heating mixtures of dry amino acids in the presence of a large excess of acidic or basic amino acids. The acidic and basic amino acids appear to protect the neutral amino acids from thermal decomposition at temperatures up to about 210°C. In a typical experiment, a dry mixture consisting of two parts of aspartic acid, two parts of glutamic acid, and one part of an equimolar mixture of 16 other amino acids was heated at 170°C for 6 hours in the absence of oxygen. Amino acid polymers were obtained in small yield along with tars and other organic compounds. Fox named the polymers *proteinoids* to distinguish them from the biologically synthesized proteins.

The proteinoids produced from mixtures of 18 amino acids have a number of interesting characteristics, including the following: (1) They contain some of each of the original 18 amino acids. (2) They possess intramolecular peptide and disulfide bonds. (3) Their molecular weights range from approximately 3000 to 10,000, and hence they are smaller in size than the great majority of proteins. (4) Their solubility characteristics are similar to those of proteins. (5) They are susceptible to digestion by proteolytic (proteo = protein; lytic = breaking down) enzymes. (6) They possess some small catalytic activity.

The catalytic properties of proteinoids have been studied in some detail. When zinc hydroxide gel is reacted with acid proteinoid, the resultant zinc proteinoid is active in the hydrolysis of ATP. Although inorganic salts of zinc are themselves capable of catalyzing the hydrolysis of ATP, the result is potentially significant since zinc adheres to the proteinoids to form a complex in which ATP-splitting activity is localized. It has been postulated that such an association could represent the initial step in the evolution of a metal-containing enzyme.

In other experiments, Fox and his coworkers studied the ability of proteinoids to accelerate the hydrolysis of *para*-nitrophenyl acetate (Fig. 8–9). The amino acid histidine can also promote this reaction, but all of about one hundred different proteinoids that were tested showed more hydrolytic activity than an equivalent amount of histidine; some proteinoids were as much as 15 times as active. In addition, the catalytic activity could be almost completely abolished by heating the proteinoids in an aqueous buffer at pH 6.8, a phenomenon which resembles protein denaturation.

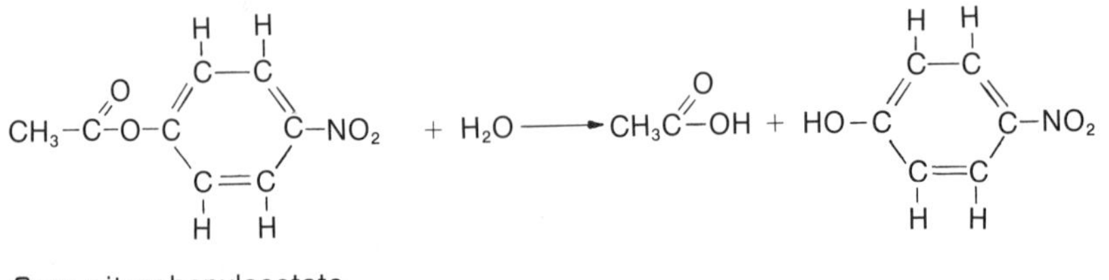

Para-nitrophenylacetate

Figure 8–9 The hydrolysis of *para*-nitrophenylacetate, shown above, is accelerated by thermally synthesized proteinoids.

Prebiotic Synthesis of Complex Organic Compounds

Structurally, proteinoids differ from naturally occurring proteins in at least one important respect. In addition to normal peptide bonds between α-carboxyl and α-amino groups, proteinoids which include aspartic acid, glutamic acid, or lysine contain peptide bonds involving side chain carboxyl and amino groups (Fig. 8–10).

It was possible to obtain enhanced yields of proteinoid if ATP, phosphoric acid, or polyphosphoric acid (PPA) were added to the amino acid mixture. In the presence of PPA (a mixture of phosphoric acid polymers, obtained by heating phosphoric acid at 200 to 350°C), appreciable yields of proteinoid were obtained when the reaction was carried out at temperatures as low as 65°C. However, since polyphosphate-containing minerals are not found in nature, PPA is not considered a valid prebiotic reagent by many workers. Nevertheless, Fox and others have argued that PPA could have been produced by the polymerization of phosphoric acid at temperatures of 300°C or less in volcanic zones.

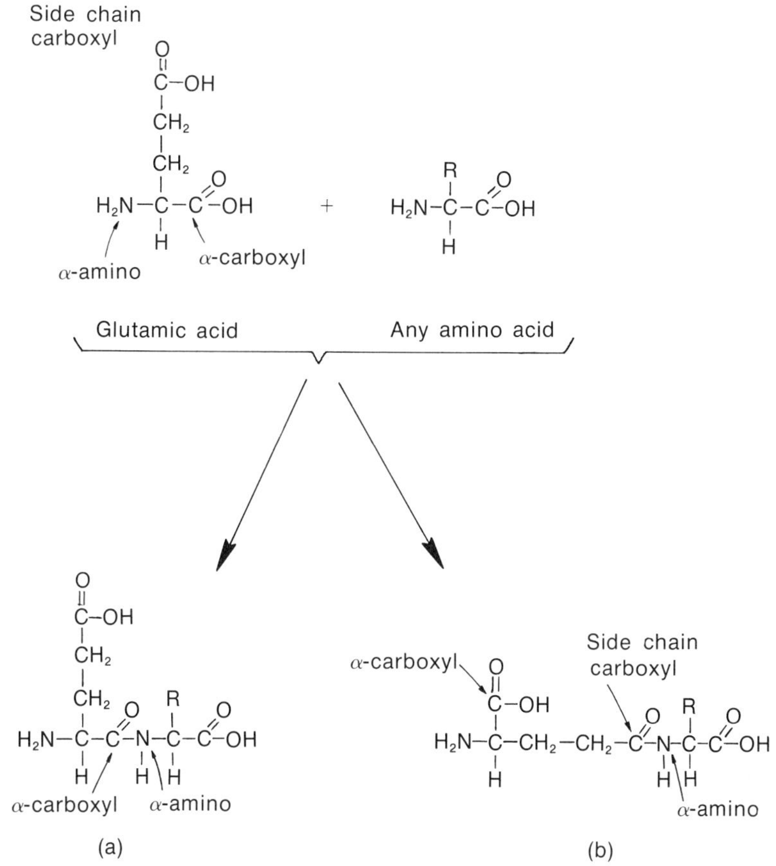

Figure 8–10 Proteinoids containing aspartic acid, glutamic acid, or lysine may possess unnatural peptide linkages, in which the side chain carboxyl or amino groups are involved in peptide bonding, as illustrated here. a, Normal peptide bond between the α-carboxyl group of one glutamic acid residue and the α-amino group of another amino acid. b, Unnatural linkage formed when the side chain carboxyl group of glutamic acid is attached to the α-amino group of a second amino acid.

Although the thermal polymerization of amino acids requires the absence of water, Fox believes that such reactions could have taken place on or near volcanoes. The importance of anhydrous thermal synthesis in prebiotic chemistry is, as we have already mentioned, a controversial issue, since many researchers feel that the opportunities for this type of synthesis would have been exceedingly limited on the prebiotic earth. Not only would amino acids have to be deposited in dry form on the edge of a volcano (perhaps by evaporation of the amino acid-containing prebiotic soup), but the temperature would have to remain within a reasonably narrow range of about 100 to 200°C—high enough to bring about polymerization but low enough to prevent thermal decomposition of the amino acids. When this acceptable temperature range is compared to a temperature of about 1200°C for molten lava, it appears that volcanic regions with conditions appropriate for amino acid condensation might be considerably restricted.

On the other hand, life may well have evolved in a local environment, and as we have pointed out many times before, it is impossible to know which chemical mechanisms were *in fact* important on the prebiotic earth. The mere fact that a particular chemical process may not have been a widespread phenomenon does not mean that it was unimportant in chemical evolution. In analyzing the relative importance of various proposed mechanisms, we must take into account a number of factors including the availability of starting materials, the plausibility of reaction conditions, the efficiency of the process, the likelihood of its occurrence, and the chemical nature of the products. The conclusions that we reach after assessing all these factors will ultimately depend to a large extent on how the significance of each factor is judged, and hence the conclusions may vary from individual to individual.

Fox has examined thermally produced proteinoids for evidence of non-random association of amino acids. This is an important point, since it bears upon the problem of whether the sequence of amino acids in present-day proteins could have arisen by some means other than nucleic acid control. In other words, do the specific chemical properties of the individual amino acid side chains influence to any degree the incorporation or sequence of amino acids into polypeptides?

To obtain information on this question, Fox compared the relative amounts of each amino acid in the proteinoid products with the relative amounts of each amino acid in the starting mixture. The results of such an analysis, shown in Table 8–2, demonstrate that the composition of proteinoids does not necessarily reflect the composition of the reactant mixture. Consequently, it appears that the incorporation of amino acids into proteinoids is not a completely random process; if it were, we would expect the proportions of different amino acids in the proteinoid and in the starting mixture to be identical.

Table 8-2* AMINO ACID COMPOSITION OF PROTEINOIDS AS COMPARED TO THE COMPOSITION OF THE REACTANT MIXTURE

	PROTEINOID[a]	REACTANT MIXTURE
Lysine	1.64	1.34
Histidine	0.95	1.43
Arginine	0.94	1.60
Aspartic acid	66.0	40.0
Glutamic acid	15.8	40.0
Proline	0.28	1.06
Glycine	1.32	0.69
Alanine	2.30	0.82
Cysteine	1.32	1.11
Valine	0.85	1.08
Methionine	0.94	1.37
Isoleucine	0.86	1.21
Leucine	0.88	1.21
Tyrosine	0.94	1.67
Phenylalanine	1.84	1.52

*Data recalculated from S. W. Fox et al. (1963). Arch. Biochem. Biophys. 102:439.
[a]The amino acid analysis was performed following acid hydrolysis; serine, threonine, and tryptophan are unstable under these conditions, and hence they are not listed.

Polymetaphosphate Ethyl Ester (PMPE) as a Condensing Agent for Amino Acids

Schramm has used PMPE in non-aqueous solvents to synthesize polypeptides from amino acids. For example, polypeptides composed exclusively of arginine (polyarginine) and possessing molecular weights of 4000 to 5000 have been obtained by the polymerization of arginine monomers in the presence of PMPE. In another synthesis, a mixture of tyrosine, alanine, glutamic acid, and PMPE yielded a polypeptide with an average molecular weight of 7300.

The relevance of using PMPE as a condensing agent in prebiotic experiments has been discussed previously.

Nitrile-Mediated Condensations of Amino Acids

Nitrile compounds have been used by a number of workers, particularly Gary Steinman, as condensing agents for peptide bond synthesis. Cyanamide and its derivatives were initially shown to promote the condensation of two amino acid units to form a dipeptide; in later experiments peptides containing two, three, or four amino acid residues were obtained by this method.

The results of experiments using nitrile compounds as condensing agents in the production of peptides have, as in the case of thermally produced proteinoids, provided evidence suggesting that amino acids are not linked to one another in a random fashion. By determining the efficiency of peptide bond formation between different amino acids in the presence of a cyanamide derivative, Steinman was able to show that a particular amino acid does not have an equal tendency to form a peptide bond with any other amino acid (Tab. 8–3, col. 1). Surprisingly, the relative tendency of two amino acids to form a dipeptide, as determined by the efficiency of cyanamide-mediated dipeptide synthesis, is directly related to the frequency with which the dipeptide occurs in a number of contemporary proteins (Tab. 8–3, col. 2).

Steinman and others have postulated that the tendency of two particular amino acids to react with one another may have been an important influence in the prebiotic development of peptides with specific amino acid sequences. According to this view, peptide bonds would form preferentially between certain amino acids because of the chemical properties of their side chain groups, and this would determine, at least to some extent, the sequence of a poly-

Table 8-3* EFFICIENCY OF DIPEPTIDE SYNTHESIS AS COMPARED WITH DIPEPTIDE FREQUENCIES IN KNOWN PROTEIN SEQUENCES

Dipeptide	VALUES (RELATIVE TO GLY—GLY)	
	Efficiency of synthesis	Frequency in proteins
Gly—Gly	1.0	1.0
Gly—Ala	0.8	0.8
Ala—Ala	0.7	0.7
Gly—Val	0.5	0.2
Gly—Leu	0.5	0.3
Gly—Ile	0.3	0.2
Gly—Phe	0.1	0.1

*Data from D.H. Kenyon and G. Steinman (1969). *Biochemical Predestination*. New York, McGraw-Hill, Inc., p. 209.

peptide. As the growing chain increased in length, interactions between amino acids, leading to folding of the chain, might bring internal amino acids near the growing end and thus put further constraints on the addition of a new amino acid to the peptide. In this way, specific amino acid sequences might have been generated prebiotically without nucleic acid intervention.

Although laboratory experiments using appropriate nitriles as condensing agents have not formed peptide products nearly as impressive as those obtained by thermal synthesis, many researchers have argued that nitrile-mediated amino acid condensations may well have been considerably more important on the primitive earth than the thermal production of proteinoids. In defense of this idea, it has been pointed out that suitable nitriles promote the condensation of amino acids *in dilute aqueous solutions at moderate temperatures,* and hence this mechanism could have occurred in the primitive ocean.

OLIGONUCLEOTIDES AND POLYNUCLEOTIDES

Many of the most successful syntheses of high molecular weight polynucleotides have employed anhydrous conditions and have been carried out in the presence of PMPE or polyphosphoric acid. For example, Schramm has synthesized polymers of high molecular weight by dissolving large amounts of a single type of nucleotide in PMPE syrup and heating the mixture at 55°C for 18 hours under anhydrous conditions. By this method, he obtained adenylic acid polymers (poly A), poly C, poly G, poly U, and poly T, varying in molecular weight from 15,000 to 50,000. Results of alkaline hydrolysis and treatment of the poly U with ribonuclease suggested a predominance of unnatural linkages between nucleotide units (such as 2'-2', 3'-3', and 2'-5' phosphoric anhydride linkages) as well as branching and crosslinking between chains.

By heating anhydrous CMP at 65°C for 2 hours with polyphosphoric acid, researchers in Fox's laboratory obtained di- and trinucleotides as well as small amounts of what appeared to be larger molecular weight polynucleotides. As when PMPE was used, chemical analysis suggested that the products were branched and that some of the nucleotides were joined together by linkages other than the natural 3'-5'.

Dinucleotides and small oligonucleotides have been obtained by using cyanic acid and various cyanamide derivatives as condensing agents to promote internucleotide bond formation in aqueous solution. In experiments of this nature, it has been observed that the polymerization of adenylic acid and adenosine to form short oligonucleotides of poly A is greatly enhanced by the presence of poly U in the incubation mixture. Poly U has no effect on the condensation of adenylic acid with uridine, cytidine, or guanosine. In a similar manner, it was shown that poly C facilitates the condensation of guanylic acid and guanosine, but does not promote the condensation of guanylic acid with uridine, cytidine, or adenosine.

It has been suggested that the specific enhancement effect of poly U and poly C is due to the intrinsic affinity between adenine and uracil, and between guanine and cytosine. In naturally occurring double-stranded nucleic acids, the polynucleotide strands are held together by specific hydrogen bonding between the nitrogenous bases; such hydrogen bonding *within* a single polynucleotide chain is also important in determining the three-dimensional configuration of single-stranded nucleic acids. With rare exceptions, adenine always forms hydrogen bonds with either uracil or thymine, and guanine always forms hydrogen bonds with cytosine (Fig. 8–11). Thus in the experiment mentioned above, adenylic acid and adenosine residues, by hydrogen bonding to uracil residues in poly U, may be held in close proximity to one another, facilitating the formation of a phosphoric anhydride linkage between them. As we shall see, the specific hydrogen bonding between bases (A to U or T, and C to G) is the key to enzyme-catalyzed nucleic acid replication. Consequently, the results above suggest a tendency of primitive nucleic acids to promote their own replication in the absence of enzymes.

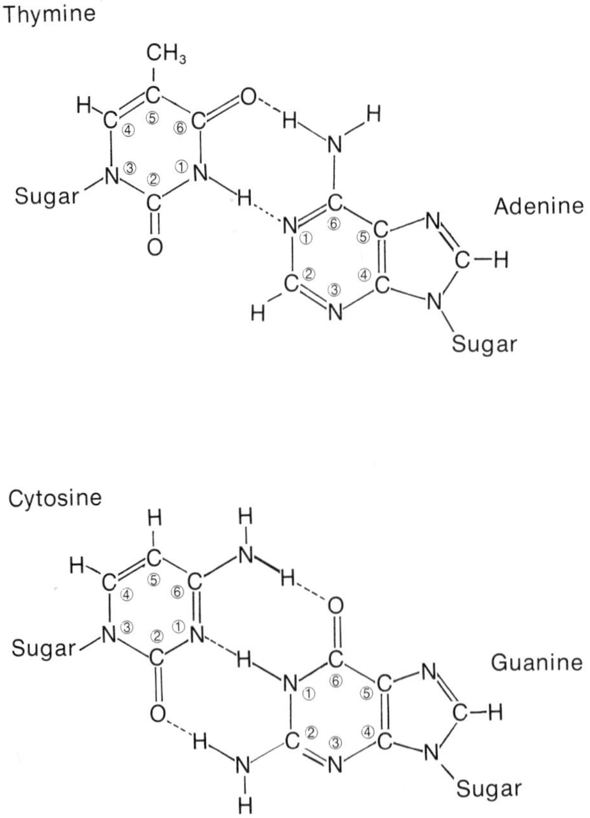

Figure 8–11 Hydrogen bonding between bases in nucleic acids. Since the structures of uracil and thymine are similar except that uracil does not have a —CH_3 group attached to Carbon 5, the hydrogen bonding pattern between the adenine and uracil is the same as that between adenine and thymine.

The poly A and poly G oligonucleotides synthesized in these experiments contain predominantly 2'-5' internucleotide bonds, and thus differ from naturally synthesized polynucleotides. Miller and Orgel have suggested that mixed 2'-5' and 3'-5' linked oligonucleotides may have occurred in early stages of chemical evolution, and that the 2'-5' linkages were eventually eliminated with the development of specific enzymes which catalyzed the replication of nucleic acids.

OLIGOSACCHARIDES AND POLYSACCHARIDES

The prebiotic formation of polysaccharides from monosaccharide units has not been studied as extensively as the formation of polypeptides or polynucleotides. The following experiments are representative of the types of approaches that have been taken.

Mora and Wood have obtained a high molecular weight glucose polymer by heating a mixture of glucose and phosphoric acid at 140 to 170°C in a vacuum. The polyglucose product is highly branched, but 1 → 6 glycosidic linkages are favored. Using PMPE in non-aqueous medium at a temperature of 55 to 60°C, Schramm and his colleagues were able to form high molecular polymers of glucose, of ribose, and of fructose.

Disaccharides have been synthesized under more reasonable prebiotic conditions by shining ultraviolet light on aqueous solutions of monosaccharides. Kaolin, a white clay, was added to the reaction mixture as a catalyst. Apparently kaolin functions by adsorbing glucose onto its surface, and thus serves as a means of concentrating the monosaccharides and promoting their condensation. Indeed, it has been postulated that the adsorption of materials to clay and other mineral surfaces may have been a general mechanism, along with freezing and evaporation, for concentrating compounds on the prebiotic earth.

FATS

Mono-, di- or triglycerides have not been identified as products in simulation experiments. As we have seen, glycerides are important constituents of cellular membranes. If it proves possible to obtain glycerol under simple prebiotic conditions, it should be reasonably easy to condense fatty acids and glycerol and to form phospholipids by phosphorylation of glycerides.

SUMMARY

In summary, then, almost all of the basic compounds important to life have been synthesized in simulation experiments. Furthermore, these experiments have demonstrated that small organic molecules of biological importance could easily have been formed on the prebiotic earth from a reducing mixture of simple gases, and that these organic compounds could well have undergone dehydration condensations to form more complex polymers. It has been shown that these chemical processes could have taken place under a variety of different environments on the primitive earth. *In fact, from what we know about conditions on the primitive earth, the formation of simple organic compounds and their more complex polymers appears to be a likely occurrence* rather than an improbable one.

In the next chapter, we will consider how organic compounds might have attained an organized state, leading ultimately to the appearance of a simple living system similar to the one we described in Figure 8–1.

REFERENCES: CHAPTER 8

Black, S. (1973). A theory on the origin of life. *Advances in Enzymology, 38*:193.
Calvin, M. (1969). *Chemical Evolution.* Oxford University Press, New York.
Fox, S. W., Ed. (1965). *The Origins of Prebiological Systems.* Academic Press Inc., New York.
Kenyon, D. H. and G. Steinman (1969). *Biochemical Predestination.* McGraw-Hill Book Co., New York.
Lawless, J. G., C. E. Folsome, and K. A. Kvenvolden (1972). Organic matter in meteorites. *Scientific American, 226*:38.
Marguilis, L. (1970). *Origin of Eukaryotic Cells.* Yale University Press, New Haven, Connecticut.
Miller, S. L. (1955). Production of some organic compounds under possible primitive earth conditions. *J. Am. Chem. Soc., 77*:2351.
Miller, S. L. and L. E. Orgel (1974). *The Origins of Life on the Earth.* Prentice-Hall, Inc., Englewood Cliffs, New Jersey.

CHAPTER **9**

From Molecules to Cells

In the last chapter, we reviewed a vast amount of experimental evidence that suggested that most classes of biologically important organic compounds could very likely have been produced on the primitive earth by abiotic means. But mixtures of organic molecules suspended in the primitive ocean or deposited as solids in dry areas as a result of evaporation are a far cry from a living cell. Indeed, one universal characteristic of living systems is that they are physically separated from their surroundings. This is not to say, of course, that a living system is completely *isolated* from its environment. On the contrary, a living system is continually exchanging materials with its surroundings, and in order to survive it must be able to respond to changes in the environment by bringing about metabolic readjustments that serve to stabilize its entire structure.

For example, a cell unable to obtain sufficient energy in the form of nutrient materials from its environment must be able to alter its metabolism to mobilize reserves. This may require the production of increased amounts of enzymes as well as other substances needed for this task. The cell is constantly involved in making metabolic adjustments of this type in its struggle for existence. In short, the living cell is a dynamic entity that owes its existence to its ability to react successfully with its environment. This is the essence of a living system: *Its stability is derived **not** from the fact that it is a static entity of fixed structure but rather from its nature as a dynamic entity of flexible structure.*

It should be obvious that it is no easy task to reconstruct the process by which living systems, so complex and dynamic in nature, developed from a collection of nonliving organic molecules. How might the transition from organic molecules to the first living cells have occurred? Although a number of different mechanisms, varying in precise detail, have been proposed to account for the process, there have been generally two main approaches to the problem.

SELF-REPLICATING MOLECULES

The first approach is to assume that life arose initially as individual molecules which were able to bring about their own replication. This theory is based upon the idea that those molecules able to replicate the fastest would have become dominant in the prebiotic soup. Hence any molecule that by some means acquired the ability to replicate faster than other molecules would have gradually replaced the more slowly replicating types. For instance, some molecules might have initially increased their replication rates by using inorganic catalysts available in the environment. Or a replicating molecule might have eventually acquired in some way the ability to control the production of polypeptides with useful catalytic properties, or to fashion molecules needed for its own replication from other organic molecules available in the environment. As self-replicating molecules thus developed the potential to direct a crude metabolism, the acquisition of a primitive limiting membrane would have become selectively advantageous; this would have allowed the accumulation of useful substances while serving to exclude useless or harmful ones. Step by step, so the theory goes, self-replicating molecules would have acquired the complex structural and functional characteristics of living cells.

POLYNUCLEOTIDES AS THE FIRST SELF-REPLICATING MOLECULES

H.N. Horowitz, L.E. Orgel, and others have held that the first self-replicating entity, the simplest form of life, was a primitive polynucleotide molecule. As Horowitz puts it, "Life arose as individual molecules in a polymolecular environment." This view has been substantiated by the discovery of the central role of nucleic acids in bearing hereditary information and in directing the metabolism of the cell through the process of protein synthesis.

Several pieces of evidence lend support to the idea that DNA- or RNA-like molecules may have been capable of independent replication, and hence may have represented the initial stage in the formation of cellular entities.

1. The simplest viruses, intracellular parasites that have many properties of living systems, consist exclusively of nucleic acid and protein. The nucleic acid of viruses is capable of directing its own replication and the production of new virus particles by using the metabolic system of the host cell to synthesize necessary enzymes. Although the hereditary information in cellular organisms is apparently stored exclusively in DNA, either RNA or DNA can serve the same function in viruses. Consequently, either RNA or DNA (or both) may have served as the first self-replicating polynucleotide.

2. It has been possible to obtain the replication of DNA and RNA *in vitro;* that is, outside the living cell. In 1955, Arthur Kornberg and his colleagues isolated from the bacterium *Escherichia coli* an

enzyme, called DNA polymerase, that is capable of catalyzing the replication of DNA in the presence of the four different kinds of deoxyribonucleoside triphosphates and magnesium ions. Since Kornberg's original experiment, other DNA polymerase enzymes have been discovered. In more recent experiments, Spiegelman and his coworkers have demonstrated the in vitro replication of the RNA of a bacterial virus (Qβ) in the presence of the four ribonucleoside triphosphates and a purified "replicase" enzyme.

3. As we mentioned in Chapter 8, there are indications that the cyanamide-mediated formation of internucleotide bonds may be influenced to some degree by specific base pairing that is important in the replication of nucleic acids within living cells. These experiments have been interpreted as evidence, albeit slight, for the ability of polynucleotides to promote their own replication without the use of enzymes.

POLYPEPTIDES AS THE FIRST SELF-REPLICATING MOLECULES

Because polypeptides are the only molecules to show catalytic activity and because amino acids and polypeptides are obtained in simulation experiments considerably more readily than polynucleotides and their precursors, other workers have suggested that the first self-replicating molecules, rather than being polynucleotides, were polypeptide in nature. These researchers contend that the role of nucleic acids evolved as a much later evolutionary development.

Direct evidence to support this position is scant, although Simon Black has constructed an impressive theoretical model that suggests that self-replicating, catalytically active proteins may have been formed as early products of chemical evolution, well before useful polynucleotides. We will consider Black's model in the final chapter.

MOLECULAR AGGREGATES

The idea that a living cell evolved from a population of self-replicating molecules has been sharply criticized by many biologists. In the first place, they contend that there is little or no evidence to suggest that polynucleotide or polypeptide molecules would have been capable of accurate self-replication in the milieu of the primitive ocean. They point out that polypeptide molecules do not now display any tendency to replicate, and that although the replication of nucleic acids can occur in vitro, appropriate enzymes are required. Indeed, we do not know of any substance that, when isolated, is capable of its own replication. *Only systems are able to replicate.*

Thus, it has been argued that it is hard to imagine how replication could have occurred, even if some primitive catalyst and necessary precursor molecules were potentially available, unless replicating molecules were able to control their own immediate environment. Much as living cells do now, replicating molecules could then accumulate in adequate concentrations the substances needed for replication, and they could exclude other substances that would adversely affect replicative activities. After all, the replication of an RNA or DNA molecule in a test tube occurs only because all necessary components are present in the immediate environment of the nucleic acid molecules—a situation contrived by the experimenter. The question is whether or not such a situation (the creation of a quasi system) could be expected to occur with reasonable frequency in the primitive environment.

An alternative theory that partially overcomes these objections has been proposed by A.I. Oparin and others. They contend that the process leading to the first living cell began initially with the formation of morphological entities consisting of molecular aggregates of organic compounds present in the prebiotic soup. These molecular aggregates thus constituted localized systems partially isolated from their aqueous surroundings. By providing a matrix for the incorporation of substances from the environment, the aggregates gradually increased the complexity of their metabolic and structural organization.

Proponents of this approach suggest that these primitive aggregates competed in a limited sense with one another for available substances. Some aggregates, by virtue of their chemical composition, were able to accumulate materials at a faster rate than others. They consequently grew more rapidly than other less "successful" types, and hence more quickly reached a size at which they tended to be broken up into smaller particles through the action of winds, waves, and other forces. As these particles continued to grow and to be divided, they would have gradually displaced other systems. By the accumulation of inorganic catalysts and of abiotically synthesized polypeptides with weak catalytic activity, some aggregates could have acquired a primitive metabolic system of a sort. Any such acquired traits would be advantageous if they enhanced the rate of mass accumulation by allowing the morphological entity to better utilize materials in its environment. Gradually, the perfection of these simple systems led to the development of a living cell.

Although this theory is hardly without problems (which we will discuss shortly), it has obtained experimental support from the observation that the types of organic compounds produced in early stages of chemical evolution exhibit natural tendencies to interact with one another to form morphological aggregates. Several different types of aggregates have been extensively studied as models for the formation of cells. Let us now briefly examine some of the work that has been done in this area.

Molecular Aggregates

MEMBRANOUS STRUCTURES

In Chapter 7, we mentioned that fatty acids and phospholipids, because of their dual hydrophilic and hydrophobic nature, showed a tendency to form films on the surface of aqueous solutions or, when surrounded by water, to orient themselves in the form of spherical micelles or bimolecular layers. We noted further that phospholipid bilayers are similar to natural membranes in dimension, and in appearance under the electron microscope.

It has been observed that artificial vesicles consisting of lipid and protein can be easily formed from a surface film. For example, if a film of oleic acid is spread on the surface of a dilute solution of egg albumin and the liquid's surface is then agitated by some mechanical means, the surface film may collapse, with the formation of elongated tubular structures possessing a diameter of about 1 to 10 μm (Fig. 9–1). As may be seen in Figure 9–1, these tubular structures consist of two layers of lipid surrounded on each side by a layer of protein. They are remarkably similar in structure to cell membranes, which appear in general to be composed of a lipid

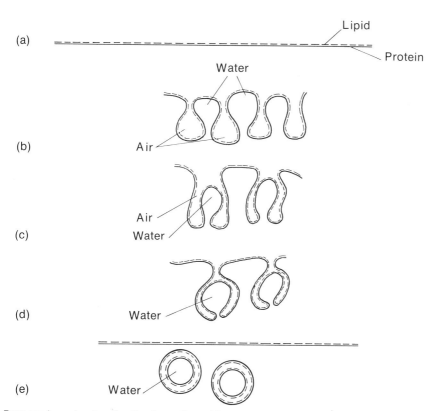

Figure 9–1 Proposed mechanism for the formation of lipoprotein vesicles. *a*, Lipid layered on a solution of protein. *b* through *e*, Successive stages in the formation of vesicles. (Modified from R.J. Goldacre *in* J.F. Danielli, K. Pankhurst, and A.C. Riddiford, Eds. [1958], *Surface Phenomena in Chemistry and Physics.* Elmsford, New York, Pergamon Press, Inc.)

bilayer in which proteins are dissolved. In related experiments, it has been shown that similar structures could well have been formed by the collapse of films produced by the accumulation of abiotically synthesized compounds on the surface of the primitive ocean.

By absorbing molecules from the surrounding environment, micellar structures and related membrane-bound vesicles similar to those described here could provide a means for spatially isolating a collection of organic compounds. They could thus have served as a matrix for the development of living cells.

MICROSPHERES

In experiments with thermally synthesized proteinoids (Chapter 8), Sidney Fox and his collaborators noticed that if proteinoids were dissolved in boiling salt solutions or water and then allowed to cool, vast numbers of small spherical structures called *microspheres* separated out of solution (Fig. 9–2). They determined that 15 mg of proteinoid dissolved in 2.5 ml of salt solution yielded between 10^8 and 10^9 microspheres. They were also able to obtain microspheres under less extreme conditions—for example, by slowly cooling a saturated solution of proteinoid from room temperature to 0°C. In some cases,

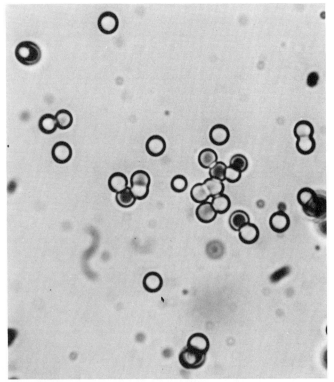

Figure 9–2 Microspheres formed by dissolving thermal proteinoid in boiling water and then allowing the preparation to cool. (*From* S.W. Fox, [1965], *The Origins of Prebiological Systems*. New York, Academic Press, Inc.)

Molecular Aggregates

microspheres appeared simply upon contact of a cold proteinoid preparation with cold water.

Since microspheres can form under such mild and general conditions, Fox has postulated that they could easily have been produced when proteinoid, synthesized in hot dry environments on the earth, was washed into the ocean by the action of tides. To test this hypothesis, a mixture of dry amino acids containing appropriate amounts of glutamic and aspartic acids was placed in a depression of a piece of lava and heated in an oven at 170°C for several hours. When the rock was then washed with a 1 per cent solution of sodium chloride, large numbers of microspheres were formed.

Microspheres possess a number of properties similar to those of living cells, and consequently they have been used as models to illustrate how cellular entities could have developed from organic compounds. Some of the characteristics of microspheres are summarized below.

Characteristics of Microspheres

SIZE AND SHAPE. Although their size varies somewhat with the conditions of preparation, microspheres are usually about 2 μm in

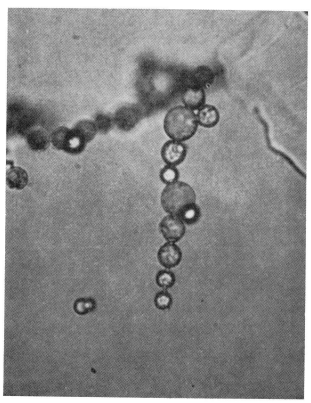

A "budding chain" of microspheres. These structures are not formed by the coalescence of individual microspheres. (*From* R.S. Young *in* S.W. Fox, Ed. [1965], *The Origins of Prebiological Systems.* New York, Academic Press, Inc.)

Figure 9-3

diameter. Fox has pointed out that they are thus similar in size and shape to the coccoid (spherically shaped) bacteria.

STABILITY. Microspheres are remarkably stable. They can be stored for several weeks without losing their structural integrity, and can be collected by centrifugation in a clinical centrifuge, again without alteration of their structure.

BUDDING. Under certain conditions, many of the microspheres appear to be undergoing division. For example, by altering the pH of the medium in which microspheres are suspended, "budding" can be induced (Fig. 9–3). This budding is not a result of coalescence of individual microspheres, as can be demonstrated by means of time lapse photography. In one set of experiments, small buds were removed by mechanical, thermal, or electrical shock and collected by centrifugation. When transferred to a solution saturated with thermal proteinoid, the buds were observed to increase in size. This growth process suggests that microspheres have the ability to incorporate materials from their surroundings.

When microspheres forming on a microscope slide are subjected to slight pressure applied to the coverslip, chains of microspheres are formed. Simple shaking of a suspension of microspheres results in fragmentation and an increase in the number of spheres.

STRUCTURE. Alteration of the pH of solutions in which microspheres have formed causes their centers to dissolve away. When microspheres treated in this way are sectioned and viewed in the electron microscope, a double-layered boundary structure can be seen (Fig. 9–4). Although this structure is superficially similar to that of cellular membranes (see Fig. 7–27), the microsphere double layer is considerably thicker than the double layer that composes the membranes found in living cells.

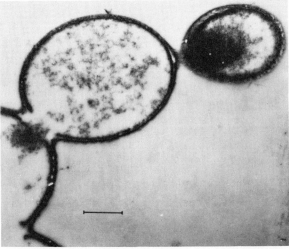

Figure 9–4 Electron micrograph of budding microspheres after their centers have been dissolved away. Notice the double-layered boundary, which superficially resembles the structure of cell membranes. (*From* S.W. Fox [1965], A theory of macromolecular and cellular origins. Nature 205:338.)

Molecular Aggregates

LOCALIZED HYDROLYTIC ACTIVITY. It is possible to produce microspheres containing zinc ions by cooling a hot aqueous solution containing proteinoid and zinc hydroxide. Although the resultant zinc-containing microspheres do not show an enhancement of ATP-hydrolyzing activity over that which is due to the zinc content alone, these results are nevertheless important because they demonstrate the localization of ATP-splitting activity in a structural unit.

STAINING REACTIONS. Like bacteria, microspheres accept Gram's stain, a crystal violet—iodine complex that is widely used by microbiologists to distinguish between different groups of bacteria. Some bacteria are classified as *Gram–negative* because Gram's stain is lost by washing the cells with 95 per cent ethanol; others, known as *Gram–positive,* do not lose Gram's stain during the wash procedure. Microspheres made from acidic proteinoid are Gram-negative, and those containing more than 35 per cent basic proteinoid are Gram-positive.

OSMOTIC PROPERTIES. When microspheres are placed in salt solutions containing a greater concentration of solute than the solution in which they were formed, the microspheres shrink in size; if they are placed in a solution containing less solute, they swell. This phenomenon, which is further evidence that microspheres interact with their environment, resembles the behavior of living cells under similar conditions. In order to understand the basis for this phenomenon, we must digress for a moment to consider the process of osmosis.

Diffusion, Osmosis and Cell Permeability. The movement of molecules from a region of high concentration to one of low concentration is called *diffusion,* and we can conveniently describe the process of diffusion in terms of chemical equilibrium. Consider, for example, a concentrated aqueous solution of sugar on which pure water is layered. The Second Law of Thermodynamics tells us that a concentration difference in two adjacent areas is an improbable, and hence an unstable, situation. Because of this, the system described above will tend to approach equal concentrations of substances in all areas of the system—its equilibrium position.

It is possible to separate two solutions, one of high and one of low solute concentration, by a *semipermeable membrane;* that is, one that allows only molecules of the solvent to pass through it. The equilibrium position of this system is still the one that approaches equal concentrations of all substances on both sides of the membrane. Consequently, water will tend to pass through the membrane from the region of low solute concentration (or high solvent concentration) to the region of high solute concentration (or low solvent concentration) in an attempt to approach the equilibrium position by equalizing the concentration of solute and solvent on both sides of the membrane. The diffusion of solvent through any membrane in response to a concentration gradient is called *osmosis.*

The plasma membrane is *selectively permeable,* meaning that it allows water to pass through it, but retards the passage of most solutes to varying degrees. Only a few substances, such as oxygen and carbon dioxide, pass through the plasma membrane as freely as does water. Because the plasma membrane is freely permeable to water, the cell demonstrates osmotic properties.

We mentioned in Chapter 3 that the solute concentration within a cell is normally maintained within an extremely narrow range. Thus, if a cell is placed in an aqueous solution containing a total solute concentration equal to the solute concentration in its own cytoplasm (an *isotonic solution*), water will diffuse into and out of the cell, but there will be no *net* transfer of water between the cell and its surroundings.* In an isotonic medium, every water molecule diffusing out of the cell will be replaced, on the average, by a water molecule diffusing into the cell. If, however, a cell is placed in a solution with a higher solute concentration than its own cytoplasm (a *hypertonic solution*), more water will tend to diffuse out of the cell than will enter it. The net effect will be a loss of water from the cell, and the cell will consequently shrink in volume. In a *hypotonic solution* (one containing a total solute concentration less than that within the cell), the net flow of water will be into the cell. The cell will swell and may eventually burst. These processes are illustrated in Figure 9–5.

Although microspheres demonstrate osmotic properties, they shrink and swell less markedly than living cells. It should also be emphasized that many kinds of membranes, including artificial membranes such as those composed of cellulose acetate, exhibit simple osmotic and permeability characteristics. Therefore, the fact that microspheres possess osmotic properties should not be accepted as evidence that they display some characteristic unique to living systems.

Although microspheres possess certain properties common to

---

*All molecules have kinetic energy, or energy of motion, except at a temperature of $-273°C$ (absolute zero), when all molecular motion ceases. Consequently, at physiological temperatures, the random motion of molecules striking the plasma membrane from both sides will propel some molecules out of the cell and some into the cell.

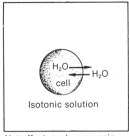

Net effect: no loss or gain of water by the cell

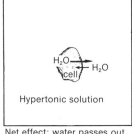

Net effect: water passes out of the cell

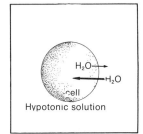

Net effect: water enters the cell

Figure 9–5 Diagrammatic illustration of osmotic effects in cells. The size of the arrow is proportional to the amount of water transferred in each direction. See text for details.

Molecular Aggregates

living cells, we should not interpret this as implying that microspheres are some kind of primitive cell. They are not. However, microspheres are valuable as cell models because they demonstrate the tendency of organic compounds to form morphological entities. Thus, as in the case of the membranous structures mentioned previously, the formation of microspheres is one means by which special, localized environments could have arisen in the prebiotic ocean. Microspheres may have represented the first step in a long process leading to the appearance of living cells.

COACERVATES

It has been known for some time that high molecular weight substances dissolved in water may aggregate at certain conditions of concentration, temperature, and pH to form viscous drops called *coacervates* (Fig. 9–6). Coacervates are visible in the light microscope, and vary in size depending on their composition and the conditions under which they are formed. They possess diameters anywhere from about 0.5 to 640 μm. Many coacervate droplets are extremely unstable and may rapidly coalesce to form a separate phase, but under appropriate conditions, others have been reported to remain as droplets sharply separated from the surrounding medium for as long as several years.

Coacervates are now generally believed to form as a result of a reduction in the hydration layer around each macromolecule; this permits a group of macromolecules to associate (Fig. 9–7). For instance, if the concentration of a salt such as sodium chloride is gradually increased in an aqueous solution of oleic acid, the ions of the salt appear to compete with the fatty acid molecules for the solvent. This brings about a decrease in the interaction between fatty acid and

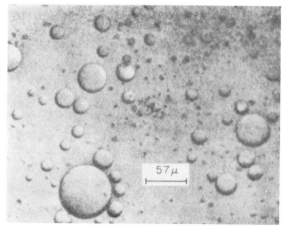

Figure 9–6 Coacervate droplets formed by the aggregation of gelatin (protein) and gum arabic (polysaccharide). (*From* A.I. Oparin [1968], *Genesis and Evolutionary Development of Life.* New York, Academic Press, Inc.)

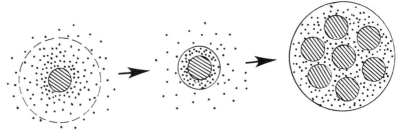

Figure 9-7 Schematic representation of the formation of coacervates. Partial dehydration of macromolecules promotes their association with the mutual exclusion of water. Water molecules are represented by dots, and the shaded circles represent macromolecules. (Modified from H.G. Bungenberg deJong [1949] in H.R. Kruyt, Ed., *Colloid Science*, Vol. II. New York, American Elsevier Publishing Co.)

solvent molecules, and results in the formation of coacervates of fatty acid.

Coacervates may be of two types. *Simple coacervates* are formed by the aggregation of identical macromolecules, as in the case of the oleic acid coacervates mentioned above. *Complex coacervates* are produced between different macromolecules of opposite charge. For example, when two solutions, one of gum arabic and one of gelatin, are mixed at a pH at which the gelatin is positively charged and the gum arabic is negatively charged, a complex coacervate of the two substances is formed. The difference in charge between the particles serves to strengthen the association between them.

Oparin has suggested that the process of coacervation may have led ultimately to the formation of cellular entities. He and others have produced coacervates from a large number of organic compounds of biological importance, including proteins, nucleic acids, polysaccharides, lipids, and porphyrin compounds, and they have studied these coacervates extensively as cell models.

Experiments on the formation of coacervates have suggested that coacervation could well have been an important mechanism for the production of morphological entities in the primitive ocean. It has been shown that coacervation may occur even when macromolecules are present in a solution at very low concentrations, as they supposedly would have been in the primitive ocean.

The most interesting property of coacervates with respect to their role as cell models is their ability to interact as open systems with the surrounding medium; that is to say, they are able to exchange materials with their environment. Coacervate droplets have been shown to absorb substances selectively from the external medium. For example, dyes such as neutral red, methylene blue, and others are concentrated in coacervates of gelatin and gum arabic by an amount many times their concentration in the initial solution. Depending on the composition of the coacervate, certain amino acids are taken up selectively. Oparin has reported that in one type of coacervate, the amino acid tyrosine is concentrated more

Molecular Aggregates

than a hundred-fold above its concentration in the surrounding medium. Tryptophan, on the other hand, is taken up only to a limited degree, reaching a concentration in the droplet about twice its concentration in the aqueous medium.

In related experiments, coacervates have been shown to be capable of absorbing enzymes from the surrounding solution, and hence acting as centers for many types of enzyme catalyzed reactions, including synthesis, hydrolysis, and electron transfer (oxidation–reduction).

The enzymatic synthesis and degradation of starch was studied in one such experiment. Oparin found that potato phosphorylase, an enzyme capable of catalyzing the formation of starch from glucose–1–phosphate (Fig. 9–8), is almost entirely taken up in coacervate droplets composed of gum arabic and histone. If glucose–1–phosphate is then added to the aqueous medium in which the droplets are suspended, starch begins to accumulate in the droplets, as may be detected by the iodine test. It was found that within about 30 minutes after the addition of starch to the medium the weight of the droplets had increased by 50 per cent. Glucose–1–phosphate is not concentrated to a significant degree by the coacervate, and throughout the experiment its concentration in the droplet remains about the same as its concentration in the surrounding medium. Consequently, glucose–1–phosphate must be continually entering the droplets from the surrounding medium as it is used up in the synthesis of starch.

If β-amylase is included in the coacervate droplets along with phosphorylase, the starch formed in the droplets is hydrolyzed to maltose, which then diffuses into the surrounding medium. The entire process is illustrated diagrammatically in Figure 9–9. If the rate of starch synthesis is greater than the rate of its hydrolysis, then starch will accumulate in the droplets, and the droplets will increase in size. This is a significant observation, since the growth of the droplets is due to a chemical reaction localized in the droplets. More is involved than simply the absorption of materials from the environment.

Oparin and his colleagues have also obtained the enzymatic synthesis and degradation of polynucleotides in coacervate drop-

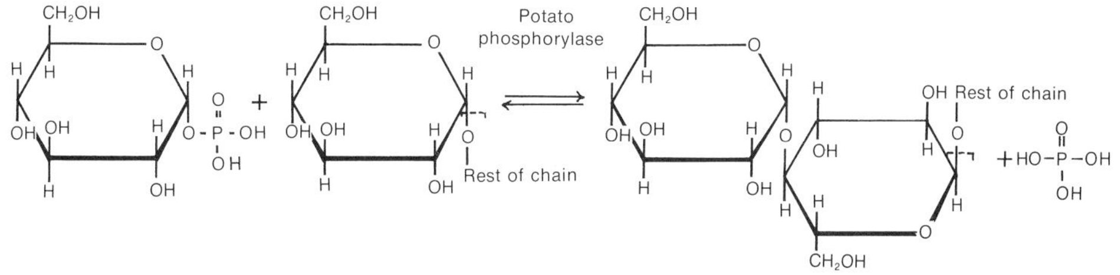

Figure 9–8 Potato phosphorylase catalyzes the formation of starch by the condensation of glucose-I-phosphate to the growing chain.

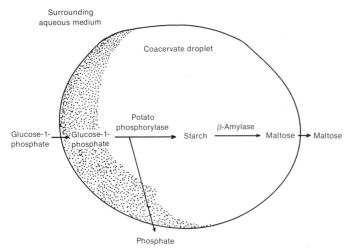

Figure 9-9 By the incorporation of glucose-l-phosphate, potato phosphorylase, and β-amylase, a coacervate of gum arabic and histone can act as a center for the synthesis and degradation of starch. (*From* A.I. Oparin [1968], *Genesis and Evolutionary Development of Life.* New York, Academic Press, Inc.)

lets. By incorporating ribonuclease (which catalyzes the hydrolysis of RNA) in coacervate droplets consisting of RNA, serum albumin, and gum arabic, the RNA is hydrolyzed within the droplets and the resultant nucleotides are released into the external medium (Fig. 9-10).

The enzymatic synthesis of polyadenylic acid (poly A) was accomplished by incorporating bacterial polynucleotide phosphorylase into coacervate droplets of histone and RNA. When ADP is then dissolved in the external medium, it enters into the droplets, where it serves as substrate for the formation of poly A (Fig. 9-11). Inorganic phosphate, formed as a byproduct of the reaction, diffuses out of the drop into the surrounding medium.

In still other experiments, Oparin demonstrated that coacervate droplets can serve as foci for electron transfer (oxidation-reduction) reactions. For example, chlorophyll incorporated into coacervate droplets can be directly involved in a light-promoted electron transfer reaction.

Coacervation, then, like the formation of microspheres and membranous structures, represents a means by which organic molecules may have interacted with one another to form morphological entities. Coacervates are not alive, nor are they close to being alive. Although they are able to interact in a limited way with their surroundings by incorporating substrates and catalysts, coacervates eventually reach a state in which there is no transfer of material or energy between them and their environment. Coacervates do not possess a metabolic system that, by virtue of its ability to respond to environmental challenges, brings about a stabilization of their struc-

Adsorption to Surfaces

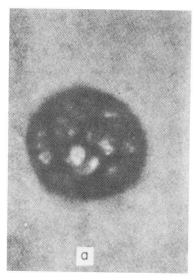 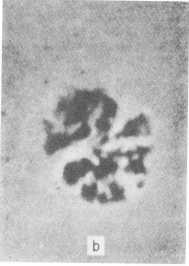

Coacervate droplet containing RNA, serum albumin, gum arabic, and ribonuclease, as seen under the electron microscope. *a*, Appearance of the droplet immediately after its formation. *b*. Its appearance 15 minutes later. The action of ribonuclease has resulted in the breakdown of RNA in the droplet. (*From* A.I. Oparin [1968], *Genesis and Evolutionary Development of Life*. New York, Academic Press, Inc.)

Figure 9–10

tures. In short, coacervates are not characterized by *dynamic* interaction with their environment. Like microspheres, they are useful to us as models serving to demonstrate the propensity of organic compounds to interact with one another. They may have been important as one link in a long chain of evolution leading to the formation of living cells.

ADSORPTION TO SURFACES

J. D. Bernal has proposed that molecular aggregates could have been formed in the primitive ocean as a result of the adsorption

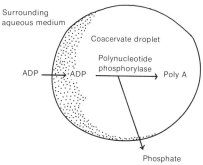

A coacervate droplet composed of histone and RNA can incorporate ADP and the enzyme polynucleotide phosphorylase, and can thus act as a center for the synthesis of polyadenylic acid. (*From* A.I. Oparin [1965] *in* S.W. Fox, Ed., *The Origins of Prebiological Systems*. New York, Academic Press, Inc.)

Figure 9–11

of organic compounds to clays and other minerals. We have already noted the potential importance of clays both in concentrating organic molecules from dilute solutions and as catalysts. Bernal assumes that the organization of molecules on clays may have led ultimately to the development of a primitive metabolic system. Of course, somewhere along the line, a boundary membrane must have evolved.

HOW DID CELLULAR ENTITIES EVOLVE?

From the foregoing discussion, it should be apparent that there is no easy answer to the question of how cellular entities have evolved. We discussed a number of quite different mechanisms, any of which might have been involved in the process; moreover, there is no reason why some mechanism involving a combination of some or all of those mentioned above could not have been important in the transition from organic molecules to cells. For example, a molecule potentially capable of replication could have become incorporated as part of a coacervate or could have been adsorbed, along with other organic compounds, onto a clay surface that provided some degree of organization. For that matter, it is quite possible that none of the mechanisms discussed so far was responsible for the establishment of the first cells.

Whether we accept the idea that a self-replicating molecule was the first manifestation of a living cell, or whether we consider that the process leading to the first living cells began initially with the formation of a morphological entity partially isolated from its surroundings, or whether we posit some entirely different mechanism to explain the transition from molecules to cells, we are still left with the problem of how the complex metabolism and structure of the cell evolved from these earliest beginnings.

The problem, of course, is that cell structure and function developed concomitantly with, not independently of, one another. That is to say, a structural unit resembling a modern cell was not somehow endowed after its formation with metabolic properties, nor did a complete metabolic system develop, later to be incorporated into a morphological unit.

The crucial point with regard to the development of living cells is how the hereditary system of the cell evolved. We have made no mention of this in the preceding discussion. We have only described hypothetical processes by which self-replicating molecules or molecular aggregates could have developed increasing complexity by accumulating materials from the environment. Although these processes superficially resemble Darwinian evolution of living systems, *this is clearly not the case.*

Natural selection, as we described it in Chapter 1, is *differential reproduction,* and requires some apparatus for transmitting hereditary information needed for constructing a new organism and for creating heritable variation. This is done in present-day cells by the storage of hereditary information in the nucleotide sequences of DNA and its translation into functional protein sequences. Nowhere in the above schemes for the evolution of molecules or molecular aggregates is there provided a means whereby the ability to use components from the environment is passed to future generations. *The missing link is the connection between the hereditary and the functional (metabolic) apparatus of the cell.* Without such a connection, the cell as we know it could not have evolved.

It is probable that many different systems for storing and utilizing hereditary information were tested and discarded during the evolutionary process before the present system was established. Unfortunately, we know very little about how the hereditary system actually developed. We will return to this point in the final chapter.

Finally, we may conclude that the transition from molecules to cells is a process about which reasonable men disagree. The available data are meager and subject to a wide variety of interpretations. Clearly much work is needed before we can say that we understand how cellular entities evolved. The topic is a fascinating one, not only because of our inherent interest in the origin of living systems but also because it illustrates the course of scientific progress so clearly. There is the quest to understand and the excitement of discovery. But most valuable is the demonstration of the principle so often lost to the beginning student: that science is questions as well as answers.

By some means we do not completely understand, cellular entities must have evolved, probably in the prebiotic soup. In the next chapters, we will consider the structure and metabolism of present-day living cells with the aim of discovering how the varied structural and metabolic patterns of living cells may have developed.

REFERENCES

Black, S. (1973). A theory on the origin of life. Advances in Enzymology 38:193.
Brooks, J. and G. Shaw (1973). *Origin and Development of Living Systems.* Academic Press, Inc. New York.
Fox, S.W. (1965). A theory of macromolecular and cellular origins. *Nature* 205:328.
Fox, S.W., Ed. (1965). *The Origins of Prebiological Systems.* Academic Press, Inc., New York.
Howland, J.L. (1968). *Introduction to Cell Physiology.* Macmillan, Inc., New York.
Kenyon, D.H. and G. Steinman (1969). *Biochemical Predestination.* McGraw-Hill, Inc., New York.
Keosian, J. (1964). *The Origin of Life.* Reinhold Publishing Corp., New York.
Oparin, A.I. (1962). Origin and evolution of metabolism. *Comparative Biochem. Physiol.* 4:371.
Oparin, A.I. (1968). *Genesis and Evolutionary Development of Life.* Academic Press, Inc., New York.
Orgel, L.E. (1973). *The Origins of Life.* John Wiley and Sons, Inc., New York.

CHAPTER **10**

The Structure of Prokaryotic Cells

It should be amply clear from the preceding discussions that although considerable progress has been made in formulating and testing a general scheme for the origin of living systems, the outline of the process is a rough one and many of the precise details, particularly in regard to the formation of the first cell-like entities from organic molecules, remain a mystery. Nevertheless, the accumulated evidence seems to justify the assumption that a minimal self-replicating cellular entity was somehow formed from abiotically synthesized organic compounds on the primitive earth, and that this entity, or one of its close relatives, evolved eventually into the first prokaryotic cell. Using this assumption as the point of departure, this chapter and the next will examine details of prokaryotic and eukaryotic cell structure. With the information that this survey will provide us, we will be in a position to consider some of the ideas that suggest how eukaryotic cells may have arisen.

THE STUDY OF CELL STRUCTURE

Before immersing ourselves in the details of cellular structure, we would do well to note some of the problems that are often encountered in considering the structure of cells. We have touched upon many of these points before, but it should nonetheless be profitable to review them briefly.

One tendency that we must overcome is the tendency to view the cell as being identical in all organisms. We must keep in mind the pitfalls of describing a "typical" cell: We do this as a convenient method of organizing vast amounts of structural information, but we must remember that, in fact, the "typical" cell does not exist. Cells, even those in the same organism, vary in structure because they are adapted to different functions.

A related pitfall is the tendency to think that a single type of cell in the same organism has a constant structure, instead of realizing

The Study of Cell Structure

that its structure may vary from one point in time to another. This kind of structural variability is a function of the dynamic aspect of a cell, a response to the cell's interaction with its environment. This is an important concept which we emphasized in the previous chapter.

There are numerous examples of cellular structural changes that reflect temporary metabolic adaptation. For example, the number of mitochondria in a cell may vary according to the energy demands of the cell. A cell may contain more ribosomes when its protein requirements are large than when they are small. The amount of storage material in a cell may fluctuate within an enormous range. Vast stores of glycogen, visible as small granules in the electron microscope, accumulate in the liver cells of a mouse prior to its birth; the depletion of glycogen reserves following birth is accompanied by an increase in the amount of endoplasmic reticulum in glycogen-rich areas of the cells. It has been shown that the endoplasmic reticulum in these liver cells possesses large amounts of glucose–6–phosphatase, an enzyme required for the breakdown of glycogen into utilizable glucose. In many plants, light is a prerequisite for chlorophyll and chloroplast formation; when grown in the dark or at low light intensities, such plants possess structures quite different from normal chloroplasts. It is thus apparent that the structural details that we see in a cell at a particular time may not be characteristic of the cell at all times.

Studying the structure of cells often conveys an impression of inactivity in yet another sense. Often, a single cell type *will* look virtually identical in structure in a number of electron micrographs taken at different times and under different environmental conditions. Such pictures, showing no indication of structural change, make the cell seem like an inanimate object, remaining essentially identical in structure from one moment to the next. The cell appears static. We speak of such a cell as "resting," but in actuality there is no such thing as a resting cell. Even in a cell that is not growing or reproducing, energy is continually being captured and utilized, compounds are continually being synthesized and broken down, and substances are being moved into the cell, around the cell, and out of the cell. What we mean by a "resting cell" is a cell in which we see no *net* change in appearance over time. But it is quite clear that metabolic activity need not necessarily be accompanied by *observable* structural change. The problem, of course, is in our methods of observation. If we could observe *living* cells in an electron microscope with greatly improved resolution, we would begin to see activity at the molecular level which we now can only infer.

Another problem that plagues students of cell structure is the interpretation of electron micrographs. Not only do these pictures represent one instant in time but they can be exceedingly misleading because each picture shows only a very thin slice of the cell. Under certain circumstances, the plane of section may be such that certain structures are not visible at all. This effect is illustrated in Figure 10–1, which shows that the same cell can appear completely

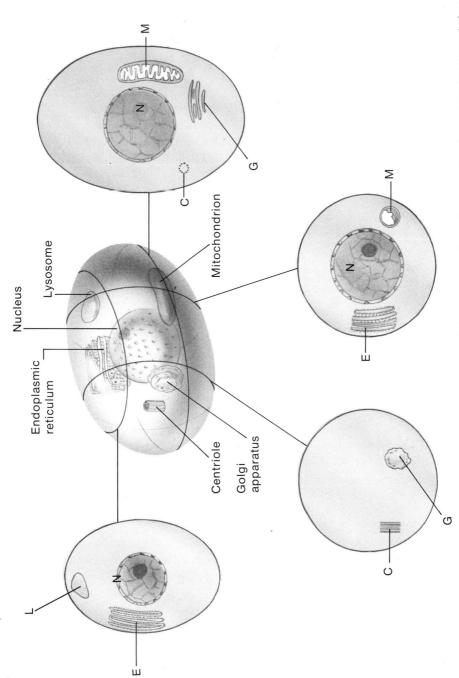

Figure 10–1 An electron micrograph may show entirely different intracellular structures depending upon the plane of section through the cell.

different in electron micrographs, depending on how the cell is sectioned. A true three-dimensional picture of a cell could be obtained only by examining *all* the sections into which it is cut. This is not easy. For instance, if a cell with a diameter of 20 μm is cut into 0.05 μm sections, some four hundred slices are obtained. It is no simple task to examine all these slices in the electron microscope. Consequently, a claim that a particular structure is missing in a cell must be substantiated by evidence of careful and extensive sampling procedures.

An additional problem with electron microscopy is that fixation and desiccation are very likely to alter the structure of the cell. Many an "intracellular structure" detected in an electron microscope has been described in detail in the scientific literature, only later to be proved an artifact of sample preparation. Thus the structures seen by electron microscopy must be interpreted with appropriate caution.

The foregoing comments are not made to disparage the importance of the electron microscope in biology, but rather to point out its limitations. The electron microscope has provided us with rich details of cellular structure, hardly imagined before its use. In proper hands, it is an extremely valuable research tool. Indeed, its widespread importance is demonstrated by the fact that cell biologists now view cellular structure primarily in terms of its electron image.

THE PROKARYOTES

In Chapter 2, we mentioned that all living organisms can be unambiguously classified as either prokaryotic or eukaryotic on the basis of their cellular structure. Bacteria and blue-green algae are prokaryotes, and all other organisms are eukaryotes. The prokaryotic cell is characterized by its relatively small size and by the fact that it lacks many of the structural features normally present in eukaryotic cells. The fundamental distinction between prokaryotes and eukaryotes has held up under rigorous examination and has been increasingly clarified, rather than clouded, as structural data have accumulated in recent years.

It has long been assumed that prokaryotes are the most ancient forms of life on earth, and that their unique cellular structure is a result of their divergence from the main line of biological evolution at a very early stage in the development of living organisms. Although the prokaryotes do share many structural and functional features in common, they are hardly a homogeneous group. For example, there are several fundamental differences between bacteria and blue-green algae. First, blue-green algae never possess flagella of any sort, whereas many bacterial strains are flagellated. Second, although there are unquestionably many biochemical similarities between the mechanisms of photosynthesis in blue-green algae and photosynthetic bacteria, the blue-green algae always produce oxygen as a byproduct of photosynthesis, whereas photosynthetic bacteria never do. In addition, the photosynthetic bacteria contain bacterial chlorophylls rather than the structurally related chlorophyll *a* which is found in all blue-green algae and higher plants.

Metabolic and structural variability among the prokaryotes exists even within the bacterial and blue-green algal groups. There are unicellular, colonial, and filamentous forms of both bacteria and blue-green algae, and these include members with a wide range of metabolic specializations.

Except for a few short paragraphs, the structure of prokaryotic cells has generally been neglected in cell biology textbooks, and emphasis has been placed almost entirely on the more complex eukaryotic cell. This is unfortunate. Such an approach gives the unjustified impression that prokaryotic cells are inherently uninteresting in structure. Moreover, the prokaryotic cell, precisely because of its relative structural simplicity, can tell us much about the evolution of simple cellular systems. As a matter of fact, much of our knowledge of cell biology has been, and will continue to be, derived from the study of prokaryotes.

PROKARYOTIC CELL STRUCTURE

THE CELL EXTERIOR

PLASMA MEMBRANE. The cytoplasm of the cell is surrounded by the *plasma membrane,* sometimes referred to as the *plasmalemma* or *cytoplasmic membrane.* (The latter term is confusing, however, since there are also a variety of membranes in the cytoplasm.) In electron micrographs, the plasma membrane is about 7.5 to 10.0 nm thick, and is seen to consist of two parallel dark (electron-opaque) lines, each about 2.0 to 2.5 nm thick, separated by a 3.5 to 5.0 nm light (electron-transparent) space (Fig. 10–2). This tripartite structure, which is characteristic of many biological membranes, has been called a "unit membrane" by J. D. Robertson, who was the first to recognize that the three lines taken together constitute one membrane.

The plasma membrane functions as the selectively permeable barrier that serves to maintain the integrity of the cell. It significantly influences (but does not completely determine) which substances enter the cell, which substances pass out of the cell, and what concentrations of substances are maintained within the cytoplasm.

We have come to realize that the plasma membrane is both functionally and structurally complex. It is not simply a passive barrier or "molecular sieve" which allows small molecules to pass through it and excludes larger ones. The size, shape, electrical charge, and solubility characteristics of a molecule all are important in determining whether or not it will pass through the membrane and at what rate.

The plasma membrane is composed primarily of lipids and proteins; however, its precise chemical composition varies from cell to

Prokaryotic Cell Structure

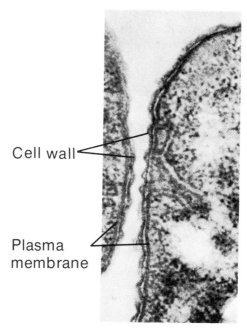

Figure 10-2

Electron micrograph of portions of two adjacent bacterial cells, showing the tripartite plasma membrane and the cell wall.

cell. Thus the plasma membrane of each type of cell contains a variety of specific molecular components. Some of these components serve as carrier substances which transport specific molecules into and out of the cell.

In addition to regulating the exchange of materials between the cell and its environment, the plasma membrane has other functions in prokaryotes. It has been reported that at least in bacteria, the plasma membrane appears to play a key role in cell division, since mutant bacteria that are unable to synthesize certain specific membrane proteins are not able to divide. In aerobic prokaryotes, many of the enzymes and other constituents essential for the production of ATP from the aerobic breakdown of nutrient substances are physically associated with the plasma membrane. In fact, several aerobic bacteria have been shown to possess small stalklike structures resembling mushrooms, attached to the cytoplasmic side of the plasma membrane. These structures are believed to represent the specific sites in which at least some of the components needed for aerobic energy production are localized.

CELL WALL. All prokaryotes with the exception of one small group of bacteria (the Mycoplasma) possess a *cell wall* surrounding the cell immediately exterior to the plasma membrane. This rigid surface coat is composed of materials secreted by the cell and may vary in thickness from about 10 nm to more than 100 nm. The cell wall determines the characteristic shape of the cell and protects the cell from physical damage and disruption. Because the cytoplasm of the cell cannot expand beyond the limits determined by the rigid cell membrane, prokaryotes are able to tolerate a wide range of solute concentration in the exterior environment. If the cell wall is removed or substantially weakened by the action of a hydrolytic enzyme such as lysozyme, the cell, called a spheroplast, will assume a spherical shape, and it will expand and eventually burst when placed in a hypotonic medium.

In recent years, a great deal of research in molecular biology has been devoted to determining the chemical structure of prokaryotic cell walls. This has been a difficult task, for their structure has turned out to be exceedingly complex. It now appears that the cell walls of all prokaryotes are composed of a similar type of structural framework in which other molecules are embedded to produce the completed cell wall; these latter molecules often differ from one type of organism to another.

The rigid structural framework of the prokaryotic cell wall is constructed of a network of polysaccharide chains. Each chain consists of alternating residues of N—acetylglucosamine and N—acetylmuramic acid joined by $\beta-(1 \rightarrow 4)$ glycosidic linkages (Fig. 10–3). Depending upon the particular organism, some or all of the N—acetylmuramic acid residues have short chains of amino acids attached to them. The length and amino acid composition of these side chains are constant for any one type of organism. Interestingly enough, some of the amino acids in the side chains may be of the D– configuration; this is one of the few exceptions to the general rule that only L–amino acids are found in the molecules of living systems.

The glucosamine–muramic acid chains are linear, and parallel chains adjacent to one another are joined to form the three-dimensional structural framework. In the cell walls of some prokaryotes, this linkage occurs by the formation of a peptide bond between

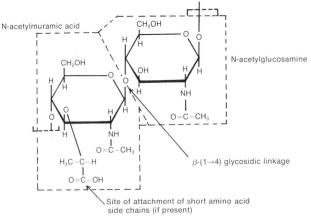

Figure 10–3 The structural polysaccharide of the prokaryotic cell wall consists of alternating units of N-acetylglucosamine and N-acetylmuramic acid, joined by β-(1 $\rightarrow$ 4) glycosidic linkages. Short side chains of amino acids are attached to some or all of the N-acetylmuramic acid residues, at the point shown above. Lysozyme, discussed in Chapter 7, catalyzes the hydrolysis of the cell wall polysaccharide between the C–1 of N-acetylmuramic acid and the C–4 of N-acetylglucosamine.

Prokaryotic Cell Structure

amino acid residues in adjacent side chains. In others, two amino acid side chains on parallel strands are connected by means of a short peptide bridge (Fig. 10–4). The structural framework consisting of crosslinked polysaccharide chains is variously referred to as *murein, mucopeptide,* or *peptidoglycan.* The substances that are embedded in the murein framework to form the completed cell wall may be lipoprotein, polysaccharide, or lipopolysaccharide, or combinations of these, depending on the type of organism.

Although the cell wall is a rigid structure, it does not possess the permeability properties of the plasma membrane, and very large molecules can pass through it. In colonial forms of blue-green algae, adjacent cells often appear to communicate with one another by means of fine cytoplasmic bridges called *plasmodesmata,* which pass through the cell walls and connect the plasma membranes of the two cells.

Much of the original research on the chemical structure of prokaryotic cell walls, particularly those of bacteria, was stimulated by the discovery that a number of antibiotics act by inhibiting cell wall synthesis in prokaryotes. Penicillin, for example, kills growing cells by preventing the attachment of the crosslinking bridges in murein. The antibacterial action of penicillin has been explained in the following way. As a normal growing cell prepares to divide, cellular enzymes known as *murein hydrolases* split the cell wall, producing open ends to which newly synthesized polysaccharide chains may attach to form additional cell wall. In actively dividing

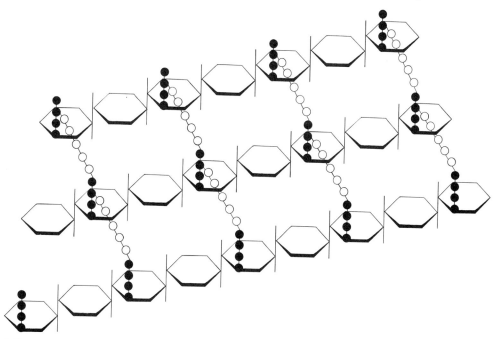

Figure 10–4 A diagrammatic model of the cell wall framework (murein) of *Staphylococcus aureus,* in which side chains of four amino acids (black dots) are linked to one another by a short connecting peptide of five glycine residues (open circles). In other organisms, such as *Escherichia coli,* the amino acid side chains are attached to one another directly, and no additional amino acids are added to bridge the chains.

cells, the murein hydrolases continue to act even in the presence of penicillin, but because the polysaccharide chains cannot be cross-linked, cell wall synthesis cannot be completed. Without the protection of its cell wall, the dividing cell will eventually burst. Because penicillin affects only murein synthesis, and because murein is present only in prokaryotes, all other organisms are insensitive to penicillin.

SLIME. Most prokaryotes secrete mucilaginous material known as *slime,* which lies exterior to the cell wall. In some organisms, slime forms only a simple amorphous coating known as a *slime layer;* in other organisms, the slime is present as a layer of uniform thickness surrounding the cell. In the latter case, the slime layer is called a *capsule* (in the case of bacteria) or a *sheath* (in the case of blue-green algae). Capsules and sheaths apparently have no metabolic function, but it is known that bacteria with capsules are not as easily destroyed by white blood cells as bacteria that are not encapsulated. Motion pictures clearly show white blood cells sliding off encapsulated bacteria they are attempting to engulf.

Capsules or sheaths are composed of polysaccharides, polypeptides, or lipoproteins, or combinations of these, and often contain traces of other substances as well. Earlier suggestions that the sheath of blue-green algae is composed partially of cellulose have not been confirmed by subsequent research.

Surface Appendages

FLAGELLA. Many bacteria have surface appendages called *flagella* (singular: *flagellum*), which confer motility on them; however, no members of the blue-green algae have any type of surface appendage. Bacterial flagella are 10 to 20 nm in diameter and may be very long, up to about 20 μm (Fig. 10–5). They are sometimes localized at one or both ends of the cell (*polar flagella*).

The prokaryotic flagellum consists of three parts: *filament, hook,* and *basal body.* The filament is the whiplike part of the flagellum that is visible in stained preparations under the light microscope. The filament is composed of parallel fibers wound about each other in a helical fashion to form a cylindrical structure with a hollow center (Fig. 10–6). Each fiber is itself composed of subunits of the globular protein, *flagellin.* Flagellin varies in structure from one bacterial species to another, but is generally characterized by the absence of cysteine, high levels of the acidic amino acids, and low levels of tryptophan, tyrosine, and phenylalanine.

If filaments are mechanically removed from the surface of bacteria, for example, by agitating a solution containing bacteria in a blender, the filaments regenerate within a few minutes. Studies of the regeneration process have shown that flagellin is synthesized in the cytoplasm, and that the completed protein leaves the cell through the hollow core of the flagellum and is added to the end of the growing filament. Because flagellin does not contain cysteine, flagella are unusually stable in both reducing and oxidizing environ-

Prokaryotic Cell Structure

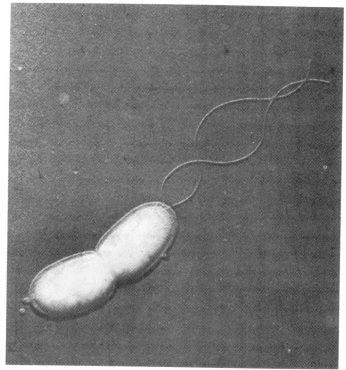

Electron micrograph showing the two polar flagella of *Pseudomonas fluorescens*. The bacterium is in the process of dividing. The excellent contrast in this photograph was obtained by metal shadowing (×22,000). (From Ambrose, E.J., and Easty, D.M. [1970], *Cell Biology*. Addison-Wesley Publishing Co., Inc.)

Figure 10-5

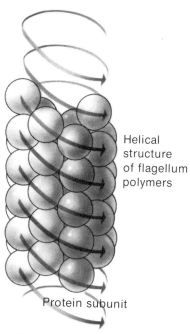

Diagrammatic representation of the structure of a flagellar filament. Individual fibers, composed of subunits of the protein flagellin (shown here as spheres), are wound about each other in a helical fashion, creating a cylindrical structure with a hollow core. (Redrawn from J. Levy, J.J.R. Campbell, and T.H. Blackburn [1973], *Introductory Microbiology*, John Wiley and Sons, Inc., New York.)

Figure 10-6

ments. Since the oxidation or reduction of cysteine affects the tertiary structure of a protein containing it (see Chapter 7), the absence of cysteine in flagellin helps to stabilize its tertiary structure under a wide range of environmental conditions.

The hook of the flagellum penetrates the cell wall and connects the filament to the basal body. Not much is known about the chemical composition or function of the hook.

The basal body binds the hook and the filament to the cell. In Gram-negative bacteria, the basal body is a structure consisting of two pairs of rings connected by a central rod. One pair of rings is attached to the cell wall, while one ring of the other pair is attached to the plasma membrane (Fig. 10–7).

The movement of flagella remains somewhat of a mystery. One widely accepted model of flagellar movement proposes that a wobbling motion of the basal body, requiring the utilization of energy, is transmitted through the hook to the relatively rigid filament, producing a gyrating motion of the filament which in turn propels the cell. This model takes account of the fact that energy is not utilized in the filament, per se.

When placed in a medium containing a concentration gradient of nutrients, a motile bacterium will move from a region of limiting nutrient concentration to one of optimum concentration. This movement, occurring in response to an external influence, is called *taxis*. Interestingly, taxis appears to be an inherited characteristic, because researchers have isolated "dumb" mutants of the bacterium *Escherichia coli* that will not move in the direction of nutrients.

PILI. Certain strains of bacteria, some flagellated and some not, have several hundred fine projections called *pili* (singular: *pilus*, the Latin word meaning "hair") or *fimbriae* (singular: *fimbria*, the Latin word meaning "fringe") covering the entire cell surface. These are considerably smaller than flagella, have diameters of 5 to 10 nm, and are as long as about 1 μm. They appear to be composed only of protein. Pili do not confer motility on bacteria that possess them. One type of pilus, called the *F-pilus*, is required for the transfer of DNA material from one bacterium to another (conjugation).

THE CELL INTERIOR

Inside the plasma membrane of the prokaryotic cell, two distinct regions are discernible in the electron microscope. The lighter (less dense) of the two regions contains DNA, the hereditary material of the cell, which forms a fibrous aggregate of irregular outline. The surrounding region, more electron-dense, is the cytoplasm.

NUCLEOID. The DNA-containing region of the cell is called the *nucleoid, chromatin body, genophore* or, especially in blue-green algae, the *nucleoplasmic region.* It is sometimes also referred to as the prokaryotic nucleus, but this is misleading, since there is no nuclear envelope.

The nucleoid consists of a mass of irregularly folded fibers, 2 to

Prokaryotic Cell Structure

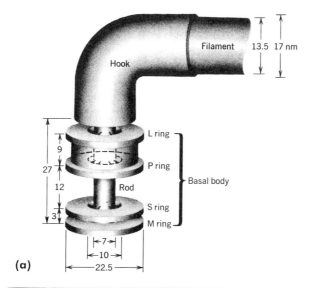

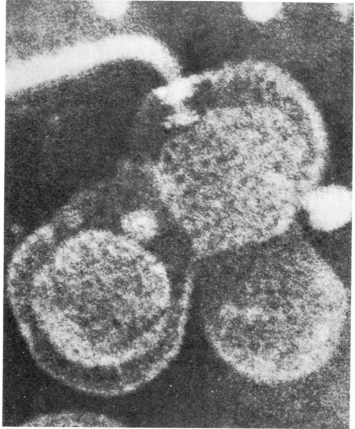

a. A model illustrating the various parts of the flagellum. The L and P rings are embedded in the cell wall and the M ring is embedded in the plasma membrane. *b.* Electron micrograph showing the attachment of the basal body to purified cell wall vesicles of *E. coli*. (From DePamphilis, M.L. and Adler, J., [1971]. J. Bacteriology *105*:384 and 396.)

Figure 10-7

5 nm in thickness (Fig. 10-8). This fibrous substance has been shown to be a single, circular molecule of double-stranded DNA. The DNA does not have protein associated with it, as it does when it forms the complex chromosomal structures characteristic of eukaryotic cells.

A growing cell may have one or more nucleoid regions, depending upon the type of cell and its rate of growth. Under certain conditions, some bacteria have been reported to have as many as seven distinct nucleoid regions.

CYTOPLASM. *Ribosomes.* The aqueous phase of the cytoplasm, the *cytoplasmic ground substance,* contains numerous ribosomes that appear as highly electron-dense granules approximately 15 to 20 nm in diameter (Fig. 10-8). Some ribosomes are also attached to the cytoplasmic side of the plasma membrane. In blue-green algae, it is common to see the ribosomes concentrated primarily in the central region of the cytoplasm in close contact with the DNA fibers, but ribosomes are also seen, in lesser concentration, in other areas of the cell.

Ribosomes are the sites of protein synthesis, and consequently their number is dependent upon the protein requirements of the cell. In a rapidly growing culture of bacteria, each cell may contain many thousands of ribosomes, constituting up to 40 per cent of the dry weight of the cell; in a starving culture, only a few hundred ribosomes may be present in a cell.

In spite of their small size, prokaryotic ribosomes are exceedingly complex structures, composed of approximately 65 per cent RNA and 35 per cent protein. The ribosomes of prokaryotes are smaller than those of eukaryotes and possess a sedimentation coefficient* of 70S. The 70S ribosome consists of two subunits. The smaller 30S subunit is composed of one molecule of single-stranded RNA, with a sedimentation coefficient of 16S, and 21 different structural proteins. The larger subunit has a sedimentation coefficient of 50S, and is composed of two RNA strands, one of 23S and one of 5S, as well as 30 to 35 different structural proteins. There is good evidence to suggest that in bacteria, the 30S and 50S ribosomal subunits associate to form 70S ribosomes only when they are involved in protein synthesis. In addition, the 70S ribosomes are frequently seen in electron micrographs to form clusters called *polyribosomes* or *polysomes.*

---

*The sedimentation coefficient, S, defines the rate of sedimentation of a molecule (or particle) in a centrifugal field. The higher the value of S, the faster the rate of sedimentation. In a given solution, the rate at which a substance sediments is dependent upon its mass, shape, and volume. Biological molecules, particularly nucleic acids and proteins but also other small particles, are often characterized by their sedimentation coefficients.

Sedimentation coefficients are not additive. Thus, for example, if a particle with a sedimentation coefficient of 30S is combined with a 50S particle, the resultant particle has a sedimentation coefficient of 70S.

Prokaryotic Cell Structure

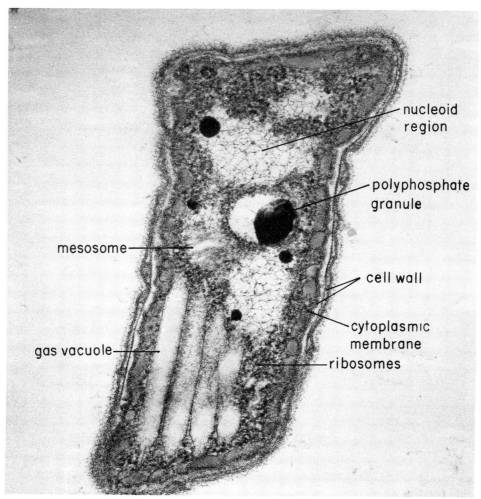

Thin section of an autotrophic bacterium, showing the DNA-containing nucleoid region and an associated mesosome, a structure formed by the infolding of the plasma membrane. Ribosomes, electron-dense granules, are seen scattered throughout the cytoplasm. Also note the gas vacuole, which provides the cell with buoyancy, and the polyphosphate granule containing a reserve source of phosphate. The cell wall and plasma membrane are also visible. See text for more detailed descriptions of structures. (From W.A. Jensen and R.B. Park [1967], *Cell Ultrastructure,* Belmont, California, Wadsworth Publishing Co., Inc., © 1967, Wadsworth Publishing Company.)

Figure 10-8

Intracellular Membranes

Although prokaryotic cells lack the complex intracellular inclusions (organelles) characteristic of eukaryotic cells, a variety of membranous structures are common in prokaryotes.

MESOSOMES. In bacteria, complex convoluted infoldings of the plasma membrane called *mesosomes* are often observed, particularly in Gram-positive bacteria, where they are quite prominent (Fig. 10-9). When present in Gram-negative bacteria, mesosomes appear to be simple infoldings, and are much less extensive than in Gram-positive cells.

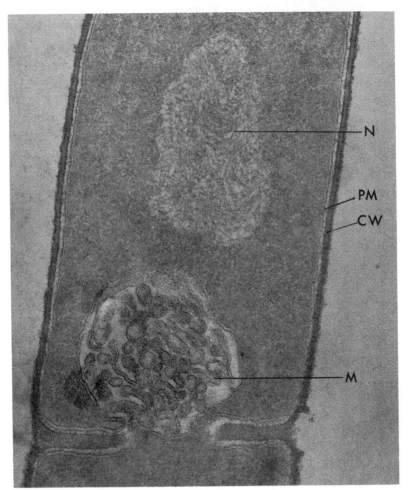

Figure 10–9 Electron micrograph of *Bacillus subtilis*, a Gram-positive bacterium, in the process of dividing. Note the mesosome (*M*), formed by infoldings of the plasma membrane at the point of septum formation; the nucleoid region (*N*), plasma membrane (*PM*), and cell wall (*CW*). From F.A. Eiserling and W.R. Romig (1962), J. Ultrastructural Res. 6:540.

Mesosomes appear to be involved in cell division, and are frequently observed at the point of septum formation, where the plasma membrane and cell wall invaginate to bring about the separation of the dividing cells into two daughter cells (see Figure 10–9). There is strong evidence that the DNA of prokaryotic cells is attached to the plasma membrane. At least in many bacteria, this attachment may occur at a mesosome, since in electron micrographs the nucleoid often appears closely associated with a mesosome (see Figure 10–8). It has been suggested that mesosomes may function to anchor the replicating DNA and thus ensure that as the cell divides, both daughter cells receive copies of the cellular DNA.

CHONDROIDS. We mentioned above that in aerobic bacteria the plasma membrane has attached to it enzymes and other compounds needed for aerobic energy generation. Some aerobic bacteria have complex infoldings of the plasma membrane, sometimes called *chondroids,* which apparently increase the amount of membranous surface area that can function in aerobic metabolism.

THYLAKOIDS. The photosynthetic prokaryotes (purple bacteria, green bacteria, and blue-green algae) have elaborate internal membranes to which photosynthetic pigments and other constituents of the photosynthetic apparatus are attached. These membranes are in the form of sacs or *thylakoids* of typical "unit membrane" structure. In photosynthetic bacteria, the thylakoids may exist in a variety of forms, including vesicles, elongated tubules, or lamellae (flattened vesicles).

In the blue-green algae, the thylakoids may be quite extensive, being visible in much of the cytoplasm (Fig. 10-10). They exist usually as lamellae, which may be arranged throughout the peripheral regions of the cell in parallel, concentric rings. The membranes are usually closely packed, and the internal space of the sacs is sometimes but not always visible. Occasionally, the lamellae are present in an irregular three-dimensional network extending throughout much of the cytoplasm.

It has been postulated that thylakoid membranes in prokaryotes are derived from the plasma membrane, and this appears to be the case in most if not all bacteria, where the thylakoids are often seen to be continuous with the plasma membrane. For blue-green algae, the evidence is much less direct. Fusion of thylakoid membranes with the plasma membrane has been reported in several instances. It has also been shown that following the degeneration of thylakoids in senescent blue-green algal cells of the genus *Oscillatoria,* new

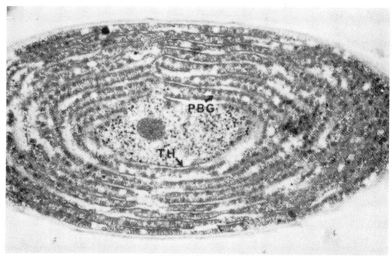

Thin section of the blue-green alga *Synechococcus lividus,* showing concentric rings of lamellar thylakoids (*TH*). Phycobilisome granules (*PBG*) are visible on the outside surface of the thylakoid membranes; phycobilisomes are rodshaped granules that contain photosynthetic pigments characteristic of blue-green algae. The nucleoid region is visible at the center of the cell, and contains a dense inclusion called a polyhedral body. Refer to the text for further details. (×42,000) (From M.R. Edwards and E. Gantt [1971], J. Cell Biol. *50*:896.)

Figure 10-10

thylakoids appear as invaginations of the plasma membrane; these invaginations then form vesicles that enlarge into thylakoid membranes. However, it is unclear whether or not this represents a general mechanism for the production of thylakoid membranes in blue-green algae.

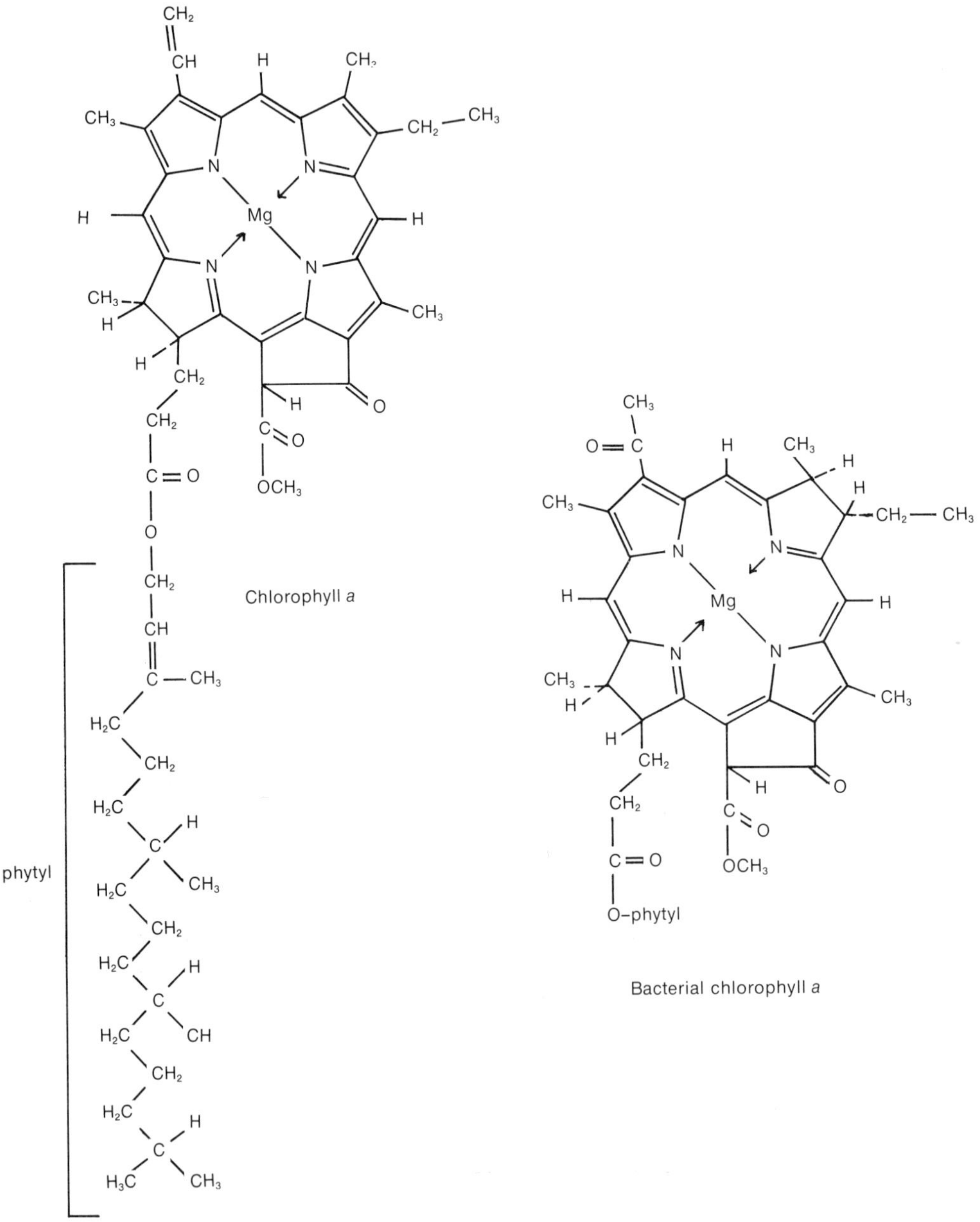

Figure 10–11 Chemical structures of chlorophyll *a* and bacterial chlorophyll *a*.

Prokaryotic Cell Structure

The thylakoids are the sites of photosynthesis. Both in bacteria and in blue-green algae, the interior of the thylakoid membranes contains light-capturing pigments as well as cytochromes and other compounds important in electron transfer during the process of photosynthesis. The membrane-associated light capturing pigments in blue-green algae are chlorophyll a (Fig. 10–11) and carotenoids (see Figure 7–29); bacteria contain one or more bacterial chlorophylls (Fig. 10–11) and carotenoid pigments. It has been shown that the chlorophylls are the pigments directly responsible for photosynthesis. The carotenoids act as accessory light-gathering pigments that transfer their absorbed light energy to chlorophyll a (in the case of blue-green algae) and to bacterial chlorophylls (in the case of photosynthetic bacteria), where the energy can be utilized in the process of photosynthesis.

In addition to the carotenoids, blue-green algae possess another class of accessory photosynthetic pigments. These are the *phycobiliproteins,* of which there are two principal kinds, the blue *phycocyanin* and the red *phycoerythrin*. One or more types of phycobiliproteins are apparently present in all forms of blue-green algae. In fact, phycobiliproteins are a major component of blue-green algae, and may constitute up to 40 per cent of the soluble protein in a cell. Like hemoglobin, the phycobiliproteins are composed of a protein to which a pigmented prosthetic group (bilin) is attached. The bilins are open-chain structures that may be formed by oxidation of closed-ring porphyrins. The probable structures of the bilin found in phycocyanin are shown in Figure 10–12; these structures should be compared with those of the porphyrin compounds heme (Fig. 7–19) and chlorophyll (Fig. 10–11).

The phycobiliproteins can be completely washed from disrupted thylakoid fragments with aqueous solutions, leaving chlorophyll and the carotenoids in the membranes. This indicates that the phycobiliproteins are *not* integral parts of the thylakoid membranes. Attempts to localize the phycobiliproteins within the cell led to the discovery of small rodshaped granules, about 35 nm in diameter, that are arranged in regular arrays on the *outer* surface of the thylakoid

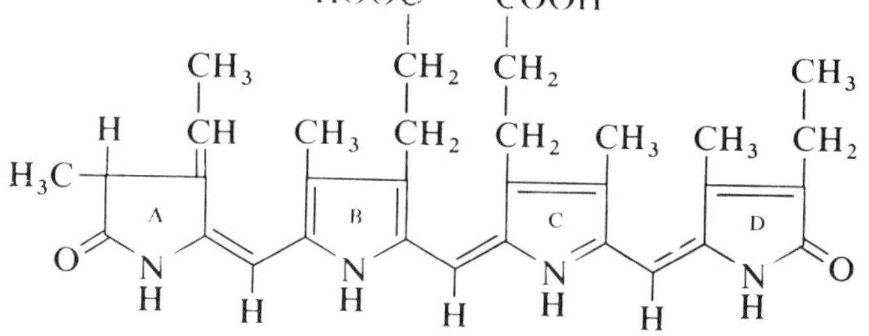

Phycocyanobilin

Figure 10–12 Chemical structure of phycocyanobilin, the prosthetic group of the protein phycocyanin.

membranes (Fig. 10–10). Since similar granules are not seen in broken cells that have been washed to remove phycobiliproteins, the granules are believed to represent macromolecular aggregates of phycobiliproteins. The granules have been termed *phycobilisomes.*

It has been postulated that phycocyanin and phycoerythyrin may have evolved as adaptations to the type of light present on the primitive earth. Harold Urey contends that methane in the primitive atmosphere would have given rise to compounds that reflected yellow and orange light. Phycocyanin, at least, absorbs light maximally at these wavelengths. It has been demonstrated that phycobiliproteins act as accessory pigments in photosynthesis. Therefore, if Urey's suggestion is correct, it would seem that blue-green algae were well adapted to carry on photosynthesis under conditions on the primitive earth.

Cytoplasmic Inclusions

Storage granules and other specialized inclusions are often seen in the cytoplasm of prokaryotic cells. Some of the inclusions most frequently encountered are described below.

GAS VACUOLES. Many photosynthetic prokaryotes, including both bacteria and blue-green algae, possess *gas vacuoles.* These are hollow, membrane-bound vacuoles that vary in size and shape among different groups of organisms.

Gas vacuoles have been most extensively studied in blue-green algae. In these organisms they are cylindrical structures, 70 to 75 nm in diameter and anywhere from about 0.2 to 2 μm in length (Fig. 10–13). The membranes of gas vacuoles are a little less than 2 nm in

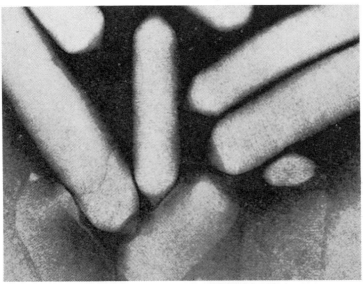

Figure 10–13 Electron micrograph of gas vacuoles isolated from the blue-green alga, *Anabaena*. Note the ribbed structure of the vacuolar membrane. (×150,000) (From A.E. Walsby [1972], Symp. Soc. Exp. Biol. *26*:233.)

thickness and have a ribbed structure. Chemical analysis of the membranes suggests that they are composed entirely of one protein. Thus both in thickness and in chemical composition, the gas vacuole membranes are fundamentally different from the typical unit membrane. The vacuole membrane is freely permeable to a large number of gases, and the gas mixture in the vacuoles has been shown to be the same as that in the aqueous medium surrounding them.

Gas vacuoles are thought to have two major functions. First, they may provide the cell with buoyancy. They probably serve to regulate the depth at which the cells float in their aqueous environment, and may serve as a mechanism by which the cell optimizes such environmental factors as light intensity and availability of nutrients. Second, by scattering light rays, gas vacuoles may shield the cell from high light intensities that might damage it. Supporting this idea is the observation that in the blue-green algae *Nostoc* and *Anabaena*, gas vacuoles develop around most of the thylakoids when cells are grown under high light intensities.

POLYPHOSPHATE GRANULES (VOLUTIN GRANULES). Large electron-dense phosphate-containing granules up to 500 nm in diameter, but lacking a surrounding membrane, are present in the cytoplasm of many prokaryotic cells (Fig. 10–8). Some evidence suggests that the *polyphosphate granules* are composed of cyclic tri- and/or tetrametaphosphate; other evidence indicates that the granules contain long open-chain polyphosphate molecules. Because the phosphate deposits are insoluble, they do not affect the osmotic pressure of the cell.

There is no evidence that the polyphosphates act as an energy source in prokaryotes. Consequently, it has been postulated that polyphosphate granules serve to store phosphate required for the synthesis of nucleic acids, phospholipids, and other phosphorus-containing compounds.

GLYCOGEN GRANULES (α–GRANULES). Glycogen is accumulated as a nutrient reserve in many prokaryotes. In blue-green algae, *glycogen granules* about 25 nm in diameter are often observed in the space between the thylakoids (Fig. 10–14). There is no membrane encasing the glycogen granules, but because of its insolubility, the glycogen does not affect the osmotic properties of the cytoplasm.

LIPID DROPLETS. Spherical droplets, less electron-opaque than polyphosphate granules and often as large as 800 nm, are frequently observed in the cytoplasm of both bacteria and blue-green algae. The droplets, containing lipid reserve material, are surrounded by a membrane about 3 nm thick.

POLYHEDRAL BODIES. Some autotrophic bacteria and many blue-green algae possess *polyhedral bodies* in the nucleoid region of the cell. These inclusions are as large as 500 nm in diameter and have a distinct polygonal shape (Fig. 10–14). In high-power electron micrographs, they appear to be composed of identical subunits arranged in a highly ordered manner.

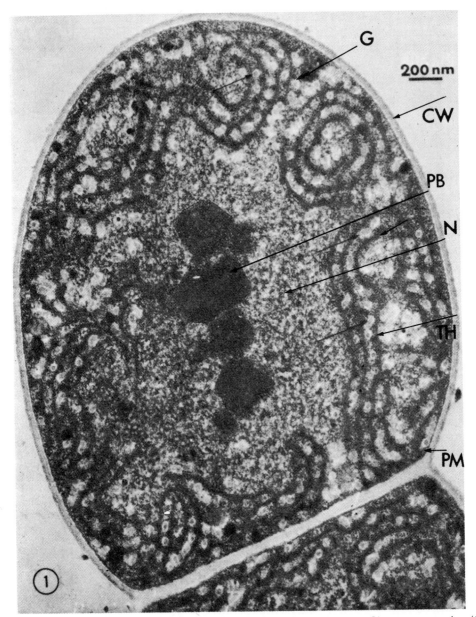

Figure 10–14 Electron micrograph of the blue-green alga, *Nostoc*. Glycogen granules (*G*) are present between the thylakoids (*TH*), and polyhedral bodies (*PB*) are located in the nucleoid region (*N*). Also notice the cell wall (*CW*) and plasma membrane (*PM*). From L. Chao and C.C. Bowen (1971), J. Bact. *105*:333.

Although polyhedral bodies were among the first inclusions identified in blue-green algae, little is known about their composition or function. Some recent work has suggested that polyhedral bodies in *Thiobacillus,* a chemoautotrophic bacterium, contain the enzyme ribulose diphosphate carboxylase, which catalyzes the addition of carbon dioxide to the five-carbon sugar, ribulose diphosphate. This reaction, known as the "fixation" of carbon dioxide into organic compounds, occurs in the manufacture of carbohydrates during photosynthesis. Consequently, the polyhedral bodies may function as sites of carbon dioxide fixation in the cell. Other researchers have contended that because polyhedral bodies are present in the nucleoid region, they may be involved in the storage of ribosomal particles.

SPECIALIZED PROKARYOTIC CELLS

In addition to the usual cells (*vegetative cells*) described above, many prokaryotes give rise to morphologically specialized cells or structures when exposed to certain environmental conditions. This process, by which relatively generalized cells become more specialized in structure and function, is called *cellular differentiation*. The most obvious example of cellular differentiation is the emergence of specialized cell types during the embryonic development of a one-celled fertilized egg into a multicellular organism. In a later chapter, we will discuss the process of cellular differentiation in some detail; here we wish merely to give a few examples of cellular differentiation among prokaryotes, in order to dispel the common impression that cellular differentiation is a process confined to eukaryotic organisms.

Endospores

Under a variety of adverse environmental conditions including, among others, low concentrations of organic or nitrogenous nutrients and the presence or absence of certain ions, some bacteria and blue-green algae will form *endospores* (Fig. 10–15). These are thick-walled structures produced within the cell that are able to survive heat, desiccation, high salt concentrations, and other extreme conditions that would adversely affect the survival of a vegetative cell. Because of their low rate of metabolism, endospores can survive for many years without a source of nutrients. When placed in a favorable environment, endospores will begin to germinate within a few minutes to form a new vegetative cell.

Akinetes

Endospores are the only differentiated cellular structure among bacteria. However, in certain strains of filamentous blue-green algae, an entire vegetative cell may develop into a type of spore called an *akinete*. During the development of an akinete, a vegetative cell forms a thickened cell wall and a surrounding envelope of carbohydrate, protein, and lipid. Cyanophycin bodies, electron-transparent granules, accumulate at the periphery of the cell. These granules contain stores of an unusual protein that is composed only of arginine and aspartic acid, and that apparently serves as a nutrient reserve. The photosynthetic thylakoids become redistrib-

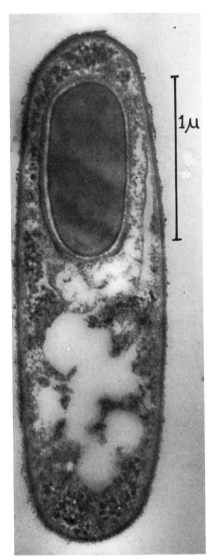

Figure 10–15 Thin section of the bacterium *Bacillus cereus* containing an endospore. (From G.B. Chapman [1956], J. Bact. *71*:348.)

uted throughout the cytoplasm, and photosynthesis continues at a gradually decreasing rate during the differentiation process, before ceasing in the mature akinete. The mature akinete retains a nucleoid region and ribosomes, and may contain polyhedral bodies and glycogen granules, although polyphosphate granules are absent. Like endospores, akinetes will germinate under favorable environmental conditions.

It should perhaps be mentioned at this point that even the vegetative cells of many species of prokaryotes that do not produce spores are remarkably resistant to extreme physical and chemical conditions. This fact, along with the ability of prokaryotes to form

Specialized Prokaryotic Cells

spores, has been taken by some workers as further evidence that prokaryotes evolved early in the earth's history, when environmental conditions would presumably have been considerably less stable than they are now.

Heterocysts

Under nitrogen-limiting conditions, many filamentous forms of blue-green algae will produce *heterocysts,* specialized cell types that differentiate from vegetative cells (Fig. 10–16).

During the differentiation of heterocysts, a thick envelope is laid down outside the wall of the vegetative cell. The original lamellar thylakoid system disintegrates and is replaced by a new system of membranes, which forms a network throughout the cytoplasm. Sometimes the new membranes form mesosome-like structures. The mature heterocyst contains a nucleoid region, ribosomes, and sometimes other small inclusions such as glycogen granules, but polyphosphate and polyhedral bodies are absent. Heterocysts have a high metabolic activity, but do not carry on carbon dioxide fixation, apparently receiving nutrient organic compounds from adjacent vegetative cells.

The primary function of heterocysts, although long in question,

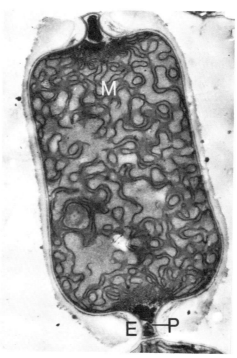

Figure 10–16 Electron micrograph of a heterocyst of the blue-green alga *Anabaena.* Note the network of intracellular membranes (*M*) throughout the cytoplasm, the thick envelope (*E*), and the pore channel (*P*), a constriction near the attachment of the heterocyst to an adjacent vegetative cell. (From P. Fay and N.J. Lang [1971], Proc. Roy. Soc. Lond. B*178*:185.)

has recently been shown to be *nitrogen fixation;* this is the process by which nitrogen gas (N_2), an extremely unreactive compound, is converted into ammonia, a combined form of nitrogen utilizable by the cell. It has been demonstrated that the differentiation of heterocysts in growing cells is controlled by a feedback system in which the availability of ammonia and other sources of combined nitrogen suppresses heterocyst formation, whereas the lack of utilizable nitrogen compounds increases heterocyst production.

Presumably all the cells of a prokaryotic species capable of sporulation or heterocyst formation carry the hereditary information for the differentiation of these structures. This potential is expressed, however, only under certain conditions, and although we know many of the environmental factors that trigger the response, we do not understand how the process actually occurs. Indeed, the mechanism of cellular differentiation, which involves the *selective expression* of certain hereditary information, is one of the major unsolved problems in biology.

OVERVIEW OF PROKARYOTIC SYSTEMS

Compared to eukaryotic cells, prokaryotic cells are relatively simple in structure, but it would be grossly incorrect to assume that because of this, prokaryotes do not function as well as eukaryotes. It does not follow that the simplest system is necessarily the least efficient one. Among prokaryotes are organisms that display a variety of structural and metabolic specializations and that collectively are adapted to a wide range of environments.

Although prokaryotic cells are less compartmentalized than eukaryotic ones, there is nevertheless considerable internal structure in prokaryotic cells that provides significant opportunities for isolating functional metabolic systems from one another. We have noted, for example, that the pigments, many of the enzymes, and other constituents of the photosynthetic apparatus are either an integral part of, or closely associated with, the thylakoid membranes, and that the metabolic reactions associated with photosynthesis occur on or near these membranes. We have also seen that the plasma membrane contains many of the enzymes necessary for aerobic metabolism, and that the mesosomes in at least some bacteria apparently serve an important, although poorly understood, role in DNA replication and cell division. Such compartmentalization can greatly increase the efficiency of metabolism by concentrating in one area all or many of the enzymes needed for a particular metabolic process, by establishing localized optimal conditions of pH, by maintaining high concentrations of substrates, and by excluding substances that may interfere with the process. We will return to this topic in the next chapter when we discuss the more developed compartmentalization of eukaryotic cells. The lesson here is that prokaryotic cells are structurally and functionally more complex than may be apparent at first glance.

Because they lack the cellular specializations and complex integration of multicellular systems, prokaryotes are simpler to study. For example, the mechanisms by which hereditary information is stored in DNA molecules and by which this information is translated into patterns of protein synthesis were discovered primarily by studying bacteria and their viruses. Sporulation and heterocyst formation in prokaryotes are being used as simple model systems for the study of cellular differentiation. Of course, if we really want to understand how eukaryotic organisms function, we must study eukaryotes. But information gained from research on prokaryotes has revealed basic biochemical mechanisms that, by analogy, have provided valuable approaches for research on eukaryotic systems.

REFERENCES

Ambrose, E.J. and D.M. Easty (1970). *Cell Biology.* Addison-Wesley Publishing Co., Inc., Menlo Park, California.

DuPraw, E.J. (1968). *Cell and Molecular Biology.* Academic Press, Inc., New York.

Dyson, R.D. (1974). *Cell Biology: A Molecular Approach,* Allyn and Bacon, Inc., Boston.

Fogg, G.E. (1956). The comparative physiology and biochemistry of the blue-green algae. Bacteriol. Rev. *20*:148.

Fogg, G.E., W.D.P. Stewart, P. Fay, and A. E. Walsby (1973). *The Blue-Green Algae.* Academic Press, Inc., New York.

Levy, J., J.J.R. Campbell, and T.H. Blackburn (1973). *Introductory Microbiology.* John Wiley and Sons, Inc., New York.

Rose, A.H. (1965). *Chemical Microbiology,* Butterworth and Company, Ltd., London.

Shively, J.M., F. Ball, D.H. Brown, and R.E. Saunders (1973). Functional organelles in prokaryotes: polyhedral inclusions (carboxysomes) in *Thiobacillus neopolitanus.* Science *182*:584.

Wolfe, S.L. (1972). *Biology of the Cell,* Wadsworth Publishing Co., Inc., Belmont, California.

Wolk, C.P. (1973). Physiology and cytological chemistry of blue-green algae. Bacteriol. Rev. *37*:32.

CHAPTER 11

The Structure and Origin of Eukaryotic Cells

There is abundant evidence testifying to the spectacular evolutionary success of prokaryotic organisms. We have seen that the prokaryotic cell can withstand relatively wide fluctuations in environmental conditions without adverse effects. As a group, prokaryotes can flourish in a broad range of environments. There are forms that thrive in the near-boiling waters of hot springs, others that exist in the subzero temperatures of the polar regions. There are prokaryotic species that grow at very high acidities (pH 1.0), and others that grow at low acidities (pH 10.0). Some prokaryotes normally exist in fresh water; others, in sea water; still others, in the concentrated solutions of salt lakes. Some are adapted to life in surface waters; some live in the high-pressure depths of the ocean.

Diversity among the prokaryotes extends to their metabolic capabilities. There are aerobic, anaerobic, and facultative forms. There are numerous heterotrophic forms that collectively are capable of metabolizing practically every organic compound synthesized by living organisms. Many prokaryotes are autotrophs, including photosynthetic organisms, which can utilize the energy in the sun's rays to synthesize organic compounds, and chemoautotrophs, which are able to extract usable energy from inorganic substances. Some prokaryotic organisms are even able to accomplish the chemically complex task of fixing atmospheric nitrogen gas into utilizable compounds.

Considering the great versatility and adaptability of prokaryotic organisms, we might ask why eukaryotic cells were selected for, and why evolution has proceeded in the direction of ever-increasing levels of organization, instead of leading to even greater diversity of prokaryotic forms of life. One answer is that eukaryotic cells must have unique properties that allowed eukaryotic organisms to compete successfully with prokaryotic forms, or to exploit environments that, for one reason or another, were not accessible to prokaryotic organisms.

PROPERTIES OF EUKARYOTIC CELLS

What are some of the properties of eukaryotic cells that may have led to their establishment? The most obvious is size. Because eukaryotic cells are much larger than prokaryotic ones, unicellular eukaryotes can obtain food by capturing and engulfing smaller prokaryotes.

Another distinguishing characteristic of eukaryotic cells is their ability to form complex multicellular systems. It is not at all clear why prokaryotic cells do not give rise to complex multicellular organisms, but it is nevertheless true that all multicellular organisms, with the exception of simple filamentous and colonial forms of bacteria and blue-green algae, are composed of eukaryotic cells. The development of multicellularity was an important breakthrough in the evolution of living forms, because it is by means of cell specialization that multicellularity can give rise to diverse forms of life able to exploit a wide range of environments. Eventually, cell specialization gave rise to forms that, by controlling their internal environment — by carrying an aqueous environment with them — were able to adapt to terrestrial life.

The very increase in size and thrust toward multicellularity that were advantageous to eukaryotic organisms created problems as well. Some of the structural features of eukaryotic cells appear to be a result of adaptation to these problems.

STRUCTURAL DESIGN OF EUKARYOTIC CELLS

SURFACE AREA TO VOLUME RATIO

There is a well-known mathematical principle stating that as the volume of a solid object increases, its surface area does not increase proportionately. Consequently, because of its larger size, a spherical eukaryotic cell has a much smaller surface area per unit of cell weight than a spherical prokaryotic cell. Since the rate at which substances pass into and out of the cell is dependent upon its surface area, the larger eukaryotic cell often needs an increased surface area in order to interact efficiently with its environment. This is particularly true when we consider that the necessity for chemical communication between cells increases in a multicellular organism.

The surface area of some eukaryotic cells is enlarged by connections between the plasma membrane and intracellular membrane systems. We shall see, for example, that the endoplasmic reticulum, a system of membranes found in most eukaryotic cells, is often observed in electron micrographs to be continuous with both the plasma membrane and the double-layered nuclear envelope that surrounds the nucleus (Fig. 11–1). Consequently, the exterior of the cell is not restricted to areas immediately adjacent to the plasma membrane, but in fact is in close contact with the nucleus and other

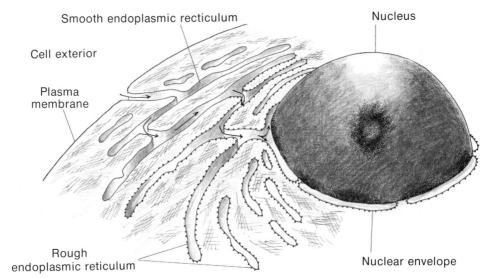

Figure 11-1 Diagram of a eukaryotic cell, showing how the endoplasmic reticulum may connect with both the plasma membrane and the double-layered nuclear envelope. Connections of this type may create channels through which materials may pass to and from the cell's surroundings without penetrating the plasma membrane. However, such channels are probably found only in cells with a highly developed system of endoplasmic reticulum.

intracellular regions that we normally think of as being far removed from the cell surface. Thus, the "surface" of the cell extends far into its interior — a conclusion that is obviously quite important, considering the interaction of a eukaryotic cell with its environment.

COMPARTMENTALIZATION

The metabolism of eukaryotic cells is considerably more complex than prokaryotic cell metabolism, primarily as a result of the elaborate cellular coordinating mechanisms in multicellular organisms. This metabolic complexity, in conjunction with the large size of eukaryotic cells, means that it is difficult to operate the metabolism of the cell efficiently without separating functional activities from one another. As a result, there is an increased need for compartmentalization—a division of labor *within* cells that allows for the establishment of optimal conditions for the operation of different functional systems.

How, specifically, does compartmentalization result in improved metabolic efficiency? In the first place, compartmentalization allows all or most of the enzymes necessary for a metabolic pathway or functional system to be concentrated in one location so that they are readily available; it also allows for the accumulation of substrates of these reactions. This is important because the rate of an enzyme-catalyzed reaction is dependent upon the concentration of enzyme and substrate (within a range of substrate concentrations,

Structural Design of Eukaryotic Cells

of course). In addition, the product of one reaction in a metabolic pathway is the substrate of the next reaction in the sequence. Because the concentrations of both enzymes and substrates can be maximized in a compartment instead of being diluted in the entire cytoplasm, the rate of metabolic reactions will be enhanced.

In all types of cells, prokaryotic as well as eukaryotic, enzymes are often associated with membranes as integral parts of the membrane structure. This arrangement appears to be crucial for the efficient operation of the metabolic pathway involved, for if the enzymes are separated from the membrane surface, they are much less effective in catalyzing the reactions of the pathway. It has been proposed that a specific arrangement of enzyme molecules in a membrane allows the passage of substrate from one enzyme to the next in the metabolic sequence without the detachment of the substrate from the membrane—a mechanism that might greatly improve the efficiency of a metabolic pathway.

Besides maintaining high concentrations of enzymes and substrates, compartmentalization also permits the establishment of such physiological conditions as may be optimum for a particular enzyme system. For example, the pH within a compartment may be regulated at a value quite different from that in the cell as a whole, and ions or other cofactors necessary for the activity of particular enzymes in a pathway may be accumulated in a compartment. At the same time, compartmentalization can act to exclude substances that might inhibit one or more reactions in the pathway. In short, each compartment is able to regulate its contents by virtue of possessing a selectively permeable membrane.

There is another potential advantage to carrying out a particular metabolic pathway within one compartment. It is quite conceivable that such an arrangement would allow for rapid and efficient regulation of the functional system. For instance, if the operation of a particular metabolic system were temporarily not needed because of an excessive accumulation of its endproduct, it is likely that "turning off" the process by some biochemical means could be accomplished more simply and quickly if all the components of the system were isolated in one region, instead of being scattered throughout the cytoplasm. Many such biochemical control mechanisms can be envisioned. As one example, accumulation of the final product of a pathway might quickly drive the final reaction, and hence the preceding reactions of the pathway, in the reverse direction. It is clear that such inhibition would operate at a faster rate in a compartment than it would if the product were diluted in the cytoplasm.

Not only is compartmentalization capable of improving the efficiency of a metabolic system; it makes possible the storage within the cell of substances that might adversely affect the cell's integrity or function. The storage of hydrolytic enzymes in membrane-bound vesicles called lysosomes is an example of this function of compartmentalization. The hydrolytic enzymes of the lysosome are needed for the digestion of nutrient material brought into the cell, but if the enzymes were not isolated from the rest of the cytoplasm, they would digest the cell itself. Ions or other substances that are required for some metabolic reactions, but are inhibitory to others, are also sometimes stored in isolated compartments.

Communication Between Compartments

For the cell to function as a viable unit, there must be communication between the separate compartments of the cell. It is easy to see, for example, how the cell's energy could be most effectively captured in a compartment devoted exclusively to that purpose, but it is equally obvious that if the energy is to be useful to the cell, there must be a mechanism for distributing it to areas in the cell that need it. Indeed, membranes not only play an important role in maintaining the specialized conditions of each compartment but they also serve as intracellular channels through which materials can be exchanged between compartments.

Thus the intracellular membranes in eukaryotic cells must not be thought of as establishing *completely separate* compartments, or as preventing movement of materials within the cytoplasm. In fact, time-lapse photography of living cells demonstrates continuous movement of the cytoplasm (*cytoplasmic streaming*) and often of cytoplasmic organelles. This activity may play an important role in the distribution of substances throughout the cytoplasm.

While studying the structure of eukaryotic cells, we must not fail to think of them as functional as well as structural units. Nor must we forget that in multicellular organisms, a cell must function both as a semi-autonomous unit and as a part of the organism as a whole, and that the structure of a eukaryotic cell has evolved concomitantly with its function. To do otherwise would be to ignore the dynamic aspects of the living cell.

EUKARYOTIC CELL STRUCTURE

Because of the enormous structural diversity among different types of eukaryotic cells, we cannot hope to describe every specialized cellular structure that occurs in eukaryotes. Therefore, in our discussions here, we will concentrate on structures that are characteristic of the great majority of eukaryotic cells or have special significance to our consideration of cell structure and function.

THE CELL EXTERIOR

Cell Wall

Almost all eukaryotic plant cells possess, exterior to the plasma membrane, a rigid cell wall that is a secreted product of the cell (Fig. 11–2). The composition and structure of the cell wall in eukaryotic plants can vary considerably in different groups of plants and in different types of cells in the same plant.

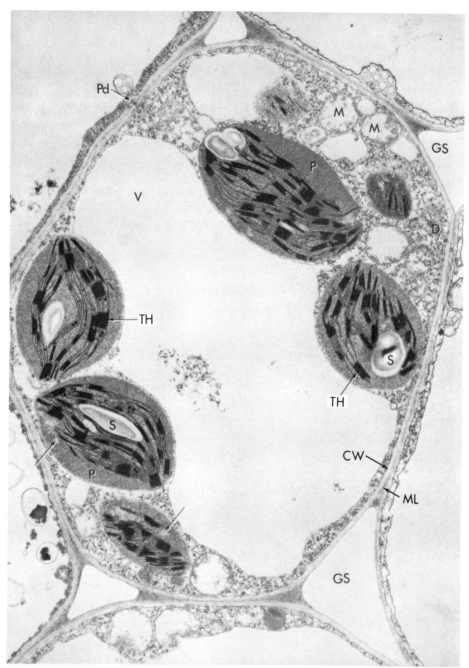

Figure 11–2 A low-power electron micrograph of a typical cell of a nonwoody higher plant (leaf tissue of timothy grass), showing the primary cell wall (CW) and middle lamella (ML). A plasmodesma (Pd) connects the cytoplasm of the central cell with an adjacent one. Also shown are five mature chloroplasts containing photosynthetic thylakoids (TH) and starch granules (S). Much of the volume of the cell is taken up by a large central vacuole (V) and a number of smaller ones; these are filled with cell sap. Between cells in the leaf tissue are gas cavities (GS) filled with carbon dioxide, oxygen, and other gases. See text for further details. From M.C. Ledbetter and K.R. Porter [1970], Introduction to the Fine Structure of Plant Cells, Springer Verlag, New York.)

During the development of the cell wall in vascular plants, a viscous mucilaginous material composed of pectin, a complex polysaccharide polymer, is laid down outside the plasma membrane. This layer, known as the *middle lamella,* is so named because it ultimately lies between the cell walls of two adjacent cells. The socalled *primary cell wall,* consisting mainly of intertwined cellulose fibers but also containing small amounts of pectin and other substances, is then secreted through the plasma membrane against the middle lamella. The primary wall consists of about 80 per cent water, and is initially quite flexible, but as increasing amounts of cellulose are deposited, the wall becomes considerably more rigid.

The cells of the soft tissues of plants contain cell walls composed only of the primary cell wall layer lying inside the middle lamella. In cells that form woody tissue, a *secondary cell wall,* again secreted through the plasma membrane, is laid down inside the primary wall as cell growth ceases. The secondary wall is composed of cellulose, lignin, pectin, and some additional non-cellulose polysaccharide substances; it also contains less water than the primary wall. The secondary wall is much harder than the primary wall, and gives woody tissues their characteristic tensile strength. After the secondary wall has been deposited, many of the cells die, leaving woody tubes formed from the cell walls. These tubes provide mechanical support for the plant, and function as channels for internal transport of water and dissolved nutrients.

The cell walls of most eukaryotic plant cells contain small circular depressions called *pits.* Many of these pits appear to have holes in them, through which *plasmodesmata* (Fig. 11–2) are formed. As in blue-green algae, the plasmodesmata connect the plasma membranes of adjacent cells and thus permit the passage of material from one cell to another.

It should be stressed that the eukaryotic cell wall in plants, although functionally similar to the cell wall of prokaryotes, has a completely different chemical composition and structure. There is no cellulose in the cell wall of prokaryotes, and there is no murein in the cell wall of eukaryotes. In fact, the composition of its cell wall is one characteristic that clearly separates a prokaryotic cell from a eukaryotic plant cell.

THE MEMBRANE SYSTEM OF THE "EXTERNAL COMPARTMENT"

The system of interconnecting cellular membranes in eukaryotic cells forms what has been called the *external compartment* of the cell. As we noted in Figure 11–1, a molecule could enter "into" the cell, or more precisely, into the inner regions of a cell, through a connection between the plasma membrane and the endoplasmic reticulum without ever passing through a cellular membrane. Once a molecule has entered into the endoplasmic reticulum or is inside one of the other intracellular membrane systems, the molecule is

Eukaryotic Cell Structure

said to be in the external compartment of the cell. In order to pass from the external compartment into the cytoplasmic matrix of the cell (the "*internal compartment*"), a substance must traverse one of the cellular membranes. As we shall see, it is convenient to describe the structure of a eukaryotic cell in terms of these two compartments. Consequently, we will first describe the structure and function of the membrane systems forming the external compartment; after this, we will consider the internal compartment.

Plasma Membrane

In the electron microscope, the *plasma membrane* or *plasmalemma* of eukaryotic cells shows the same "unit membrane" structure as the prokaryotic plasma membrane. However, the eukaryotic plasma membrane does not have enzymes of aerobic metabolism associated with it, and consequently does not possess the small projections detected on the inner side of the plasma membrane of some aerobic bacteria. Although the plasma membrane in eukaryotes may vary in composition from one type of cell to another, it is composed primarily of lipids and proteins plus, in some instances, carbohydrates.

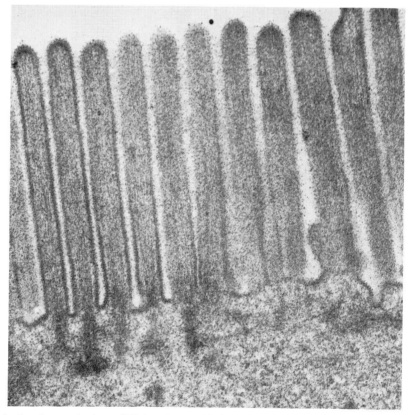

Figure 11-3 Electron micrograph showing part of a cell lining the small intestine of a rat. The plasma membrane is folded into a number of microvilli forming the brush border of the cell (×44,500).

The main function of the plasma membrane in eukaryotes, as in all cells, is to maintain the integrity of the cell by forming a selective barrier between the cell and its surroundings.

In many types of eukaryotic cells, the plasma membrane, instead of stretching over the surface of the cell, will increase its sur-

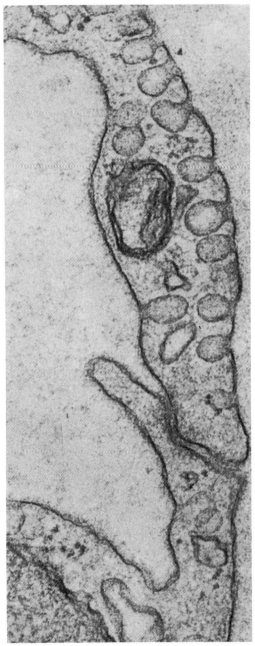

Figure 11-4 Invaginations of the plasma membrane in an elongated part of a cell forming the wall of a blood capillary. These invaginations pinch off from the plasma membrane to form vesicles that enter the cytoplasm (see Figure 11–5).

face area by forming specialized structures. For example, in cells whose main function is the absorption of materials from their surroundings, such as the cells forming the inner lining of the intestine (intestinal epithelial cells), the plasma membrane is folded to produce numerous cylindrical projections called *microvilli;* these fingerlike projections are often known collectively as a *brush border* (Fig. 11-3). Since the rate at which substances can pass through the plasma membrane is dependent upon its surface area, microvilli considerably increase the absorptive capacity of intestinal cells or any cell type possessing them.

In addition to permanent structures such as microvilli, the plasma membrane of a great many cell types has been shown to form small temporary depressions (invaginations) and projections at the cell surface; these resemble sacs open at one end, and can be clearly seen in the electron microscope (Fig. 11-4). A number of studies have demonstrated that the cell uses these structures to transport material into the cell. Electron micrographs show that material external to the cell is entrapped in a membrane-bound vesicle when the open end of the sac closes by the fusion of the plasma membrane across the opening. The vesicle then detaches from the inner surface of the plasma membrane and moves into the cytoplasm. This process, by which material is transported into cells by the formation of vesicles, is called *endocytosis* (Fig. 11-5). When liquid material, perhaps containing organic molecules or other nutrient substances in solution, is entrapped in a vesicle, the process is referred to more specifically as *pinocytosis* (cell drinking), and the vesicles that are formed are known as *pinocytotic vesicles* or, less commonly, as *pinosomes.* In some instances, most notably in the case of amoebae or scavenging white blood cells, particulate material enters the cell by endocytosis. This process is called *phagocytosis* (cell eating), and the vesicles are referred to as *phagocytotic vesicles* or *phagosomes.*

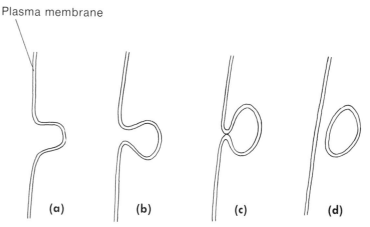

Figure 11-5

Diagram illustrating successive stages in the formation of an endocytotic vesicle from an invagination of the plasma membrane.

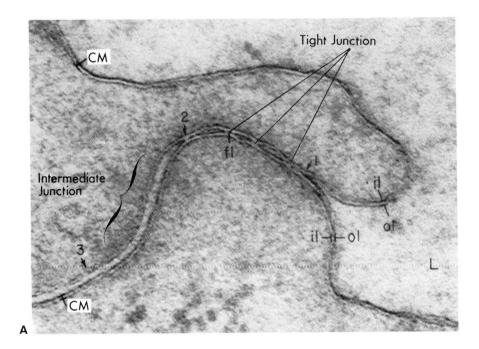

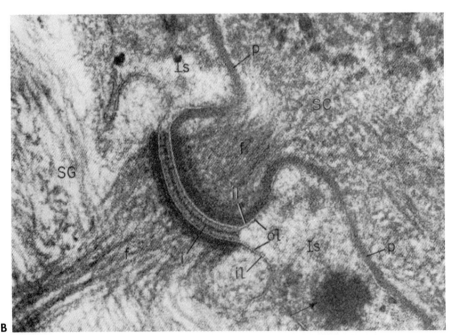

Figure 11–6 Types of junctions between cells in a tissue. *a.* Intermediate (loose) junction and tight junction. *b.* Desmosome. (From: [*a*] M.G. Farguhar and G.E. Palade [1963], J. Cell Biol. *17*:375; [*b*] M.G. Farguhar and G.E. Palade [1965], J. Cell Biol. *26*:263.)

Continued on the opposite page

Eukaryotic Cell Structure

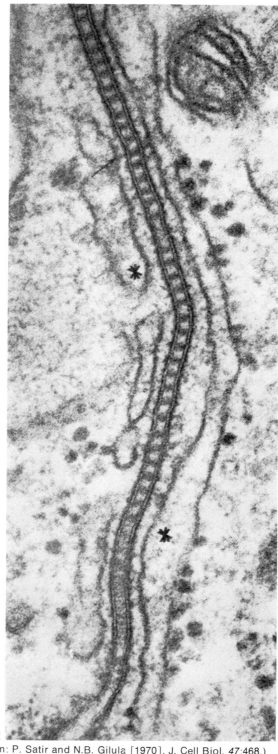

c. Septate desmosome. (From: P. Satir and N.B. Gilula [1970], J. Cell Biol. *47*:468.)
Illustration continued on the following page.

Figure 11-6

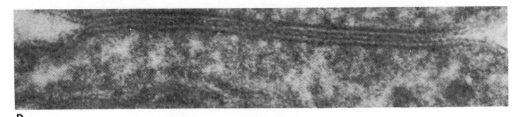

D

Figure 11-6 *d*. Gap junction. (From F.J. Silverblatt and R.E. Bulger [1970], J. Cell Biol. 47:514.)

An endocytotic vesicle, once inside the cell, may fuse with a lysosome to form what is known as a *secondary lysosome* or *digestive vacuole*. The enzymes of the lysosome then hydrolyze the ingested material to small organic molecules that diffuse through the membrane of the vacuole into the cytoplasm. Residual undigestible material, along with the vacuole, may be extruded from the cell by *exocytosis,* a process that is essentially the reverse of endocytosis. In fact, exocytosis is the principal mechanism used by cells in the secretion of hormones and other substances: vesicles containing the substance to be secreted fuse with the plasma membrane and release their contents to the surroundings.

Although phagocytosis has been known since biologists first began to study amoebae many years ago, the importance of endocytosis and exocytosis as general mechanisms for transporting substances into and out of cells was greatly underestimated until recently, because the smaller vesicles involved in transport were not visible in the light microscope. It was not until the use of the electron microscope several decades ago that small endocytotic and exocytotic vesicles were noted in many different types of cells.

The plasma membrane is also involved in the attachment of cells to one another in the tissues of multicellular animals. In most instances of cell adhesion, the plasma membranes of two adjacent cells are closely apposed to one another, and are separated by a distance of about 10 to 20 nm. This type of junction between two cells is called an *intermediate junction,* or sometimes a *loose junction* (Fig. 11-6a). The space between the two membranes is reportedly filled with an amorphous colloidal material, which some investigators have assumed to represent an "extracellular cement." However, the data derived from research in this area have been somewhat contradictory, and it is difficult to conclude whether or not an extracellular cement actually exists.

Localized regions of two adjacent plasma membranes often show differentiation in structure. These regions are believed to be specialized sites where the attachment between plasma membranes is stronger or more permanent than in the intermediate junction.

The most complex of these differentiated structures is the *desmosome* (desmos = binding). In this structure, apposing cell membranes appear thickened and are separated by an enlarged inter-

Eukaryotic Cell Structure

cellular space, often 20 to 50 nm in diameter, which is filled with an electron-dense material that forms a distinct layer in the middle of the intercellular space. In addition, fine cytoplasmic filaments radiate from the cytoplasmic side of the thickened cell membranes into the surrounding cytoplasm (Fig. 11–6b).

In other cases, the adjacent cell membranes appear to fuse for a short distance so that there is no intercellular space detectable in electron micrographs. This type of structure has been called a *tight junction* (Fig. 11–6a).

There are two other types of specialized junctional structures sometimes noted between plasma membranes. The *septate desmosome* is formed when two adjacent membranes, approximately 20 to 30 nm apart, are joined by short electron-dense structures which give the junction the appearance of a zipper (Fig. 11–6c). The *gap junction* is formed by two closely apposed membranes which are separated by small electron-dense particles about 2 nm in diameter (Fig. 11–6d).

There is evidence that suggests that the movement of materials between two cells occurs at tight junctions, septate desmosomes, and gap junctions, leading some workers to conclude that these junctions contain minute cytoplasmic bridges. However, not all communication between cells necessarily occurs only at such junctions. In some instances two animal cells have been shown by electron microscopy to be joined by conspicuous cytoplasmic bridges.

Endoplasmic Reticulum

The *endoplasmic reticulum* is a system of membranes present in most eukaryotic cells. It consists of a number of interconnecting membranous sacs which may be in the form of large flattened vesicles (known as *cisternae*), inflated vesicles, or tubules (Fig. 11–7). These elements form a cavernous system of membrane-bound channels which divide the cell into two distinct areas: the channels themselves and the rest of the cell. Thus membranes forming the endoplasmic reticulum have an outer surface which borders on the cytoplasm and an inner surface which borders on the channels. Membranes of the endoplasmic reticulum have a "unit membrane" structure, but are often less distinctly trilaminar than the plasma membrane and are also frequently somewhat narrower in width, usually about 5 to 6 nm. Like the plasma membrane and other cell membranes, the membranes of the endoplasmic reticulum are selectively permeable, allowing certain molecules, but not others, to pass through them. At least in some cases, as we have noted previously, the membrane system of the endoplasmic reticulum appears to be continuous with the plasma membrane and with the nuclear envelope.

There are two distinct forms of endoplasmic reticulum which may, but need not, be present in a single cell, and which are continuous with one another. One of these forms usually appears as a system of flattened cisternae and has ribosomes attached on its outer

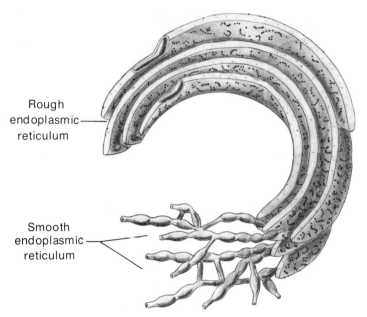

Figure 11-7 Diagram of the three-dimensional structure of the endoplasmic reticulum. Some of the endoplasmic reticulum may contain ribosomes attached to its outer surface; it is called rough endoplasmic reticulum. Other sections, known as smooth endoplasmic reticulum, do not contain ribosomes.

cytoplasmic surface. Because of the ribosomes, the membranes of this type of endoplasmic reticulum have a granular or "rough" appearance under the electron microscope; consequently, this form is known as *rough endoplasmic reticulum* or *rough ER* for short. The other form is more tubular in appearance, and because it lacks ribosomes, it is called *smooth endoplasmic reticulum* or *smooth ER* (Fig. 11–8).

The extent and complexity of the endoplasmic reticulum may vary from one type of cell to another. In some cells it may fill most of the cytoplasm; in some it may consist of only a few elements; in some it may not be present at all. Its extent and its structure may also vary within a particular cell type, depending upon the developmental and functional state of the cell. Rough ER predominates in cells that are actively synthesizing large amounts of protein—an observation that is not particularly surprising, considering the presence of ribosomes on rough ER. On the other hand, smooth ER is more characteristic of cells that are involved in the synthesis of nonprotein substances such as steroid hormones, phospholipids, and glycolipids (glyco = carbohydrate). However, rough ER can convert to smooth ER, and vice versa, depending upon the metabolic requirements of the cell. Indeed, the endoplasmic reticulum

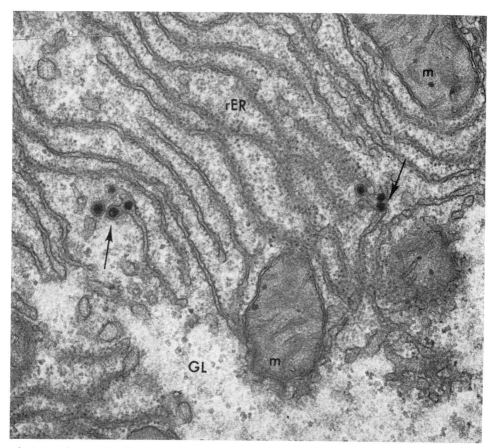

An electron micrograph of a mammalian liver cell showing connections between lamellae of the rough endoplasmic reticulum (rER) and tubules of the smooth endoplasmic reticulum (arrows). Lipoprotein granules are present in the smooth ER. Also shown are mitochondria (m) and large electron-transparent areas of particulate glycogen (GL). (×49,500) (From A. Claude [1970], J. Cell Biol. 47:748.)

Figure 11-8

is capable of continual reorganization. Like the cell as a whole, the endoplasmic reticulum is a dynamic, not a static, system.

In addition to its synthetic functions, the smooth endoplasmic reticulum has been implicated in the metabolism of glycogen and lipids, including the mobilization of glycogen to glucose-1-phosphate and the initial steps in the breakdown of fatty acids. In muscle cells, there are large amounts of smooth endoplasmic reticulum. Here it is known as *sarcoplasmic* (sarco = muscle) *reticulum,* and is involved in the concentration of calcium ions by a process requiring the utilization of ATP (energy-requiring). The calcium ions are stored in the sarcoplasmic reticulum and released following the stimulation of muscle by nerve impulses, hormones, or other means. The release of calcium ions leads to muscular contraction.

The endoplasmic reticulum also has a general role in the intracellular transport and storage of proteins and sometimes lipids. For example, in experiments with starved rats that were fed corn oil, droplets of corn oil were detected initially in pinocytotic vesicles below the brush border of the intestinal cells. About an hour after feeding, lipid was found distributed in the rough and smooth endoplasmic reticulum, in the nuclear envelope, and in the Golgi apparatus. It is important to realize, however, that since many cells do not contain extensive systems of endoplasmic reticulum, it is quite possible that such a transport system is important primarily in cells that have major absorptive or secretory functions.

The role of the endoplasmic reticulum in the synthesis and transport of proteins has been extensively studied. Many of the proteins synthesized on the ribosomes of the rough ER enter into the cisternae of the endoplasmic reticulum. Although it is not understood how the proteins pass through the membranes of the rough ER, some researchers have proposed that a small channel exists through the membrane near the site of ribosome attachment. There is good evidence to suggest that the newly synthesized proteins, once inside the cisternae, are transported to an area of smooth ER where they are accumulated in vesicles that then migrate to the Golgi apparatus (see below).

It should be stressed that not all proteins synthesized in the cell are accumulated in the cisternae of the endoplasmic reticulum for transport to the Golgi apparatus. There are many ribosomes in the cytoplasm that are not attached to endoplasmic reticulum, and it is undoubtedly true that proteins synthesized on these cytoplasmic ribosomes, as well as some of the proteins synthesized on ribosomes of the rough ER, are released directly into the cytoplasm.

Little is known about the development of the endoplasmic reticulum. It has been suggested by some workers that the endoplasmic reticulum may be budded from the nuclear envelope. Others, however, contend that both the ER and the nuclear envelope are formed by the invagination of the plasma membrane. There is some evidence to support both viewpoints.

Golgi Apparatus

Functionally, the endoplasmic reticulum is intimately associated with the *Golgi apparatus,* a system of membranes named for its discoverer, Camillo Golgi. The Golgi apparatus (often referred to as the *Golgi body*) is distinctly different in morphology from the endoplasmic reticulum. It consists essentially of a stack of closely spaced flattened sacs, about 20 nm apart, and a number of associated vesicles (Fig. 11–9). The sacs are composed of 6 nm unit membranes which, like the membranes of the ER, are not distinctly trilaminar. Each sac forms a disclike structure with a convex and a concave side; the membranes in the central region of each disc are about 15 to 20 nm apart, but at the edges of the disc, the membranes are swollen apart to about 60 to 80 nm. Numerous vesicles

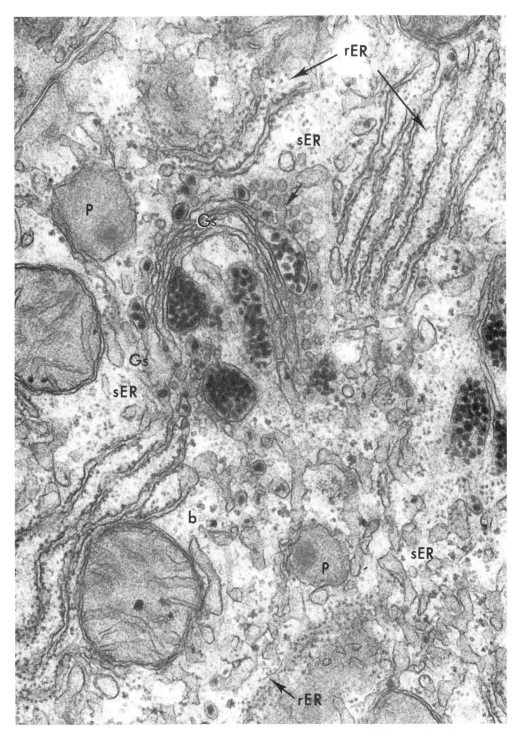

Figure 11-9 Portion of a mammalian liver cell similar to the one shown in Figure 11-8. The rough endoplasmic reticulum (rER) and smooth endoplasmic reticulum (sER) are involved in the synthesis of lipoprotein granules. These are shown in the process of being transferred to the Golgi apparatus (Gs), where they are concentrated into membrane-bound vesicles. The small arrow near the Golgi apparatus points to the fusion of an element of the smooth ER with a concentrating vesicle. Also visible is a peroxisome (P). This electron micrograph conveys an impression of the dynamic aspect of cell structure. (From A. Claude [1970], J. Cell Biol. 47:753.)

(roughly 20 to 100 nm in diameter) are visible, usually at the inner concave face of the membranous stacks and around the edges of the discs; the vesicles appear to be budded from the margins of the discs.

In plant cells, Golgi bodies (usually called *dictyosomes* by botanists) may be scattered throughout the cytoplasm; in animal cells, it is more typical for the Golgi bodies to be near the nucleus. The number of Golgi bodies in cells is highly variable, and depends upon both the cell type and the species. There are usually on the order of twenty Golgi bodies per cell, but many thousands have been reported in some cells, especially those involved in significant secretory activity.

Although the function of the Golgi apparatus is not well understood, it appears to play an important role in the storage, packaging, and secretion of certain cell products. It is involved in the formation of lysosomes and other enzyme-containing cellular inclusions, and in the formation of secretory granules in cells such as those found in the pancreas, the pituitary and mammary glands, and the mucus-secreting glands of the intestine, and in many other cell types.

Most of the proteinaceous substances secreted by the Golgi apparatus are glycoproteins; these are formed in the Golgi apparatus by the attachment of carbohydrate to the protein products of the endoplasmic reticulum. The secretion of substances by the Golgi apparatus is thought to occur in the following way: After enzymes are synthesized and accumulated in vesicles of the endoplasmic reticulum, the vesicles migrate to the Golgi apparatus, where they fuse with the Golgi membranes. In the Golgi apparatus, the newly synthesized products of the ER are concentrated, and some of the proteins are modified by the addition of carbohydrates or other prosthetic groups. The concentrated products then accumulate in the edges of the Golgi discs, where they are packaged into small vesicles that bud off the edges of the discs and are released into the cytoplasm. Here, smaller vesicles may fuse into larger ones. If the vesicles contain a secretory product, they will migrate to the plasma membrane, where they release their products into the surrounding environment by exocytosis. In other cases, lysosomes or other specialized membrane-bound structures that remain intact in the cytoplasm may be formed. This entire process is summarized in Figure 11–10.

The Golgi bodies have been implicated in the synthesis of many of the polysaccharides they secrete. For example, pectin and other mucilaginous substances of the plant cell wall are synthesized in the Golgi and are packaged in vesicles for secretion. There is convincing evidence to suggest that the carbohydrate portions of the glycoproteins secreted by Golgi bodies are probably synthesized within the Golgi itself.

One particularly intriguing aspect of Golgi function that remains a complete mystery is how the selectivity of packaging by the Golgi apparatus is determined. For instance, in a single cell, what mechanism determines that certain hydrolytic enzymes are packaged in lysosomes, and that secretory products are packaged separately in

Eukaryotic Cell Structure

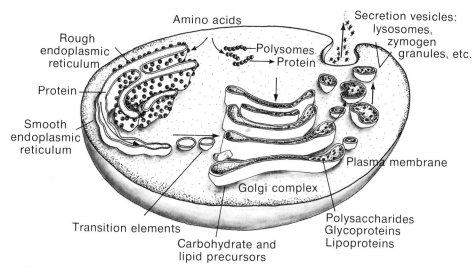

Figure 11-10 The role of the endoplasmic reticulum and the Golgi apparatus in the synthesis of membrane-bound products is summarized in this diagram. (From S.L. Wolfe [1972], *Biology of the Cell*, Belmont, California, Wadsworth Publishing Co., Inc.)

secretory granules? It is apparent that some kind of sorting process must take place.

There have been reports that in electron micrographs Golgi bodies occasionally appear to be connected with smooth endoplasmic reticulum. It is not clear, however, whether this is generally true; in many cases, the endoplasmic reticulum is seen to form vesicles which appear to migrate to the Golgi apparatus, suggesting that no structural continuity exists between the two systems. There is no reason, of course, why there could not be physical continuity between the endoplasmic reticulum and Golgi apparatus in some cells, but not in others.

Many workers have contended that Golgi bodies arise by differentiation of the endoplasmic reticulum. A highly speculative, but nevertheless intriguing, process of "membrane flow" has been proposed to account for the dynamic turnover of intracellular membranes. Because Golgi membranes are continually used up in the formation of vesicles, it has been suggested that Golgi membranes are constantly replaced by material from the endoplasmic reticulum which, in turn, is proliferated from the nuclear envelope.

Nuclear Envelope

At least in some cells, the nuclear envelope is continuous with the endoplasmic reticulum and therefore constitutes part of the external compartment. The nuclear envelope consists of two unit membranes about 10 nm thick, each of which resembles very closely the structure of the plasma membrane. The two membranes of the nuclear envelope are separated by a 10 to 30 nm space known as the *perinuclear space* (Fig. 11-11). The outer surface of the nuclear envelope is often lined with a layer of ribosomes.

The Structure and Origin of Eukaryotic Cells

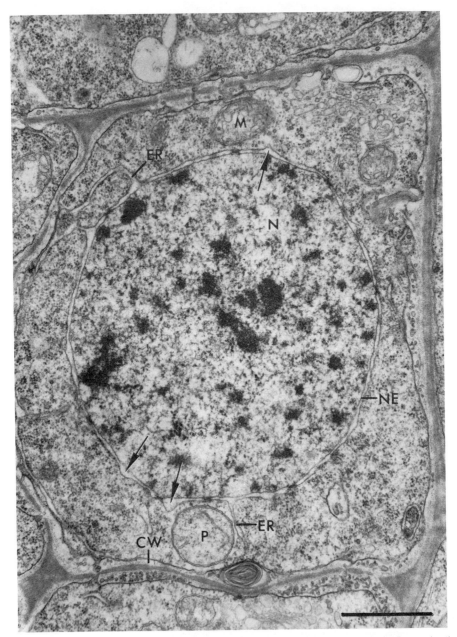

Figure 11–11 This electron micrograph of a cell from a liverwort (Bryophyte plant) shows the double-layered nuclear envelope (NE) and connections between it and the endoplasmic reticulum (ER). (From Z.B. Carothers [1972], J. Cell Biol. 52:273.)

The outer and inner membranes of the nuclear envelope fuse in many places to form numerous openings called "*pores,*" which are completely separated from the perinuclear space. The "pores" appear to be octagonal, with diameters of about 65 to 70 nm (Fig. 11–12).

The "pores" of the nuclear envelope are not, however, simple

Eukaryotic Cell Structure

Figure 11-12 "Pores" in the nuclear envelope as shown by freeze-etching. In this technique, a small piece of tissue is frozen to −100°C and then splintered. The lines of fracture tend to run through the cells along areas of weakness, particularly along the surface of membranes. The fractured preparation is then etched by evaporating water from it under vacuum, and a shadowed surface replica of the sample is prepared for viewing in the electron microscope. (×48,500) (From Bronk, J.R. [1973] *Chemical Biology*, New York, The Macmillan Co., Inc.)

open channels. They are filled with a somewhat electron-dense proteinaceous substance called the *annular material*. The annular material forms a "plug" which extends through the opening a short distance on either side of the nuclear envelope; because the diameter of the annular material is about 120 nm, it obscures the edge of the "pores" when the surface of the nuclear envelope is viewed. In the electron microscope, the annular material appears less dense toward the center of the "pore" opening, and consequently the annular material is often described as, although in fact it is not, a hollow cylinder of material that extends around and through the opening. The entire structure—the "pore" formed by the fusion of the membranes of the nuclear envelope, and the annular material that surrounds and fills the opening—is called the *pore complex* (Fig. 11-13). The structure of the pore complex is strikingly similar

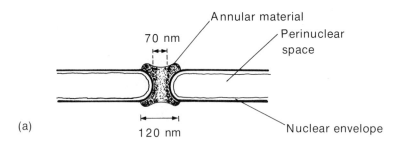

(a)

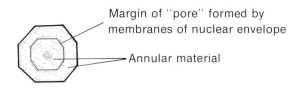

(b)

Figure 11–13 a. Cross-sectional diagram of the pore complex, showing the relationship of the annular material to the "pore" in the nuclear envelope. b. A pore complex, as it would appear to an observer looking down on the surface of the nuclear envelope if the annular material were transparent.

in cells of different species, but the number of pores in the nuclear envelope may vary greatly from species to species.

Material passing from the nucleus to the cytoplasm, or in the opposite direction, must traverse the nuclear envelope. We know, for example, that RNA- and protein-containing complexes that are precursors to cytoplasmic ribosomes are produced within the nuclear region, and that these complexes must pass through the nuclear membrane in order to enter the cytoplasm. However, studies on the transport of ions between the nucleus and cytoplasm have indicated that, at least at times, the pore complex is closed to free diffusion of ions and other molecules. One piece of evidence supporting the latter conclusion is the observation that nuclei exhibit osmotic properties; that is, they swell when placed in a hypotonic solution and shrink in a hypertonic solution. However, the permeability of the nuclear envelope is variable and differs according to the metabolic activity and developmental stage of the cell. Surprisingly, changes in permeability are not accompanied by observable morphological changes in the pore complex.

In still other studies, Carl Feldherr injected gold particles into the cytoplasm of living amoebae and was able to demonstrate by electron microscopy that particles as large as about 14.5 nm could pass from the cell's cytoplasm into the nucleus. When this occurred, the gold particles were visible only in the center of the pore complex. Since particles larger than 14.5 nm could not enter the nucleus, it was assumed that a central channel of about that diameter existed in the pore complex, which, under certain conditions, allowed materials to pass through it.

Somehow, then, the annular material must regulate the passage of substances through the pore complex. Many researchers believe that the pore complexes are special structures whose function is to permit the exchange of very large molecules between the nucleus and the cytoplasm while maintaining the integrity of the nuclear compartment. Some recent evidence suggests that the nuclear pores may be able to distinguish between different types of RNA molecules, and may allow only some of these to pass into the cytoplasm while restricting others to the nucleus. The mechanism by which this may occur is, however, unknown.

The passage of large molecules or molecular aggregates through the pore complex probably occurs by some energy-requiring process. Studies on the localization of enzymes within regions of the cell have shown that an enzyme catalyzing the hydrolysis of ATP is present in relatively high concentrations in the regions of the pore complexes. Whatever the mechanism of transport, substances that pass through the pore complexes are probably restricted to an area of about 14.5 nm in the center of the complexes.

In Chapter 18, we will return to a discussion of the role of the nuclear envelope and pore complex in the transport of macromolecules between the nucleus and the cytoplasm.

INTRACELLULAR INCLUSIONS OF THE INTERNAL COMPARTMENT

Nucleus

The *nucleus* was discovered by the Scottish botanist Robert Brown in 1831, and is now recognized as the structure that contains the hereditary material of the cell. Most of the nucleus is filled with fibrils which may vary in diameter from about 3 to 30 nm, depending on the method of preparation; these fibrils constitute the *chromatin strands* of the nucleus. Chromatin has been shown by chemical analysis to consist of double-stranded DNA (two hydrogen-bonded polynucleotide strands; see Chapter 7) to which large amounts of protein and a small amount of RNA are bound. The proteins are primarily histones, a class of strongly basic proteins (positively charged at neutral pH), although a few acidic proteins are also present. The DNA molecule has a diameter of about 2 nm, and with the attached RNA and protein, a single chromatin fiber is 3.0 to 3.5 nm in thickness. Chromatin tends to coil spontaneously to form strands 10 to 30 nm in diameter, and it is these strands that are visible in some electron micrographs (Fig. 11–14). Occasionally, the chromatin fibers are seen anchored to the inside of the nuclear envelope at the edges of the annular material.

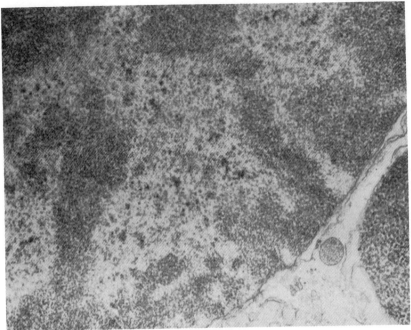

Figure 11-14 Electron micrograph of part of a nucleus, showing relatively electron-dense areas of chromatin fibers. In many cases, the fibers are cut in cross-section and appear as dots. (×48,500) (From S.L. Wolfe [1972], *Biology of the Cell,* Belmont, California, Wadsworth Publishing Co., Inc.)

In an *interphase cell,* that is, one that is not in the process of dividing, the chromatin may exist in two states. Some of it may remain in an extended, dispersed state called *euchromatin;* under the electron microscope, euchromatin appears as relatively electron-transparent masses of 10 nm fibers. Other chromatin, known as *heterochromatin,* is tightly coiled, and in electron micrographs appears as extremely electron-dense aggregates of material (Fig. 11–15).

In the nuclear ground substance, or *nucleoplasm,* of the interphase nucleus in many cells, several types of small granules and fibrils, distinct from the chromatin, are also visible. These particles include *perichromatin granules,* which are 40 to 45 nm in diameter and closely associated with chromatin, and *interchromatin granules,* which are smaller particles (20 to 25 nm in diameter) present in regions between the chromatin. Fibrillar structures known as *perichromatin fibers* and *coiled bodies* are often visible. By chemical analysis, all these intranuclear structures have been shown to be composed of RNA complexed with protein, but their precise identity and function are not known.

When a cell begins to divide, the chromatin strands lose their attachments to the nuclear envelope, the nuclear envelope breaks up, and the chromatin strands condense into a number of discrete bodies which, after staining, are easily visible in the light microscope. These bodies are termed *chromosomes* (chromo = colored;

Eukaryotic Cell Structure

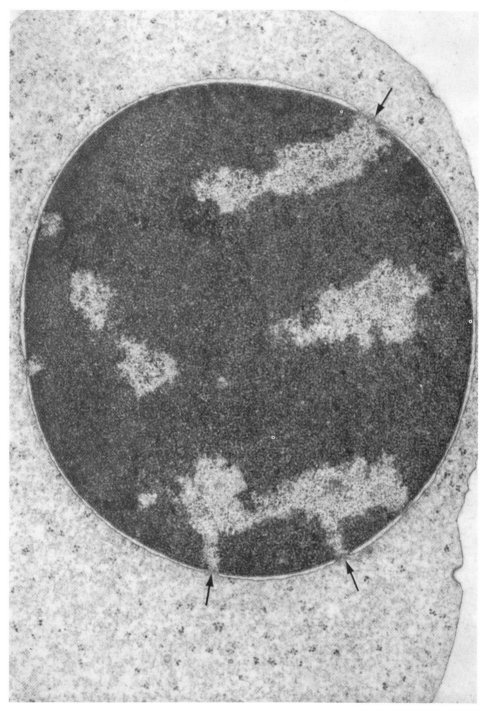

The nucleus of a developing red blood cell from guinea pig bone marrow. Most of the chromatin is aggregated into extremely dense masses of heterochromatin; there are a few areas of less densely coiled chromatin (euchromatin) mostly near the pore complexes (arrow). (×17,000) (From D.W. Fawcett, [1966], *The Cell: An Atlas of Fine Structure,* Philadelphia, W.B. Saunders Co.)

Figure 11-15

somes = bodies). Once the chromatin is condensed into chromosomes, it can be seen that each chromosome has replicated during interphase and consists of two identical strands called *chromatids,* which are united at one point called the *centromere.* During cell division, chromosomes are distributed among daughter cells. This process is termed *mitosis* when it involves the division of any type of cell *except* sex cells; it is called *meiosis* when eggs and sperm are being formed by the division of sex cells. Mitosis and meiosis are extremely important cellular processes, and we will describe them in detail in Chapter 17.

Nucleolus

Inside the nucleus are one or more small bodies of variable size called *nucleoli* (singular: *nucleolus,* meaning "small nucleus"). The nucleolus has no membrane of its own, but is considerably more dense than the surrounding nucleoplasm and hence is distinctly visible. It consists of areas of granules (15 to 20 nm particles, somewhat smaller than ribosomes) and fibrils (about 5 to 10 nm in diameter) suspended in a proteinaceous ground substance (Fig. 11–16). Areas of chromatin are also visible within the nucleolus; the chromatin material can be recognized because it is considerably less electron-dense than the fibrillar region of the nucleolus.

In most cells, the chromatin material in the nucleolus consists of a section of chromatin called the *nucleolar organizer region.* The chromatin of the nucleolar organizer becomes associated with fibrillar and granular material to form the nucleolus. During cellular reproduction, when the chromatin is condensed into discrete chromosomes, nucleoli usually disappear, but they re-form later in the interphase nuclei of the new daughter cells. There may be more than one nucleolar organizer region among the chromatin strands in a nucleus, and hence more than one nucleolus may be present in a single cell.

The nucleolus is known to be the cellular site for the synthesis of ribosomal RNA, the RNA components of the ribosomes. The chromatin material of the nucleolar organizer carries the information that directs the formation of the nucleolus and of the ribosomal RNA. In the nucleolus, the newly synthesized ribosomal RNA (probably the fibrillar material of the nucleolus) combines with proteins, which are apparently synthesized in the cytoplasm and transported to the nucleolus, to form particles (the nucleolar granules) that are precursors to cytoplasmic ribosomes. The granules pass from the nucleolus to the nucleoplasm and then through the pores of the nuclear envelope into the cytoplasm. Thus, unlike prokaryotes, in which ribosome formation occurs in the cytoplasmic ground substance, eukaryotes manufacture ribosomes in the nucleolus.

The nucleolus is a highly dynamic structure. Cells active in protein synthesis, with a consequent large requirement for ribosomes, may have large nucleoli and/or a number of nucleoli. Nucleoli are absent in certain specialized cell types (such as sperm) that exhibit

Eukaryotic Cell Structure

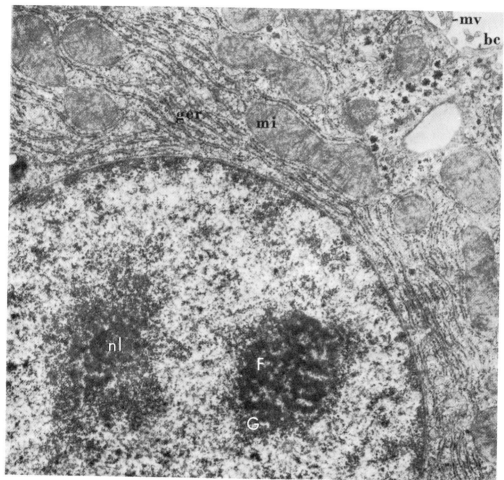

Portion of a liver cell, showing two nucleoli (*nl*) within the nucleus. Note the electron-dense fibrillar regions (*F*) and the granular regions (*G*) of each nucleolus. Also shown are mitochondria (*mi*) and rough endoplasmic reticulum (*ger*). Glycogen granules appear as rosettes towards the periphery of the cell. (From E.B. Sandborn [1970], *Cells and Tissues by Light and Electron Microscopy,* New York, Academic Press, Inc.)

Figure 11–16

little or no protein synthesis, although the nucleolar organizer region is, of course, present.

In some cells, particularly in the developing egg cells (oocytes) of amphibians, the rate of protein synthesis is unusually high, and there may be many nucleoli. It has been reported that some amphibian oocytes may possess more than a thousand nucleoli per nucleus. Since the chromatin of a single nucleus contains, at most, only a few nucleolar organizer regions, each of these nucleoli cannot be associated with one of the original nucleolar organizers. Instead, the nucleolar organizer region is replicated many times to produce small pieces of DNA, and one of these pieces is contained in each nucleolus. In this way, each nucleolus is built around a copy of the nucleolar organizer region.

Ribosomes

The ribosomes of eukaryotic cells are not only found attached to the endoplasmic reticulum but may be free in the cytoplasm as single units or as polyribosomes. Eukaryotic ribosomes are larger than those in prokaryotes, and are about 50 per cent RNA and 50 per cent protein. They consist of a 40S and a 60S subunit, which together make up an 80S ribosome. The 40S subunit is composed of a strand of 18S RNA in addition to a number of structural proteins; the 60S subunit contains a strand of 28S RNA, a strand of 5S RNA, and associated proteins. The structural proteins of eukaryotic ribosomes have not been well studied, and it is not known precisely how many different proteins are present in each ribosomal subunit and how much variability exists in the ribosomal proteins of different species. However, it is generally assumed that the number of proteins composing ribosomes in eukaryotes is probably about the same as the number in prokaryotic ribosomes.

Lysosomes

Lysosomes are membrane-bound vesicles, anywhere from about 0.15 to 0.8 μm in diameter, which appear dense and finely granular in the electron microscope; at times, the material in a lysosome crystallizes to form regular arrays. Lysosomes are a combined product of the endoplasmic reticulum and the Golgi apparatus, and contain a large number of hydrolytic enzymes that collectively are capable of hydrolyzing every class of biological macromolecule.

Interestingly, lysosomes were discovered initially by Christian deDuve in 1952, not by microscopy, but in centrifuged extracts of the cytoplasm. DeDuve, studying enzyme levels in various parts of the cell, noticed that when cells were gently disrupted and then centrifuged, a fraction of the cytoplasm could be obtained that sedimented more slowly than mitochondria and released a hydrolytic enzyme (acid phosphatase) upon standing for several days. It thus seemed probable that the acid phosphatase was normally present in some type of membrane-bound structure and was released only when the structure was disrupted. In 1955, Alex Novikoff identified lysosomes in rat liver cells by electron microscopy.

Lysosomes are particularly abundant in leukocytes (white blood cells), amoebae, and other cell types that ingest large quantities of extracellular material. They are also abundant in damaged cells or degenerating cells, and in cells undergoing metabolic (and structural) readjustments. For example, when a tadpole's tail is resorbed during metamorphosis of the tadpole into an adult frog, the cells of the tail have large numbers of lysosomes.

It has become clear that, in addition to their role in the digestion of extracellular material (mentioned in our discussion of endocytotic vesicles), lysosomes are important in the "recycling" of materials both within the cell and between cells of a multicellular organism. For example, a cell that, because of reduced energy requirements, no longer needs as many mitochondria as it contains, is

Eukaryotic Cell Structure

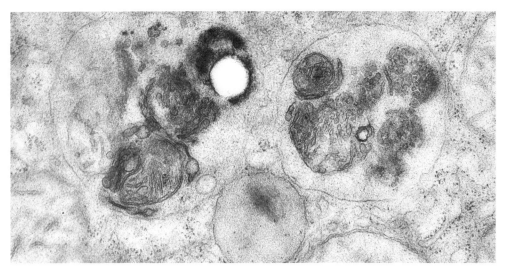

Figure 11-17 Mammalian liver cell lysosomes containing partly digested mitochondria. (×26,500) (From J.T. Dingle and H.B. Fell [1969], *Lysosomes in Biology and Pathology*, Vol. II, New York, American Elsevier Publishing Co.)

able to digest the excess mitochondria with lysosomal enzymes and use the resultant materials for other purposes (Fig. 11–17). Thus, internal restructuring of a cell is aided by lysosomes. In the same way, the self-digestion of damaged cells or of senescent cells in multicellular organisms provides raw materials that can be reused by the organism. The observation that lysosomes often release their contents into the cytoplasm previous to or following death has prompted deDuve to refer to lysosomes as "suicide bags."

Microbodies

In addition to lysosomes, there are other membrane-bound vesicles, called *microbodies,* which are present in the cytoplasm of many eukaryotic cells and are a product of the endoplasmic reticulum—Golgi system. Two common types of microbodies are *peroxisomes* and *glyoxysomes.*

Peroxisomes are involved in the oxidation of a variety of substances in what appears to be a two-step reaction (Fig. 11–18). In the first step, oxidase enzymes present in peroxisomes catalyze the

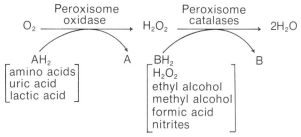

Figure 11-18 Cellular oxidations carried out by peroxisomes, as suggested by C. DeDuve and P. Baudhuin (1966), Peroxisomes, Physiological Rev. *46*:323.

oxidation of amino acids, uric acid, lactic acid, and other substances, using molecular oxygen as the oxidizing agent and forming hydrogen peroxide (H_2O_2) in the process. In the next step, hydrogen peroxide in turn is used as an oxidizing agent to oxidize another group of substances in reactions catalyzed by the enzyme catalase. These latter substances may include another molecule of H_2O_2 (which is oxidized to O_2) or of ethyl alcohol, methyl alcohol, nitrites, and formic acid (among others).

Peroxisomes have been found in yeasts, in protozoans, and in mammalian liver and kidney cells, as well as in other cell types (Fig. 11–19). It has been estimated that as much as 20 per cent of the oxygen (O_2) consumed in liver cells is utilized in peroxisomes. The function of the oxidation–reduction reactions that take place in peroxisomes is unclear; there is no evidence that oxidation in peroxisomes is coupled to energy capture. One possible role of these reactions may be removal of H_2O_2 formed in other cellular metabolic processes; H_2O_2 is known to have destructive effects on cells. Another possibility is that they may serve to regulate the concentration of certain metabolites in the cell.

Glyoxysomes are often found in microorganisms and in higher plants. In addition to catalase and oxidase enzymes, glyoxysomes contain enzymes of the glyoxylate cycle, a metabolic pathway that converts fats into carbohydrates.

Vacuoles

Both plant and animal cells may contain a number of different kinds of membrane-bound *vacuoles* (large vesicles). For example, fresh-water protozoans are hypertonic to their environment and possess *contractile vacuoles* to remove water, which tends to diffuse into the cell through the plasma membrane. Excess water is collected from the cytoplasm in the contractile vacuole and is then expelled from the cell through a small surface pore in the vacuole by a rapid contraction of the vacuole. Many plants have a large central vacuole (Fig. 11–2) which occupies most of the volume of the cell,

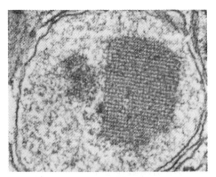

Figure 11–19 Peroxisome of mammalian kidney containing regular arrays of paracrystalline material. (From Z. Hraban and M. Rechcigl [1969], *Microbodies and Related Particles,* New York, Academic Press, Inc.)

often as much as 90 per cent; it contains cell sap, which consists of various organic molecules and ions dissolved in water.

Mitochondria

Almost all eukaryotic cells, except for a few highly specialized cell types that do not obtain their energy by aerobic metabolism, contain one or more *mitochondria* (singular: *mitochondrion*). The function of mitochondria is to generate ATP by oxidizing selected nutrient molecules in the presence of molecular oxygen. Consequently, mitochondria are the source of almost all of the ATP produced in a nonphotosynthetic cell and have therefore been called the "powerhouses" of the cell. Cells capable of photosynthesis also possess mitochondria, but a large proportion of the ATP utilized by these cells is produced by capturing energy from the rays of the sun — a process that does not involve mitochondria.

The shape, size, and number of mitochondria in a cell are extremely variable and depend upon both the type of the cell and its metabolic state. Mitochondria are typically rod-shaped, about 0.2 to 1.0 μm in diameter and 3 to 10 μm long. However, some mitchondria may be very much larger than this; for example, mitochondria 50 to 60 μm in length have been reported. There are spherical, helical, ringshaped, and starshaped mitochondria, as well as some with other unusual morphology. As a general rule, cells that expend a great deal of energy have many large mitochondria, whereas the mitochondria of cells with low energy requirements are small and less numerous. Mitochondria are also observed in regions of the cell where there is greatest energy demand. For example, in sperm cells the mitochondria surround the axial filament of the flagellum and hence provide energy for flagellar motion; in skeletal muscles, the mitochondria are in close contact with the contractile fibers. Whatever their form, concentration, or position within the cell, mitochondria are dynamic structures. The mitochondria of living cells observed under the phase contrast microscope may change shape frequently and may be seen to branch, divide, and coalesce.

Structurally, mitochondria are composed of two membranes: a smooth *outer membrane* which covers the entire surface, and an *inner membrane*. The inner membrane is most frequently convoluted to form a large number of folds called *cristae* (= ridges), although in the mitochondria of some cells, the inner membrane may form a system of tubules (Fig. 11–20).

The membranes of the mitochondrion separate it into two distinct compartments. An *outer compartment,* sometimes called the *intermembrane space,* is formed between the outer and inner membranes. An *inner compartment,* also known as the mitochondrial *matrix,* is bounded by the inner membrane. Enzymes and other substances needed for aerobic metabolism are localized on the mitochondrial membranes or in the compartments of the mitochondrion.

High-resolution electron microscopy of preparations fixed with

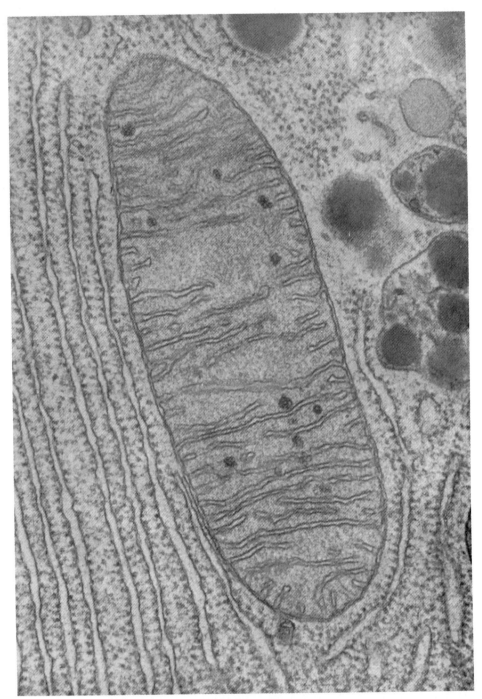

Figure 11–20 *a.* Mitochondrion from a cell of the bat pancreas. Note the smooth outer membrane and the highly folded inner membrane which forms the cristae. Intramitochondrial granules of unknown function are also visible. (×95,000)

Continued on the opposite page.

Eukaryotic Cell Structure

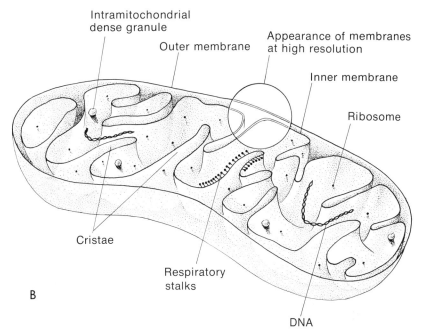

b. A three-dimensional sketch of a typical mitochondrion, showing the relationship of the inner membrane to the outer membrane. (From: [a] D.W. Fawcett [1966], *The Cell: An Atlas of Fine Structure,* Philadelphia, W.B. Saunders Co.; [b] B. Tandler and C.L. Hoppel [1972], *Mitochondria,* New York, Academic Press, Inc.)

Figure 11-20 (*Continued*)

osmium or permanganate shows the mitochondrial membranes to be about 6 nm in thickness and to have a distinctly granular appearance, quite different from that of a typical unit membrane. This granularity has been interpreted to mean that the mitochondrial membranes consist of arrays of spherical particles about 5 nm in diameter. Each particle appears to consist of an electron-transparent core surrounded by an electron-dense region.

By the use of negative staining,* the inner surface of the cristae are seen to be covered with small projections, each of which consists of a spherical particle (8 to 9 nm in diameter) connected to the membrane by a short stalk (Fig. 11–21). These structures are similar to those seen on the inner side of the plasma membrane of some aerobic bacteria. The spherical particles have been called "*elementary particles*" or "*respiratory assemblies*," and have been shown to contain enzymes involved specifically in the formation of ATP. Because the elementary particles are not usually visible by ordinary staining and fixation methods, some investigators contend that they are not normally present on the surface of the cristae, but rather are extruded from within the membrane by phosphotungstic acid during negative staining.

In addition to soluble enzymes associated with aerobic metabolism, the matrix of the mitochondrion often contains large spherical

---

*In negative staining techniques, an unstained specimen is embedded in an electron-dense substance, usually phosphotungstic acid. The specimen is visible in the electron microscope because of the contrast between it and the dense background.

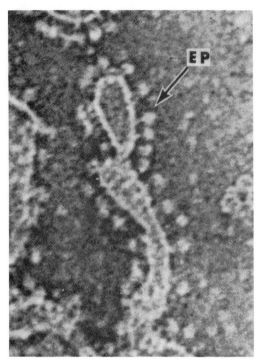

Figure 11-21 Mitochondrial cristae of a beef heart mitochondrion as seen in the electron microscope by negative staining. Projections called elementary particles are clearly visible on the membrane surface. (×420,000) (From H. Fernández-Morán, T. Oda, P.V. Blair, and D.E. Green [1964]. J. Cell Biol., 22:63.)

granules, 25 to 35 nm in diameter, known as *intramitochondrial granules* (Fig. 11-20). The function of these granules is not known. Also present in the matrix are ribosomes, RNA, and a small quantity of DNA. The ribosomes in mitochondria are smaller than cytoplasmic eukaryotic ribosomes and more nearly the size of prokaryotic ribosomes. It has been shown that the mitochondrion is capable of protein synthesis and that the DNA it contains codes for the synthesis of some, *but not all,* of the mitochondrial proteins. Mitochondria have also been shown to grow and reproduce by binary fission at a rate that is not necessarily synchronized with cell division.

Because mitochondria are able to replicate and to synthesize some of their own proteins, mitochondria are thought of as semi-autonomous cellular inclusions. However it is important to realize that they are not independent structures. They are a component of the cell, and although they possess DNA and can replicate, their structure and functional activities are determined to a considerable degree by information coded within the nuclear chromatin.

Recent research on a particular strain of the yeast *Saccharomyces* has produced some extremely interesting results with regard to the structure of mitochondria. Electron micrographs of *Saccharomyces* appear to show several mitochondria in one cell. However, if consecutive serial sections of a single cell are examined in the electron microscope, and a three-dimensional picture of the

Eukaryotic Cell Structure

cell is reconstructed from the individual electron micrographs, it can be shown that this particular strain of *Saccharomyces* possesses one giant, branched mitochondrion per cell. Thus individual electron micrographs of thin sections of *Saccharomyces* give the *impression* of several mitochondria only because branches of the single mitochondrion are cut in cross section. Although other workers have since shown that the number of mitochondria in *Saccharomyces* is dependent upon the growth conditions of the cell, and that under some conditions 50 to 100 small mitochondria may be visible in a single cell of *Saccharomyces,* the original observation raises the possibility that what appear to be many mitochondria in at least some eukaryotic cells may actually be parts of a single very large mitochondrion.

Plastids

Eukaryotic plants contain a heterogenous group of cell inclusions called *plastids,* which vary in morphology and function. Some plastids are colorless and are called *leucoplasts.* These are specialized to store reserve material such as starch, proteins, and lipids. Other plastids, termed *chloroplasts,* are green because they contain chlorophyll; still others, the *chromoplasts,* contain carotenoid pigments which are a brilliant red or yellow.

All three types of plastids differentiate from small, spherical amoeboid structures, 0.4 to 0.9 μm in diameter, called *proplastids.* In addition, chloroplasts may develop from leucoplasts, and chromoplasts, which are considered end forms of plastid differentiation, may develop from either leucoplasts or chloroplasts. A scheme summarizing plastid differentiation is shown in Figure 11–22. Since proplastids can differentiate into one of three types of plastids and since, in certain cases, one type of plastid can differentiate into another, it has been generally assumed that all plastids are essentially the same structure, with the ability to differentiate in various ways, depending upon the requirements of the cell.

Because it is the site of photosynthesis in eukaryotic cells, the chloroplast is the most widely studied of the plastids. Chloroplasts are generally oval bodies approximately 4 to 10 μm long and 2 to 3 μm thick, and in higher plants there are usually 20 to 40 per cell. However, in many algae there may be only one or a few chloroplasts, and these may be in the shape of a star or a spiral ribbon, or may have other exotic forms.

Like all plastids, chloroplasts are bounded by two concentric

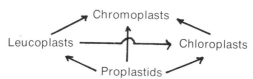

Figure 11–22 Scheme summarizing plastid differentiation. The manner in which a plastid differentiates apparently depends upon the metabolic requirements of the cell.

membranes about 6 nm thick. The two membranes enclose an inner region called the *stroma,* in which a vast network of membranes is suspended. These membranes, consisting of flattened vesicles called *thylakoids,* are not usually continuous with the inner chloroplast membrane. The thylakoids enclose an interior space known as the *intrathylakoidal space.* All the enzymes and other substances associated with the process of photosynthesis are either bound to the membranes of the chloroplast or are soluble in the stroma.

There may be great variation in the arrangement of thylakoids in different cell types. In lower eukaryotic plants, the thylakoids may occur singly in the stroma or may be fused in vast parallel arrays. Particularly in the chloroplasts of higher plants, and in some green algae as well, thylakoid membranes may form stacks of disclike structures termed *grana,* which are interconnected by extensions of the thylakoids, called *frets* or sometimes *stromal lamellae* (Fig. 11–23). Although it was originally believed that photosynthetic pigments are located only within the grana, there is now evidence that all the membranes of the thylakoid system contain photosynthetic pigments.

The ultrastructure of the thylakoid membrane has not been described in detail, and much contradictory evidence has accumulated. Nevertheless, the following tentative picture emerges. Arrays of spherical particles about 11 nm in diameter may form the framework of the membrane. Embedded on the surface of this membrane, or interspersed among the particles of the membrane framework, are other particles called *quantasomes* (Fig. 11–24). These are roughly rectangular (18.0 nm $\times$ 15.5 nm $\times$ 10.0 nm) and appear to consist of four subunits. The quantasomes are composed of about equal amounts of lipid and protein, and contain chlorophyll, carotenoids, and other components of the photosynthetic apparatus. It has been suggested that one part of the photosynthetic process—the conversion of light energy into chemical energy—may take place exclusively within the quantasomes.

In addition to dissolved salts and enzymes of photosynthesis, the stroma of the chloroplast, like the matrix of the mitochondrion, contains RNA, DNA, and ribosomes, and is capable of carrying on protein synthesis. The chloroplast ribosomes are the same size as ribosomes in prokaryotes. Chloroplasts are also like mitochondria in being semi-autonomous. They can grow and divide, and their DNA contains a portion of the genetic information needed for the synthesis of chloroplast proteins. Hence the structural and functional properties of chloroplasts are only partially determined by information stored in the nuclear chromatin.

Filamentous Structures of the Cytoplasm

FILAMENTS. Filaments of indefinite length appear in many cell types. These generally fall into two groups..

Microfilaments are 5 to 7 nm in diameter and are often found in clusters immediately beneath the plasma membrane (Fig. 11–25).

Eukaryotic Cell Structure

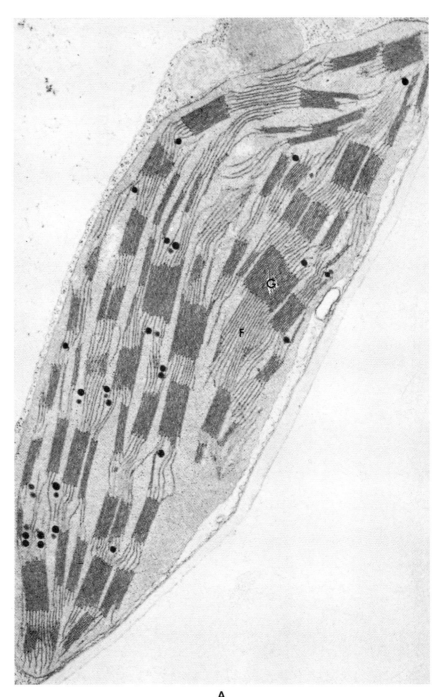

A

a. Electron micrograph of a mature chloroplast in a leaf cell of timothy grass. The grana (G) are visible as stacks of thylakoid membranes which are interconnected by membranes known as frets (F) or stromal lamellae. The internal membrane system is suspended in the stroma, which contains numerous ribosomes, visible as very small dots, as well as larger dense particles, called osmiophilic granules, whose function is not known. (×22,000) (Photo by William P. Hergin, Department of Agriculture, Beltsville, Maryland. Reproduced by permission.)

Figure 11–23

Illustration continued on the following page

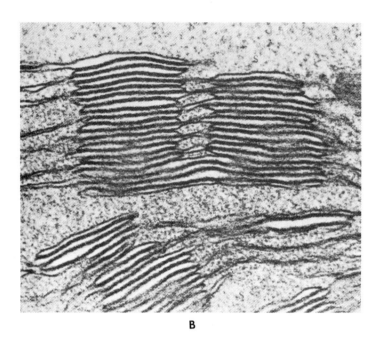

B

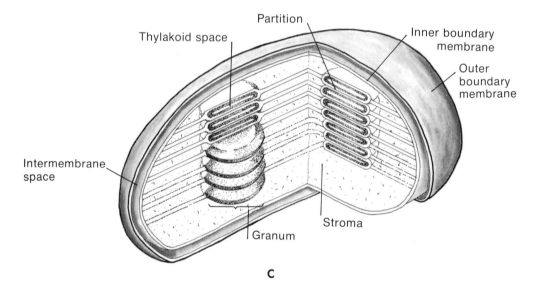

C

Figure 11-23 *b*. High-power electron micrograph of thylakoid membranes in the sunflower chloro-
(*Continued*) plast. Note the lamellar structure of the grana and the interconnecting frets. (×93,000) (Courtesy of Dr. Harry J. Horner, Jr., Iowa State University. From S.D. Gerking [1974]. *Biological Systems,* 2nd Ed., Philadelphia, W. B. Saunders Co.) *c*, A three-dimensional sketch of a chloroplast in a higher plant. (Adapted from Wehrmeyer [1964] Planta *62*: 272.)

Eukaryotic Cell Structure

An array of quantasomes in a thylakoid disc of a spinach chloroplast, as revealed by freeze-etching. Fracturing has removed part of the outer membrane, exposing quantasomes underneath. (From: W.A. Jensen and R.B. Park [1967], *Cell Ultrastructure*, © 1967 by Wadsworth Publishing Co., Inc., Belmont, California. Reprinted by permission of American Scientist, Journal of Sigma XI, The Scientific Research Society of North America.)

Figure 11-24

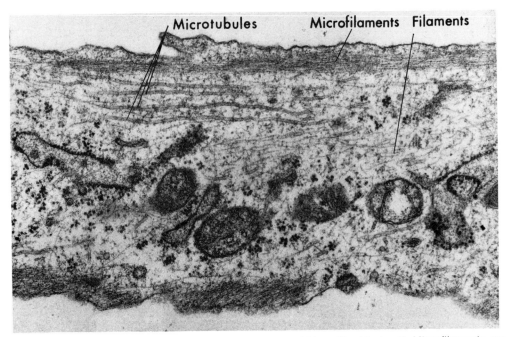

Figure 11-25 Longitudinal section through a part of a kidney fibroblast cell. Microfilaments are visible in bundles beneath the cell membrane. Also evident are microtubules and 100 Å filaments. (From R.D. Goldman [1971], J. Cell Biology 51:752.)

Some cell types such as neurons possess larger filaments 8 to 10 nm in diameter. A microfilament is composed of a single strand of identical globular protein subunits; these subunits are extremely similar if not identical to the protein *actin,* which composes thin filaments in muscle cells. There is indirect evidence to suggest that microfilaments are involved in movement. For example, during cell division the separation of the cytoplasm into two cells begins with the formation of a *cleavage furrow* around the dividing cell; the cleavage furrow becomes deeper and deeper as it divides the cytoplasm. It has been shown that furrow formation is inhibited by cytochalasin B, an antibiotic that destroys the microfilaments. Hence it has been proposed that microfilaments are intimately involved in the movement associated with furrow formation. Many other cellular processes, including cytoplasmic streaming in some plant cells and cell migration during early stages of embryonic development, are inhibited by cytochalasin B, suggesting that these processes may be brought about by microfilaments. The mechanism by which microfilaments produce movement of this type is largely unknown.

MICROTUBULES. Microtubules are long, cylindrical structures, usually about 25 nm in diameter (Fig. 11–25). In most cases the cylinder has a hollow core, although the core sometimes appears to contain electron-dense material. Electron micrographs of microtubules from a variety of different cell types demonstrate that the wall of the cylinder is composed of thirteen smaller filaments, called *protofilaments.* Each protofilament is 5 nm in diameter, and is itself a polymer of globular protein subunits termed tubulins (Fig. 11–26).

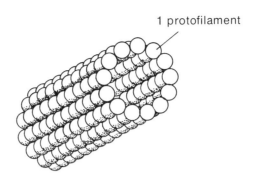

Model of microtubule structure. **Figure 11-26**

However, the protein subunits composing microfilaments and the tubulins composing the protofilaments of microtubules are different in amino acid composition, amino acid sequence, and molecular weight, and therefore the two types of filaments are not identical.

Two general classes of microtubules are recognized: *labile microtubules,* which are easily disrupted by treating the cell with specific inhibitors (e.g., colchicine or vinblastine), by changing the temperature, and by applying other treatments; and *stable microtubules,* which are not easily disrupted. The stable microtubules are components of centrioles, basal bodies, cilia, and flagella (see below). Labile microtubules are seen in many cells distributed at random throughout the cytoplasm, and also form the spindle fibers of mitosis. Spindle fibers attach to the duplicated chromosomes and are responsible for the separation of chromosomes during mitosis so that a complete set of chromosomes is incorporated into each daughter cell. We will discuss the role of spindle fibers in detail when we consider mitosis in Chapter 17.

It is becoming increasingly clear that microtubules are involved in a wide variety of cellular processes. In addition to their role as spindle fibers and as important structural components of centrioles, basal bodies, cilia, and flagella, microtubules are also responsible for determining the shape of certain specialized cells by providing internal support. For example, the biconcave disc shape of circulating red blood cells is maintained by a "cytoskeleton" (cyto = cell) of microtubules. Microtubules have also been implicated in the intracellular transport of materials, such as the release of insulin from pancreatic β-cells, the movement of melanin granules in melanocytes, and many other transport processes.

In spite of the large amount of research which is currently being done on microtubules, we do not understand how microtubules bring about their various effects. Nor are we completely sure how microtubules are formed. Some very recent evidence suggests that growth of a new microtubule occurs by addition of tubulin subunits to a growth point at the distal end of the microtubule, but we still do not know how the formation of microtubules is regulated by the cell.

Centrioles and Basal Bodies

Near the nucleus of most animal cells and some algae are two or more hollow, cylindrical structures called *centrioles.* These are about 0.2 μm in diameter and 2.0 μm in length, and are usually found lying at right angles to one another in pairs (Fig. 11–27a).

Centrioles consist of nine sets of microtubules forming the edge of the cylinder. There are three microtubules in each set, and the sets are joined by fine, fiberlike connections. (Fig. 11–27b).

As we shall see in our discussion of mitosis, centrioles appear to play a template role in the formation and organization of spindle fibers during cell division. Centrioles are to some extent semi-autonomous structures, and in this respect are similar to mitochondria and chloroplasts. There is some debate as to whether centrioles contain their own DNA, but in any case they are clearly able to direct their own replication. In addition to their role in cell division, centrioles will replicate repeatedly under certain conditions to give rise to identical structures that migrate toward the plasma membrane of the cell and form *basal bodies,* units from which cilia and flagella develop. Note that the basal bodies of eukaryotic cells are much larger than the basal bodies of prokaryotes and are also completely different in structure.

Cilia and Flagella

Eukaryotic cells possess two types of motile projections: *cilia* and *flagella.* Cilia and flagella have identical structures; cilia are simply small flagella. Although cilia are commonly about 2 to 20 μm

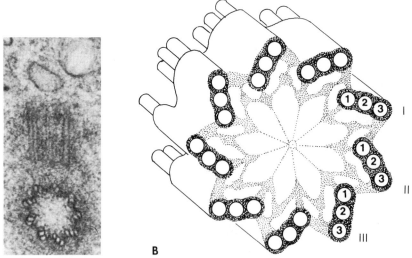

Figure 11–27 a. Two centrioles lying at right angles to one another; one is cross-sectioned, the other is sectioned longitudinally. (×73,000) (From J.B. Rattner and S.E. Philips [1973], J. Cell Biol. *57*:368. b. Diagram of the cross-sectional structure of a centriole. There are nine groups of three microtubules. (From S.L. Wolfe [1972], *Biology of the Cell,* Belmont, California, Wadsworth Publishing Co., Inc.)

Eukaryotic Cell Structure

in length, flagella may be hundreds, or in some cases, even thousands of micrometers long. A flagellated eukaryotic cell usually possesses only one or two flagella, but ciliated cells possess many cilia. In unicellular eukaryotes, cilia and flagella are important in cell motility and often in capturing food. In multicellular organisms, flagella are involved in the movement of various specialized cell types such as sperm, and cilia are often used to move substances past stationary cells. The classic example of this latter function occurs in lung tissue, where ciliated cells move dust and other small particles out of the lung.

Flagella and cilia are extensions of basal bodies and are usually covered by outfoldings of the cell's plasma membrane (Fig. 11-28). The internal structure of motile appendages is strikingly similar in all

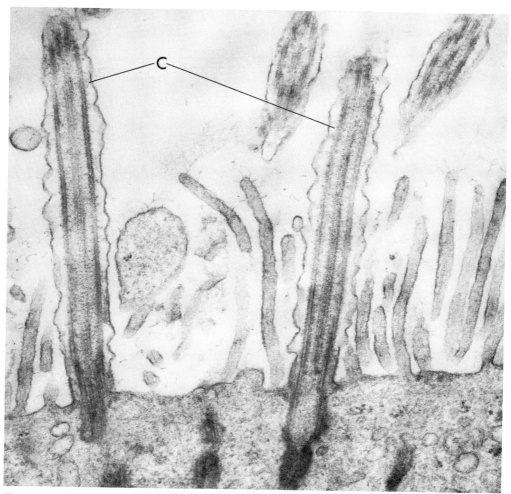

Figure 11-28 Electron micrograph of a portion of a ciliated cell lining the trachea. The cilia (C) are surrounded by a loose sheath formed by an outfolding of the plasma membrane, and contain microtubules arranged in a regular pattern. The small projections at the surface of the cell are microvilli, which have irregular filamentous tufts at the apexes. (×60,000) (From E.B. Sandborn [1970], *Cells and Tissues by Light and Electron Microscopy*, New York, Academic Press, Inc.)

eukaryotes; in fact, the characteristic structure of eukaryotic cilia and flagella is one morphological feature that clearly distinguishes prokaryotes from eukaryotes, and has given rise to the hypothesis that all eukaryotes are derived from an ancestral flagellated eukaryotic cell.

In cross-section under the electron microscope, all cilia and flagella are seen to consist of nine pairs of microtubules peripherally arranged, usually around two central microtubules (Fig. 11–29). This

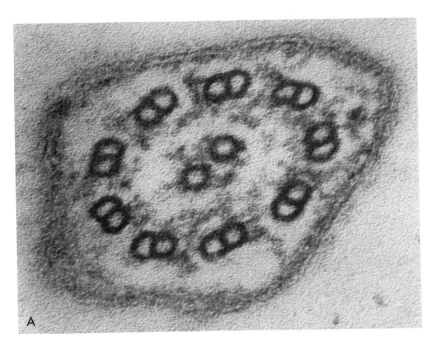

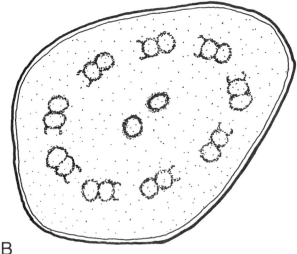

Figure 11–29 *a.* Electron micrograph showing a cross-section of a flagellum of the green alga *Chlamydomonas.* (×260,000) (From J.R. Warr, et al. [1966], Genetics Research 7:3.) *b.* Schematic drawing of the flagellum shown in (*a*). One microtubule of each peripheral pair possesses small arms.

Eukaryotic Cell Structure

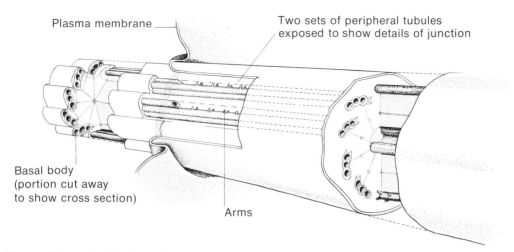

Diagram illustrating the three-dimensional structure of a flagellum. (From S.L. Wolfe [1972], *Biology of the Cell,* Belmont, California, Wadsworth Publishing Co., Inc.)

Figure 11–30

is the socalled (9 + 2) configuration—nine peripheral *groups* of microtubules surrounding two central microtubules—which is distinct from the (9 + 0) configuration of centrioles and basal bodies (see Fig. 11–27b). There are exceedingly few exceptions to the (9 + 2) configuration of eukaryotic cilia and flagella; these exceptions occur notably in some sperm cells in which only a single central filament is present.

The longitudinal structure of cilia and flagella is shown diagrammatically in Figure 11–30. As can be seen, each of the peripheral pairs of microtubules traverses the length of the structure; these microtubules are extensions of two tubules of each basal body triplet. In addition, one of the two tubules of each peripheral pair characteristically possesses small arms attached to it. The arms may also be seen in cross-section in Figure 11–29.

Because the peripheral tubules are invariably present, it has been assumed that these structures are involved in the movement of cilia and flagella and that the central microtubules are not necessary for the motile function. This assumption has obtained support from the finding that only the peripheral tubules contain enzymes associated with the hydrolysis of ATP (ATPases). These enzymes are undoubtedly necessary for the release of energy needed for flagellar motion. More recently, it has been shown that the arms of the microtubules consist primarily of an ATPase. In addition, evidence shows that ATP can move up the shaft of the peripheral tubules.

The mechanism of ciliary and flagellar motion is by no means well understood, but a number of workers have speculated that a "sliding filament" mechanism, which has been shown to be operative in muscular contraction, may be involved in the movement associated with cilia and flagella. (The sliding filament model has also been proposed as a mechanism for the movement of microfilaments and other microtubular structures.) In the sliding filament model, it is envisioned that the sliding of microtubules on one side of the flagellar shaft, relative to those on the other side of the shaft, would cause the flagellum to bend, as illustrated in Figure 11–31. This type of mechanism is supported by the observation that the microtubules do not contract or buckle when movement of cilia or flagella occurs.

The cilia and flagella of eukaryotes should not be confused with those of prokaryotes. Eukaryotic cilia and flagella are not only considerably larger, but possess a different and more complex structure (see Chapter 10).

SUMMARY OF EUKARYOTIC CELL STRUCTURE

Two main points emerge from our consideration of eukaryotic cell structure. In the first place, the eukaryotic cell is highly compartmentalized, with various metabolic functions of the cell occurring in relative isolation from one another. Second, the metabolic reactions that take place in separate compartments must be coordinated with one another so that the cell will function as a viable unit and, in multicellular organisms, as an integral part of the whole organism. The necessity for functional integration, both

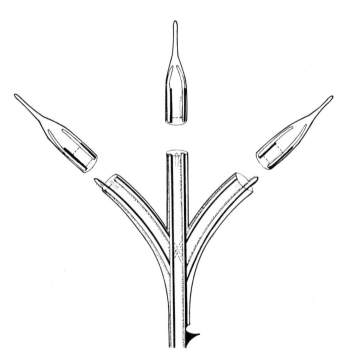

Figure 11–31 Diagram demonstrating the sliding filament model of ciliary and flagellar motion. According to this hypothesis, bending of the flagellum, for example, to the left, is caused by the sliding of filaments on the right side of the flagellum relative to those on the left side. (Redrawn after P. Satir, [1968], J. Cell Biol. 39:77.)

within the cell and between cells in the same organism, accounts for the dynamic aspect of cell structure. Therefore, compartments are not immutable; on the contrary, they change in structure and degree of function in response to the requirements of the cell or of the organism.

THE ORIGIN OF EUKARYOTIC CELLS

As we have pointed out, a good deal of evidence supports the widely held assumption that organisms not unlike present-day prokaryotes were among the first forms of life on earth. In Chapter 2, we proposed the following scheme for the subsequent evolution of living organisms during the Precambrian era: The first prokaryotes were anaerobic heterotrophs. As utilization of abiotically synthesized organic compounds began to deplete the resources available to these organisms, natural selection favored the establishment of anaerobic autotrophs—the chemosynthetic bacteria and the photosynthetic bacteria and blue-green algae. Eventually, one group of the photosynthetic blue-green algae evolved the capacity to release molecular oxygen into the environment. As oxygen began to accumulate in the atmosphere, systems of aerobic metabolism were established, perhaps as early as about 2.7 billion years ago. When the concentration of oxygen reached very high levels, aerobic heterotrophs and aerobic autotrophs became the dominant cell types. The fossil record suggests that eukaryotic cells arose about 1.5 billion years ago. These aerobic cells subsequently became the structural and functional unit of all protists, animals, plants, and fungi.

By what process did eukaryotic cells evolve? The history of this question is extremely long and complex, for the subject has been hotly debated, particularly in recent years. In general, three quite different theories have been proposed to explain the origin of eukaryotic cells:

1. Eukaryotic cells arose completely independently of prokaryotic ones; that is, eukaryotic cells and prokaryotic cells arose as separate and unrelated events in the origin of living forms from non-living matter.

2. Eukaryotic cells evolved from a single ancestral prokaryote by continuous evolution. According to this idea, which follows from classical evolutionary theory, the step-by-step accumulation of selectively advantageous mutations would result in the eventual appearance of eukaryotic cells.

3. Eukaryotic cells arose from prokaryotic cells, not by a gradual process of evolutionary change but in a relatively sudden process by the establishment of specific symbiotic* associations. Thus, this theory proposes that eukaryotic cells were derived from several ancient prokaryotes rather than one. For example, it is en-

---

*Symbiosis is the process by which two organisms live together to their mutual benefit.

visioned that mitochondria may be descended from a free-living aerobic bacterium that was engulfed, but not digested, by a larger anaerobic heterotrophic prokaryote, and that the two cells formed a stable symbiotic association. It is assumed that chloroplasts of eukaryotes may be derived from symbiotic blue-green algae by the same mechanism, and that centrioles, basal bodies, and the cilia and flagella of eukaryotes may have evolved from symbiotic spirochaete-like motile prokaryotes. Because all eukaryotic cells are aerobic and contain mitochondria (except for those cells that have clearly undergone subsequent evolutionary change), the symbiotic theory contends that the first step in the evolution of eukaryotic cells was the establishment of the symbiotic event that led to the evolution of mitochondria, and that symbiotic associations leading to the evolution of flagellated cells and photosynthetic cells must have occurred at a later time.

The major features of each of the three theories of eukaryotic cell origin are summarized in Figure 11–32. The first of these theories—the separate and unrelated origin of prokaryotic and eukaryotic cells—has been rejected by practically all biologists on the ground that the remarkable similarities of the metabolic mechanisms in prokaryotes and eukaryotes suggest a common origin for all living forms. As we pointed out in Chapter 1, it is hard to imagine that metabolic systems so similar to one another could have evolved independently. Indeed, the absence of eukaryotes in the early fossil record would seem to suggest that if prokaryotes and eukaryotes are unrelated, eukaryotes must have arisen at a much later time than prokaryotes. This also seems unlikely, since a newly evolving form would be required to compete with well-established prokaryotes (unless, of course, eukaryotes arose from non-living matter in some highly isolated environment in the absence of prokaryotes—another possibility that does not seem very likely).

Serious debate centers around the second and third theories, which we will refer to as the "classical* theory" and the "symbiotic theory," respectively. Interestingly enough, the symbiotic theory was first suggested in the late nineteenth century, but had been rejected until quite recently in favor of the classical theory. Since the late 1960s, the symbiotic theory has obtained renewed and substantial support, particularly as a result of a comprehensive and persuasive treatise written by Lynn Margulis in 1970 in defense of the symbiotic theory. Only during the mid-1970s has the symbiotic theory been seriously challenged again.

THE SYMBIOTIC THEORY

Interest in the symbiotic theory was rekindled about a decade ago by the discovery of DNA in chloroplasts by H. Ris and W. Plaut

---

*By use of the word classical, we mean to imply not that the theory is outmoded but rather that it is derived from considerations of classical evolutionary theory.

The Origin of Eukaryotic Cells

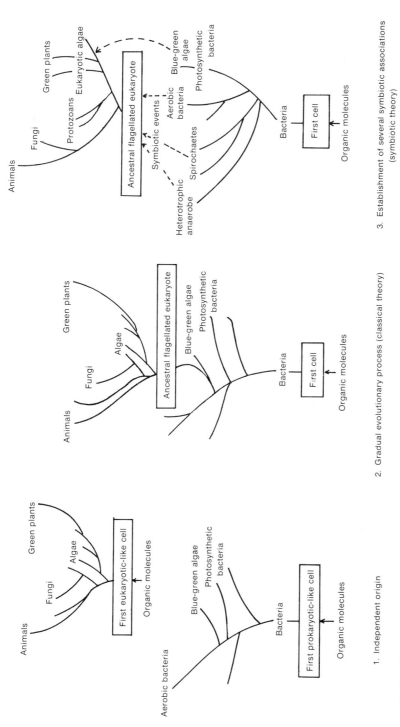

Figure 11-32 Summary of theories describing the origin of eukaryotic cells. 1. Independent origin of eukaryotic cells from that of prokaryotes. 2. Gradual process of evolution of eukaryotic cells from ancestral prokaryotes by mutation and natural selection. 3. Formation of eukaryotic cells by the establishment of symbiotic associations between several types of prokaryotes. (Adapted from L. Margulis [1971], Am Sci. 59:231.)

(1962) and in mitochondria by M. Nass and J. Nass (1963). The excitement of these discoveries stimulated a great deal of research on the origins of mitochondria and chloroplasts. Particularly, biologists were interested in the role of mitochondrial and chloroplast DNA. They wondered whether these DNAs carried the information that determines the structure and function of mitochondria and chloroplasts, and if so, whether these DNAs might be evolutionary relics of the DNAs of ancient intracellular symbiotic prokaryotes. The result of this research has been the accumulation of a large amount of circumstantial evidence that appears to support the symbiotic theory, at least with regard to the origin of mitochondria and chloroplasts. (Except for their ability to direct their own replication, there is little evidence to reinforce the contention that centrioles, basal bodies, and their related structures may have evolved from intracellular symbiotic prokaryotes.) Some of the evidence that is frequently mentioned in support of the symbiotic origin of mitochondria and chloroplasts is the following:

1. The structural discontinuity between prokaryotic and eukaryotic cells—that is, the lack of cell types intermediate in morphology to prokaryotic and eukaryotic cells—suggests that eukaryotes did not evolve from prokaryotes by a gradual evolutionary process involving the step-by-step acquisition of selectively advantageous traits. If they had, then organisms representing transition types might be expected to exist.

2. Intracellular symbiosis is known to occur in large numbers of eukaryotic organisms. For example, blue-green algae are frequent endosymbionts (endo = within) of amoebae, diatoms, flagellated protists, fungi, and "green" algae that lack chloroplasts; in all these cases, the blue-green algae endow their hosts with photosynthetic ability (Fig. 11–33). It has been known for many years that *Paramecium aurelia* may possess several types of endosymbiotic Gram-negative bacteria whose adaptive value, however, is not clear. Nitrogen-fixing bacteria (*Rhizobium*) are endosymbionts of many leguminous plants.

3. We noted in our discussions of cell structure that both mitochondria and chloroplasts are semi-autonomous; that they grow and reproduce by simple binary fission, as do bacteria; and that their rate of division is not always synchronized with that of the cell in which they are present. Both mitochondria and chloroplasts contain their own DNA and separate machinery for protein synthesis, including ribosomes that differ in structure from cytoplasmic ribosomes. The DNA in each organelle contains information that directs the synthesis of some of its own components. However, the structure and function of mitochondria and chloroplasts are determined to a large extent by information stored in the nuclear DNA. For example, the structural proteins of mitochondrial ribosomes appear to be synthesized on cytoplasmic ribosomes and transported into the mitochondrion. Many other proteins important to the function of mitochondria and chloroplasts are synthesized on cytoplasmic ribo-

The Origin of Eukaryotic Cells

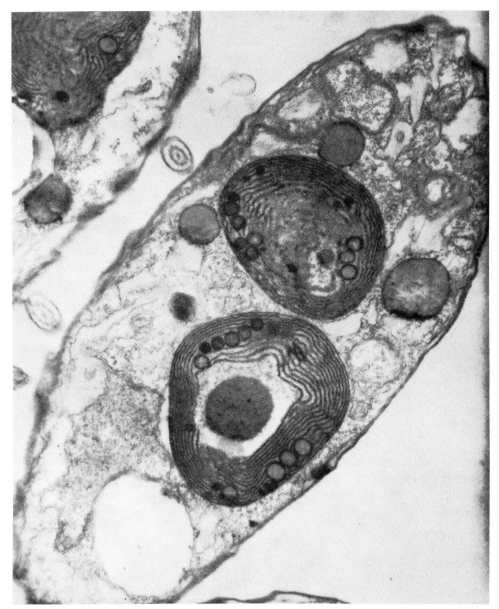

An example of intracellular symbiosis. This electron micrograph shows *Cyanophora paradoxa,* a protozoan with a blue-green algal symbiont. Notice that the blue-green algae do not have cell walls; this is presumably an adaptation to the intracellular environment. (From W.T. Hall and G. Claus [1963], J. Cell Biol. *19*:551.)

Figure 11-33

somes from information coded in the nuclear DNA. Thus control of the structure and function of chloroplasts and mitochondria is shared by the DNA of the organelle and of the nucleus.

4. Structurally, mitochondria appear to resemble bacteria, and chloroplasts resemble blue-green algae in a number of respects. Typical mitochondria and chloroplasts are, respectively, similar in size to bacteria and blue-green algae. The plasma membrane of bacteria is often convoluted, as is the inner membrane of mitochon-

dria; the elementary particles on mitochondrial cristae are similar in structure to the particles seen on the plasma membrane of some aerobic bacteria, and both types of particles have been implicated in reactions associated with aerobic metabolism. Both the thylakoids of the blue-green algae and those of chloroplasts contain light-capturing pigments as an integral part of their structure. The ribosomes of mitochondria and chloroplasts are smaller in size than those in eukaryotes, and more closely resemble the ribosomes of prokaryotes. The DNA of chloroplasts, mitochondria, and prokaryotes consists of double-stranded molecules that are not complexed with proteins and RNA, as is the DNA in the chromosomes of eukaryotes. The naked DNA is attached to the inside of the plasma membrane in prokaryotes and to the inner side of the inner membrane of mitochondria and chloroplasts. Although the DNA in prokaryotes, chloroplasts, and the mitochondria of animal cells exists in the closed circular form, the mitochondrial DNA of plants, fungi, and protists is frequently linear. It has been suggested, however, that these linear DNAs may well be fragments of circular molecules.

5. There are a number of remarkable functional similarities between mitochondria, chloroplasts and prokaryotes. The most striking of these is the resemblance between ribosomes of mitochondria and chloroplasts and those of prokaryotes in their response to certain antibiotics. For instance, chloramphenicol (Chloromycetin) inhibits protein synthesis on the ribosomes of mitochondria, chloroplasts, and prokaryotes, but not on cytoplasmic eukaryotic ribosomes; cycloheximide, on the other hand, inhibits protein synthesis on eukaryotic ribosomes, but does not affect the synthesis of proteins on prokaryotic or organelle ribosomes.

In summary, then, mitochondria and chloroplasts resemble prokaryotic organisms in a number of ways, and these similarities in structure and function may indicate that the two organelles were derived from prokaryotic organisms. Symbiotic associations of the type described here would give obvious selective advantage to both organisms. The prokaryotic endosymbiont would exist in a relatively protected and stable environment in which it could easily obtain necessary nutrients, and the host organism would acquire increased biochemical flexibility. Proponents of the symbiotic theory explain the dependence of mitochondria and chloroplasts on products of nuclear DNA as being a result of mutual adaptation. Thus they contend that duplicate functions, originally performed both by nuclear and by organelle DNA, would have been eliminated as the symbiotic association became firmly established.

THE CLASSICAL THEORY

The symbiotic theory has been criticized recently, especially by R. A. Raff and H. R. Mahler (1972) and by T. Uzzell and C. Spolsky (1974), who contend that the evidence in support of the classical theory for the origin of eukaryotic cells is by far the more compel-

The Origin of Eukaryotic Cells

ling. They argue that the symbiotic theory has two major flaws that greatly decrease its plausibility. In the first place, they contend that it is unreasonable to assume the existence of a large anaerobic prokaryote capable of engulfing smaller aerobic bacteria or blue-green algae. Such an organism, with its primitive anaerobic metabolism, would have had great difficulty competing with more efficient aerobic and autotrophic prokaryotes. Consequently, if it existed, it would most probably have been restricted to localized anaerobic environments where aerobic and photosynthetic prokaryotes would not have thrived. As a result, there would have been little opportunity for symbiotic associations to be established.

Second, the symbiotic theory implicitly requires the transfer of DNA from the endosymbiotic prokaryote to the DNA of the host organism, and it is not at all clear how this could have occurred. Symbiotic organisms are known to eliminate duplicate functions, *but they retain functions that are not duplicated.* Proponents of the symbiotic theory assume that duplicate functions of the endosymbiont and the host were eliminated, with the result that the control of mitochondrial and chloroplast structure and function was divided between the nuclear and organellar DNA. But, if this assumption is valid we must wonder why the DNA of the host contained information that determined metabolic functions identical to those of tion is valid, we must wonder why the DNA of the host contained information that determined metabolic functions identical to those of the endosymbiont, and if it did, why there was any selective advantage to the establishment of a symbiotic relationship between the two organisms.

In addition to the foregoing two major objections to the theoretical aspects of the symbiotic theory, biologists who support the classical theory argue that much of the evidence purporting to show a close relationship between prokaryotes on the one hand, and mitochondria and chloroplasts on the other, has been exaggerated. For example, the ribosomes of animal mitochondria are considerably smaller (about 55S) than bacterial ribosomes, and therefore these two types of ribosomes may be no more similar to each other in structure than are prokaryotic and eukaryotic ribosomes. Some support for this conclusion comes from the observation that 5S RNA is a component of prokaryotic and eukaryotic ribosomes, but is not present in mitochondrial ribosomes, although it is present in chloroplast ribosomes that are identical in size to those of prokaryotes. Furthermore, fusidic acid, an antibiotic that inhibits both bacterial and eukaryotic protein synthesis, has been reported to have no effect on mitochondrial protein synthesis. It is also argued that the apparent structural similarities in the membrane systems of chloroplasts and blue-green algae, and of mitochondria and aerobic bacteria, do not necessarily prove that chloroplasts and mitochondria arose as endosymbionts. Indeed, it is reasonable to assume that increased surface area on membranes is necessary for aerobic metabolism and for photosynthesis, and that the proliferation and folding of membranes to increase their surface area would have adaptive value, and therefore would be selected for in any metabolic

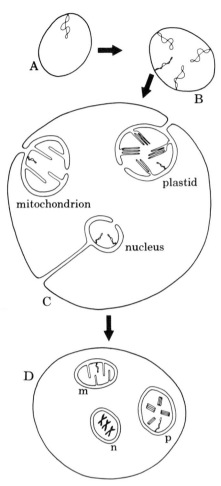

Figure 11-34 A model showing how nuclei, mitochondria, and chloroplasts may have been formed by a classical evolutionary process involving DNA duplication and invagination of the plasma membrane in a prokaryotic cell ancestral to present-day eukaryotes. *a.* Prokaryotic cell. *b.* Duplication of prokaryotic DNA without cell division. *c.* Invagination of the plasma membrane to form double-membrane structures. *d.* Differentiation of the structures and elimination of duplicate functions. The mechanism is described in detail in the text. (From T. Uzzell and C. Spolsky [1974], Am. Sci. *62*:338.)

structure. To put it simply, mitochondria and chloroplasts evolved as membranous structures because of the requirements of their metabolic functions.

As an alternative to the symbiotic theory, Uzzell and Spolsky have proposed an interesting model, similar to one developed by Raff and Mahler, to explain the evolutionary origin of mitochondria and chloroplasts. Their model is based upon the widely held assumption that the nuclear envelope of eukaryotic cells evolved originally in an ancestral prokaryote as an invagination of the plasma membrane. Since there is no evidence to support the symbiotic origin of the nucleus, the development of the nuclear envelope by a gradual evolutionary process of mutation and natural selection is accepted by most biologists. Uzzell and Spolsky argue that if the

The Origin of Eukaryotic Cells

nuclear envelope could evolve by such a mechanism, certainly the same mechanism could be used in the formation of mitochondria and chloroplasts. A diagram of their model, illustrating how this process might have occurred, is shown in Figure 11–34. They envision that the process leading to the formation of the nucleus, mitochondrion, and chloroplast would have begun in an ancestral *aerobic and photosynthetic* prokaryote with the replication of the prokaryotic DNA without cell division. It seems quite reasonable that this could have occurred, since many prokaryotes possess more than one copy of their DNA. The multiple DNA copies would be attached to the inner surface of the plasma membrane, as is DNA in prokaryotes now, and invagination of the membrane would occur at these attachment sites. This would result initially in the formation of bodies enclosed by double membranes, with an identical strand of DNA attached to the inner membrane of each body.

Thus, in the beginning stages of the process, all these bodies would have been very similar in structure and function, and they all may have possessed membrane-bound components of photosynthesis and aerobic metabolism derived from the plasma membrane of the original cell. Eventually, the bodies would have become increasingly differentiated, with the end result that the nuclear envelope would have lost its metabolic functions, and the membrane-bound components of photosynthesis and aerobic metabolism would have been partitioned in chloroplasts and mitochondria, respectively. In the process of this differentiation, the DNA in mitochondria and chloroplasts would have lost duplicate functions to the nucleus. It has been proposed that those functions that were retained by the DNA in chloroplasts and mitochondria may have directed the synthesis of proteins that, because of insolubility or other chemical or physical characteristics, could not have entered the organelle through its membrane. Uzzell and Spolsky point out that if a mechanism such as the one described here was operative in the evolution of mitochondria and chloroplasts, it is clear why no morphological transition types between prokaryotes and eukaryotes are found: They never existed. That is, the inner and outer membranes of mitochondria and chloroplasts were formed in one step.

THE SYMBIOTIC THEORY VERSUS THE CLASSICAL THEORY

At the present time, it is difficult to conclude which of these two hypotheses more clearly represents the mechanism by which eukaryotic cells arose. Neither theory has all the answers. A great deal of research is being done in this area, and undoubtedly there will be important breakthroughs in the next few years that will place the weight of evidence strongly behind one or the other mechanism.

REFERENCES

Ambrose, E. J. and D. M. Easty (1970). *Cell Biology.* Addison-Wesley Publishing Co., Inc., Menlo Park, California.
Brokaw, C. J. (1972). Flagellar movement: a sliding filament model. Science *178*:455.
Bryan, J. (1974). Microtubules. Bioscience *24*:701.
Cohen, S. S. (1970). Are/were mitochondria and chloroplasts microorganisms? American Scientist *58*:281.
Cohen, S. S. (1973). Mitochondria and chloroplasts revisited. American Scientist *61*:437.
DeDuve, C. (1969). The peroxisome: a new cytoplasmic organelle. Proc. Roy. Soc. London B*173*:71.
DuPraw, E. J. (1968). *Cell and Molecular Biology.* Academic Press, Inc. New York.
Dyson, R. D. (1974). *Cell Biology: A Molecular Approach.* Allyn and Bacon, Inc., Boston.
Gibbs, M., ed. (1971). *Structure and Function of Chloroplasts.* Springer Verlag, Berlin.
Giese, A. C. (1973). *Cell Physiology,* W. B. Saunders Co., Philadelphia.
Hoffmann, H. P. and C. J. Avers (1973). Mitochondrion of yeast: ultrastructural evidence for one giant, branched organelle per cell. Science *181*:749.
Howland, J. L. (1968). *Introduction to Cell Physiology: Information and Control.* Macmillan, Inc., New York.
Margulis, L. (1970). *Origin of Eukaryotic Cells.* Yale University Press, New Haven, Connecticut.
Racker, E. (1968). The membrane of the mitochondrion. Scientific Am. *218*:32.
Raff, R. A. and H. R. Mahler (1972). The non-symbiotic origin of mitochondria. Science *177*:575.
Raven, P. H. (1970). A multiple origin for plastids and mitochondria. Science *169*:641.
Sager, R. (1972). *Cytoplasmic Genes and Organelles.* Academic Press, Inc., New York.
Tedeschi, H. (1974). *Cell Physiology: Molecular Dynamics.* Academic Press, Inc., New York.
Uzzell, T. and C. Spolsky (1974). Mitochondria and plastids as endosymbionts: a revival of special creation? American Scientist *62*:334.
Wolfe, S. L. (1972). *Biology of the Cell.* Wadsworth Publishing Co., Inc. Belmont, California.

CHAPTER **12**

The Evolution of Cellular Metabolism

The term *cellular metabolism* describes all the chemical transformations that take place within living cells. These transformations traditionally have been divided into two categories: (1) those processes referred to as *catabolic,* which involve the degradation of chemical substances, and (2) those processes referred to as *anabolic* or *biosynthetic,* which are responsible for the synthesis of cellular materials needed for growth, repair, and reproduction. Anabolic processes require energy, and this energy, in the form of ATP, is supplied by catabolic processes. (The exception is photosynthesis, in which anabolic processes that result in the synthesis of carbohydrates are driven by energy captured from the sun's rays.)

Of course, there are many cellular activities which require energy, but which do not result in the synthesis of cellular molecules. These include processes that involve cell motility and muscular contraction (mechanical work), the transport of substances into and out of the cell (transport work), and the conduction of electrical impulses (electrical work), as well as many others (Fig. 12–1). These too are metabolic activities, and they use energy in the form of ATP supplied by catabolic processes. For the purposes of our present discussion, however, we will use the term cellular metabolism in a restricted sense to include only those processes involving energy production and/or biosynthesis. In this and the next three chapters we will consider the mechanisms by which energy is captured as ATP (catabolism), and the relation of catabolic processes to biosynthetic ones. In later chapters, we will discuss other forms of cellular work.

It would be a serious mistake to think of anabolic processes as being separate from catabolic ones; it is because of the interrelatedness of the two types of processes that we have chosen to discuss them together. In fact, there is no clear distinction between anabolic and catabolic processes, and although such categorization is often useful for purposes of discussion, it creates the erroneous impression that the two types of processes constitute entirely separate metabolic subsystems that relate to one another only with

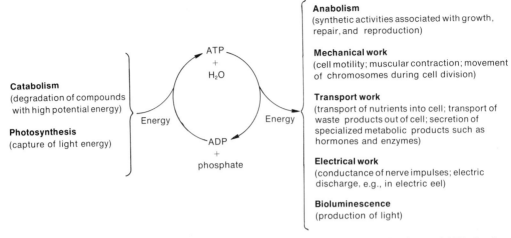

Figure 12-1 Catabolic processes and photosynthesis capture energy in the form of ATP; the hydrolysis of ATP releases free energy that can be used to drive a large number of endergonic biological processes. (Adapted from N.A. Edwards and K.A. Hassall [1971], *Cellular Biochemistry and Physiology*, McGraw-Hill, p. 132.)

regard to the transfer of energy between them. Many metabolic pathways are involved in both the energy-yielding degradation of nutrient molecules and in the biosynthesis of cellular constituents. For example, a compound produced in an energy-yielding pathway may be shunted from the pathway to be used in the synthesis of another needed compound. A cell's metabolic system constitutes an interconnecting, integrated series of chemical reactions that may operate in any one of many directions, depending upon the physiological requirements of the cell.

Thus, whereas catabolic pathways provide the necessary energy for anabolic ones, the two types of pathways are not necessarily separate, but are often linked by common intermediate compounds (Fig. 12-2). This is hardly surprising, considering that the metabolic machinery of the cell evolved as a single system. The individual metabolic pathways of the cell are interrelated in the same way as are its individual structural components. On a higher level, moreover, the total structural and metabolic systems of the cell evolved concurrently, and are therefore inextricably interwoven. While discussing any particular metabolic pathway, we must always remember that any single sequence of reactions is merely a part of an integrated whole.

Discussions of the metabolism of living cells often turn out to be the most confusing and by far the dullest part of a course in cell biology. This is unfortunate, for cellular metabolism is interesting, and potentially exciting. In no other aspect of cell biology are the intricacy and precision of living systems more evident, nor is the outcome of the evolutionary process more immediately apparent. The design of

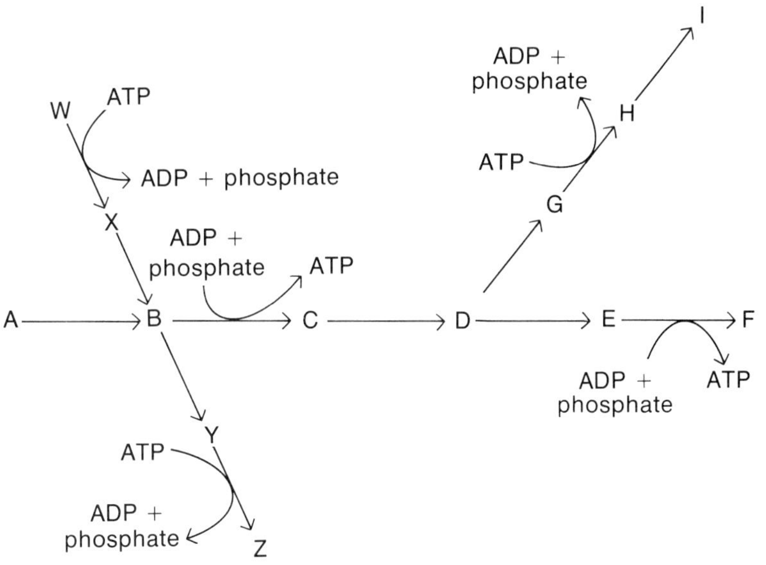

Basic Principles of Metabolism

There is not always a clear distinction between catabolic and anabolic pathways, as illustrated in this diagram.

Figure 12-2

cellular metabolic systems in different types of organisms becomes far easier to rationalize when we stop viewing these systems as interminable lists of metabolic reactions—the descriptive approach—and start thinking of them as evolutionary solutions to environmental challenges. A particular type of metabolic system became established because of the selective advantage that certain hereditary changes (mutations) gave a particular type of organism in a changing environment. Thus it is only in the context of environmental change that the diversity of metabolic systems apparent in present-day living organisms can be adequately explained.

BASIC PRINCIPLES OF METABOLISM

What are some of the general characteristics common to all metabolic systems, and what are the fundamental principles by which cellular metabolism operates?

All biological work requires energy, and a major role of metabolism is to provide an adequate source of energy for the cell. We have already mentioned the ADP–ATP cyclic energy transfer system, and we have noted that cells use energy primarily in the form of ATP, which is hydrolyzed to ADP with the release of free energy. In order to understand the process by which cells capture energy, we must understand how cells extract energy from nutrient substances, and how this energy is used to synthesize ATP from ADP (the process known as the phosphorylation of ADP).

ENERGY CAPTURE IN THE LIVING CELL

The process of energy capture by living cells is essentially one of oxidation and reduction. You will recall from Chapter 6 that *oxidation* is the removal of electrons from a substance and that *reduction* is the gain of electrons by a substance. Energy capture in metabolic systems takes advantage of the fact that *the removal of electrons from a substance* (oxidation) *releases energy.* We know from our previous discussion of oxidation and reduction that electrons cannot be removed from a substance unless they are accepted by another substance, which is called the oxidizing agent or the electron acceptor. The amount of energy that is released as a result of the oxidation of a substrate molecule, and is thus potentially available for the phosphorylation of ADP to ATP, is dependent upon the oxidation state of the substrate and of the electron acceptor.

The tendency of a substance to accept or to donate electrons in a chemical reaction with another substance is defined by its *oxidation-reduction potential* (also known as its *redox potential*), which is usually expressed in volts. A knowledge of the redox potentials of chemical substances is extremely useful, because redox potentials tell us the direction in which a particular oxidation-reduction reaction, or sequence of oxidation-reduction reactions, will proceed. The greater the tendency of a substance to donate an electron (the greater its reducing properties), the greater its *negative* redox potential. Thus, electrons can be transferred spontaneously from one substance to another substance that has a greater tendency to accept electrons—a more positive redox potential. When such an electron transfer (oxidation-reduction) occurs, energy is released, because electrons move to a state of lower energy.

In fact, the greater the gap in redox potential between the donor and the acceptor (the greater the difference in energy of a given electron in the donor and in the acceptor molecules), the greater the yield of energy. Conversely, to transfer electrons from one substance to another that has a more negative redox potential requires energy, and the greater the difference in redox potential between the two substances, the greater the amount of energy required. In subsequent discussions, we will refer to substances with high negative redox potentials, relative to those of other biologically important substances, as *strong reductants* (strong reducing agents), and to compounds with comparatively high positive redox potentials as *strong oxidants* (strong oxidizing agents).

In the living cell, then, the oxidation of a substance is coupled to the phosphorylation of ADP to ATP. More precisely, the energy used to phosphorylate ADP is derived from the spontaneous transfer of an electron from one substance to a less reduced substance. This leads us to an important conclusion. *All reactions that bring about the synthesis of ATP or other high-energy compounds in living cells have one feature in common: cells capture energy by moving electrons to a state of lower energy.* The difference between the energy of an electron in the donor substance and in the acceptor substance

Basic Principles of Metabolism

represents the amount of energy available for manufacturing ATP by phosphorylating ADP (Fig. 12–3).

Later, we will see that anaerobic oxidation of an organic molecule is much less efficient in terms of energy capture than aerobic oxidation of the same molecule. For example, much less ATP is produced during the anaerobic oxidation of glucose than is produced when glucose is oxidized under aerobic conditions. This is because an electron in glucose is transferred to a much lower energy state when it is accepted by molecular oxygen (in aerobic oxidation) than when it is accepted by an organic molecule, as in anaerobic oxidation. In other words, during aerobic oxidation the substrate is oxidized to a greater extent than it is during anaerobic oxidation, and therefore more energy is available for ATP synthesis. The same point can be made still another way: Oxygen is a stronger oxidant than is a reduced organic compound.

The foregoing discussion represents a greatly simplified view of the principle of energy capture within living cells, and tells us nothing about the specific chemical transformations that occur during energy capture; nevertheless, electron transfer is the fundamental aspect of the process. We will see that pathways of energy capture in living cells are designed specifically to allow the transfer of electrons from substrate to acceptor molecules under the physiological conditions (temperature, pH, concentration of substances, among others) existing within the cell.

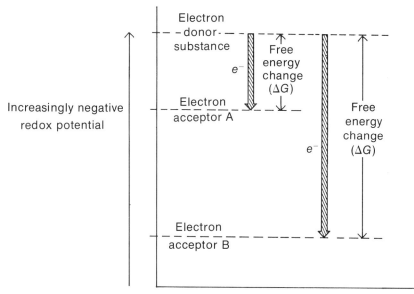

Figure 12-3

The amount of free energy released upon the transfer of an electron from a given electron donor to an electron acceptor depends upon the redox potential of the electron acceptor. Consequently, a greater amount of free energy is released when an electron is transferred to electron acceptor B than when it is transferred to A.

STEPWISE NATURE OF METABOLISM

There is another characteristic of cellular metabolism that is related to the principle of energy transfer mentioned above. *Rather than occurring by a one-step chemical reaction, metabolic processes tend to proceed by a multistep metabolic pathway.* For example, in the aerobic oxidation of glucose, electrons are not transferred directly from glucose to molecular oxygen. Rather, they are transferred in a series of reactions, each catalyzed by a specific enzyme, and each occurring with a relatively small free energy change.

Such a design has several advantages. In the first place, most metabolic processes could not occur in one step under physiological conditions. If glucose were to react in one step directly with oxygen, a great deal of energy would be released as heat, resulting in a fatal rise in intracellular temperature. However, the transfer of electrons from glucose to oxygen by way of a series of reactions, each of which involves a small free energy change instead of a large one, results in the controlled release of small amounts of energy. Second, a multistep oxidative pathway can potentially capture a far greater amount of energy in the form of ATP than could be captured if the oxidative reaction occurred in one highly exergonic step. Since the phosphorylation of one molecule of ADP occurs by virtue of the fact that it is coupled in some way to another reaction that releases energy, it follows that a greater yield of ATP can be obtained if the substrate is oxidized in a series of reactions, each possessing a modest free energy change. This is because more individual reactions can be coupled to the phosphorylation of ADP, and a good deal of the free energy released in any one step may potentially be conserved in phosphorylation (providing, of course, that at least enough free energy is released in a particular step to phosphorylate one molecule of ADP). Figure 12–4 illustrates the efficiency of energy capture in a one-step transformation as compared to the efficiency of energy capture in the same transformation occurring in a multistep pathway.

In addition to increasing the efficiency of energy capture, multistep pathways can be controlled easily and sensitively, so that metabolic activity can be well coordinated with the physiological requirements of the cell. It follows that the greater the number of steps in a pathway, the more opportunities there are for metabolic controls to influence the speed and direction of the pathway.

Multistep pathways also provide intermediate compounds that may be used in other pathways, thereby enhancing the integration of metabolic processes. This characteristic of metabolic pathways is clearly advantageous to the cell from the perspective of energy requirements. The greater the coordination between individual reactions, the less energy is wasted; the final result is a cell metabolism that is more efficient than otherwise, and consequently appears to have greater selective advantage from an evolutionary standpoint. For instance, when an anabolic and a catabolic process are integrated by virtue of the existence of chemical intermediates common to both processes, not only may a particular nutrient be used for several

Basic Principles of Metabolism

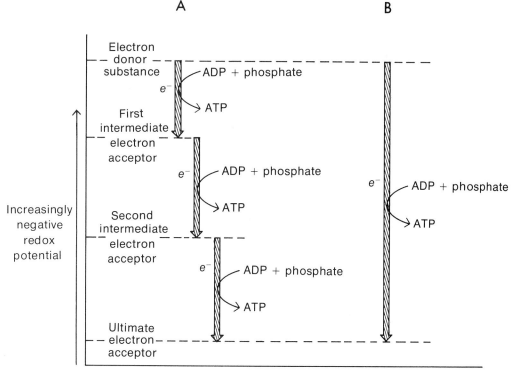

Each separate electron transfer reaction may be coupled to the phosphorylation of one molecule of ADP. Consequently, more ADP can be phosphorylated, and hence more energy can be captured when the overall electron transfer (from electron donor to ultimate electron acceptor) is broken down into a number of separate electron transfer reactions, as in (A), than when it occurs in one step, as in (B).

Figure 12-4

quite different metabolic purposes instead of being wasted but also that the number of enzymes or individual metabolic reactions required for a functioning metabolic system is reduced. Nutrients are not always available in excessive amounts, and even when they are, it takes energy to transport them into the cell. Consequently, the fewer materials that need be captured from the environment, the more completely each nutrient is utilized, and the simpler the metabolic machinery of the cell, the greater will be the efficiency of the system.

DEFINITIONS

The study of metabolism is often made unduly confusing by variations in the way certain terms are employed by different sources, so it is well to clarify our terminology at the outset. *Anaerobes* are defined as organisms that live only in the absence of oxygen; *aerobes,* as organisms that require the presence of oxygen in order to live; and *facultative anaerobes* (or *facultative aerobes*), as organisms that can exist in either the presence or the absence of oxygen. *Anaerobic metabolism* operates in the absence of oxygen, and *aerobic metabolism* operates in the presence of oxygen. More specifically, we will use the term *anaerobic respiration* to describe the energy-generating oxidation of reduced compounds by the use of organic substances or inorganic substances other than molecular oxygen as electron acceptors; the term *aerobic respiration* will be used to describe the oxidation of reduced compounds by the use of molecular oxygen as the final electron acceptor.*

It should be emphasized that the difference between aerobic and anaerobic metabolism is not always as clearcut as it might seem from these definitions. Identical metabolic pathways exist in both anaerobes and aerobes, and a particular pathway may often operate as well in the presence of oxygen as in its absence. Sometimes, particularly in respiratory metabolism, aerobic and anaerobic pathways may differ only with respect to end products and energy yields. This is an important point, because it emphasizes that anaerobic metabolism does not use a system of pathways entirely different from that of aerobic metabolism; rather, the two types of metabolism are very similar to one another. Indeed, it is well to remember that there are many basic features common to all metabolic systems.

THE DEVELOPMENT OF THE FIRST METABOLIC PATHWAYS

It has been generally assumed that the first organisms obtained all their essential nutrients and energy (including condensing agents) from among organic compounds synthesized abiotically on the primitive earth. It also seems likely that the first metabolic pathways to develop were those directly associated with reproduction. These first pathways probably involved primitive forms of nucleic acid and protein synthesis, membrane synthesis, and nucleic acid replication, and operated essentially by the condensation of preformed monomers present in the prebiotic soup (see Figure 8–1). Apart from their ability to carry out such condensation reactions, the first primitive cells are thought to have been entirely dependent on the prebiotic soup for necessary organic compounds—these first cells possessed no biosynthetic capabilities, nor did they have the capacity to capture energy in usable form.

A number of researchers have proposed theoretical models to explain how the most primitive biosynthetic and energy-generating pathways might have developed in early cells. Essentially, all these models are based on the contention that metabolic pathways

---

*The term *respiration* is sometimes restricted to oxidations involving oxygen. In yet other terminology, *anaerobic respiration* is used to refer only to the oxidation of organic compounds, using inorganic substances as electron acceptors. We have defined the terms anaerobic respiration and aerobic respiration in a broader sense to emphasize the fact that all energy-generating oxidations are similar in basic design, although they differ in detail.

evolved as a consequence of the competition of living organisms for scarce organic components of the prebiotic soup.

Under conditions in which there was an ample supply of essential nutrients and high-energy compounds in the environment, it does not seem likely that there would have been any selection pressure for the establishment of biosynthetic and energy-generating pathways. In fact, such metabolic capabilities would most likely have been *disadvantageous* at this stage in the evolution of living systems because they would have been superfluous. Any unnecessary metabolic activities represent a needless expenditure of energy—a wasteful process, both in terms of the "cost" of maintaining and operating the added metabolic machinery (such as enzymes) required for such tasks, and in terms of the energy requirements of the syntheses themselves.

However, scarcity or depletion of one or more essential compounds in the prebiotic broth would have represented selection pressure in favor of those organisms able to manufacture the necessary compounds from other chemically related compounds available in the environment. It has been proposed that, at least in some cases, the first enzymes may have evolved as catalysts to reactions that proceeded inefficiently in the primitive environment. For example, if a particular prebiotic synthesis contained a step that was *rate limiting* (proceeding at a rate much slower than other steps in the pathway), then a primitive enzyme that catalyzed this particular step might have eventually evolved. Thus, according to this idea, the first metabolic pathways would have duplicated pathways of chemical synthesis that occurred on the prebiotic earth. Miller and Orgel have concluded, however, that such a mechanism may have operated in the development of a few pathways, but was not the general mechanism by which metabolic pathways developed.

The conclusion of Miller and Orgel is supported by two pieces of evidence. First, the great majority of metabolic pathways clearly do not follow any sequence of reactions that conceivably could have operated on the primitive earth. Many reactions in present-day metabolic pathways require highly specific and complex enzymes in order to proceed at anything more than negligible rates; it is extremely unlikely that these enzymes were present in the prebiotic soup. Second, many prebiotic syntheses occur only in the atmosphere, and could not proceed in the aqueous environment characteristic of living organisms.

In 1945, H.N. Horowitz suggested another mechanism to explain how metabolic pathways may have developed, and his hypothesis is now widely accepted by evolutionary biologists. The Horowitz hypothesis proposes that early metabolic pathways developed in the direction opposite to that in which they now operate; this occurred by the sequential acquisition in living organisms of the ability to synthesize necessary enzymes.

Suppose, for example, that the supply of a required compound (A), perhaps an amino acid, were low or exhausted. Any mutant then would have clear selective advantage if it possessed the ability, by virtue of synthesizing a specific enzyme, to manufacture compound A from another compound (B), a chemically related amino acid readily available in the environment. Eventually, as a population of organisms able to convert B to A became established, compound B in turn would become scarce. Under such conditions, an additional mutation that endowed an individual in this population with the ability to synthesize B from another compound (C), would be selectively advantageous. Because the ability to synthesize the enzyme that catalyzed the conversion of B to A had already been established in the population, only one new enzyme (the one catalyzing the conversion of C to B) would be required to convert C into A.

This process could be repeated, giving rise to the step by step elaboration of a complex, multistep metabolic pathway (Figure 12–5). Of course, this description of the development of metabolic pathways does not include an explanation of how the entire process may actually occur at the level of the genetic material of the cell; we will discuss this point in Chapter 18.

Notice that Horowitz's scheme assumes that as soon as the ability to synthesize an enzyme arises, the enzyme can perform a useful function. This is an extremely important point, because *evolution operates without planning.* Short of the chance *simultaneous* acquisition of the ability to synthesize all the required enzymes of a pathway, there is no obvious way that a multistep pathway such as

$$E \xrightarrow{\text{Enzyme D}} D \xrightarrow{\text{Enzyme C}} C \xrightarrow{\text{Enzyme B}} B \xrightarrow{\text{Enzyme A}} A \text{ (required compound)}$$

could evolve in the forward direction (from E to A). It is clear that the ability to synthesize a single enzyme at the beginning (left side) or in the middle of the pathway (for example, Enzyme C) could not be selected for, because such an ability alone would provide no selective advantage without the later enzymes in the pathway. The probability that several enzymes would appear by simultaneous mutation is very small, and the larger the number of enzymes involved, the smaller would be the probability of such an event.

$$B \xrightarrow{\text{Enzyme A}} A \quad \text{1st synthetic enzyme develops}$$

$$C \xrightarrow{\text{Enzyme B}} B \xrightarrow{\text{Enzyme A}} A \quad \text{2nd synthetic enzyme develops}$$

$$D \xrightarrow{\text{Enzyme C}} C \xrightarrow{\text{Enzyme B}} B \xrightarrow{\text{Enzyme A}} A \quad \text{3rd synthetic enzyme develops}$$

$$E \xrightarrow{\text{Enzyme D}} D \xrightarrow{\text{Enzyme C}} C \xrightarrow{\text{Enzyme B}} B \xrightarrow{\text{Enzyme A}} A \quad \text{4th synthetic enzyme develops}$$

Figure 12–5 A metabolic pathway leading to the biosynthesis of an important cellular constituent (A) may develop in reverse of the direction in which it first operates. (Adapted from L.E. Orgel [1973]. *The Origins of Life,* John Wiley and Sons Inc., p. 174.)

Horowitz's hypothesis has been extended by the suggestion that biosynthetic pathways, in many cases, may represent the reverse of pathways by which biologically important organic molecules were spontaneously degraded in the primitive broth. For example, the spontaneous decomposition of a compound (X) will produce Z, which is closely related structurally to X. Hence, if an organism acquires the ability to synthesize an enzyme capable of catalyzing this conversion, and if the conversion can be linked to an exergonic process such that Z is converted into X, this may lead to the production of a metabolic pathway. At least some of the metabolic pathways that occur in living systems today may have evolved in this way. The pathway of purine synthesis, for instance, does follow quite closely the expected pathway for the spontaneous degradation of purines in aqueous solutions.

It is quite possible, indeed probable, that other mechanisms besides those described above may have been important in the development of cellular metabolic systems. We can easily conceive of more complex mechanisms; those described here have the virtue of simplicity.

It is important to remember that the development of metabolic pathways involved not only those pathways that provided for the manufacture of necessary structural components but also those that were concerned with energy capture. Although free energy for biosynthesis and other cellular activities was probably initially derived from the hydrolysis of high-energy compounds in the environment, metabolic pathways providing alternative free energy sources would have been needed when these sources of free energy became scarce. These pathways, resulting in the synthesis of high-energy compounds such as ATP, could have evolved by mechanisms like those described above.

THE COURSE OF METABOLIC DEVELOPMENT DURING BIOLOGICAL EVOLUTION

In attempting to trace the course by which metabolic systems developed during biological evolution, we must keep in mind that all such reconstructions are, and must be, tentative. The evolutionary process that occurred during the first 4.6 billion years of the earth's history cannot be duplicated, and we therefore must use circumstantial rather than direct evidence to reconstruct the sequence of events. In many cases, the validity of the evidence itself is questionable. Geologic data, for example, give a very imperfect picture of environmental conditions present during the Precambrian Era. In addition, the fossil record is hardly unequivocal. But by combining physical geologic data and data from the fossil record with a comparative study of metabolic patterns in present-day organisms, we can put together a tentative outline of how metabolic systems may have developed during the evolution of living forms. It is well to emphasize that most of the discussion in this chapter concerns the development of metabolic systems in prokaryotic organisms, because it is generally thought that all major types of metabolism had developed in prokaryotes before the evolution of eukaryotes about 1.5 billion years ago.

As A.I. Oparin has pointed out, an examination of metabolic systems in contemporary organisms suggests that during the course of biological evolution, an increasing number of metabolic pathways became linked with one another. The fundamental metabolic mechanism of living organisms appears to be that of anaerobic heterotrophy. All animals, most protozoa, most bacteria, and all fungi are heterotrophs, using organic compounds not only as sources of energy for cellular activities but also as precursors for the synthesis of cellular components. But even in most autotrophs, heterotrophic pathways are present, as well.

For example, many normally autotrophic forms of algae, both prokaryotic and eukaryotic, are able to convert to heterotrophic metabolism when growing in polluted water rich in organic compounds. In higher plants, only the chlorophyll-containing photosynthetic cells are autotrophic, and even these cells metabolize heterotrophically in the absence of light. All other cells of higher plants use as nutrients organic compounds that have been manufactured in their photosynthetic cells. Indeed, although many autotrophs are able to function as heterotrophs when organic nutrient sources are available, the reverse situation is not seen; that is, normally heterotrophic organisms do *not* have the ability to convert to autotrophic metabolism under conditions in which they are deprived of organic compounds.

Similarly, evidence derived from the study of present-day cells suggests that the original heterotrophic organisms were also anaerobes. Although most contemporary organisms are obligate aerobes, many aerobic cells can survive short periods in the absence of oxygen (mammalian muscle cells are a classic example). Most anaerobes, on the other hand, cannot survive even short periods in the presence of oxygen. When we compare metabolic systems in present-day aerobes and anaerobes, we find that the basic design of energy metabolism in all modern aerobes is the same as that in anaerobic organisms, and that oxidation reactions using molecular oxygen as the final electron acceptor are supplementary to pathways of anaerobic oxidation found in almost all organisms.

This evidence gives credence to the theory that present-day metabolic systems developed by adding links to the metabolic chain already established in anaerobic heterotrophs. Had aerobic mechanisms evolved first, we would expect to find the existing relationship between anaerobic and aerobic pathways in contemporary organisms reversed. That is, we would expect that metabolism in most organisms would be based on an aerobic system, and that this system would be supplemented in some organisms by an anaerobic mechanism. The same reasoning would lead us to conclude that the

The Course of Metabolic Development During Biological Evolution

wide distribution of heterotrophic as compared to autotrophic metabolic pathways suggests that heterotrophic metabolism was the first to appear.

Before beginning a discussion of the probable course of the development of metabolic systems, we should emphasize several points with regard to biological evolution. A particular type of organism is adapted to the environment in which it thrives. Thus, as we are discussing different kinds of metabolic systems, we must remember that each type of organism has selective advantages in a particular environment; if it did not, it would have been eliminated by more successful competitors. It is grossly inaccurate, for example, to say that anaerobes are not well adapted to life on earth. Although it is true that they are not adapted to life in the presence of oxygen, they are very well adapted to live in anaerobic environments, where aerobes cannot survive. *Therefore, to speak of adaptation without reference to a particular environment is meaningless.*

We must also remember that the development of a new type of metabolic system did not necessarily lead to the extinction of previous metabolic types, although this may have occurred in some cases. For example, the development of autotrophic metabolism did not bring about the extinction of anaerobic heterotrophs, which we have good reason to believe were the first types of living organisms. Indeed, anaerobic heterotrophs still exist on the earth in local environments. Autotrophic metabolism probably became increasingly widespread as abiotically synthesized sources of organic compounds continued to diminish because of the metabolic demands of anaerobic heterotrophs. It was the changing environment — in this example, an increasing scarcity of organic compounds — that provided selective advantage to developing forms possessing autotrophic mechanisms of metabolism.

One of the marvels of biological evolution is the ability of living organisms to utilize a wide variety of nutrient sources by making only minor modifications in the design of metabolic systems that had previously evolved. This point will be amply illustrated in the discussion that follows.

PATHWAYS OF ANAEROBIC HETEROTROPHY

As we have noted, the earliest anaerobic pathways of metabolism were undoubtedly simple ones. As nutrients became exhausted, biosynthetic pathways were selected for. In addition, the depletion of energy sources led eventually to the establishment of multistep pathways of anaerobic energy generation in which organic compounds were oxidized, using other organic compounds as electron acceptors. The free energy released in the process was used to phosphorylate ADP.

A diagram outlining the main features of this type of anaerobic respiration is shown in Figure 12-6. Electrons are passed spontaneously from an electron donor (AH_2) to an electron acceptor (B), both organic compounds. This exergonic electron transfer from one substance to another substance with a more positive redox potential is coupled to the endergonic phosphorylation of ADP to ATP. Of course, it must be emphasized that the scheme illustrated in Figure 12-6 does not provide any of the details—the chemical pathway—by which such reactions actually occur, nor does it give any indication of the stepwise nature of all metabolic pathways. We will discuss these pathways further in Chapter 13. The reason for describing pathways of anaerobic heterotrophy in abbreviated form (as in Figure 12-6) is to emphasize the fact that the fundamental mechanism of energy capture involves the transfer of electrons from donor to acceptor molecules, as we discussed in the initial part of this chapter.

PHOTOSYNTHESIS AND THE ELIMINATION OF OXYGEN

The next phase in the evolution of living forms is thought to have occurred when easily utilized organic compounds began to become scarce in the primitive broth as a result of the metabolic processes of the anaerobic heterotrophs. This situation obviously represented a new and important environmental challenge to the existing forms of life. The ability to obtain energy by means other than by extraction from organic compounds, and the ability to synthesize organic compounds from carbon dioxide, suddenly possessed great selective advantage. One alternative source of energy consisted of reduced inorganic compounds; another far more abundant source

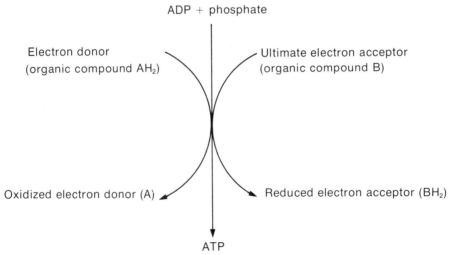

Figure 12-6 A diagrammatic representation of one type of anaerobic respiration in which the electron donor and the ultimate electron acceptor are both organic compounds. Energy released in the spontaneous transfer of electrons from AH_2 to B is used to drive the phosphorylation of ADP to ATP.

was the sun. Two groups of autotrophic organisms evolved to exploit these energy sources. One group consisted of *anaerobic chemoautotrophs* (presently a small and relatively unimportant group of bacteria), which were able to extract energy from reduced inorganic compounds. The other group, the *anaerobic photosynthetic bacteria*, were able to utilize the energy of the sun's rays to synthesize energy-rich organic compounds.

Contemporary anaerobic autotrophs possess the ability to manufacture all essential organic compounds from carbon dioxide (CO_2) —a process known as *autotrophic carbon dioxide fixation*.* Because organic cell constituents are more reduced than carbon dioxide, the pathways of autotrophic CO_2 fixation require, in addition to an energy source (ATP), a strong reductant capable of reducing carbon dioxide.

In anaerobic chemoautotrophs, ATP derived from the oxidation of *inorganic* compounds is used to generate the strong reductant needed for fixation of carbon dioxide. In photosynthetic organisms, light energy is used both to generate ATP *and* to generate a strong reductant in a series of reactions illustrated in outline form in Figure 12–7. Although the details of photosynthesis may vary considerably among different types of organisms, in most cases the initial elec-

---

*Carbon dioxide fixation, by which CO_2 is incorporated into organic compounds, occurs in one form or another in all organisms. However, heterotrophic organisms can fix CO_2 only to a very limited extent, and they cannot derive all, or even a significant proportion of, the organic compounds they need from CO_2 because they do not possess the necessary enzymes. CO_2 fixation in heterotrophs is known as *heterotrophic carbon dioxide fixation* to distinguish it from the autotrophic type with which we are primarily concerned here.

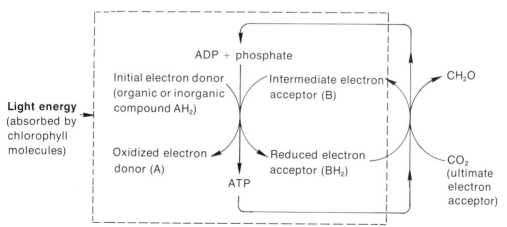

Figure 12–7

A greatly simplified diagram summarizing the major chemical events in photosynthesis. Light energy is absorbed by chlorophyll and used to drive the endergonic reactions that lead to the production of ATP and a strong reductant (XH_2), both of which are required to fix CO_2 into organic compounds. The CO_2 fixation reactions do not require light. Hydrogen sulfide (H_2S), thiosulfate ($S_2O_3^{-2}$), H_2, or organic compounds are used as electron donors in bacterial photosynthesis; H_2O is used as an electron donor in photosynthesis in blue-green algae and all eukaryotes. See text for further details. (Adapted from J.R. Bronk [1973], Chemical Biology, Macmillan, Inc., p. 316.)

tron donor compounds employed in photosynthetic systems are not sufficiently strong reductants to be used in the phosphorylation of ATP or in the reduction of carbon dioxide. In the photosynthetic process, light energy absorbed by chlorophyll molecules is, in effect, used to raise electrons in the electron donor compounds to a state of higher potential energy (an excited state). Such excited electrons have sufficient energy to bring about the phosphorylation of ADP, and to be accepted by an electron acceptor compound to produce a strong reductant that *is* capable of reducing carbon dioxide. Thus, we see that photosynthetic systems use the same basic mechanism as do heterotrophic ones: Energy is transferred by moving electrons.

The first photosynthetic organisms most likely used hydrogen sulfide (H_2S), thiosulfate ($S_2O_3^{-2}$), hydrogen (H_2), or organic compounds as electron donors in photosynthesis, as do present-day photosynthetic bacteria. However, as these electron donor compounds became increasingly scarce, selection pressure favored the establishment of organisms that used the ubiquitous substance water as an electron donor—a process that is energetically very difficult and results in the release of molecular oxygen

$$2\ H_2O \longrightarrow 4\ e^- + 4\ H^+ + O_2$$

Organisms using water as an electron donor were presumably ancestral to present-day blue-green algae, the only contemporary prokaryotes that eliminate molecular oxygen as a byproduct of photosynthesis. All photosynthetic eukaryotes also use water as an electron donor in photosynthesis, and consequently they also produce oxygen.

It is not known whether the ability to phosphorylate ADP by using energy derived from inorganic compounds (in the case of anaerobic chemoautotrophs) or from the sun's rays (in the case of photosynthetic organisms) developed before or after the metabolic pathways required for carbon dioxide fixation. Some biologists have contended that the coupling of the phosphorylation of ADP to processes other than the oxidation of organic compounds probably preceded the evolution of pathways of carbon dioxide fixation, because a supply of ATP may have permitted the transformation of simple nonutilizable organic compounds (such as acetic acid) into more complex compounds. In this way, a large number of organic compounds that formerly were not useful to existing organisms might have become utilizable in biosynthetic pathways by organisms capable of generating ATP. In addition, these researchers have contended that an organism possessing the metabolic machinery for synthesizing organic compounds from carbon dioxide, but not for producing ATP, would have been unable to function unless some adequate source of ATP were available. Pathways of autotrophic carbon dioxide fixation, however, occur in some strictly anaerobic nonphotosynthetic prokaryotes that are thought to have existed dur-

The Course of Metabolic Development During Biological Evolution

ing early stages in the evolution of living systems. Because of this, other researchers contend that metabolic pathways for carbon dioxide fixation must have been established early in the evolution of living forms, at least before photosynthetic systems had evolved. Although the available evidence does not allow us to choose between these alternatives, the exact sequence of events is unimportant to our scheme. The crucial point is this: Organisms capable of obtaining energy elsewhere than from organic compounds and capable of synthesizing all their organic compounds from carbon dioxide eventually evolved, some time before significant amounts of molecular oxygen had accumulated in the atmosphere as a result of photosynthesis.

Whether the development of pathways of carbon dioxide fixation preceded or followed the development of mechanisms that allowed the extraction of usable energy from sources other than organic compounds, the development of photosynthesis must have been preceded by acquisition of the ability to synthesize porphyrin compounds (see Figure 7–19). Porphyrin derivatives are found in all organisms, with the exception of a few anaerobes. In the form of chlorophylls, porphyrins act to absorb the energy of visible radiation that can be converted into the energy of chemical compounds during the process of photosynthesis. The cytochromes, iron-containing porphyrins, are present in all photosynthetic systems, including strictly anaerobic photosynthetic bacteria, and in all aerobes.

It is also most likely that prokaryotes that possessed the ability to synthesize porphyrins were additionally capable of synthesizing the simpler isoprenoid compounds (see page 180) from acetic acid. The carotenoids and the phytol tail of the chlorophylls are isoprenoid derivatives; both carotenoids and chlorophylls are present in all photosynthetic systems. Margulis has pointed out that a good deal of evidence exists to support the hypothesis that isoprenoid compounds initially evolved to protect cellular organic compounds from photochemical oxidation (photo-oxidation).

At about the same time that anaerobic autotrophs were evolving, the scarcity of utilizable organic compounds probably also led to the establishment of a small group of anaerobic bacteria that could generate high-energy phosphates from the organic end-products (reduced electron acceptors) of the metabolism of anaerobic heterotrophs (see Figure 12–6). These organisms, the carbon dioxide reducers, the sulfate reducers, and the nitrate reducers, used the inorganic substances carbon dioxide (CO_2), sulfate (SO_4^{-2}), and nitrate (NO_3^-), respectively, as electron acceptors in the oxidation of reduced carbon compounds* (Fig. 12–8). This was possible because

---

*The carbon dioxide, sulfate, and nitrate reducers are a very specialized group of anaerobic bacterial heterotrophs that use *inorganic* compounds as electron acceptors when oxidizing organic compounds, thus differing from the anaerobic heterotrophs believed to be the first living forms. The latter, like contemporary heterotrophs, base their metabolism entirely upon the use of other *organic* compounds as electron acceptors. Consequently, the use of *organic* electron acceptors in heterotrophic metabolism will be assumed in subsequent discussion unless a specific note to the contrary is made.

carbon dioxide, sulfate, and nitrate have redox potentials more positive than those of many small organic compounds, and therefore carbon dioxide, sulfate, and nitrate reducers were able to make use of small organic compounds not usable by other organisms. Indeed, the ability of many present-day prokaryotes to use nitrate in place of oxygen as an alternative electron acceptor in the normal oxidation of organic compounds has been interpreted as evidence that the nitrate reducers were direct ancestors of aerobic prokaryotes.

Obviously, the pathways of energy metabolism found in the carbon dioxide, sulfate, and nitrate reducers are simply modifications of the basic metabolic design of anaerobic heterotrophs. Here again, we see striking chemical similarities between metabolic systems that superficially appear quite different.

PATHWAYS OF NITROGEN UTILIZATION

Reduced nitrogen compounds are needed for the synthesis of amino acids, nucleotides, and other cell constituents containing nitrogen. The depletion of easily utilized reduced carbon compounds in all likelihood paralleled the depletion of available reduced nitrogen compounds, increasing the selection pressure for the establishment of metabolic pathways resulting in the production of ammonia. All photosynthetic bacteria, many genera of blue-green algae, and some heterotrophic bacteria, including both aerobes and anaerobes, are capable of *nitrogen fixation:* the process by which nitrogen gas (N_2) is reduced to ammonia, and hence made available for biosynthesis. Because of the widespread occurrence of nitrogen fixing capability in prokaryotes, particularly anaerobic ones, some

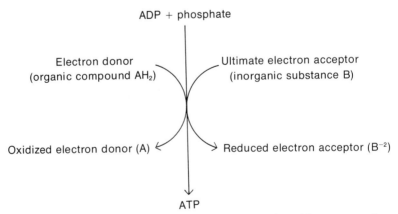

Figure 12–8 A type of anaerobic respiration in which electrons are transferred from an organic compound (AH_2) to an inorganic electron acceptor (B). B may be CO_2, SO_4^{-2}, or NO_3^-; CO_2 is reduced to methane or acetic acid, SO_4^{-2} is reduced to H_2S, and NO_3^- is reduced to NO_2^- (nitrite) or other reduced forms of nitrogen including N_2O, N_2, and NH_3.

workers believe that the pathway of nitrogen fixation must have been one of the earliest metabolic pathways to evolve. This would suggest that atmospheric ammonia was depleted at a very early point in the evolution of living systems, considerably before molecular oxygen accumulated. Perhaps ammonia was destroyed photochemically.

Other workers, however, contend that ammonia might not have completely disappeared before free oxygen appeared in the atmosphere. If this was the case, nitrogen may have been present in the more oxidized form of nitrate (NO_3^-) or nitrite (NO_2^-), and these substances may have been used initially as nitrogen sources by means of metabolic reduction to ammonia. Indeed, all plants utilize either nitrate or nitrite as a source of nitrogen for biosynthesis, and possess enzymes for the reduction of these ions to ammonia (which exists primarily as ammonium ion, NH_4^+). Thus, the latter pathways may have evolved before the pathways leading to the fixation of nitrogen gas.

PATHWAYS OF AEROBIC RESPIRATION

The third stage in the evolution of living systems, which is continuing even today, presumably resulted from the success of organisms capable of oxygen-eliminating photosynthesis. The period beginning with the first appearance of free oxygen in the atmosphere was increasingly inhospitable to anaerobic life, and, although some organisms continued to exist in localized anaerobic environments, selection pressure strongly favored those forms that were able to adapt to the adverse effects of molecular oxygen.

It is not entirely clear why oxygen gas is poisonous to modern anaerobes. At least in some cases, it is apparently poisonous because molecular oxygen is reduced to hydrogen peroxide (H_2O_2) within the cell. Anaerobic bacteria lack a mechanism for destroying H_2O_2, a strong, toxic oxidizing agent that inhibits their growth. Conversely, facultative aerobes and strict aerobes, besides possessing the ability to use oxygen in their metabolic activities, possess the enzyme catalase, an iron–porphyrin protein that degrades H_2O_2 almost as quickly as it is formed.

Whatever causes the sensitivity of anaerobes to oxygen, the accumulation of oxygen in the environment must have led to the establishment of a variety of new, selectively advantageous, metabolic functions. The most successful of these was the development of aerobic metabolism, in which molecular oxygen could be used as the final electron acceptor in energy-generating pathways.

It has been suggested by McElroy and others that bioluminescence—the emission of light by living organisms—originally developed in anaerobic bacteria as a mechanism for detoxifying oxygen. In the process of bioluminescence, a substrate called luciferin

336 **The Evolution of Cellular Metabolism**

(which varies in structure in different types of organisms) is oxidized by molecular oxygen in the presence of ATP and an enzyme called luciferase to form an excited molecule that fluoresces (emits light and some heat) on spontaneous return to the ground state. Because molecular oxygen is reduced to water in this reaction, anaerobic bacteria may have used bioluminescence as a way of eliminating oxygen from their intracellular environment. It is interesting to note that the reduction of one molecule of oxygen in the bioluminescence reaction requires one molecule of ATP, and therefore is very costly from an energy standpoint.

Bioluminescence is observed in a wide range of modern orga-

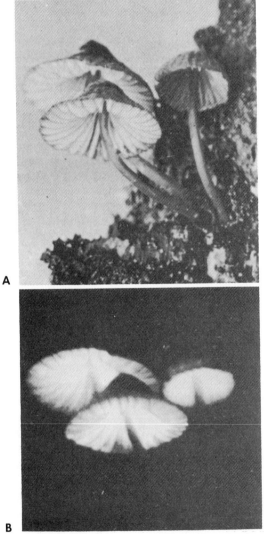

Figure 12-9 Bioluminescent fungi, photographed (a) in daylight and (b) in the dark. (*From* E.O. Wilson et al. [1973], *Life on Earth,* Sinauer, p. 544.)

nisms, including some bacteria, some fungi, and many animals (Fig. 12-9). In order to account for the diversity of organisms in which bioluminescence now occurs, McElroy and Seliger have proposed that although bioluminescence was one of the earliest oxygen detoxifying mechanisms, it lost this function in the great majority of organisms as the ability to synthesize catalases developed, and as metabolic pathways that utilized oxygen were established. The presence of fluorescence in advanced organisms such as fireflies has been explained by some biologists as the retention of an ancient mechanism now used for other functions. For example, fireflies use bioluminescence as an adjunct to mating—a function that has developed relatively recently in evolution.

In addition to the toxic effect of molecular oxygen on anaerobes, the accumulation of oxygen in the atmosphere led to a drastic change in the chemistry of the primitive earth. Not only would the establishment of an ozone shield (see Chapter 6) have eliminated ultraviolet radiation, an abundant source of energy previously available for prebiotic synthesis, but prebiotic synthesis would no longer have been possible in the presence of oxygen. In addition, organic compounds are oxidized to carbon dioxide by oxygen, and consequently all living forms incapable of autotrophic carbon dioxide fixation became dependent on autotrophic organisms for a source of organic compounds.

One class of organism that may have evolved during the period of transition from a reducing to an oxidizing atmosphere is the *aerobic chemoautotrophs*. These organisms, presumably established at a time when there was an insufficiency of readily utilizable organic compounds, are able to generate ATP by oxidizing a variety of reduced inorganic substances in the earth's crust. Unlike the anaerobic chemoautotrophs, the aerobic chemoautotrophs possess the ability to use molecular oxygen as an electron acceptor (Fig. 12-10). Some of the ATP generated by the oxidation of inorganic sub-

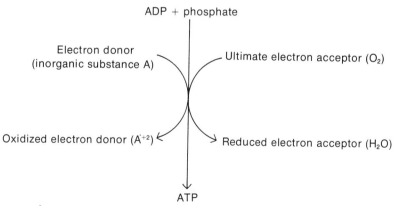

Figure 12-10 Summary of energy-generating metabolism in aerobic chemoautotrophs. The transfer of electrons from an inorganic electron donor (A) to oxygen is coupled to the phosphorylation of ADP. Electron donor A may be NH_4^+, NO_2^-, Fe^{++}, CO, H_2, H_2S, $S_2O_3^{-2}$, S, or other inorganic substances. ATP is used to generate a strong reductant needed for CO_2 fixation.

stances is used to produce a strong reductant; ATP and the strong reductant are then used to drive carbon dioxide fixation.

As the concentration of oxygen in the atmosphere continued to increase, the chemical substances at the surface of the earth presumably became increasingly oxidized. This process would have greatly diminished the supply of reduced inorganic compounds, leading to a considerable restriction in the distribution of aerobic chemoautotrophs, which became dependent for nutrients primarily on the reduced *inorganic* end-products formed by the carbon dioxide reducers, sulfate reducers, nitrate reducers, and anaerobic chemoautotrophs.

As a consequence of the success of the chemosynthetic autotrophs and the oxygen-eliminating photosynthetic relatives of present-day blue-green algae, organic matter probably began once again to accumulate in the oceans. This, in turn, may have created selection pressure for the development of aerobic heterotrophs. Use of oxygen as an electron acceptor in respiratory metabolism allowed aerobic organisms to extract amounts of energy from organic compounds considerably greater than were extractable when organisms used compounds more reduced than oxygen (i.e., organic compounds) as electron acceptors (Fig. 12–11). The development of an efficient oxygen-utilizing form of energy metabolism in prokaryotes set the stage for the evolution of eukaryotes, which are exclusively aerobic forms.

AN OVERVIEW OF METABOLIC DEVELOPMENT

In summary, three general stages can be distinguished in the development of metabolic systems during the evolution of living forms: (1) development of metabolism associated with anaerobic heterotrophs, (2) development of photosynthesis and elimination of

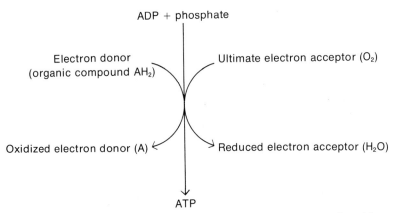

Figure 12–11 Summary of respiration in aerobic heterotrophs. Electrons are transferred from an organic electron donor (AH_2) to molecular oxygen, and the energy released in the process is used to phosphorylate ADP.

oxygen, and (3) development of aerobic metabolism. We will consider these stages in Chapters 13, 14, and 15. We must keep in mind, however, that it is not possible to make clear distinctions between processes occurring in the three stages. As is apparent from the previous discussion, the metabolic systems of living forms did *not* develop in stages, but developed continuously, and therefore any division of the process is wholly artificial.

The continuity of evolution is illustrated by the observation that the metabolic processes that presumably developed initially in anaerobic heterotrophs provide a fundamental framework on which other metabolic processes are built. Not only do all metabolic systems capture energy by moving electrons, but many *specific* metabolic pathways occur in similar or identical form in a wide variety of organisms possessing diverse types of metabolic systems. For example, the pathway of glucose oxidation in anaerobes occurs in identical form, using identical enzymes and identical steps, in *all* organisms that degrade glucose. Glucose oxidation in aerobic organisms has, in addition to the basic pathway of glucose oxidation found in all organisms, reactions associated with the use of molecular oxygen as a terminal electron acceptor. In a similar way, the main pathway of autotrophic carbon dioxide fixation uses many of the same enzymes and chemical reactions as occur in an alternate pathway of glucose degradation which probably developed initially in anaerobic heterotrophs (the pentose phosphate pathway; to be discussed in Chapter 13). However, for carbon dioxide fixation, the pathway runs essentially in the reverse direction. By thus comparing metabolic systems in a wide range of organisms, we can see many additional examples of the continuity of evolution.

The purpose of this chapter has been to provide an overview of the design and development of metabolic systems, in order to help you understand why metabolic systems are the way they are. This information should provide you with a valuable perspective for learning the details of cellular metabolism presented in the next three chapters.

REFERENCES

Bronk, J. R. (1973). *Chemical Biology*. Macmillan, Inc., New York.
Howland, J. L. (1968). *Introduction to Cell Physiology*. Macmillan, Inc., New York.
Krebs, H. A., and H. L. Kornberg (1957). *Energy Transformations in Living Matter*. Springer Verlag.
Lehninger, H. L. (1965). *Bioenergetics*. W. A. Benjamin, Inc., New York.
Levy, J., J. J. R. Campbell, and T. H. Blackburn (1972). *Introductory Microbiology*. John Wiley and Sons, Inc., New York.
Margulis, L. (1970). *Origin of Eukaryotic Cells*. Yale University Press, New Haven, Connecticut.
Miller, S. L., and L. E. Orgel (1974). *The Origins of Life on the Earth*. Prentice-Hall, Inc., Englewood Cliffs, New Jersey.
Oparin, A. I. (1968). *Genesis and Evolutionary Development of Life*. Academic Press, Inc., New York.
Orgel, L. E. (1973). *The Origins of Life*. John Wiley and Sons, Inc., New York.
Rose, A. H. (1965). *Chemical Microbiology*. Butterworths, London.

CHAPTER **13**

The Metabolism of Anaerobic Heterotrophs

It would be an extensive and doubtless unprofitable task to attempt to discuss in detail all the known metabolic pathways that occur in present-day cells. One of our main purposes in studying metabolism is to confirm the basic similarity of the metabolic systems of all organisms. We will therefore restrict our attention primarily, though by no means exclusively, to carbohydrate metabolism, which includes the energy-yielding pathways involved in the oxidation of glucose and related compounds (respiration), as well as the anabolic pathways in autotrophs that result in the synthesis of carbohydrates from carbon dioxide.

This approach, although admittedly limited, is nevertheless valid. Carbohydrates in various forms are used as an energy source (though not necessarily the only energy source) in the great majority of organisms, both anaerobes and aerobes, and the synthesis of carbohydrates that takes place in autotrophic carbon dioxide fixation is the ultimate source of organic compounds for all living organisms. In addition, the major pathways of glucose degradation connect with many other catabolic and anabolic pathways at points involving common intermediates. Indeed, carbohydrate metabolism, in one form or another, is so important to living organisms that its pathways might be described as the "core" of cellular metabolism.

THE ANAEROBIC OXIDATION OF GLUCOSE

THE EMBDEN-MEYERHOF PATHWAY

In most anaerobic heterotrophs, energy is derived primarily from the anaerobic oxidation of glucose (or its storage compounds, glycogen or starch) to lactic acid, ethyl alcohol, or various other organic end-products. This was probably one of the first energy-capturing processes to develop. It is frequently called *glycolysis* (glycogen breakdown), although the term *fermentation* is often used to refer to the same process, especially when ethyl alcohol

The Anaerobic Oxidation of Glucose

is the end-product. Whether the process is called glycolysis or fermentation, the pathway is nearly identical in all organisms, differing primarily with respect to end-products. Confusion caused by use of both the terms "glycolysis" and "fermentation" to refer to the same process tends to obscure the fact that its pathway is fundamentally the same in all organisms where it occurs. We therefore prefer to refer to the chemical transformations involved in the process of glucose oxidation by the term *Embden-Meyerhof pathway*, a name that honors two of the scientists who first described them.

Before describing the individual reactions that compose the Embden-Meyerhof pathway, we should clarify several points. The beginning student is often confused by the fact that the Embden-Meyerhof pathway operates in both anaerobic and aerobic organisms. Thus, although the Embden-Meyerhof pathway brings about the oxidation of glucose in anaerobic organisms, it can, and does, function in the presence of oxygen, *but it does not use oxygen.* It is an anaerobic pathway, even in aerobes. In aerobic organisms, the Embden-Meyerhof pathway is supplemented by additional pathways that do use oxygen; as we will see, these supplementary processes serve to complete the oxidation of the end-products of the Embden-Meyerhof pathway to yield carbon dioxide and water.

The presence of the Embden-Meyerhof pathway in almost all organisms that degrade glucose is good evidence that this pathway developed early in the evolutionary process. Additional evidence supports this contention. For instance, the *aerobic* reactions of glucose oxidation apparently take place in localized regions of the cell. In aerobic prokaryotes, at least some of the enzymes of aerobic glucose oxidation appear to be associated with the plasma membrane; in eukaryotes, all these enzymes are located in the mitochondria. In direct contrast to the enzymes of aerobic glucose oxidation, however, the enzymes of the Embden-Meyerhof pathway are soluble in the cytoplasm, and are not apparently associated with any particular intracellular structure. The fact that the operation of the Embden-Meyerhof pathway does not require a special structural organization is additional evidence that the pathway developed at an early stage in the evolution of living forms, presumably before intracellular structure had developed to any great extent.

Because of the central importance of the Embden-Meyerhof pathway in cellular metabolism, and because the design of the Embden-Meyerhof pathway illustrates principles that are employed in other metabolic pathways, we will describe its individual reactions in detail.

Reactions of the Embden-Meyerhof Pathway

Before glucose can be oxidized by the Embden-Meyerhof pathway, it must first enter the cell. This seemingly obvious point is mentioned here because it is usually overlooked. The point is important because the uptake of glucose by a cell often requires the expenditure of energy; that is, glucose is brought into the cell by energy-

Embden-Meyerhof pathway

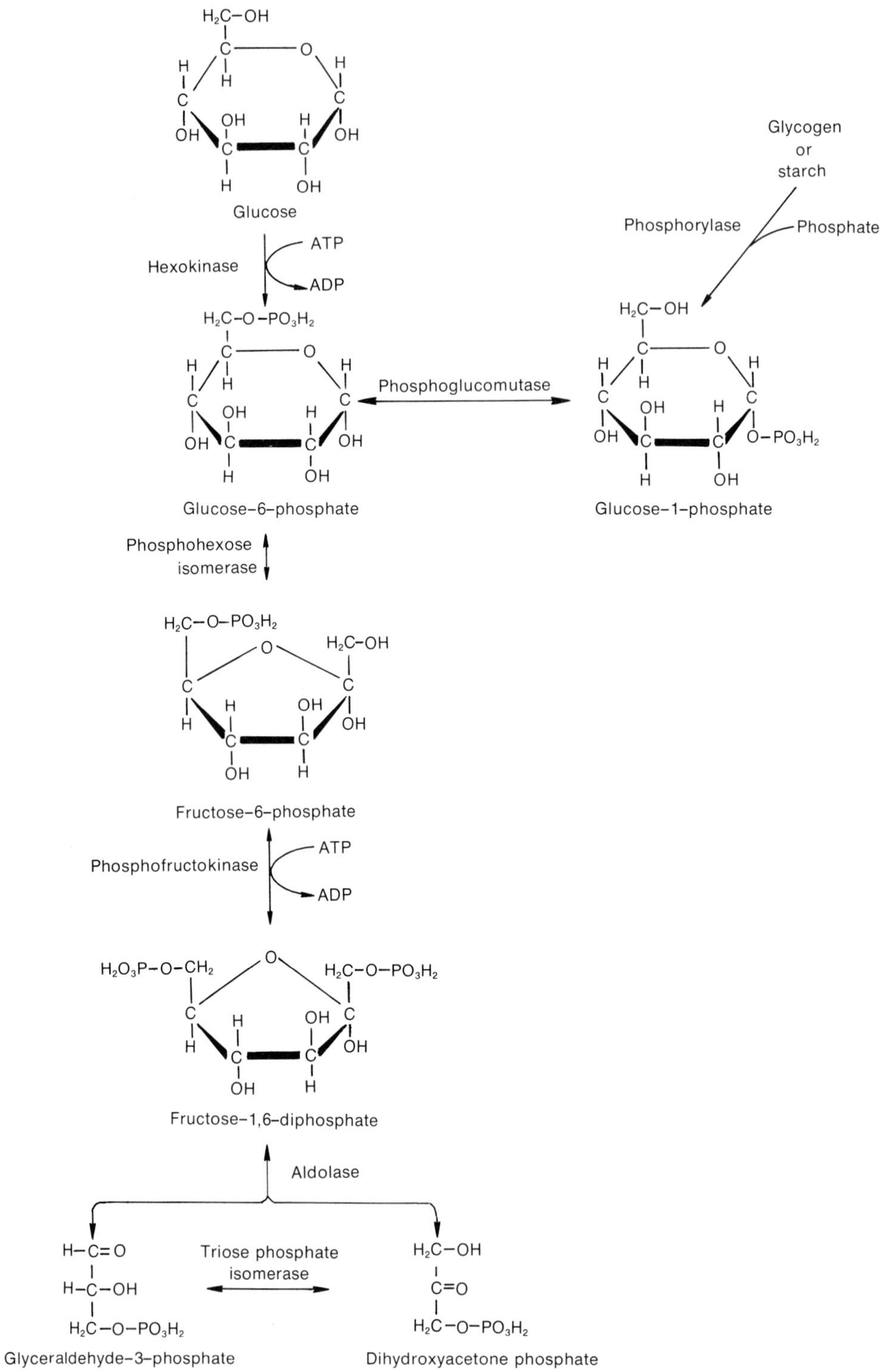

Figure 13–1 The Embden-Meyerhof pathway. At the pH existing within the cell (about pH 7), the —COOH and $—PO_3H_2$ groups would be ionized to $—COO^-$ and $—PO_3^{-2}$, respectively. All components are shown here in their nonionized forms so that interconversions are easier to follow. *(Continued on next page)*

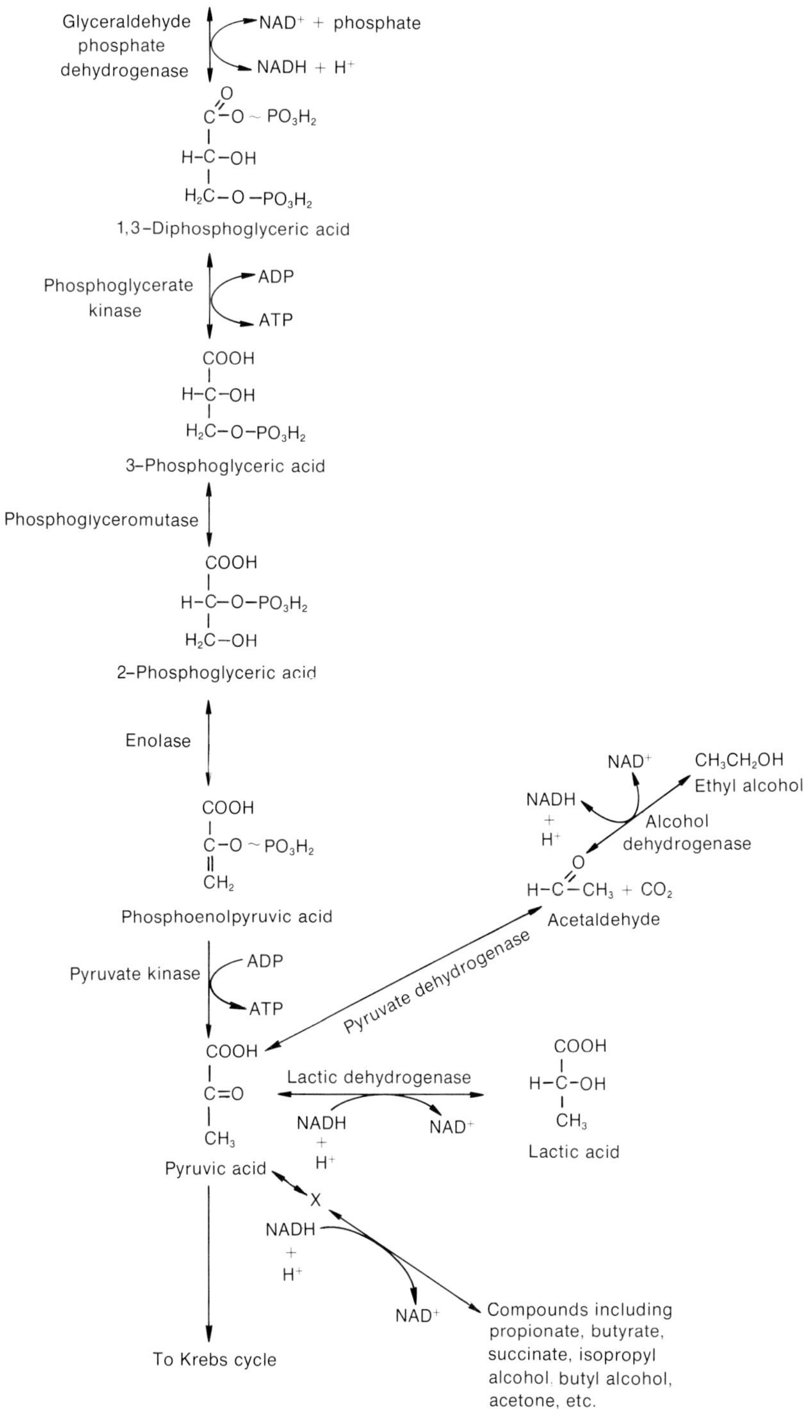

The Anaerobic Oxidation of Glucose

Glyceraldehyde phosphate dehydrogenase ⇅ NAD$^+$ + phosphate / NADH + H$^+$

$$\underset{\text{1,3-Diphosphoglyceric acid}}{\begin{array}{c} O \\ \| \\ C-O \sim PO_3H_2 \\ | \\ H-C-OH \\ | \\ H_2C-O-PO_3H_2 \end{array}}$$

Phosphoglycerate kinase ⇅ ADP / ATP

$$\underset{\text{3-Phosphoglyceric acid}}{\begin{array}{c} COOH \\ | \\ H-C-OH \\ | \\ H_2C-O-PO_3H_2 \end{array}}$$

Phosphoglyceromutase ⇅

$$\underset{\text{2-Phosphoglyceric acid}}{\begin{array}{c} COOH \\ | \\ H-C-O-PO_3H_2 \\ | \\ H_2C-OH \end{array}}$$

Enolase ⇅

$$\underset{\text{Phosphoenolpyruvic acid}}{\begin{array}{c} COOH \\ | \\ C-O \sim PO_3H_2 \\ \| \\ CH_2 \end{array}}$$

Pyruvate kinase ⇅ ADP / ATP

$$\underset{\text{Pyruvic acid}}{\begin{array}{c} COOH \\ | \\ C=O \\ | \\ CH_3 \end{array}}$$

Pyruvate dehydrogenase → Acetaldehyde $H-\overset{O}{\overset{\|}{C}}-CH_3 + CO_2$

Alcohol dehydrogenase: NADH + H$^+$ / NAD$^+$ → CH$_3$CH$_2$OH Ethyl alcohol

Lactic dehydrogenase: NADH + H$^+$ / NAD$^+$ →

$$\underset{\text{Lactic acid}}{\begin{array}{c} COOH \\ | \\ H-C-OH \\ | \\ CH_3 \end{array}}$$

X: NADH + H$^+$ / NAD$^+$ → Compounds including propionate, butyrate, succinate, isopropyl alcohol, butyl alcohol, acetone, etc.

To Krebs cycle

Figure 13-1 (*Continued*)

requiring transport processes (we will discuss these in Chapter 16). Consequently, energy may have to be expended by the cell *before* the oxidation of glucose can begin.

The sequence of reactions involved in the anaerobic oxidation of glucose by the Embden-Meyerhof pathway is shown in Figure 13-1. The pathway can be broken down into the following steps.

1. In the first reaction of the pathway, glucose is phosphorylated to glucose–6–phosphate. The phosphorylation of glucose requires energy, and is coupled to the hydrolysis of ATP:

Glucose + H_3PO_4 ⇌ glucose–6–phosphate + H_2O $\Delta G° = +3.9$ Kcal/mole
ATP + H_2O ⇌ ADP + H_3PO_4 $\Delta G° = -7.3$ Kcal/mole

Net reaction:
Glucose + ATP ⇌ glucose–6–phosphate + ADP $\Delta G° = -3.4$ Kcal/mole

Despite the impression conveyed by the above equations, phosphorylation does not actually occur in two steps. Instead, the phosphate group is transferred directly from ATP to glucose in a reaction catalyzed by the enzyme hexokinase. *Kinases* are a class of enzymes that catalyze the transfer of phosphate from ATP to an acceptor molecule (in this case, glucose). Like all enzymes, kinases catalyze specific reactions in both directions.

In this phosphorylation reaction, as in the uptake of glucose by some cells, energy must be expended. The formation of glucose–6–phosphate from the glucose molecule is often referred to as the "activation" of glucose, and has been likened to pump-priming. In the same way that fluid must be used to prime a pump before the pumping process can begin, so energy must be expended before a greater amount of energy can be extracted from the glucose molecule.

Breakdown of Stored Glucose. An alternate source of glucose–6–phosphate is the breakdown of glycogen or starch, the cellular storage forms of glucose. This process, which is shown in Figure 13-1, occurs in two steps. Initially, glycogen or starch is broken down with the formation of glucose–1–phosphate in a reaction that requires inorganic phosphate, and is catalyzed by the enzyme phosphorylase. Unlike the direct phosphorylation of glucose, this phosphorylation reaction occurs spontaneously and does not require energy in the form of ATP. In the next step of the process, glucose–1–phosphate is converted to glucose–6–phosphate. This reaction is catalyzed by phosphoglucomutase. *Mutase enzymes* are responsible for the internal transfer of a phosphate group from one hydroxyl group to another on the same molecule. The mutases are one type of a more general class of enzymes called *isomerases,* which catalyze a redistribution of atoms, or groups of atoms, within a molecule.*

---

*The original molecule and the new molecule formed by an isomerization reaction are called *structural isomers* of one another. In contrast to optical isomers, structural isomers are molecules having the same molecular composition, but differing in at least one chemical or physical property. Structural isomerism results from the different order in which atoms are bonded to one another. For example, butane (CH_3—CH_2—CH_2—CH_3) and *iso*butane (CH_3—CH—CH_3) are structural isomers.
$\qquad\qquad\qquad\qquad\qquad\qquad\qquad\qquad\quad |$
$\qquad\qquad\qquad\qquad\qquad\qquad\qquad\qquad\; CH_3$

The Anaerobic Oxidation of Glucose

Thus, ATP is not required for the formation of glucose-6-phosphate if glycogen or starch, present as a reserve polysaccharide, is used by the cell. Consequently, one less ATP molecule is expended in the Embden-Meyerhof pathway if the starting point of the pathway is glycogen or starch instead of glucose. However, utilizing stored starch does not amount to getting something for nothing, because the intracellular synthesis of glycogen or starch requires energy, and the energy requirement for this synthesis is the equivalent of two molecules of ATP per glucose unit.

2. Following the production of glucose-6-phosphate, another isomerase enzyme, glucose phosphate isomerase, catalyzes the conversion of glucose-6-phosphate to its structural isomer, fructose-6-phosphate.

3. In the next step, fructose-6-phosphate is phosphorylated at the first carbon to produce the compound fructose-1,6-diphosphate, an essentially symmetrical molecule possessing a phosphate group at either end. This phosphorylation reaction, like the initial phosphorylation of glucose, occurs through the expenditure of ATP.

Up to this point in the pathway, no energy has been generated. In fact, two molecules of ATP have been expended for each molecule of glucose entering the pathway. The energy-capturing reactions of the Embden-Meyerhof pathway all occur in subsequent steps.

4. Fructose-1,6-diphosphate is now split into two three-carbon molecules, the triose phosphates glyceraldehyde-3-phosphate, and dihydroxyacetone phosphate. These two compounds are not identical, but are nevertheless readily interconvertible by another isomerase enzyme, triose phosphate isomerase.

The cell utilizes glyceraldehyde-3-phosphate in the next step of the pathway. Because dihydroxyacetone phosphate is convertible to glyceraldehyde-3-phosphate, the cleavage of one molecule of fructose-1,6-diphosphate may be thought of as producing *two* molecules of glyceraldehyde-3-phosphate for subsequent reactions in the pathway. In other words, as glyceraldehyde-3-phosphate is used up in the next step of the pathway, it is replaced by conversion of dihydroxyacetone phosphate. Thus, *for each molecule of glucose oxidized,* all subsequent reactions of the pathway after the triose phosphate isomerase reaction must be considered as operating twice. This is an important consideration in calculating the yield of ATP derived from each molecule of glucose oxidized by the Embden-Meyerhof pathway.

The aldolase reaction, catalyzing the conversion of fructose-1,6-diphosphate to the two triose phosphates, is significant in yet another respect. The reaction is highly endergonic; that is, the equilibrium of the reaction is strongly in favor of the production of fructose-1,6-diphosphate rather than the production of the two triose phosphates. Consequently, the Embden-Meyerhof pathway could not proceed in the forward direction past the aldolase reaction, were it not for the removal of glyceraldehyde-3-phosphate in the next step. The aldolase reaction is a good example of the importance of thermodynamically coupled reactions in metabolism. It is the free energy released in the subsequent reactions in the pathway that drives the aldolase reaction in the direction of triose phosphate production by virtue of the removal of glyceraldehyde-3-phosphate.

5. The next step in the pathway is the oxidation of glyceraldehyde-3-phosphate in the presence of inorganic phosphate and a coenzyme molecule (NAD^+) to form 1,3-diphosphoglyceric acid. Oxidation of glyceraldehyde-3-phosphate involves the removal of two electrons and two protons (the equivalent of two hydrogen atoms). Since all oxidations are accompanied by reductions, the oxidation of glyceraldehyde-3-phosphate must be linked to the reduction of another substance. In this case, as in most biological oxidations not involving molecular oxygen, the electrons are transferred to a coenzyme molecule that acts as an electron carrier. The coenzyme molecule is capable of being alternately oxidized and reduced, and therefore can donate or accept electrons in biological reactions.

The coenzyme molecule involved in this reaction is nicotinamide adenine dinucleotide (often abbreviated as NAD*). The structures of the oxidized and reduced forms of NAD are shown in Figure 13-2. The oxidized form of NAD has a net positive charge, and hence is commonly represented as NAD^+. When NAD^+ is reduced, two electrons and one proton are added to the nicotinamide part of the molecule, and one proton is liberated into the aqueous medium. The reduced form of NAD is denoted as NADH. The reduction of NAD^+ (or, in the reverse direction, the oxidation of NADH) may be summarized by the equation

$$NAD^+ + 2 H^+ + 2 e^- \rightleftarrows NADH + H^+$$

The enzyme glyceraldehyde phosphate dehydrogenase, which catalyzes the oxidation of glyceraldehyde-3-phosphate in the presence of NAD^+ and inorganic phosphate, serves to combine the coenzyme molecule and the substrate (glyceraldehyde-3-phosphate) on its surface. Hence, the enzyme mediates the transfer of electrons between donor (in this case, the substrate) and acceptor (in this case, NAD^+), without itself becoming oxidized or reduced. NAD is the most frequently encountered electron carrier; it interacts with a number of different enzymes and serves as a coenzyme in many intracellular reactions. Within the cell, NAD^+ exists in equilibrium with its reduced form (NADH), and thus there is normally a pool of both oxidized and reduced NAD within the cell.

The net result of this step in the pathway may be summarized by the equation

$$\text{Glyceraldehyde-3-phosphate} + H_3PO_4 + NAD^+ \rightleftarrows$$
$$\text{1,3-diphosphoglyceric acid} + NADH + H^+$$

---

*NAD is sometimes referred to by its older names, diphosphopyridine nucleotide (DPN) or coenzyme I.

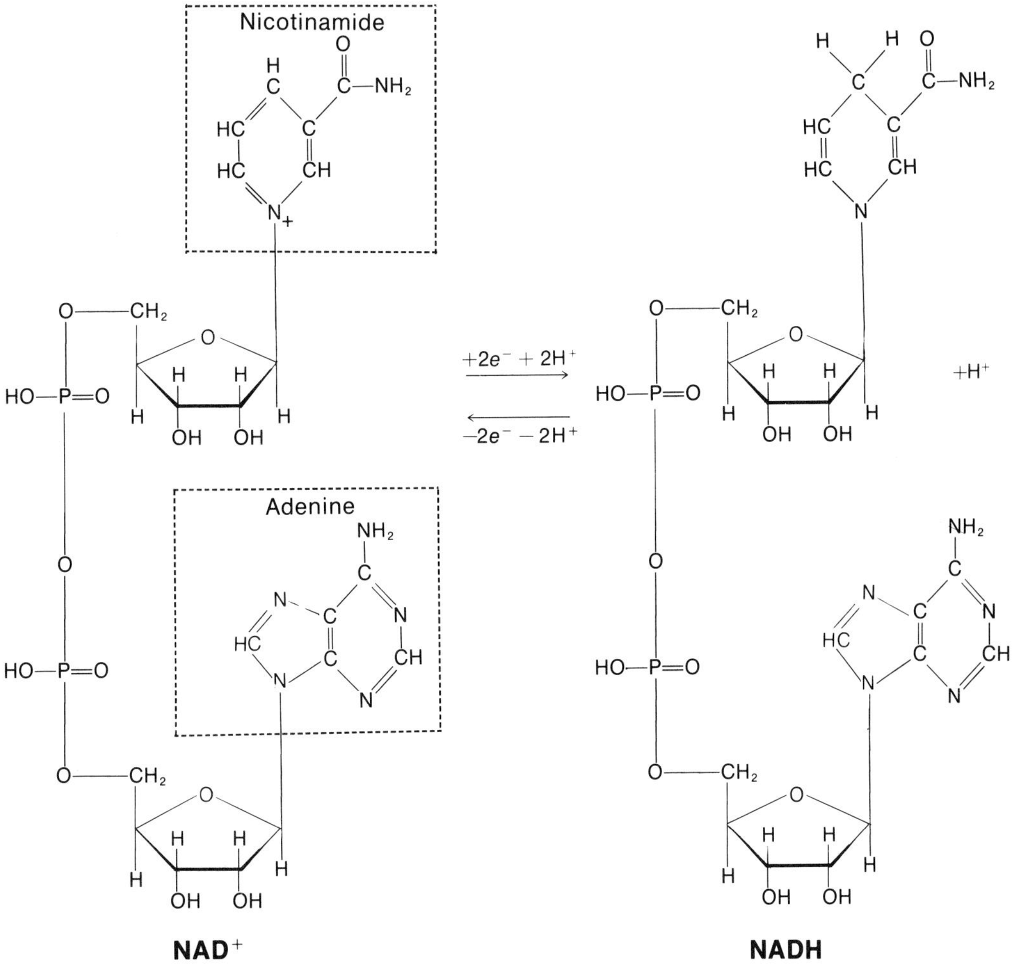

The Anaerobic Oxidation of Glucose

Figure 13-2 The oxidized (*left*) and reduced (*right*) forms of nicotinamide adenine dinucleotide (NAD). In the reduction of NAD^+, 2 electrons and 1 proton are added to the nicotinamide ring; the other proton is released into the solution.

Notice that glyceraldehyde-3-phosphate is phosphorylated in this reaction, and that the phosphate is derived from inorganic phosphate and not from ATP. Indeed, the oxidation of glyceraldehyde-3-phosphate to 1,3-diphosphoglyceric acid is a strongly exergonic reaction, and most of the energy released in the oxidation is stored in NADH. As we will see, NADH is an exceptionally strong reductant (electron donor), and its oxidation can in turn release energy.

Up to this point, the reactions of the Embden-Meyerhof pathway have utilized two molecules of ATP, and have produced two molecules of the strong reductant NADH and two molecules of 1,3-diphosphoglyceric acid for every molecule of glucose oxidized.

6. You will recall that a phosphate group attached to an atom that also bears a double bond has a high energy of hydrolysis. This configuration is characteristic of 1,3-diphosphoglyceric acid, which is therefore a high-energy compound.

At this point in the pathway, the enzyme phosphoglycerate kinase catalyzes the transfer of the high-energy phosphate group in 1,3-diphosphoglyceric acid to ADP, resulting in the formation of 3-phosphoglyceric acid and ATP. As a result of this reaction, much of the energy released by removal of the high-energy phosphate group from 1,3-diphosphoglyceric acid is captured in the form of ATP. In terms of energy, this reaction may be summarized by the equations

1,3-Diphosphoglyceric acid + H_2O ⇌
$\qquad$ 3-phosphoglyceric acid + H_3PO_4 $\quad \Delta G° = -14.1$ Kcal/mole
$\qquad$ ADP + H_3PO_4 ⇌ ATP + H_2O $\quad \Delta G° = +7.3$ Kcal/mole

---

Net reaction:
1,3-Diphosphoglyceric acid + ADP ⇌
$\qquad$ 3-phosphoglyceric acid + ATP $\quad \Delta G° = -6.8$ Kcal/mole

The production of a high-energy phosphate derivative of a substrate, followed by formation of ATP by the group transfer of the high-energy phosphate group to ADP, is called *substrate-level phosphorylation* to distinguish it from other types of ATP formation encountered in the processes of photosynthesis and aerobic respiration.

Because two molecules of 1,3-diphosphoglyceric acid result from the oxidation of each molecule of glucose, two molecules of ATP are produced in this step of the pathway. Therefore, up to this point, two molecules of ATP have been consumed and two molecules have been produced.

7. In the next several reactions in the pathway, another high-energy phosphate compound is produced in two steps. In the first step, 2-phosphoglyceric acid is formed from 3-phosphoglyceric acid. This transfer of the phosphate group from the third to the second carbon is catalyzed by the mutase enzyme phosphoglyceromutase. During the second step, 2-phosphoglyceric acid is dehydrated to form phosphoenolpyruvic acid, a reaction catalyzed by the enzyme enolase. The result of the latter reaction is the formation of

The Anaerobic Oxidation of Glucose

phosphoenolpyruvic acid, a compound that possesses a high-energy phosphate configuration.

8. In the next step of the pathway, another group transfer reaction occurs, catalyzed by the kinase enzyme, pyruvate kinase. The high-energy phosphate group in phosphoenolpyruvic acid is transferred to ADP, forming ATP and pyruvic acid. Here, as in the conversion of 1,3-diphosphoglyceric acid to 3-phosphoglyceric acid, two molecules of ATP are formed for each glucose molecule oxidized, adding up to a net gain of two ATP molecules up to this point in the oxidation of glucose.

9. Under anaerobic conditions, pyruvic acid is usually converted into either lactic acid or ethyl alcohol (see Figure 13-1), depending on the particular organism involved. The reducing agent for these transformations is NADH that was formed earlier during the oxidation of glyceraldehyde-3-phosphate. The effect of the latter conversions of pyruvic acid is to regenerate NAD^+ that can be used in the oxidation of another molecule of glyceraldehyde-3-phosphate (Fig. 13-3). In fact, if NAD^+ were not regenerated, the Embden-Meyerhof pathway could not continue to operate.

Similarity of Anaerobic Glucose Oxidation in All Organisms

Although the anaerobic decomposition of glucose to either lactic acid or ethyl alcohol occurs most frequently in living forms, several types of anaerobic prokaryotes (of which there are very few present-day forms) use various other organic compounds as terminal electron acceptors (X), as shown in Figure 13-1. However, no matter what the terminal electron acceptor, if the Embden-Meyerhof pathway is present at all, it is identical in all organisms to the point of pyruvic acid formation; the differences begin thereafter. As long as NAD^+ is regenerated at the end of the pathway, it is not critical what the terminal electron acceptor is. The other steps of the pathway, however, depend on the formation of precise chemical configurations, enabling the energy released in the oxidation of glucose to be coupled to the phosphorylation of ADP. It is possible to envision other ways that energy could be captured by the oxidation of glucose, but significantly different pathways would have had to develop in order to accomplish this. Oparin has suggested that the Embden-Meyerhof pathway must have high biological efficiency, because since its development (presumably at an early stage in the evolution of life), it has survived essentially unchanged through subsequent evolution. Moreover, it occurs in prokaryotes that we believe

Glyceraldehyde-3-phosphate NAD^+ Lactic acid

1,3-Diphosphoglyceric acid $NADH + H^+$ Pyruvic acid

An illustration of the cyclic role of NAD in the Embden-Meyerhof pathway.   Figure 13-3

to be of ancient origin as well as in the metabolism of higher plants and animals.

The essence of the anaerobic Embden-Meyerhof pathway may be summarized as follows: By use of the intermediate electron carrier NAD, electrons are in effect transferred spontaneously from glucose to a terminal electron acceptor, to produce lactic acid, ethyl alcohol, or some other organic compound. Some of the energy released in these electron transfers is captured in the form of ATP.

In aerobic organisms, pyruvic acid is oxidized in the presence of oxygen to carbon dioxide and water, a process that results in the capture of a great deal of energy. We will discuss aerobic respiration in Chapter 15.

Efficiency of the Anaerobic Embden-Meyerhof Pathway

The net yield of useful energy from the oxidation of one molecule of glucose by the Embden-Meyerhof pathway under anaerobic conditions is two molecules of ATP. If we assume that the formation of each mole of ATP requires 7.3 Kcal (the standard free energy change for the phosphorylation of one mole of ADP to ATP), then 14.6 Kcal (7.3 Kcal × 2) represents the amount of energy captured in the form of ATP for each mole of glucose oxidized. Since the complete oxidation of one mole of glucose to carbon dioxide and water proceeds with a $\Delta G°$ of -686.0 Kcal, the Embden-Meyerhof pathway under anaerobic conditions has an efficiency of energy capture of about two per cent.

$$\frac{14.6}{686.0} \times 100 = 2.1\%$$

In actuality, the efficiency of the process is probably somewhat greater than this equation indicates. As we have noted previously, the ΔG value for the phosphorylation of one mole of ADP to ATP *under intracellular conditions* is probably greater than 7.3 Kcal (estimates range from 8.5 to 14.5 Kcal/mole).

Most of the energy available in glucose is retained in the end products of the Embden-Meyerhof pathway (such as lactic acid and ethyl alcohol). For example, the oxidation of glucose to two molecules of lactic acid proceeds with a $\Delta G°$ of -47.4 Kcal/mole. This means that 638.6 Kcal/mole (686.0 Kcal $-$ 47.4 Kcal) cannot be extracted from glucose under anaerobic conditions in which lactic acid is not oxidized further. It is this energy that is available for capture under aerobic conditions.

Cellular Control of the Embden-Meyerhof Pathway

The ratio of oxidized to reduced NAD is probably the most important factor in regulating the rate at which the entire Embden-

The Anaerobic Oxidation of Glucose

Meyerhof pathway operates. NAD^+ is required for the oxidation of glyceraldehyde-3-phosphate, and is itself reduced to NADH in this reaction. NADH is, in turn, reoxidized to NAD^+ during the reduction of pyruvate. Under anaerobic conditions, where NADH cannot be oxidized by molecular oxygen as it is in aerobic cells, there is a tight linkage between the formation and utilization of NADH (or NAD^+); this linkage keeps the beginning and end of the pathway operating at coordinated rates.

Of course, in addition to NAD-mediated control, the availability of glucose, inorganic phosphate, and ADP also influences the rate at which the Embden-Meyerhof pathway operates. Another important control is the availability of enzymes utilized in the pathway; this is determined by the rate of cellular enzyme synthesis and by the concentration of intracellular substances that may act as enzyme inhibitors.

Yet another control on the rate of glucose oxidation by the Embden-Meyerhof pathway is related to the integration of catabolic and anabolic processes, mentioned in Chapter 12. The Embden-Meyerhof pathway provides not only a supply of ATP for the cell but also starting materials for cellular biosynthesis. This is apparent in Figure 13–4, which shows the relationship of the Embden-Meyerhof pathway to a number of biosynthetic pathways. At various stages during the breakdown of glucose by the Embden-Meyerhof pathway, intermediate compounds, instead of being oxidized further, may be shunted toward the synthesis of a particular cellular component. For example, dihydroxyacetone phosphate is an intermediate in the synthesis of lipids (see Figure 13–4). A cellular demand for a particular lipid compound may cause an increased breakdown of glucose, not for the purpose of generating ATP, but for the purpose of generating dihydroxyacetone phosphate needed for the biosynthesis. Various metabolic pathways are, therefore, in competition with one another for common intermediate compounds, and this competition is important in the regulation of metabolic activity. This type of regulation depends upon the fact that a low concentration of a particular compound may pull the pathway in the direction of increased synthesis of that compound (Fig. 13–5).

THE PENTOSE PHOSPHATE PATHWAY

Other pathways besides the Embden-Meyerhof pathway are important to anaerobic carbohydrate metabolism. Krebs and Kornberg have speculated that the *pentose phosphate pathway*, found in most present-day organisms, is one such pathway that may have developed initially in anaerobic prokaryotes.* The pentose phosphate pathway is usually not the main pathway of carbohydrate oxidation; instead, it appears to supplement the Embden-Meyerhof pathway

---

*The pentose phosphate pathway, like the Embden-Meyerhof pathway, can operate under both anaerobic and aerobic conditions.

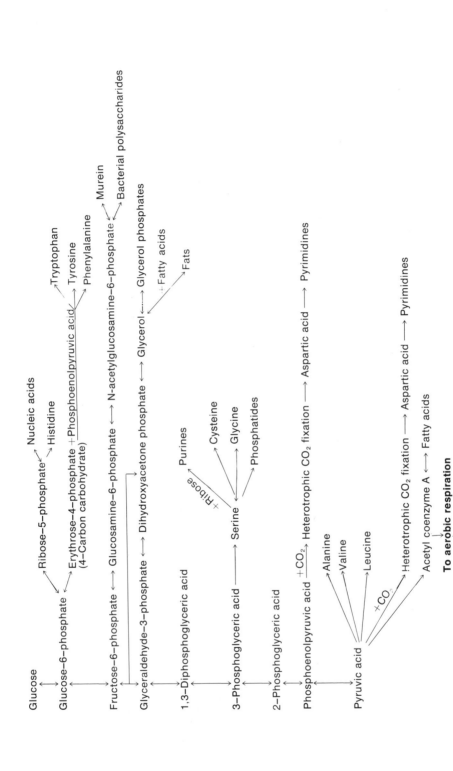

Figure 13–4 A diagram illustrating the relationship between the Embden-Meyerhof pathway and various other catabolic and anabolic pathways of the cell. (Adapted from Levy, J., J.J.R. Campbell, and T.H. Blackburn [1973]. *Introductory Microbiology*, John Wiley and Sons, Inc., New York.)

The Anaerobic Oxidation of Glucose

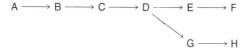

Figure 13-5

Control of cellular metabolism as a result of competition between pathways for common intermediate compounds. A biochemical pathway normally proceeds in the direction of A to F. However, if H were drastically depleted, the pathway might be shifted in the direction of increased production of H by conversion of the common intermediate compound (D) to G, rather than to E.

in the majority of cell types. The main functions of the pentose phosphate pathway apparently are to provide the cell with ribose needed for nucleotide synthesis and to supply it with reduced coenzymes needed for reductive biosynthesis.

Details of the pentose phosphate pathway are shown in Figure 13-6. Although this pathway looks complicated, it is actually quite simple. Three molecules of glucose-6-phosphate, produced by the phosphorylation of glucose, are oxidized and decarboxylated (carbon dioxide is removed) to three molecules of ribulose-5-phosphate, a 5-carbon sugar. Through a series of reactions involving four-carbon, five-carbon, six-carbon, and seven-carbon intermediates, the three molecules of ribulose-5-phosphate are converted into two molecules of glucose-6-phosphate and one molecule of glyceraldehyde-3-phosphate. The latter compound may enter the Embden-Meyerhof pathway for further oxidation. One of the five carbon intermediates of the pentose phosphate pathway is ribose-5-phosphate, which may be shunted into the synthesis of nucleic acid components. Because certain reactions of the Embden-Meyerhof pathway are bypassed in the oxidation of glucose by this scheme, the pentose phosphate pathway is also known as the *pentose phosphate shunt.*

Notice that in the oxidation and decarboxylation of glucose-6-phosphate to ribulose-5-phosphate, a compound designated as $NADP^+$ acts as an electron acceptor. This compound is the coenzyme *nicotinamide adenine dinucleotide phosphate.** It is very similar to NAD in structure, except that NADP possesses an additional phosphate group (Fig. 13-7). NADP was originally discovered at the time that the pentose phosphate pathway was elucidated.

NADP functions as an electron carrier, and, like NAD, is capable of being alternately oxidized and reduced. In general, NAD is associated with reactions involved in energy transfer (energy capture and utilization), whereas NADP is involved in many cellular biosyntheses. NADPH, the reduced form of NADP, possesses a high negative redox potential, similar to that of NADH, and is usually used as the electron donor whenever a reduction occurs in a biosynthetic pathway.

The *net result* of the pentose phosphate pathway is the oxidation of one molecule of glucose to one molecule of glyceraldehyde-3-phosphate; this result may be expressed by the equations

---

*NADP was formerly called triphosphopyridine nucleotide (TPN) or coenzyme II.

354 The Metabolism of Anaerobic Heterotrophs

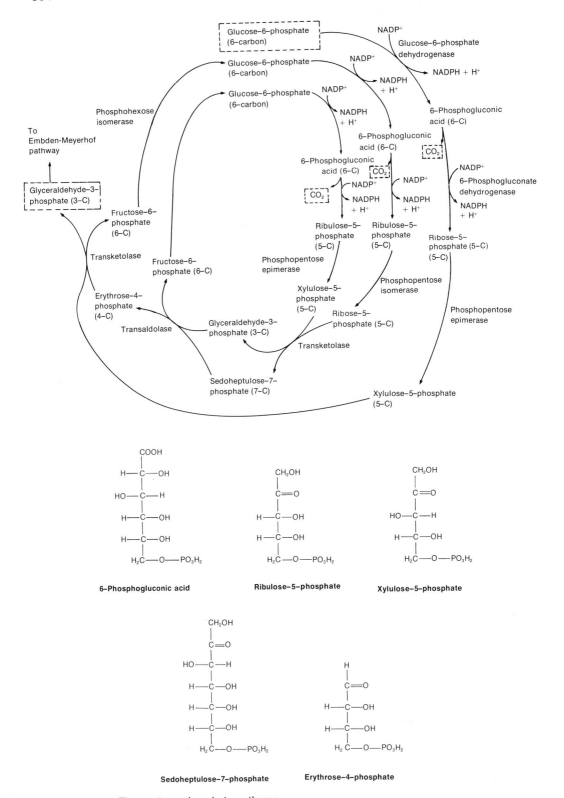

Figure 13-6 The pentose phosphate pathway.

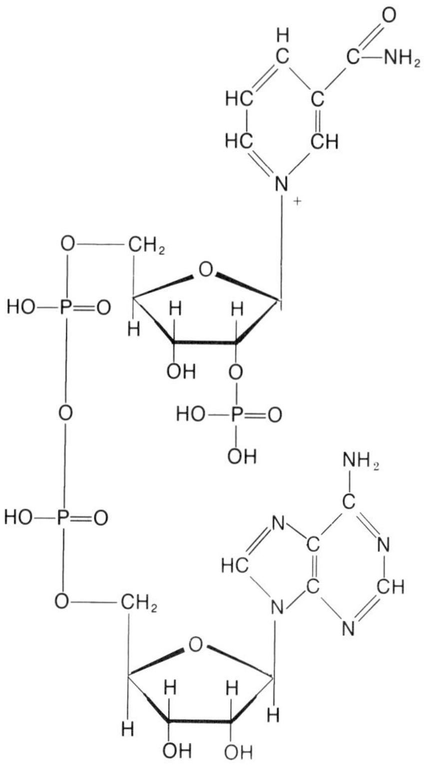

The Anaerobic Oxidation of Glucose

Figure 13-7 The chemical structure of the oxidized form of nicotinamide adenine dinucleotide phosphate (NADP). The reduction of $NADP^+$ occurs in a manner identical to the reduction of NAD^+ (Fig. 13-2).

3 Glucose-6-phosphate + 6 $NADP^+$ ⟶
    2 glucose-6-phosphate + 1 glyceraldehyde-3-phosphate
    + 6 NADPH + 6 H^+ + 3 CO_2

or

Glucose-6-phosphate + 6 $NADP^+$ ⟶
    1 glyceraldehyde-3-phosphate + 6 NADPH + 6 H^+ + 3 CO_2

The overall yield of ATP in the pentose phosphate pathway is only one molecule of ATP for each glucose molecule oxidized. (One ATP is utilized in the phosphorylation of glucose to glucose-6-phosphate, and two ATPs are produced in the oxidation of one molecule of glyceraldehyde-3-phosphate to pyruvic acid.) Consequently, the pentose phosphate pathway is only half as efficient as the Embden-Meyerhof pathway in terms of the capture of energy in the form of ATP. Since its yield of ATP is low, the pentose phosphate pathway appears to function mainly to produce NADPH, as well as ribose and other carbohydrate intermediates important in cellular metabolism. For example, the four-carbon intermediate erythrose–4–phosphate is a very early precursor in the synthesis of the amino acids tryptophan, tyrosine, and phenylalanine (see Figure 13–4).

The pentose phosphate pathway has much in common with the Embden-Meyerhof pathway. Both pathways operate under anaerobic as well as aerobic conditions, and the enzymes of both pathways are soluble in the cytoplasm instead of being tightly bound to a particular intracellular structure. As in the case of the Embden-Meyerhof pathway, these characteristics suggest that the pentose phosphate pathway developed during early stages of biological evolution. Kornberg and Krebs have postulated that the pentose phosphate pathway probably arose as a result of the depletion of ribose in the environment, some time after the Embden-Meyerhof pathway had developed. This conclusion is supported by the observation that although the pentose phosphate pathway possesses individual reactions that also occur in the Embden-Meyerhof pathway, suggesting that one pathway evolved from the other, the pentose phosphate pathway is now the less widely distributed of the two.

One reason for focusing on the pentose phosphate pathway is that it further illustrates the evolutionary continuity of living forms. The main pathway of autotrophic carbon dioxide fixation possesses many reactions identical to those in the pentose phosphate pathway; and in addition, it contains several reactions found in the Embden-Meyerhof pathway. However, in the pathway of carbon dioxide fixation, these reactions operate in the reverse direction, with the result that organic compounds are built up, rather than degraded. Since the pentose phosphate pathway does not require complex organic compounds such as porphyrins in order to operate, some workers have suggested that the pentose phosphate pathway arose earlier than the metabolic pathways associated with photosynthesis. Given the existence of the Embden-Meyerhof and the pentose phosphate pathways, only two additional reactions in the pathway of autotrophic carbon dioxide fixation, plus the processes associated with the capture of light energy (which use the porphyrin molecule, chlorophyll), are required for photosynthesis. If we assume that, as a general rule, metabolism has acquired increasing complexity throughout evolution, it also seems reasonable to assume that photosynthesis developed in organisms in which the pentose phosphate pathway had already been established. This assumption has strong support, but it is not universally accepted.

The metabolic pathways that we have discussed in this chapter are representative of the ways in which evolution proceeded, still without free oxygen in the environment, and in the presence of a diminishing supply of abiotically synthesized organic compounds. We will now turn to a detailed discussion of the development of photosynthesis and pathways of autotrophic carbon dioxide fixation.

REFERENCES

Bronk, J. R. (1973). *Chemical Biology.* Macmillan, Inc., New York.
Howland, J. L. (1968). *Introduction to Cell Physiology.* Macmillan, Inc., New York.
Krebs, H. A., and H. L. Kornberg (1957). *Energy Transformations in Living Matter.* Springer-Verlag, New York.
Levy, J., J. J. R. Campbell, and T. H. Blackburn (1973). *Introductory Microbiology.* John Wiley and Sons, Inc., New York.
Oparin, A. I. (1968). *Genesis and Evolutionary Development of Life.* Academic Press, Inc., New York.
White, A., P. Handler, and E. L. Smith (1973). *Principles of Biochemistry.* McGraw-Hill Book Co., Inc., New York.

CHAPTER **14**

Photosynthesis and the Buildup of Oxygen in the Atmosphere

In the previous chapter, we described several anaerobic pathways of energy metabolism and biosynthesis that probably evolved in the earliest heterotrophs as a result of a gradual decline in the supply of abiotically synthesized organic nutrients and ATP (or other high-energy compounds). Our tentative scheme describing the development of metabolism during biological evolution suggests that as readily utilizable organic compounds became severely depleted, selection pressure led to the establishment in anaerobic bacterial heterotrophs of alternative mechanisms of energy capture, as well as pathways by which essential organic compounds could be synthesized from carbon dioxide. One of these new mechanisms was the ability to extract energy from inorganic compounds, as practiced by the anaerobic chemoautotrophs. But by far the most important occurrence from an evolutionary standpoint was the development of the light-driven mechanism of photosynthesis. This chapter describes the metabolic pathways associated with photosynthesis, the process that now supports, directly or indirectly, all life on earth.

THE UNIQUE PROCESS OF PHOTOSYNTHESIS

Most of the initial research on photosynthesis was concentrated on higher plants, and the fundamental chemical mechanisms of photosynthesis consequently were long overlooked. It was this early work that unfortunately led to the description of photosynthesis as "the process by which carbon dioxide and water are converted to glucose and molecular oxygen in the presence of light energy and chlorophyll." The process according to this definition is described by the equation

$$6\ CO_2 + 6\ H_2O \xrightarrow[\text{chlorophyll}]{\text{light}} C_6H_{12}O_6 + 6\ O_2 \quad \Delta G° = +686.0\ \text{Kcal/mole} \quad (1)$$

The Unique Process of Photosynthesis

Subsequent research has established that this definition, although valid in a general sense, is inaccurate in a number of specific respects. In the first place, it applies only to photosynthesis occurring in blue-green algae and eukaryotic plant cells. In photosynthetic bacteria, for example, hydrogen sulfide, molecular hydrogen, and other reduced compounds are used in place of water in the process of photosynthesis. Thus molecular oxygen, which is derived from water during photosynthesis in eukaryotes and blue-green algae, is never released in bacterial photosynthesis. In addition, not all organisms that are able to derive energy from the rays of the sun are also capable of autotrophic carbon dioxide fixation; that is, some of these organisms use an organic compound such as acetate as a carbon source in synthesizing other organic compounds. Since the distinguishing characteristic of photosynthesis is not autotrophic carbon dioxide fixation but rather the ability to use the energy of the sun to synthesize organic compounds, it seems reasonable to classify these latter organisms as photosynthetic. This designation is, however, at odds with the definition of photosynthesis stated in Equation 1.

As we have come to realize that the process generically defined as photosynthesis is not identical in all organisms, it has also become apparent that the photosynthetic process in all organisms *does* have a common fundamental characteristic: conversion of the energy from the rays of the sun into the energy present in covalent bonds. In other words—*in photosynthesis light energy is converted into chemical energy.* Indeed, photosynthesis is the only significant metabolic process that uses energy other than the energy in chemical bonds.

The conventional definition of photosynthesis summarized in Equation 1 is misleading in yet another sense: It obscures the complex nature of the photosynthetic process. The production of organic compounds from simpler substances does not occur in one step, as Equation 1 would imply. Photosynthesis, like the process of anaerobic glucose oxidation, is a multistep process—it is as misleading to describe photosynthesis by a single equation as it would be to describe the anaerobic Embden-Meyerhof pathway as the oxidation of glucose to lactic acid, without mentioning the existence of intermediate steps in the pathway.

Actually, photosynthesis involves two quite separate processes: (1) the photochemical process that requires light energy, and (2) the synthesis of carbohydrates or other organic compounds. The photochemical process, also called the *light reactions* of photosynthesis, consists of a series of light-driven reactions by which ATP and reduced coenzymes are generated. Once these substances have been produced, the synthesis of carbohydrates proceeds in another series of reactions known as the *dark reactions.* The dark reactions are so named because they do not require light (although they may occur in the presence of light); instead, the dark reactions are driven by the ATP and reduced coenzymes formed in the light reactions. This dual aspect of photosynthesis has been demonstrated in all photosynthetic organisms. However, specific details of the light and the dark reactions may vary among different types of organisms.

We have mentioned before, but would do well to stress again, that there is nothing unique about the chemical mechanisms used for the process of carbon dioxide fixation, nor about the manufacture of more complex organic compounds from simpler organic compounds. These are energy-requiring processes, and as long as enzymes are available to catalyze these reactions, and as long as there is an adequate supply of ATP and necessary reducing agents, these reactions may occur by chemical mechanisms found in most metabolic pathways. For example, autotrophic carbon dioxide fixation in most chemoautotrophs occurs by a pathway that is identical to that used in photosynthetic organisms. In chemoautotrophs, ATP and reducing agents required to drive the pathway of autotrophic carbon dioxide fixation are supplied by the oxidation of inorganic substrates, and no light energy is involved in the process. Heterotrophs are incapable of autotrophic carbon dioxide fixation because they do not possess all the enzymes required for the process. Indeed, it would be energetically absurd for heterotrophs to extract energy by degrading carbohydrates in order to provide the energy necessary to synthesize the very same carbohydrates. The Second Law of Thermodynamics tells us that such a process would be a losing proposition; that is, the amount of energy required for the synthesis of one molecule of glucose is greater than the amount that can be derived from the oxidation of one molecule of glucose.

The foregoing discussion leads us to an important conclusion: The unique aspect of photosynthesis is not the "synthesis" but the "photo" — the utilization of light energy to drive chemical processes. The evolutionary success of photosynthetic organisms is due to the fact that they have a virtually inexhaustible source of energy: the sun.

Because all photosynthetic bacteria are anaerobes, it is reasonable to assume that they were the first photosynthetic organisms to evolve, and that the oxygen-eliminating blue-green algae appeared at a later time. Photosynthetic eukaryotes probably evolved, in turn, from these organisms. Consequently, we shall begin this chapter by discussing photosynthetic mechanisms in bacteria, and then consider photosynthesis in blue-green algae and eukaryotic plant cells.

THE LIGHT REACTIONS OF BACTERIAL PHOTOSYNTHESIS

PIGMENTS

All photosynthetic bacteria (the green bacteria *Chlorobacteriaceae* and the purple bacteria *Thiorhodaceae* and *Athiorhodaceae*) contain several types of pigments that are associated with the thylakoid membranes, and are important in photosynthesis. These pigments always include *carotenoids* and one or more of the *bacte-*

The Light Reactions of Bacterial Photosynthesis

rial chlorophylls (either bacterial chlorophyll a or one of its close derivatives). (See Figure 10-11, page 250.) The carotenoids are of two types: *carotenes*, which are orange-yellow oxygen-free hydrocarbon compounds, and *xanthophylls*, which are oxygenated derivatives of carotenes. (The chemical structure of β-carotene is shown in Figure 7-29, page 182.)

Each photosynthetic pigment has a characteristic *absorption spectrum;* that is, it absorbs light of certain wavelengths, and it transmits (allows to pass through it) or reflects light of other wavelengths. The transmitted and reflected wavelengths combine to produce the characteristic color of the pigment. For example, most leaves are green because the chlorophyll they contain masks other pigments that are present in lesser concentrations. Chlorophyll of green plants absorbs visible light of short wavelengths (violet and blue) and long wavelengths (orange and red), and it transmits or reflects green and yellow light of intermediate wavelengths (Fig. 14-1).

Most purple bacteria contain bacterial chlorophyll a (Bchl a) as their major chlorophyll pigment. Bchl a has an absorption maximum in the near infrared approximately between 850 and 890 nm, as well as a minor peak at 800 nm; it also significantly absorbs light in the blue regions of the spectrum. The green bacteria possess mainly a chlorophyll known as *Chlorobium* chlorophyll, which absorbs light maximally between 735 and 755 nm; they also have a small amount of bacterial chlorophyll a. The carotenoids are present in all types of photosynthetic bacteria, and absorb maximally in the region of approximately 400 to 500 nm.

The type of light absorbed by the majority of strains of purple bacteria and green bacteria is shown in Figure 14-2. The curves in Figure 14-2 represent the absorption spectra of the combined pigments *in the intact organism,* and thus do not indicate the absorption spectrum of each individual pigment. The absorption spectra of the photosynthetic pigments are given in this way because it has been demonstrated that bacterial chlorophyll a displays an entirely different absorption spectrum when isolated in organic solvents than when it is in an intact organism. This is presumably due to the fact that bacterial chlorophyll a is complexed to other substances within the cell, and therefore the isolation of bacterial chlorophyll a disrupts such associations. Indeed, one of the major problems in

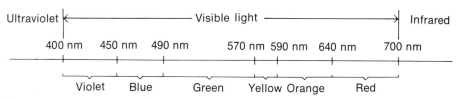

Figure 14-1 The relationship between wavelength and color of light. Visible light encompasses a range of wavelengths between about 400 and 700 nm; ultraviolet light falls immediately below this zone, and infrared falls above. The energy of a photon is inversely proportional to its wavelength.

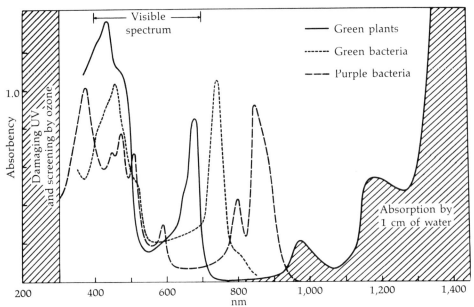

Figure 14-2 Absorption spectra of the combined photosynthetic pigments in the purple bacteria and the green bacteria. (From *Molecular Physics in Photosynthesis* by Roderick K. Clayton. © 1965, by Xerox Corporation. Used with permission of Ginn and Company [Xerox Corporation].)

using absorption spectroscopy as a tool for studying biological systems is that purified substances isolated in a test tube often show quite different absorption properties than the same substances in vivo.

LIGHT ABSORPTION

Before light can be used to drive the chemical processes of photosynthesis, it must be absorbed by the cellular pigment molecules. The discussion in Chapter 6 states that the absorption of light by a molecule brings about a transition in the energy level of one of the electrons surrounding an atomic nucleus. The molecule, after absorption, is said to be in an *excited state* because it has more energy than it does in its ground state. Because the energy present in an absorbed photon must equal the amount of energy required for the electronic transition, and because the energy of a photon is inversely proportional to the wavelength of the light, only certain wavelengths can be absorbed by a particular molecule. Chapter 6 also points out that an excited molecule can return to the ground state by giving up its absorbed energy in a number of different ways. With respect to the process of photosynthesis, two types of transitions from the excited to the ground state are important: (1) The excited molecule may transfer its excitation energy to another molecule, or (2) the excited molecule may enter into a chemical reaction.

The Light Reactions of Bacterial Photosynthesis

It has been demonstrated that bacterial chlorophyll a (Bchl a) is the only pigment that is *directly* involved in the chemical events of photosynthesis.* The carotenoids apparently have a dual function within the cell. Although they do not participate directly in the chemical events of photosynthesis, they are capable of transferring their excitation energy of absorbed light to Bchl a. Consequently, they act rather like antennae, helping to gather light energy for use in photosynthesis. Because of this role, carotenoids are termed *accessory pigments*. By absorbing light at wavelengths other than those absorbed by Bchl a, the accessory pigments expand the range of light that may be used for photosynthesis. In addition to functioning as accessory photosynthetic pigments, carotenoids are probably involved in shielding certain organic compounds of the cell from photo-oxidation.

The actual mechanism by which energy is transferred from carotenoids to Bchl a has not been clearly elucidated. One mechanism that has been proposed is that of *inductive resonance*. Simply expressed, this theory proposes that the vibration of an excited electron in one molecule can induce a similar vibration in an adjacent molecule, causing the adjacent molecule to become excited, and the original molecule to return to its ground state. However, each pigment molecule can only transfer energy to another molecule that absorbs light at the same or at longer wavelengths. Energy transfer in the reverse direction cannot occur, because the energy present in an excited molecule is not sufficient to induce an excited state in a molecule that absorbs at a shorter wavelength, and hence requires more energy. Consequently, energy always migrates among pigment molecules to those that absorb light at the longest wavelength (at the lowest energy).

In the early 1950s, Louis Duysens, working with purple bacteria, discovered a special long-wavelength form of Bchl a that absorbed light maximally at 890 nm.† Subsequently, it was shown that energy absorbed by other pigment molecules, including carotenoids and the ordinary short-wavelength form of Bchl a, is transferred to this special form of Bchl a called P890 (P = pigment). It is P890 that is directly involved in the chemical reactions of photosynthesis. The P890 initiates an electron transfer reaction (oxidation-reduction) with adjacent molecules. This reaction is referred to as the *initial photochemical event* of photosynthesis, for it is at this point that the energy absorbed by pigment molecules initiates the chemical events of photosynthesis. Because of its central role in initiating the chemical events in the photosynthetic process, the P890 molecule is called the *reaction center*.

---

*Most of the research on photosynthetic bacteria has been done on the purple bacteria, and therefore we will concentrate on the purple bacteria in this discussion. The photosynthetic process may differ somewhat in other types of bacteria, but it apparently is similar in principle.

†The absorption maximum of the special long-wavelength form of Bchl a apparently varies somewhat in different types of bacteria.

There is ample evidence to suggest that the pigments associated with photosynthesis in bacteria are present in functional units that consist of carotenoid pigments, about forty molecules of ordinary Bchl *a*, and one molecule of P890, which forms the reaction center of the unit. This functional grouping is called the *photosynthetic unit* or *pigment system,* and there are, of course, numerous such units embedded in the thylakoid membranes. Thus the thylakoid membranes may be thought of as forming an extended area of light-collecting pigments with an occasional reaction center.

Not all energy absorbed by pigment molecules is used in photosynthesis. Because the excited state of the pigment molecules is not stable, an excited pigment molecule may theoretically return to its ground state within about 10^{-9} seconds, with re-emission of the absorbed energy. This re-emission of energy occurs in the form of light and heat. It is called *fluorescence* when there is only a short delay between absorption and re-emission, and when the wavelength of light re-emitted is only slightly longer than that of the light absorbed. The spectrum of fluorescent light emitted by a molecule is characteristic of the molecule under defined conditions. The chlorophyll found in higher plants, for example, always gives off light in the red wavelengths (fluoresces in the red).

The study of fluorescence has provided considerable information on the mechanism of the light reactions of photosynthesis. One particularly significant observation, which led to the discovery of energy transfer between pigment molecules, is that carotenoids are never observed to fluoresce within the cell, although they possess characteristic absorption and fluorescence spectra when extracted from the cell. In fact, only Bchl *a* fluoresces inside the bacterial cell. This demonstrates that light energy absorbed by carotenoids is transferred to Bchl *a*. Absorbed energy reaching Bchl *a* may be re-emitted as fluorescence if the energy is not transferred within about 10^{-9} seconds to P890, or if all the energy reaching P890 cannot be utilized in the slower chemical events of photosynthesis. The latter would occur if the rate of light energy capture were greater than the rate of energy utilization.

What, specifically, is involved in the initial photochemical event? You will remember that the energy-capturing mechanisms of photosynthesis, like all cellular processes involving energy transfer, take place by means of the transfer of electrons. The energy of the excited electron in P890 must somehow be conserved if it is to be available to drive the chemical reactions of photosynthesis. This is accomplished by using the electron in a chemical reduction, in which the energy of the excited electron is conserved by transferring the electron to an electron acceptor substance. The initial photochemical event is believed to occur by a two-step process that may be summarized as follows:

$$D/P890^*/A \longrightarrow D/P890^+/A^- \longrightarrow D^+/P890/A^-$$

The Light Reactions of Bacterial Photosynthesis

In the first step, the photoexcited P890 (P890*) reacts initially with an electron acceptor (A), transferring its excited electron to A; in the process, A is reduced and P890 is oxidized. In the second step, the oxidized P890 (P890$^+$) regains an electron from an electron donor (D). The net effect of the primary photochemical event is thus to form an initial oxidizing agent (D$^+$) and an initial reducing agent (A$^-$).

Regardless of what the earliest mechanism of autotrophic carbon dioxide fixation may have been, the two pathways of carbon dioxide fixation that are known to operate in present-day autotrophic bacteria require substantial amounts of ATP and NADH.* For example, the most commonly occurring of these pathways, the *Calvin cycle* (also known as the *reductive pentose phosphate pathway*) requires eighteen molecules of ATP and twelve molecules of NADH to fix six molecules of carbon dioxide into glucose. The production of ATP and NADH occurs in reactions subsequent to the initial photochemical event.

The light reactions of photosynthesis that follow the initial photochemical event are not identical in all types of photosynthetic bacteria. However, the following overall scheme does emerge. In bacteria, ATP and NADH are apparently synthesized in separate pathways. Thus, electrons transferred to an initial electron acceptor either can be used to reduce NAD$^+$ to NADH, or their energy can be used to drive the phosphorylation of ADP to ATP in a process known as *cyclic photophosphorylation*.

PHOTOPHOSPHORYLATION: THE LIGHT-DRIVEN PHOSPHORYLATION OF ADP

A tentative pathway for cyclic photophosphorylation in photosynthetic bacteria is shown in Figure 14-3. In essence, excited electrons from P890, which have been transferred to an initial electron acceptor in the primary photochemical event, are passed spontaneously downhill through a series of electron carriers. The electron carriers, which are bound to the thylakoid membranes in close association with the photosynthetic pigment systems, form what is called an *electron transport chain*. Each of the electron carriers in the chain accepts one or more electrons from the carrier preceding it, and passes those electrons to the carrier following it. Because this electron transfer occurs spontaneously, we know that each of the electron carriers in the chain is a stronger reducing agent than the one that follows it. Thus, as shown in Figure 14-3, electrons are transferred from the initial electron acceptor (of unknown chemical composition) to a quinone compound, from the quinone to a type of cytochrome *b*, and from cytochrome *b* to a type of cytochrome *c*. Fi-

---

*NADH, rather than NADPH, is required for carbon dioxide fixation in bacterial photosynthesis; all other organisms use NADPH in these reactions.

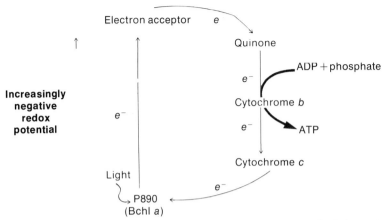

Figure 14-3 A tentative scheme for cyclic photophosphorylation in photosynthetic bacteria. Arrows denote the transfer of an electron from one compound to another. The identity of the electron acceptor is unknown; it is also uncertain whether one or two molecules of ATP are produced in each turn of the cycle. Details of the pathway may vary somewhat in different bacterial strains. (Adapted from Vernon, L. P. [1968], Bact. Rev. 32:243.)

nally, the electrons, which now have substantially less energy, are transferred from cytochrome c to P890, perhaps through several other unidentified electron carriers. In this way, an electron is returned to chlorophyll at its ground state. The absorption of another photon by P890 is then required to re-initiate the process. The fundamental characteristic of the process of photophosphorylation is that energy released during the transfer of electrons down the electron transport chain is used to phosphorylate one or perhaps two molecules of ADP. The process is referred to as cyclic photophosphorylation because the electron is returned to the same molecule of P890 from which it had been ejected (hence the process is cyclic), and because energy originally derived from light is used to phosphorylate ADP (hence the process is one of *photo*phosphorylation).

The types of electron carriers encountered in the pathway of bacterial cyclic photophosphorylation are important in other metabolic pathways involving energy capture, and deserve brief mention. The *cytochromes,* of which there are a number of different molecular types, are enzymes that possess an iron–porphyrin prosthetic group, very similar in structure to the heme group in hemoglobin. The iron in the center of the porphyrin ring can exist in either the ferrous (Fe^{+2}) or ferric (Fe^{+3}) state, depending on whether it is, respectively, reduced or oxidized. Thus one electron is carried by each cytochrome in an electron transport chain. The various types of cytochromes (*b, c*) differ from one another in their amino acid sequences, in the nature of the side groups on their iron–porphyrin rings, and in the way their porphyrin groups are attached to the protein moiety.

Quinones are another important type of electron carrier. There are a number of different kinds, all of which contain a quinone group that is capable of existing in reduced and oxidized states by gaining or losing one or two electrons.

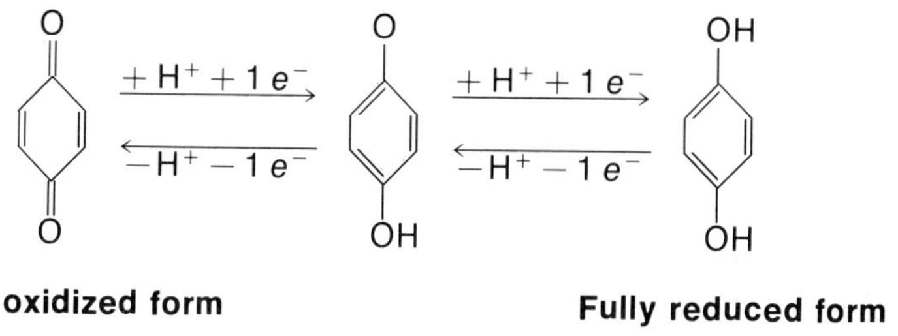

The Light Reactions of Bacterial Photosynthesis

[Quinone redox structures shown: Fully oxidized form ⇌ semiquinone ⇌ Fully reduced form, each step $\pm H^+ \pm 1e^-$]

Fully oxidized form   **Fully reduced form**

The most common quinones involved in electron transport in photosynthetic bacteria are the *ubiquinones* (Fig. 14-4). Ubiquinones are also known as *coenzyme Q*, a name that was assigned before the structures of the ubiquinones had been elucidated. It is technically inaccurate, because the ubiquinones are not coenzymes.

There are other electron carriers that are important in bacterial photosynthesis. In addition to the coenzyme NAD, one coenzyme that has been implicated in many photosynthetic systems is *ferredoxin*, a protein that possesses a non-heme iron–sulfur complex as a prosthetic group that may be alternately oxidized and reduced.

Some enzymes employ a riboflavin (vitamin B_2) derivative as an electron-carrying prosthetic group; this may be either *flavin adenine dinucleotide* or *flavin mononucleotide*. The combination of a flavin nucleotide with protein is known as a flavoprotein. The prosthetic group of a flavoprotein undergoes oxidation or reduction by losing or gaining two hydrogen atoms. The oxidized forms of flavin adenine dinucleotide and flavin mononucleotide are designated, respectively, FAD and FMN, and their reduced forms are designated $FADH_2$ and $FMNH_2$ (Fig. 14-5).

THE PHOTOREDUCTION OF NAD^+

Carbon dioxide fixation in bacteria requires not only ATP as a source of energy but also a reducing agent in the form of NADH. In many bacteria, the production of NADH is thought to occur by a *noncyclic* pathway in which excited electrons of P890, transferred to an initial electron acceptor, are passed spontaneously through several electron carriers, including ferredoxin and a flavoprotein, to NAD^+ (Fig. 14-6). This pathway of NAD^+ photoreduction is noncyclic because the electrons that pass from P890 to NAD^+ are ultimately used to reduce carbon dioxide in the dark reactions of photosynthesis, and consequently are not returned to P890. Instead, electrons are donated to P890 from reduced sulfur compounds such as

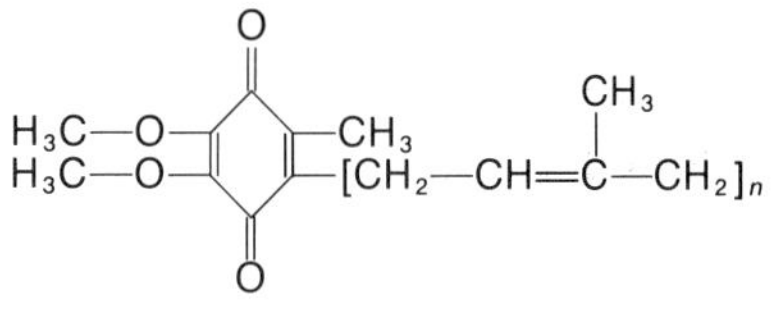

[Chemical structure of ubiquinone shown with H_3C-O- groups, $-CH_3$ substituent, and isoprenoid side chain $-[CH_2-CH=C(CH_3)-CH_2]_n$]

The chemical structure of the ubiquinones; n may vary from 6 to 10.   **Figure 14-4**

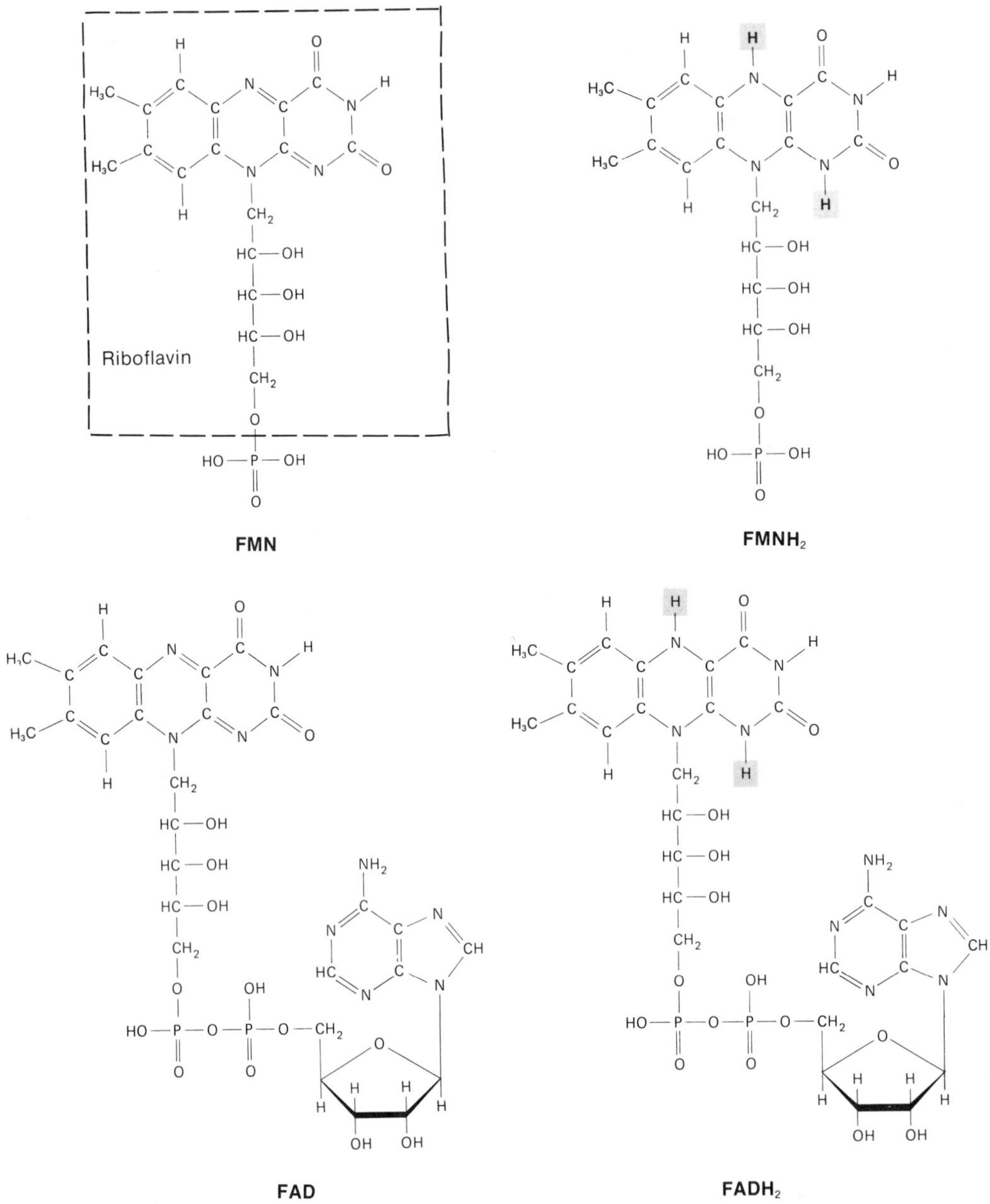

Figure 14-5 Chemical structures of the oxidized and reduced forms of flavin mononucleotide and flavin adenine dinucleotide.

hydrogen sulfide (H_2S) or thiosulfate ($S_2O_3^{-2}$), or from organic compounds. The *net* result, then, of the noncyclic pathway of NAD^+ reduction is that electrons in weak reducing agents (H_2S, $S_2O_3^{-2}$ and organic compounds) are raised to a high enough energy level, through the photoexcitation of P890, to reduce NAD^+ to NADH, a strong reducing agent.

In some photosynthetic bacteria, reduced NAD may be generated by a process that apparently does not involve the noncyclic

Light Reactions in Blue-Green Algae and Photosynthetic Eukaryotes

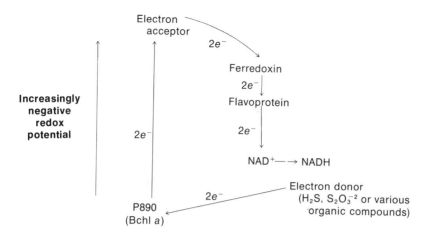

Tentative scheme of NAD^+ photoreduction in bacterial photosynthesis. The pathway is noncyclic because electrons are transferred from an electron donor to NAD^+. (Adapted from Levy, J., J.J.R. Campbell, and T.H. Blackburn [1973], *Introductory Microbiology*, John Wiley and Sons, Inc., New York.)

Figure 14-6

pathway of NAD^+ reduction. Instead, ATP generated in cyclic photophosphorylation is used to "drive" electrons to the potential at which they can be accepted by NAD^+, in a reversal of the process by which NADH is oxidized in aerobic cells (see Chapter 15).

LIGHT REACTIONS IN BLUE-GREEN ALGAE AND PHOTOSYNTHETIC EUKARYOTES

Because there was no ozone screen in the primitive atmosphere prior to the buildup of oxygen, the earliest photosynthetic organisms must have existed well below the surface of the ocean, where damaging ultraviolet radiation could not penetrate. In this environment, there may well have been insufficient amounts of reduced soluble compounds to serve as electron donors in photosynthesis; hence, photosynthetic organisms could not operate at optimal efficiency. If this was the case, selection pressure would have favored the establishment of organisms able to use other compounds as electron donors in photosynthesis. Water was the most readily available electron donor, and eventually an organism evolved that was ancestral to present-day blue-green algae—an organism capable of using water as an electron donor in photosynthesis.

The use of water in this reaction, however, led to the production of molecular oxygen, which was toxic to anaerobes, initially including the organisms that produced it. During the initial period in the development of oxygen-eliminating photosynthesis, however, most free oxygen was quickly removed by reaction with reduced substances present on the primitive earth, and hence did not accumulate in high enough concentrations to harm the organisms that produced it. Eventually, of course, organisms arose that could use oxygen in their metabolic processes. Curiously, it would appear as if the establishment of a process that allowed early photosynthetic organisms to be unaffected by curtailments in the supply of electron donors in turn gave rise to the Oxygen Revolution and the subsequent development of aerobic organisms.

THE DUAL PIGMENT SYSTEMS

Photosynthesis in present-day blue-green algae is very similar to photosynthesis in eukaryotes, and both resemble bacterial photosynthesis in principle, although they differ in details. In blue-green algae and eukaryotes, the cyclic and noncyclic electron transport systems of photosynthesis are coupled, instead of being separate as they are in photosynthetic bacteria.

Initial studies of photosynthesis in eukaryotic plant cells led to the idea that ATP, reduced coenzymes (NADPH, in the case of organisms other than bacteria), and molecular oxygen were all produced in a single pathway as a result of the absorption of light by chlorophyll. However, research by Robert Emerson in the 1940s and 1950s led to a reappraisal of this idea. Emerson demonstrated that when a plant was illuminated with visible light at wavelengths of both 700 nm and 650 nm (or with light both of 700 nm and of a wavelength shorter than 650 nm) simultaneously, the efficiency of photosynthesis was considerably greater than was the sum of the efficiencies when the plant was illuminated at each wavelength separately. Even more unexpected was Emerson's finding that an enhanced efficiency was also obtained if the two wavelengths of light were flashed alternately, at intervals of no more than several seconds. This phenomenon has been termed the *Emerson enhancement effect.* It has been interpreted as indicating that the light reactions of photosynthesis in blue-green algae and eukaryotes involve two separate photochemical events that cooperate to produce ATP, NADPH, and oxygen.

Evidence has accumulated to support this idea, and it now appears that each "single" photochemical event is actually a combination of reactions occurring in each of two separate pigment systems, and that each pigment system absorbs light of a somewhat different wavelength. The fact that enhancement of the efficiency of photosynthesis was obtained by alternating light of short and long wavelengths suggests that the photochemical events in the two pigment systems are linked by some chemical process. Were this not the case, alternating light at intervals of several seconds would not be expected to enhance the efficiency of photosynthesis, for light energy absorbed by one pigment system would be dissipated as fluorescence within a period of about 10^{-9} second, and would not be available to cooperate with light absorbed by the other pigment system a second or so later. However, if the light energy absorbed by both pigment systems were converted into chemical intermediates, the two pigment systems could cooperate chemically even if energy were absorbed by one of the pigment systems about a second before the other.

Evidence that the photosynthetic systems in blue-green algae

Light Reactions in Blue-Green Algae and Photosynthetic Eukaryotes

Table 14-1 RELATIVE AMOUNTS OF EACH PIGMENT IN THE TWO PIGMENT SYSTEMS OF BLUE-GREEN ALGAE AND PHOTOSYNTHETIC EUKARYOTES

PIGMENT SYSTEM I	PIGMENT SYSTEM II
Most carotenes	Most xanthophylls
Some chlorophyll b	Most chlorophyll b
	Phycobiliproteins (only in blue-green and red algae)
Chlorophyll a	Chlorophyll a
P700	P680

and eukaryotic plants consist of two pigment systems, often referred to as *pigment systems I* and *II* (or *photosystems I* and *II*), has thus accumulated. As in bacteria, the pigment systems are embedded within the thylakoid membranes. Each pigment system consists of a number of different photosynthetic pigments, and has its own reaction center, in which initial oxidizing and reducing agents are produced by a mechanism fundamentally similar to that which occurs in bacterial photosynthesis.

By treating chloroplasts with detergents, two separate kinds of pigment–protein complexes can be extracted; one of these contains components of pigment system I and the other contains components of pigment system II. Pigment systems I and II both contain chlorophyll a, chlorophyll b (a close structural relative of chlorophyll a), and carotenoids, although the distribution of these pigments in the two systems varies. (See Figure 10–11, page 250, and Table 14–1.) Of the carotenoids, carotenes are more abundant in pigment system I than they are in pigment system II; xanthophylls, on the other hand, predominate in pigment system II. Pigment system II contains considerably more chlorophyll b than pigment system I. Chlorophyll a is the primary photosynthetic pigment in both systems; however, the reaction center of pigment system I consists of a form of chlorophyll a that absorbs light maximally at 700 nm (P700). The reaction center of pigment system II is another special form of chlorophyll a, P680, which absorbs maximally at 680 nm.* In blue-green algae and red algae, the phycobiliproteins (primarily phycoerythrin and phycocyanin) are present as additional accessory pigments primarily in pigment system II, although the precise structural relationship between phycobilisomes and the pigment–protein complex of pigment system II has not been worked out. (Phycobilisomes are discussed in Chapter 10.)

Absorption spectra of purified photosynthetic pigments ex-

---

*In at least some cases, the reaction center of pigment system II appears to be P690, a type of chlorophyll a that absorbs maximally at 690 nm. For simplicity, however, we will refer to the reaction center of pigment system II as P680 throughout this discussion.

tracted from blue-green algae and eukaryotic plant cells are shown in Figure 14–7. Within the cell, chlorophyll a and chlorophyll b have absorption peaks at about 670 and 645 nm, respectively; both chlorophylls also absorb strongly at wavelengths below approximately 500 nm. The phycobiliproteins absorb light in the region between 500 and 650 nm, and consequently are able to capture light energy at longer wavelengths than the carotenoids can.

In each of the two pigment systems, as in the pigment system of bacterial photosynthesis, absorbed light energy is transferred, probably by inductive resonance, from accessory pigments to the reaction center; these transfers, of course, always proceed from one pigment to another that absorbs light at a longer wavelength. Thus, in the chloroplasts of eukaryotic cells, for example, energy absorbed by the carotenoids in one of the pigment systems is passed to chlorophyll a, which in turn transfers its excitation energy to the long-wavelength reaction-center pigment of that pigment system. The excitation energy of chlorophyll b is transferred directly to chlorophyll a, from which it is transferred to the reaction center. In pigment system II of the blue-green and eukaryotic red algae, the

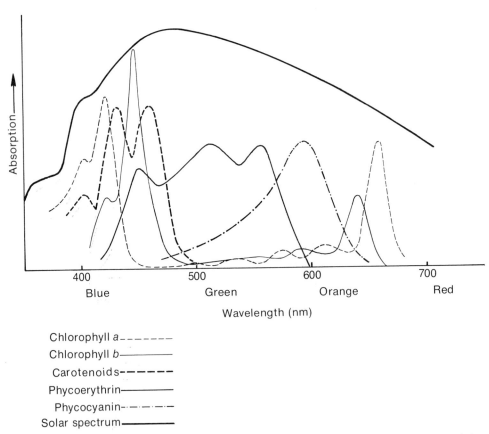

Figure 14–7 Absorption spectra of chlorophyll a, chlorophyll b, carotenoids, phycoerythrin, and phycocyanin dissolved in an organic solvent. The absorption spectra of these pigments in vivo are very similar to those shown here. (From Govindjee and R. Govindjee [1974], Sci. Am. 231:6, p. 72.)

Light Reactions in Blue-Green Algae and Photosynthetic Eukaryotes

Figure 14-8
Transfer of absorbed energy between photosynthetic pigments in blue-green and red algae. In eukaryotes other than red algae, carotenoids transfer their energy directly to chlorophyll a, since these organisms do not possess the phycobiliproteins, phycoerythrin and phycocyanin. (Adapted from Govindjee and R. Govindjee [1974], Scientific American, December, 1974, 231:77. Copyright © 1968 by Scientific American, Inc. All rights reserved.)

only organisms that possess phycobiliproteins, energy is passed from carotenoids to phycobiliproteins, to chlorophyll a, and then to the reaction center. Of course, light may be absorbed by any of these pigments, and the energy passed through longer-wavelength pigments to the reaction center. For example, light energy absorbed directly by chlorophyll a is transferred to the reaction center without the participation of carotenoids or phycobiliproteins. A summary of the transfer of energy between photosynthetic pigments is given in Figure 14-8.

It has been estimated that 300 pigment molecules of ordinary chlorophyll a and 1 molecule of the reaction center pigment (P700 in pigment system I and P680 in pigment system II) are present, along with accessory pigments, in each pigment system. The figure of 300 molecules of chlorophyll a has been derived from results of experiments suggesting that 2400 molecules of chlorophyll a are required to produce 1 molecule of oxygen or to fix 1 molecule of carbon dioxide in photosynthesis. Since eight photochemical events are required to generate one molecule of oxygen, it has been assumed that each pigment system contains 300 molecules of chlorophyll a.

PHOTOSYNTHETIC ELECTRON TRANSPORT: PHOTOPHOSPHORYLATION AND THE PHOTOREDUCTION OF NADP$^+$

The transfer of electrons in the dual pigment systems of blue-green algae and eukaryotic plants is summarized in Figure 14-9. Light energy, absorbed by P680 in pigment system II, raises an electron to an excited state. The excited electron is transferred to an unidentified electron acceptor, called Q. Q is so named because in the oxidized state it quenches (extinguishes) the fluorescence of pigment system II by accepting the excited electron. The primary photochemical event in pigment system II thus results in the production of an initial reducing agent (Q) and a hypothetical oxidizing agent (Z). Electrons lost from Z by transfer to P680 are ultimately replaced by electrons derived from water.

Experiments have shown that one molecule of oxygen is evolved for every four electrons donated by Z to P680; these four electrons are acquired from the oxidation of two molecules of water.

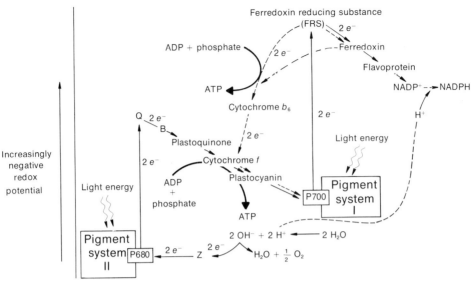

Figure 14-9 Transfer of electrons during the light reactions of photosynthesis in blue-green algae and eukaryotic plant cells. Electrons may flow in a noncyclic pathway (solid black arrows) from water to $NADP^+$, producing NADPH; energy released in the process is also used to phosphorylate ADP to ATP. A cyclic electron flow (dashed lines) may also occur between FRS (and possibly ferredoxin) and P700, resulting in the phosphorylation of ADP, but not in NADPH production. The yield of ATP in both the noncyclic and cyclic pathways is in question. See text for further details. (Adapted from J.R. Bronk [1973]. *Chemical Biology,* Macmillan, Inc., New York.)

The transfer of electrons from water to Z may well occur in a number of steps involving one or more electron carriers. In a sense, then, the photochemical production of an initial oxidizing agent, Z, brings about the splitting of water molecules—a reaction often referred to as the "photolysis" of water. This terminology is unfortunate, since the term "photolysis" (photo = light; lysis = breakdown) implies that the water molecules are split by the direct action of light, a process that clearly does not occur. Instead, the splitting of water is actually an oxidation of water by Z. We know very little about how this occurs, although it appears to require the presence of Mn^{++} ions.

The electrons accepted by Q are spontaneously transferred to pigment system I through a series of electron carriers. Although the identity and sequence of these carriers have not been unequivocally determined, they appear to involve (in order) an unidentified substance called B, a plastoquinone, cytochrome *f,* and plastocyanin. Plastoquinones are quinone compounds that are closely related structurally and functionally to the ubiquinones found in many systems of bacterial photosynthesis (and in aerobic respiration, as well). The quinone group in plastoquinones is oxidized and reduced in the same manner as is the quinone group in ubiquinones. Plastocyanin, the last electron carrier in the sequence, is a copper-containing protein.

A photon absorbed by P700 initiates the photochemical reactions in pigment system I, and an excited electron is transferred from P700 to an acceptor substance of unknown structure that has been termed *ferredoxin reducing substance* (FRS). The electron lost by P700 is recovered from plastocyanin, which becomes oxidized. Hence the primary photochemical event in pigment system I involves the production of oxidized plastocyanin (an oxidizing agent) and reduced FRS (a reducing agent). Electrons from reduced FRS are transferred, again spontaneously, through ferredoxin and a flavoprotein enzyme (ferredoxin–$NADP^+$ reductase) to $NADP^+$, which is reduced to NADPH. Two electrons are required for the reduction of each molecule of $NADP^+$.

Thus, in this series of reactions, electrons in effect are transferred from the weak reducing agent water to $NADP^+$, creating the strong reducing agent, NADPH, which in turn is capable of reducing carbon dioxide. In this process, electrons from water are raised in a series of light-driven reactions to a potential high enough so that they may be accepted by $NADP^+$. To produce one molecule of oxygen, four electrons must be removed from water, resulting in the production of two molecules of NADPH; the electron transfer reactions involved in this process require eight photons of light energy (four photons absorbed by pigment system II and four absorbed by pigment system I).

The production of NADPH is not, however, the only result of the electron transfer reactions summarized in Figure 14–9. The spontaneous transfer of electrons from Q to plastocyanin is coupled to the phosphorylation of ADP to ATP. It is not known precisely to which step in the electron transport chain this phosphorylation reaction is coupled, nor is it certain how many ATPs are produced for each pair of electrons that passes down the chain of carriers. Evidence suggests that at least one ATP is produced during the downhill transfer of each pair of electrons from Q to plastocyanin, and it is quite likely that two molecules of ATP are produced.

In addition to ATP production during electron transfer between pigment system II and pigment system I, ADP may also be phosphorylated during a cyclic flow of electrons from reduced FRS, or possibly from reduced ferredoxin, back to P700. In this case, electrons are not used for the reduction of $NADP^+$, and the process as a whole resembles photophosphorylation in bacterial photosynthesis. As is shown in Figure 14–9, the cyclic electron flow might proceed via cytochrome b_6 to cytochrome f, to plastocyanin, and back to the reaction center chlorophyll, although the precise pathway has not been clearly established. There is some evidence that the phosphorylation site in cyclic electron flow is not the same as in the electron transfer reactions from Q to plastocyanin.

Cyclic photophosphorylation is important because it provides a source of ATP that is independent of the production of NADPH. Many workers have suggested that the rate of cyclic electron flow may be determined by the availability of ATP. Thus, for example, if the amount of ATP produced during the flow of electrons from water to $NADP^+$ is inadequate to drive the dark reactions of photosynthesis, cyclic electron flow can provide a supplementary source of ATP.

The precise chemical mechanism involved in photophosphorylation is not known, although experiments by André Jagendorf have shed some light on the process. It is known that the thylakoid sacs will accumulate H^+ ions and pump out K^+ and Mg^{++} ions by a process that depends on photosynthetic electron transport. Thus, ATP is apparently required for this process. Jagendorf has demonstrated that H^+ movements across the thylakoid membranes can be used to form ATP in the dark.

By equilibrating isolated intact thylakoid preparations in an acid solution, and then rapidly bringing the solution to an alkaline pH, Jagendorf was able to synthesize ATP. Thus the outward passage of protons appears to be linked to the phosphorylation of ADP (Fig. 14–10). This is at least energetically feasible, because the accumulation of protons requires ATP, and therefore the reverse of this process is spontaneous and could be coupled to ADP phosphorylation. Peter Mitchell has proposed a detailed mechanism to explain how ATP synthesis might actually be coupled to the collapse of a proton gradient, as demonstrated by Jagendorf. In Chapter 15 and Appendix 4, we discuss Mitchell's hypothesis in connection with ADP phosphorylation during electron transport in mitochondria.

EFFICIENCY OF THE LIGHT REACTIONS OF PHOTOSYNTHESIS

Researchers have not yet worked out the light reactions of bacterial photosynthesis in as much detail as those of photosynthesis in blue-green algae and eukaryotic plants. Sufficient data do exist to allow us to estimate the efficiency of the light reactions in these latter organisms, and to draw useful general conclusions from these approximations.

In blue-green algae and eukaryotes, eight photons of light produce two molecules of NADPH and four molecules of ATP (as-

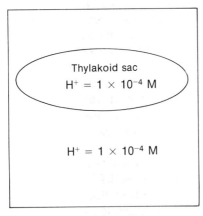 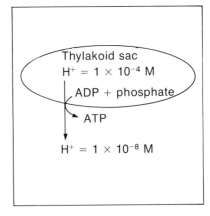

A   B

Figure 14–10 The phosphorylation of ADP by collapse of a proton (H^+) gradient, as demonstrated by André Jagendorf. (A) Intact thylakoids are equilibrated in a medium at pH 4.0 ($H^+ = 1 \times 10^{-4}$ M). (B) When the pH of the medium in which the thylakoids are suspended is rapidly changed to 8.0 ($H^+ = 1 \times 10^{-8}$ M), protons pass spontaneously across the thylakoid membranes. This translocation of protons is coupled to the phosphorylation of ADP.

The Dark Reactions of Photosynthesis

suming that two ATPs are produced for each pair of electrons transferred between Q and plastocyanin). The reduction of $NADP^+$ by water and the phosphorylation of ADP to ATP occur with the following $\Delta G°$ values:

$$NADP^+ + H_2O \rightleftarrows NADPH + \frac{1}{2} O_2 \quad \Delta G° = +52.6 \text{ Kcal/mole} \quad (2)$$

$$ADP + H_3PO_4 \rightleftarrows ATP + H_2O \quad \Delta G° = +7.3 \text{ Kcal/mole} \quad (3)$$

Remember that 1 mole equals 6.02×10^{23} particles (in this case, molecules). Consequently, because the production of two molecules of NADPH and four molecules of ATP requires eight photons, the production of 2 moles of NADPH and 4 moles of ATP requires $8 \times 6.02 \times 10^{23}$ photons. At a wavelength of 700 nm (the wavelength of light absorbed by P700), 6.02×10^{23} photons possesses 40.8 Kcal of energy. Therefore, the production of 2 moles of NADPH and 4 moles of ATP *utilizes* 8×40.8 Kcal or 326.4 Kcal. On the other hand, the production of 2 moles of NADPH *captures* 2×52.6 Kcal or 105.2 Kcal, and the production of 4 moles of ATP *captures* 4×7.3 Kcal or 29.2 Kcal.

Thus, the efficiency of light energy capture in the light reactions of photosynthesis is

$$\frac{105.2 \text{ Kcal} + 29.2 \text{ Kcal}}{326.4 \text{ Kcal}} \times 100 = \frac{134.4}{326.4} = 44.5\%$$

A value of this magnitude is extremely high compared to the efficiency of manmade machines (the internal combustion engine is about 20 per cent efficient). Of course, the efficiency of the light reactions may be even greater than 44.5 per cent if the hydrolysis of ATP yields more free energy than 7.3 Kcal/mole under intracellular conditions, a likely possibility. Alternatively, the efficiency will be reduced somewhat if only one molecule of ATP, not two, is produced during the transfer of each pair of electrons between Q and plastocyanin. It is also clear that cyclic photophosphorylation in the dual pigment systems of blue-green algae and eukaryotes captures much less energy than it would if the electrons were also used for reducing $NADP^+$.

THE DARK REACTIONS OF PHOTOSYNTHESIS

The separation of the light reactions from the fixation of carbon dioxide was first demonstrated in 1937 by Robert Hill, who observed that a leaf homogenate containing intact chloroplasts was able to reduce ferric potassium oxalate (an artificial electron acceptor) with the concomitant evolution of oxygen, even though the fixation of carbon dioxide in such preparations occurred at negligible rates. It has since been shown that a large number of substances may be reduced in the presence of light by intact chloroplasts. The reduction of an artificial electron acceptor under these conditions is now known as the *Hill reaction,* and may be summarized by the equation

$$H_2O + A \underset{}{\overset{\text{light, chloroplasts}}{\rightleftarrows}} \tfrac{1}{2}O_2 + AH_2$$

(A = Electron acceptor; AH_2 = Reduced electron acceptor)

In subsequent work with disrupted chloroplasts, primarily done by R.B. Park and N.G. Pon, the thylakoid membranes were separated from the stroma of the chloroplasts, and preparation containing only thylakoid membranes were able to carry out the Hill reaction and to phosphorylate ADP. The stromal fraction, on the other hand, was shown to contain soluble enzymes associated with carbon dioxide fixation, although the actual fixation of carbon dioxide by the stromal fraction required the addition of the thylakoid fraction. It was concluded from these experiments that carbon dioxide fixation required the ATP and reducing agent produced by the thylakoid fraction.

THE CALVIN CYCLE

In most autotrophic organisms, the fixation of carbon dioxide occurs in a series of reactions known as the *Calvin cycle,* or the *reductive pentose phosphate pathway.* The details of this pathway are shown in Figure 14–11. In essence, carbon dioxide is added to ribulose–1,5–diphosphate, and the resulting six-carbon intermediate is converted into two molecules of 3-phosphoglyceric acid. This is then reduced to glyceraldehyde–3–phosphate in a reaction requiring NADPH (NADH, in bacteria) and ATP. Through a series of sugar phosphate interconversions, ribulose–1,5–diphosphate is regenerated from ribulose–5–phosphate in an ATP-requiring reaction. Ribulose–1,5–diphosphate can then react with an additional molecule of carbon dioxide to repeat the cycle.

The net result of the Calvin cycle is that one molecule of carbon dioxide is incorporated into carbohydrate with the expenditure of two molecules of NADPH (NADH in bacteria) and three molecules of ATP. Thus, the Calvin cycle can be thought of as producing one molecule of glucose from six molecules of carbon dioxide in reactions utilizing twelve molecules of NADPH and eighteen molecules of ATP. Glucose should not, however, be considered as the only product of the Calvin cycle; rather, the Calvin cycle brings about the reduction of carbon dioxide into carbohydrates, and any of the intermediate compounds in the cycle can be withdrawn from the cycle as they are required in other metabolic reactions. For example, 3-phosphoglyceric acid can enter the Embden-Meyerhof pathway and lead to the production of pyruvic acid, used in many biosyntheses; ribose–5–phosphate may be used for DNA and RNA synthesis; erythrose–4–phosphate is important in the synthesis of a number of amino acids; and so on. Of course, the quantity of intermediate compounds withdrawn from the cycle must not exceed the quantity

The Dark Reactions of Photosynthesis

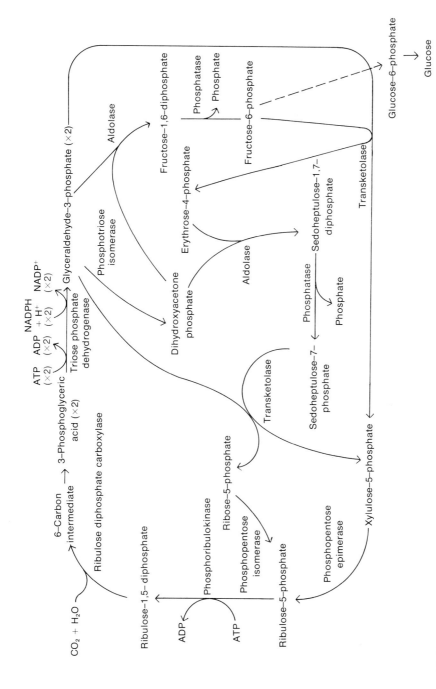

Figure 14-11 The Calvin cycle, the dark reactions of photosynthesis. See Figure 13-6 for chemical structures of carbohydrate intermediate compounds. (Adapted from J.A. Bassham and M. Calvin [1957]. *The Path of Carbon in Photosynthesis*, Prentice-Hall, Inc., New York.)

necessary to regenerate ribulose–1,5–diphosphate and thus to sustain the operation of the cycle.

As is apparent, the Calvin cycle has many individual steps that are identical to steps in the pentose phosphate pathway (see Figure 13–6, page 354). It is for this reason that the Calvin cycle is often referred to as the *reductive* pentose phosphate pathway—carbon dioxide is reduced to carbohydrate in the Calvin cycle, whereas carbon dioxide is released as a byproduct of the oxidative pentose phosphate pathway. Two of the enzymes of the Calvin cycle are unique to that pathway. These are *phosphoribulokinase,* which catalyzes the phosphorylation of ribulose–5–phosphate to ribulose–1,5–diphosphate at the expense of ATP, and *ribulose diphosphate carboxylase,* which is involved in the actual fixation of carbon dioxide by catalyzing the reaction between ribulose–1,5–diphosphate and carbon dioxide to yield two molecules of 3–phosphoglyceric acid. The other enzymes used in the Calvin cycle are found in most organisms, where they catalyze reactions of either the oxidative pentose phosphate pathway or the Embden-Meyerhof pathway. It is the similarity of the Calvin cycle to the oxidative pentose phosphate pathway that has led some biologists to suggest that the Calvin cycle developed from the pentose phosphate pathway in those organisms that acquired the ability to synthesize phosphoribulokinase and ribulose diphosphate carboxylase.

The Calvin cycle appears to be present in most, if not all, autotrophic organisms. However, there are other, additional pathways of carbon dioxide fixation that operate in some organisms. Interestingly enough, carbon dioxide fixation in some anaerobic bacteria may proceed essentially by a reversal of the decarboxylation reactions that occur in the anaerobic pathway of pyruvic acid oxidation (i.e., the Krebs cycle; see Chapter 15).

EFFICIENCY OF THE DARK REACTIONS OF PHOTOSYNTHESIS

The dark reactions of photosynthesis constitute one of the most efficient metabolic systems known. The production of 1 mole of glucose in the dark reactions of photosynthesis requires 18 moles of ATP and 12 moles of NADPH (NADH in bacteria*). The energy input (see Equations 2 and 3) is consequently

18 moles × 7.3 Kcal/mole + 12 moles × 52.6 Kcal/mole = 762.6 Kcal

Since the complete oxidation of one molecule of glucose to carbon dioxide and water proceeds with a $\Delta G°$ of −686.0 Kcal/mole, the efficiency of energy capture in the synthesis of glucose by the dark reactions of photosynthesis is

$$\frac{686.0}{762.6} \times 100 = 89.9\%$$

---

*The oxidation of NADH to NAD^+ proceeds with a $\Delta G°$ that is almost identical to that of the oxidation of NADPH to $NADP^+$.

A COMPARISON OF PHOTOSYNTHESIS IN BACTERIA, BLUE-GREEN ALGAE, AND EUKARYOTES

LOCALIZATION OF PHOTOSYNTHESIS

In all photosynthetic organisms, whether prokaryotic or eukaryotic, the pigment molecules, enzymes, and other constituents associated with the light reactions of photosynthesis are bound to the thylakoid membranes. If the integrity of the thylakoid membranes is destroyed, the phosphorylation of ADP and the reduction of coenzymes cease. This suggests that the individual components of the light reactions must be arranged in a highly organized fashion in order to operate.

It has been proposed that the components of the light reactions in chloroplasts may be localized in the quantasomes (Chapter 11). Quantasomes are tightly bound to or embedded within the thylakoid membrane of chloroplasts. They have been shown to consist of a lipid and a protein fraction; the lipid fraction contains chlorophyll, carotenoids, and plastoquinones, whereas the protein fraction has been shown to contain several cytochromes, plastocyanin, and ferredoxin.

The enzymes needed for the dark reactions of photosynthesis are present in the stroma of chloroplasts, in solution, or in aggregates loosely bound to the thylakoid membranes. In bacteria and blue-green algae, which do not contain chloroplasts, the enzymes of the dark reactions are dissolved, or loosely bound to structures, in the cytoplasm of the cell.

THE LIGHT REACTIONS

It is important to remember that the electron transport scheme shown in Figure 14–9 occurs only in blue-green algae and eukaryotic plants. There is no evidence to indicate that an enhancement effect of two different wavelengths of light exists in bacterial photosynthesis. The lack of such evidence tends to confirm that bacterial photosynthesis has only a single pigment system. A number of researchers have postulated that the second pigment system in blue-green algae and eukaryotic plants may have evolved as an adaptation to the use of water as an ultimate electron donor in photosynthesis, because of the increased energy requirements involved in using water as an electron donor. (Water is a considerably weaker reducing agent than the electron donors used in bacterial photosynthesis.) Others, however, have argued that it is not likely that photosynthesis in bacteria differs so fundamentally from that in blue-green algae and eukaryotes, and that it is consequently possible that a second pigment system exists in bacteria but as yet has not been discovered.

Although blue-green algae, like eukaryotic algae and other plants, normally use water as the ultimate electron donor in photosynthesis, blue-green algae and certain other eukaryotic algae are capable under some conditions of utilizing other electron donors in photosynthesis. For example, several strains of blue-green algae, along with the green alga *Scenedesmus,* are able to use hydrogen gas as a reductant in photosynthesis under anaerobic conditions. There have been some reports that, in some strains of blue-green algae, hydrogen sulfide can be used as a hydrogen donor either exclusively or simultaneously with water. The ability of some blue-green and green algae to use electron donors other than water suggests that the photosynthetic systems in these organisms and in bacteria are closely related.

The separation of ATP synthesis and coenzyme reduction in photosynthetic bacteria means that these organisms can produce either substance independently of the other if only one is needed. The flexibility imparted by this arrangement appears to have been conserved in the cyclic electron transport flow that occurs in pigment system I of blue-green algae and eukaryotes.

THE DARK REACTIONS

The dark reactions of the Calvin cycle are interesting from an evolutionary standpoint because they are practically identical in all organisms in which they occur. The one difference is that NADH is required in the dark reactions of bacterial photosynthesis, whereas NADPH is used in the dark reactions of blue-green algae and eukaryotes. The fact that the Calvin cycle occurs in chemoautotrophic as well as photosynthetic organisms suggests that this pathway of carbon dioxide fixation arose independently of the photochemical mechanisms of photosynthesis. This is an extremely important point, for it emphasizes the fact that it is only the photochemical process that is unique to photosynthetic organisms.

REFERENCES

Bacon, K. E. (1973). The primary electron acceptor of photosystem I. Biochem. Biophys. Acta *301*:1.

Fogg, J. E., W. D. P. Stewart, P. Fay, and A. E. Walsby (1973). *The Blue-Green Algae.* Academic Press, Inc., New York.

Frenkel, A. W. (1970). Multiplicity of electron transport reactions in bacterial photosynthesis. Biol. Revs. *45*:569.

Gibbs, M., Ed. (1971). *Structure and Function of Chloroplasts.* Springer-Verlag, Berlin.

Giese, A. C. (1973). *Cell Physiology.* W. B. Saunders Co., Philadelphia.

Govindjee and R. Govindjee (1974). The absorption of light in photosynthesis. Sci. Am. *231*:68.

Levine, R. P. (1969). The mechanism of photosynthesis. Sci. Am. *221*:58.

Levy, J., J. J. R. Campbell, and T. H. Blackburn (1973). *Introductory Microbiology.* John Wiley and Sons, Inc., New York.

Sager, R. (1972). *Cytoplasmic Genes and Organelles.* Academic Press, Inc., New York.

Tedeschi, H. (1974). *Cell Physiology: Molecular Dynamics.* Academic Press, Inc., New York.

Vernon, L. P. (1968). Photochemical and electron transport reactions of bacterial photosynthesis. Bact. Rev. *32*:243.

White, A., P. Handler, and E. L. Smith (1973). *Principles of Biochemistry.* McGraw-Hill Book Co., Inc., New York.

Whittingham, C. P. (1970). The mechanism of photosynthesis and the structure of the chloroplast. Progress in Biophysics and Mol. Biology *21*:125.

CHAPTER 15

Aerobic Metabolism

The accumulation of oxygen in the atmosphere as a result of photosynthesis in organisms ancestral to present-day blue-green algae led to the establishment of metabolic systems that could use oxygen as an electron acceptor in the oxidation of nutrient molecules. This development was of extreme importance to the subsequent evolution of living forms, for it not only allowed organisms to overcome the toxic effects of oxygen but also allowed them to extract a far greater amount of usable energy from nutrient molecules than could be extracted by anaerobic respiration.

Because oxygen has a high positive redox potential, electrons transferred from organic compounds to oxygen can release far more energy than can electrons transferred to an organic electron acceptor. For example, pyruvic acid is used as an electron acceptor in the anaerobic oxidation of glucose by the Embden-Meyerhof pathway. When the pyruvic acid is reduced to lactic acid, ethyl alcohol, or other organic compounds, NADH produced by the oxidation of glyceraldehyde−3−phosphate is reoxidized to NAD^+. Thus, by using pyruvic acid as an electron acceptor, the cell is able to generate NAD^+, which is required for continued operation of the pathway (see Figure 13–3, page 349). In aerobic respiration, NADH may be reoxidized by oxygen, and pyruvic acid, freed of its role as electron acceptor, may be completely oxidized to carbon dioxide and water.

FORMATION OF ACETYL COENZYME A

The oxidation of glucose in aerobic organisms proceeds by way of the Embden-Meyerhof pathway as far as pyruvic acid; however, in aerobic cells, organic end-products such as lactic acid are not formed from pyruvic acid except under conditions of severe cellular oxygen deprivation. Rather, pyruvic acid undergoes oxidative decarboxylation to a two-carbon acetyl $\left(CH_3C\overset{\displaystyle O}{\diagup}\!\!\!- \right)$ fragment that becomes attached to a coenzyme molecule, called *coenzyme A* (HS—CoA). This process is summarized in the following equation:

$$\underset{\text{(Pyruvic acid)}}{\underset{\substack{|\\ \text{COOH}}}{\overset{\substack{\text{CH}_3 \\ |}}{\text{C}}}=\text{O}} + \underset{\text{(Coenzyme A)}}{\text{HS—CoA}} + \text{NAD}^+ \underset{\longleftarrow}{\overset{\substack{\text{pyruvate} \\ \text{dehydrogenase} \\ \text{complex}}}{\longrightarrow}} \underset{\text{(Acetyl coenzyme A)}}{\text{CH}_3\text{—}\overset{\overset{\text{O}}{\|}}{\text{C}}\sim\text{S—CoA}} + \text{NADH} + \text{H}^+ + \text{CO}_2 \quad (1)$$

The oxidative decarboxylation of pyruvic acid actually occurs in a number of different steps, and is accomplished by three separate enzymes and four different cofactors, known collectively as the *pyruvate dehydrogenase complex*. One of these cofactors is coenzyme A, which acts as a carrier of acetyl groups, in the same way that NAD functions as an electron carrier. The chemical structure of coenzyme A, with and without an attached acetyl group, is shown in Figure 15–1. Coenzyme A consists of adenine, ribose, three phosphate groups, the vitamin pantothenic acid, and a sulfur-containing constituent, β–mercaptoethylamine. When coenzyme A carries an acetyl group, the acetyl is linked to the sulfhydryl group (—SH) of coenzyme A to form *acetyl coenzyme A* (*acetyl*$\sim$S—CoA). Acetyl coenzyme A is a high-energy compound, and it is the two-carbon acetyl group of acetyl coenzyme A that is oxidized to carbon dioxide and water in subsequent reactions.

It should be emphasized at this point that pyruvic acid is *not* the only cellular source of acetyl groups. Fats and amino acids are also oxidized by the cell to two-carbon fragments that combine with molecules of coenzyme A to form acetyl coenzyme A. Thus, coenzyme A clearly plays a central role in the aerobic oxidation of various types of organic compounds.

The oxidative decarboxylation of pyruvic acid requires two other cofactors in addition to NAD^+ and coenzyme A. One of these is *thiamine pyrophosphate,* a phosphate derivative of thiamine (vitamin B_1); the other is a derivative of *lipoic acid,* a sulfur-containing compound. Thiamine pyrophosphate and lipoic acid are required as cofactors in other cellular decarboxylation reactions, as well.

Two molecules of acetyl coenzyme A are formed from each molecule of glucose oxidized by the Embden-Meyerhof pathway (since two pyruvic acid molecules are produced for each oxidized glucose molecule). The net result of the degradation of glucose up to this point in the oxidative sequence of reactions is summarized in Table 15–1.

The enzymes of the Embden-Meyerhof pathway, remember, are soluble components of the cytoplasm in all cell types. In aerobic prokaryotes, the enzymes and other compounds associated with the aerobic oxidation of pyruvic acid are either soluble in the cytoplasm or bound to the plasma membrane. Thus, the conversion of pyruvic acid to acetyl coenzyme A occurs in the cytoplasm of prokaryotes, as do the subsequent reactions involving the oxidation of the acetyl fragment of acetyl coenzyme A to carbon dioxide and water. However, in aerobic eukaryotes, these reactions occur in the mitochondria, where the necessary enzymes are located. In these cells,

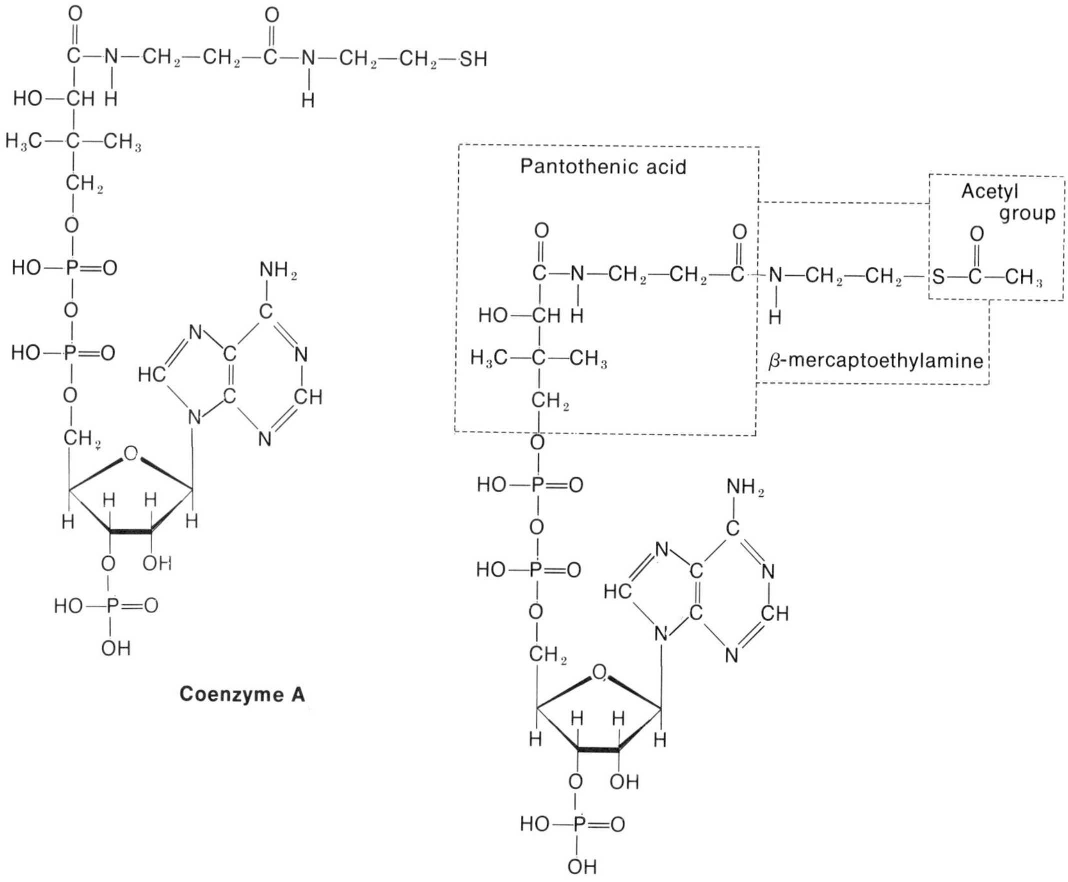

The Krebs Cycle

Chemical structures of coenzyme A and acetyl coenzyme A.   **Figure 15-1**

pyruvic acid produced in the cytoplasm by the Embden-Meyerhof pathway must enter the mitochondria in order to be oxidized further.

THE KREBS CYCLE

The acetyl group carried by coenzyme A is completely oxidized (in mitochondria of eukaryotic cells or in the cytoplasm of aerobic prokaryotes) by a cyclic series of reactions known variously as the *Krebs cycle,* in honor of Hans Krebs, who first described the pathway in the 1930s; as the *citric acid cycle,* because citric acid is an intermediate in the pathway; or as the *tricarboxylic acid (TCA) cycle,* because several intermediates have three carboxyl groups. It is important to remember that the acetyl groups oxidized in the Krebs cycle may be derived either from glucose (via pyruvic acid), from fats, or from amino acids; therefore, the Krebs cycle is the degrada-

Table 15-1 OXIDATION OF GLUCOSE TO ACETYL COENZYME A

REACTION	ATP YIELD

Embden-Meyerhof Pathway *(see Figure 13-1)*

Glucose + 2 ADP + 2 phosphate + 2 NAD$^+$ ⟶ 2 pyruvic acid + 2 ATP + 2 NADH + 2 H$^+$	2

Oxidative decarboxylation of pyruvic acid *(see Equation 15-1)*

2 pyruvic acid + 2 coenzyme A + 2 NAD$^+$ ⟶ 2 acetyl coenzyme A + 2 CO$_2$ + 2 NADH + 2 H$^+$	

Net reaction (glucose to acetyl coenzyme A)

Glucose + 2 ADP + 2 phosphate + 4 NAD$^+$ + 2 coenzyme A ⟶ 2 acetyl coenzyme A + 2 CO$_2$ + 2 ATP + 4 NADH + 4 H$^+$	2

tive pathway for oxidation of other substances besides carbohydrates.

Details of the Krebs cycle are summarized in Figure 15-2. The reactions of the cycle are initiated by the fusion of an acetyl group from acetyl coenzyme A to oxaloacetic acid (a four-carbon compound) to form citric acid (a six-carbon compound)

$$CH_3-\overset{O}{\underset{\|}{C}}\sim S-CoA + \underset{\underset{COOH}{|}}{\underset{CH_2}{|}}\overset{COOH}{\underset{|}{C}}=O \xrightleftharpoons[\text{citrate synthetase}]{} HO-\underset{\underset{CH_2-COOH}{|}}{\overset{CH_2-COOH}{\underset{|}{C}}}-COOH + HS-CoA \quad (2)$$

(Acetyl coenzyme A) (Oxaloacetic acid) (Citric acid) (Coenzyme A)

The subsequent reactions of the cycle involve a series of structural rearrangements leading to oxidations and decarboxylations, and resulting in the conversion of citric acid back to oxaloacetic acid; the cycle is then reinitiated by the reaction of oxaloacetic acid with another molecule of acetyl coenzyme A.

The Krebs Cycle

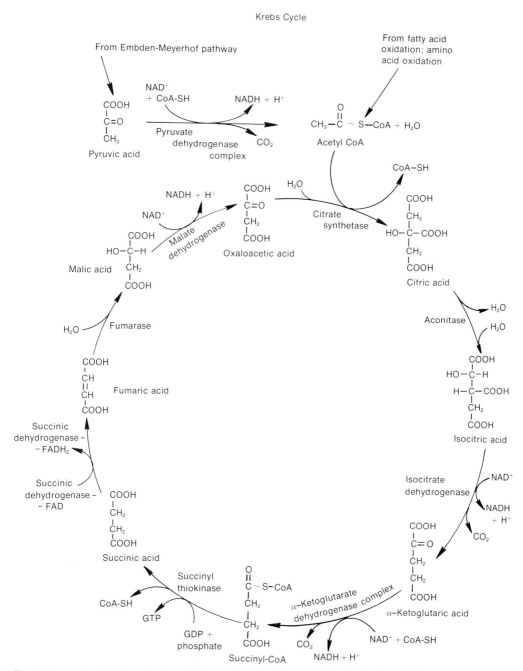

Figure 15-2

The Krebs cycle. At intracellular pH, the carboxyl groups would be ionized to COO$^-$; they are shown here as nonionized in order to make the transformations of the cycle easier to follow. (Adapted from J.R. Bronk [1973], *Chemical Biology: An Introduction to Biochemistry*. New York, Macmillan, Inc.)

The Krebs cycle consists of the following steps, as shown in Figure 15-2.

1. Citric acid, formed by the fusion of an acetyl group from acetyl coenzyme A to oxaloacetic acid, is converted to isocitric acid, a structural isomer of citric acid, in a reaction catalyzed by the enzyme aconitase.

2. Isocitric acid undergoes oxidative decarboxylation to α-ketoglutaric acid. This reaction is catalyzed by isocitrate dehydrogenase, an enzyme that uses NAD as a coenzyme.* The oxidation of isocitric acid is therefore linked to the reduction of NAD^+.

3. A further oxidative decarboxylation results in the conversion of α-ketoglutaric acid to succinyl coenzyme A in a multistep reaction catalyzed by three enzymes collectively known as the α-ketoglutarate dehydrogenase complex. This reaction is analogous to that occurring in the oxidative decarboxylation of pyruvic acid (Equation 1). The reactions catalyzed by the α-ketoglutarate dehydrogenase enzyme complex, like those promoted by the pyruvate dehydrogenase complex, require NAD^+, coenzyme A, thiamine pyrophosphate, and lipoic acid. The similarity of the mechanisms by which these two reactions are carried out illustrates the economy of cellular metabolism; that is, entirely different processes are often accomplished by using identical chemical mechanisms.

4. The high-energy compound succinyl coenzyme A reacts with GDP (guanosine diphosphate) and inorganic phosphate to form GTP (guanosine triphosphate) and succinic acid. The phosphorylation of GDP in this reaction is a substrate-level phosphorylation analogous to the phosphorylation of ADP that occurs in the Embden-Meyerhof pathway. GTP, formed in this way, may be used as a high-energy donor in certain cellular reactions. Alternatively, ATP may be synthesized from GTP by a simple group transfer reaction that occurs as follows:

$$\text{GTP} + \text{ADP} \xrightarrow{\text{nucleoside diphosphokinase}} \text{GDP} + \text{ATP} \qquad (3)$$

5. Succinic acid is oxidized to fumaric acid, a reaction catalyzed by the enzyme succinic dehydrogenase. Succinic dehydrogenase is a component of the mitochondrial inner membrane, and is a flavoprotein enzyme that uses FAD as a prosthetic group. Thus the oxidation of succinic acid is linked to the reduction of the FAD of succinic dehydrogenase. This reaction is the only oxidative step in the Krebs cycle that does not employ NAD^+ as an electron acceptor.

---

*Animal cells contain two different isocitrate dehydrogenase enzymes that catalyze the conversion of isocitric acid to α-ketoglutaric acid. One of these uses NAD as a coenzyme; the other uses NADP. The NAD-specific enzyme is present exclusively in the mitochondrion, and the NADP-specific enzyme is found throughout the cell and in the mitochondria. The NAD-linked dehydrogenase is the enzyme that operates in the Krebs cycle; the NADP-linked enzyme functions primarily in the nucleus and cytoplasm to provide a supply of NADPH required for various biosyntheses.

The Krebs Cycle

6. The enzyme fumarase catalyzes the addition of one molecule of water to fumaric acid to form malic acid.

7. Malic acid is oxidized to oxaloacetic acid. This reaction is catalyzed by the NAD-linked enzyme malate dehydrogenase, and results in the reduction of NAD^+. No decarboxylation occurs in this reaction, and hence it differs from the other two oxidations in the Krebs cycle that involve NAD.

Oxaloacetic acid, formed in this reaction, may now condense with another molecule of acetyl coenzyme A, reinitiating the cycle.

To summarize: for each revolution of the Krebs cycle, one acetyl group is oxidized to two molecules of carbon dioxide. In the process, one molecule of GTP (or, indirectly, ATP) is formed, as are three molecules of NADH and one of $FADH_2$. The net result of the cellular oxidation of one molecule of glucose up to this point is shown in Table 15-2. Remember that *two* acetyl fragments are produced for each molecule of glucose oxidized by the Embden-Meyerhof pathway.

The primary function of the Krebs cycle is to generate ATP needed by the cell, but the substrate-level phosphorylation that occurs in the cycle is not the only, or even the principal, source of ATP produced as a consequence of its operation. It is obvious that the Krebs cycle cannot operate for any length of time without a mechanism for reoxidizing the NADH and $FADH_2$ produced in the

Table 15-2 OXIDATION OF GLUCOSE TO CARBON DIOXIDE BY THE EMBDEN-MEYERHOF PATHWAY AND THE KREBS CYCLE

REACTION	ATP YIELD
Glucose to acetyl coenzyme A *(see Table 15-1)*	
Glucose + 2 ADP + 2 phosphate + 4 NAD^+ + 2 coenzyme A $\longrightarrow$ 2 acetyl coenzyme A + 2 CO_2 + 2 ATP + 4 NADH + 4 H^+	2
Krebs Cycle *(see Figure 15-2)*	
2 Acetyl coenzyme A + 6 H_2O + 2 ADP + 2 phosphate + 2 FAD + 6 $NAD^+ \longrightarrow$ 2 coenzyme A + 4 CO_2 + 2 ATP[a] + 2 $FADH_2$ + 6 NADH + 6 H^+	2
Net reaction (glucose to CO_2)	
Glucose + 6 H_2O + 4 ADP + 4 phosphate + 2 FAD + 10 $NAD^+ \longrightarrow$ 6 CO_2 + 4 ATP + 2 $FADH_2$ + 10 NADH + 10 H^+	4

[a]The GTP produced in the Krebs Cycle is convertible to ATP.

oxidative steps of the cycle. In fact, most of the energy released in the four oxidation reactions of the Krebs cycle is stored in NADH and in the FADH$_2$ of succinic dehydrogenase: *It is the subsequent oxidation of these substances by molecular oxygen that captures a great deal of energy for the cell by a coupled process in which the released energy is used to phosphorylate ADP to ATP.* NADH and FADH$_2$ are oxidized in a series of reactions called the *respiratory chain* or the *electron transport system*, in which electrons from NADH and FADH$_2$ are transferred to molecular oxygen. The phosphorylation of ADP that occurs during electron transfer in the respiratory chain is entirely analogous to the phosphorylations that occur in electron transfers during the light reactions of photosynthesis.

THE RESPIRATORY CHAIN

The components of the respiratory chain in aerobic prokaryotes vary among different types of organisms, and have not been well studied. Consequently, this discussion concentrates on the mitochondrial respiratory chain in eukaryotes. It should be understood, however, that although each individual component of the respiratory chain may not be the same in prokaryotes as in the mitochondria of eukaryotes, the fundamental principle involved in the coupling of electron transport to phosphorylation is identical in all aerobic organisms, prokaryotic or eukaryotic.

The respiratory chain as it is thought to exist in mitochondria is shown in Figure 15–3. Each component of the chain is capable of accepting and donating electrons. Electrons are transferred spontaneously down the chain; each component of the chain accepts one or more electrons from the electron carrier immediately preceding it, and donates electrons to the carrier immediately following it.

The sequence of electron carriers shown in Figure 15–3 is stated tentatively. It has been deduced indirectly by comparing redox potentials of the individual components of the chain. Because each carrier presumably follows another that possesses a more negative redox potential, it is possible to determine a tentative sequence by arranging the carriers in order from the most negative to the least negative (or most positive). Determining the sequence of carriers by this method may be inaccurate, however, since the redox potentials that are used to derive the sequence are calculated by using *purified* electron carriers.

In the cell, the respiratory chain is tightly bound to the inner mitochondrial membrane (or to the plasma membrane in prokaryotes), and, with the exception of cytochrome *c*, the individual carriers cannot be removed from the membrane easily. Because of this factor, the redox potential of a particular carrier may be greatly altered by stripping it from the membrane. Redox potentials calculated by using purified carriers may therefore by quite different from the

The Respiratory Chain

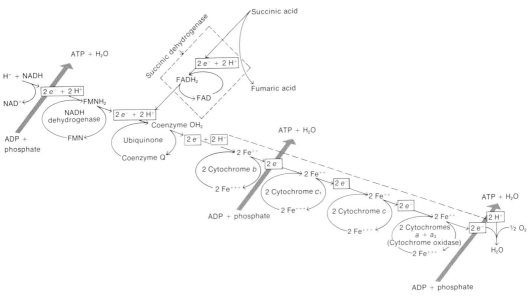

The mitochondrial respiratory chain.

Figure 15-3

redox potentials of the same carriers as they actually occur within the cell.

By comparing redox potentials with other types of experimental data, a probable sequence of electron carriers acceptable to most biochemists has been established. It is, however, quite possible that alternative arrangements may later be proved correct. Moreover, because it is difficult to remove carriers from membranes and to purify these carriers in order to identify them, it is quite likely that additional carriers, as yet undetected, are components of the mitochondrial respiratory chain.

The respiratory chain in mitochondria begins with NAD^+, which, like other components of the chain, is bound to the inner mitochondrial membrane. The NAD-linked dehydrogenases (for example, those in the Krebs cycle such as isocitrate dehydrogenase, α-ketoglutarate dehydrogenase, and malate dehydrogenase, as well as others involved in fat metabolism and other metabolic processes) reduce the membrane-bound NAD^+ of the respiratory chain; that is, NADH formed as a result of substrate oxidation in the mitochondria transfers its electrons to the membrane-bound NAD^+ of the respiratory chain. The resultant NADH of the respiratory chain passes its electrons to the next component of the chain, a flavoprotein enzyme (*NADH dehydrogenase**) that possesses FMN as a prosthetic group. The reduced flavoprotein then passes its electrons to *coenzyme Q* (a ubiquinone).

---

*Do not be confused by the terminology. NADH dehydrogenase is a flavoprotein that catalyzes the transfer of electrons from NADH of the respiratory chain to its own prosthetic group, FMN. NADH dehydrogenase does *not* use NADH as a coenzyme.

There is another branch of the respiratory chain involved with succinic acid oxidation. *Succinic dehydrogenase,* which is situated on the inner mitochondrial membrane with other components of the respiratory chain, is an FAD-linked enzyme. Succinic dehydrogenase passes its electrons, derived from the oxidation of succinic acid, from its $FADH_2$ prosthetic group directly to coenzyme Q, bypassing the earlier steps of the chain. Hence there are two flavoproteins at the substrate end of the respiratory chain: One of these, NADH dehydrogenase, catalyzes the oxidation of respiratory chain NADH, which is produced as a result of various NAD-linked dehydrogenase reactions; the other, succinic dehydrogenase, is both a component of the respiratory chain and a catalyst that brings about the oxidation of succinic acid (see Figure 15–3).

From coenzyme Q, electrons are passed to molecular oxygen through a series of cytochrome enzymes including, in order, cytochrome b, cytochrome c_1, cytochrome c, and a complex of cytochromes a and a_3. The cytochrome $a + a_3$ complex is known as *cytochrome oxidase,* because it is capable of reducing molecular oxygen. Cytochrome oxidase contains copper ions as well as the iron–porphyrin cytochromes, but it is not yet known how cytochrome oxidase actually brings about the reduction of oxygen.

In addition to the components shown in Figure 15–3, there are at least three areas in the chain where a nonheme iron protein (a protein that contains iron *not* bound within a porphyrin ring) is located. One such protein is associated with NADH dehydrogenase; another with succinic dehydrogenase; a third may be located in the respiratory chain between cytochromes b and c_1. These nonheme proteins apparently assist in some way with electron transfer.

Hydrogen ions as well as electrons are considered to flow along parts of the respiratory chain. Two electrons as well as two hydrogen ions flow from NADH (one hydrogen ion is supplied by the medium) to the FMN of NADH dehydrogenase and from the flavin prosthetic groups of NADH dehydrogenase or succinic dehydrogenase to coenzyme Q. Since the cytochromes carry only electrons, hydrogen ions from coenzyme Q are released into the medium, to be accepted by oxygen during its reduction to water. It is also evident that NAD, the flavoproteins, and coenzyme Q are oxidized or reduced by the removal or addition of two electrons, whereas each of the cytochromes carries only one electron. This has been interpreted as implying either that the cytochromes are oxidized twice as fast as the other components of the chain, or that there are twice as many cytochromes, perhaps acting in parallel, as there are molecules of NAD, flavoprotein, and coenzyme Q.

The respiratory chain functions only when the inner mitochondrial membrane, to which the components of the respiratory chain are attached, is intact. Thus the spatial orientation of the electron carriers to one another appears to be crucial to their proper functioning. It has been assumed that a precise spatial orientation is necessary in order that electrons may be transferred from one car-

The Respiratory Chain

rier to another at maximal efficiencies without migration of the carriers from their positions in the membrane.

OXIDATIVE PHOSPHORYLATION

Phosphorylation of ADP to ATP takes place during the transfer of electrons along the respiratory chain—a process that is referred to as *oxidative phosphorylation.* The energy originally stored in reduced coenzymes by the oxidation of substrate molecules is, in turn, released during the oxidation of the coenzymes, and its release is coupled to the endergonic phosphorylation of ADP. The exact points in the respiratory chain where electron transfer is linked to phosphorylation have not been unambiguously determined. It is, however, known that three molecules of ATP are produced for each molecule of NADH oxidized in the respiratory chain, and that consequently there must be three places in the chain where electron transport is coupled to phosphorylation. It is generally, but not universally, agreed by biochemists that phosphorylation is linked to (1) the transfer of electrons from NADH to the FMN prosthetic group of NADH dehydrogenase, (2) the transfer of electrons from two molecules of cytochrome b to two molecules of cytochrome c_1, and (3) the transfer of electrons from two molecules of cytochrome oxidase to oxygen (see Figure 15–3).

Since the oxidation of NADH by molecular oxygen has a $\Delta G°$ of -52.6 Kcal/mole, and three moles of ATP are produced for each mole of NADH oxidized, the capture of energy in the form of ATP by electron transport has an efficiency of approximately

$$\frac{3 \times 7.3 \text{ Kcal/mole}}{52.6 \text{ Kcal/mole}} \times 100 = 41.1\%$$

Although the oxidation of one molecule of NADH yields three molecules of ATP, the oxidation of the $FADH_2$ prosthetic group of one molecule of succinic dehydrogenase yields only *two* molecules of ATP. This is because the first phosphorylation site in the respiratory chain (between NADH and the FMN of NADH dehydrogenase) is bypassed when electrons are transferred from $FADH_2$ of succinic dehydrogenase to coenzyme Q. It is important to consider this fact when calculating the efficiency of energy capture in the oxidation of glucose.

OXIDATION OF CYTOPLASMIC NADH

NADH produced in the cytoplasm by oxidation of glyceraldehyde–3–phosphate in the Embden-Meyerhof pathway is also oxidized in the respiratory chain. A supply of NAD^+ needed for the operation of the pathway is thus maintained. This presents no problem in aerobic prokaryotes, whose respiratory chain components are located on the plasma membrane in contact with the cytoplasm of the cell. However, in eukaryotes, where the respiratory chain is located within the mitochondria, NADH would have to enter the mitochondria in order to be oxidized. This is not always possible, since the mitochondria of many cells are relatively impermeable to NADH. In such cases, the oxidation of cytoplasmic NADH apparently occurs not by the direct entry of cytoplasmic NADH into the mitochondrion but by the transfer of electrons to other substances that *can* freely penetrate the mitochondria. Thus a type of "shuttle" is set up between the cytoplasm and the mitochondria, in which some substrate that is reduced by cytoplasmic NADH traverses the mitochondrial membrane, is reoxidized in the mitochondrion, and is then returned to the cytoplasm (Fig. 15–4).

ANAEROBIC METABOLISM IN AEROBES

In multicellular animals, actively metabolizing cells (for example, muscle cells that are contracting rapidly) may consume oxygen faster than the bloodstream can supply it. Under such conditions of severe oxygen deprivation, the respiratory chain, and therefore the Krebs cycle, is inhibited, and pyruvic acid is reduced to lactic acid, providing a source of NAD^+ to keep the Embden-Meyerhof pathway functioning. The lactic acid produced in this reaction does not go to waste. It is usually excreted by the cell and carried by the blood-

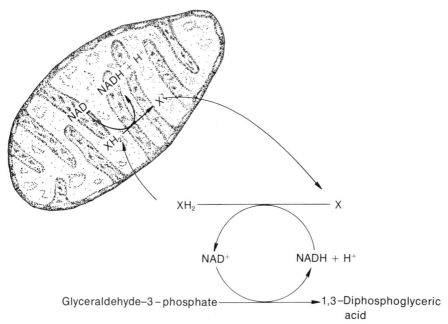

Figure 15–4 A schematic representation of the mechanism by which electrons of NADH are transferred from the cytoplasm to the mitochondria. Cytoplasmic NADH, which does not easily penetrate mitochondria, transfers electrons to a compound (X) that can pass freely in and out of mitochondria. Reduced X (XH_2) is oxidized by NAD^+ within the mitochondrion, and then X reenters the cytoplasm. The net result is that cytoplasmic NADH is oxidized to NAD^+.

stream to the liver where, in the presence of an ample supply of oxygen, it may be oxidized to pyruvic acid. The pyruvic acid may then enter the Krebs cycle for further oxidation or be converted into glucose and stored as glycogen. Although the energy of lactic acid is lost to the cell under oxygen-limiting conditions, the lactic acid is retained by the organism. The conservation of energy attained through this mechanism has obvious selective advantage.

The entire preceding discussion of energy generation in aerobic cells further emphasizes the interrelationship of anaerobic and aerobic pathways of metabolism. The Embden-Meyerhof pathway will operate, to different ends, both in the presence and in the absence of oxygen. The Krebs cycle, although it does not directly require molecular oxygen for its operation, does require a functioning respiratory chain to maintain a supply of oxidized coenzymes. It is thus apparent that the distinction between aerobic and anaerobic pathways of metabolism is not a clear one—a situation that is not surprising in view of the way metabolic systems probably evolved.

EFFICIENCY OF THE AEROBIC OXIDATION OF GLUCOSE

The total yield of ATP from the complete oxidation of glucose by the Embden-Meyerhof pathway and the Krebs cycle, followed by the oxidation of coenzymes in the respiratory chain, is shown in Table 15–3. Since 38 ATP molecules are produced by the aerobic oxidation of one molecule of glucose, compared to two ATP molecules for every molecule of glucose oxidized anaerobically in the Embden-Meyerhof pathway, aerobic respiration captures nineteen times as much energy for the cell as does anaerobic respiration. Thus, in an aerobic environment the ability to respire aerobically has great selective advantage in terms of energy capture.

MECHANISMS OF OXIDATIVE PHOSPHORYLATION

In normally functioning mitochondria, the rate of respiration is determined by the ATP requirements of the cell. For example, unless a supply of ADP and inorganic phosphate is present within the cell, substrates are not oxidized and oxygen is not consumed. This coupling between energy utilization and energy capture operates as a metabolic control mechanism: If the concentrations of ADP and inorganic phosphate in the cell are low, the rate of respiration is low; as the concentrations of ADP and inorganic phosphate (the hydrolysis products of ATP) increase, the rate of respiration increases. Numerous experiments have shown that even though phosphorylation is tightly coupled to electron transport in the respiratory chain, phosphorylation and electron transport are distinctly separate processes that may be caused to operate independently under experimental conditions. For example, certain chemical substances, called *uncoupling agents,* act as specific inhibitors of phosphorylation without interfering with electron transport. One

Table 15-3 COMPLETE OXIDATION OF GLUCOSE BY THE EMBDEN-MEYERHOF PATHWAY AND KREBS CYCLE, FOLLOWED BY OXIDATION OF COENZYMES IN THE RESPIRATORY CHAIN

REACTION	ATP YIELD
Glucose to CO_2 by Embden-Meyerhof pathway and Krebs cycle (see Table 15-2)	
Glucose + 6 H_2O + 4 ADP + 4 phosphate + 2 FAD + 10 NAD^+ $\longrightarrow$ 6 CO_2 + 4 ATP + 2 $FADH_2$ + 10 NADH + 10 H^+	4
Respiratory chain (see Figure 15-3)	
2 $FADH_2$ + 4 ADP + 4 phosphate + O_2 $\longrightarrow$ 2 FAD + 4 ATP + 2 H_2O	4
and	
10 NADH + 10 H^+ + 30 ADP + 30 phosphate + 5 O_2 $\longrightarrow$ 10 NAD^+ + 30 ATP + 10 H_2O	30
Net reaction (complete oxidation of glucose and oxidation of coenzymes)	
Glucose + 38 ADP + 38 phosphate + 6 O_2 $\longrightarrow$ 6 CO_2 + 6 H_2O + 38 ATP	38

uncoupling agent that is frequently used to study the relationship between phosphorylation and electron transport is 2,4-dinitrophenol (DNP).

Although much research has been performed in attempts to elucidate the mechanism of oxidative phosphorylation, we still do not understand how phosphorylation and electron transport are coupled. That is, we do not yet know how the energy released in electron transport is channeled for use in the phosphorylation of ATP. Three basic mechanisms have been proposed in attempts to explain how phosphorylation is coupled to electron transport. These are discussed in Appendix 4. Each of these mechanisms is supported by some experimental evidence, and it is not as yet possible to determine which, if any, of the three mechanisms actually operates within the cell.

THE RELATIONSHIP BETWEEN CARBOHYDRATE METABOLISM AND OTHER METABOLIC PROCESSES

OXIDATION OF SUBSTANCES OTHER THAN CARBOHYDRATES

We have mentioned that cells can use fats and proteins, in addition to carbohydrates, as sources of energy. In fact, fats are the

Carbohydrate Metabolism and Other Metabolic Processes

primary energy reserve in most animal cells. The oxidation of fats and proteins occurs primarily in the Krebs cycle, and their oxidation therefore shares a major pathway with carbohydrate oxidation.

OXIDATION OF FATS

In the first step in their oxidation, fats are hydrolyzed in the cytoplasm of the cell to glycerol and fatty acids. The glycerol is converted to dihydroxyacetone phosphate at the expense of ATP, and then may enter the Embden-Meyerhof pathway for further oxidation. In eukaryotic cells, mitochondria are the major site of fatty acid oxidation. The long-chain fatty acid residues enter the mitochondrion, where they are sequentially broken down into two-carbon acetyl units linked to coenzyme A. The first stage in this process is the attachment of the fatty acid molecule to coenzyme A. This occurs in a two-step reaction in which one molecule of ATP is hydrolyzed to AMP, thus utilizing the energy in two high-energy phosphate bonds

$$R-CH_2-CH_2-\overset{O}{\underset{\|}{C}}-OH + HS-CoA + ATP \underset{\text{Fatty acid thiokinase}}{\overset{Mg^{++}}{\rightleftharpoons}}$$

(Fatty acid)

$$R-CH_2-CH_2-\overset{O}{\underset{\|}{C}}\sim S-CoA + AMP + \text{Pyrophosphate} \quad (4)$$

(Fatty acyl coenzyme A)

$$HO-\overset{O}{\underset{\|}{\underset{OH}{P}}}-O-\overset{O}{\underset{\|}{\underset{OH}{P}}}-OH + H_2O \longrightarrow 2\ HO-\overset{O}{\underset{\|}{\underset{OH}{P}}}-OH \qquad \Delta G° = -8\ \text{Kcal/mole} \quad (5)$$

(Pyrophosphate)   (Phosphoric acid)

This mechanism is seen in many reactions, particularly biosynthetic ones. The high-energy bond in the fatty acyl coenzyme A molecule has about the same $\Delta G°$ of hydrolysis as the high-energy phosphate bond in ATP. Consequently, the first reaction (Equation 4) would proceed only if the reactants remained at a higher concentration than the products, were it not for the fact that the reaction is coupled to the hydrolysis of pyrophosphate (Equation 5). Since this latter reaction proceeds with a $\Delta G°$ of -8 Kcal/mole, the entire process is strongly exergonic.

After this preliminary expenditure of energy, no further energy is required for the oxidation of the entire fatty acid molecule, and consequently the attachment of the fatty acid molecule to coenzyme A is usually referred to as the *preliminary activation step*. Following activation, acetyl coenzyme A is cleaved from the fatty acyl coenzyme A molecule in a four-step pathway that yields one molecule of NADH and one molecule of FADH$_2$, and regenerates a new fatty acyl coenzyme A molecule that possesses two less carbon atoms than the original fatty acyl coenzyme A molecule

$$R-CH_2-CH_2-\overset{O}{\underset{\|}{C}}\sim S-CoA + FAD + NAD^+ + HS-CoA + H_2O \longrightarrow$$
(Fatty acyl coenzyme A with n carbon atoms)

(6)

$$R-\overset{O}{\underset{\|}{C}}\sim S-CoA + CH_3-\overset{O}{\underset{\|}{C}}\sim S-CoA + FADH_2 + NADH + H^+$$
(Fatty acyl coenzyme A with $n-2$ carbon atoms)

FADH$_2$ and NADH may be oxidized in the respiratory chain for a yield of five ATPs for each acetyl coenzyme A fragment split from the fatty acyl coenzyme A. The oxidation of the acetyl coenzyme A fragment in the Krebs cycle yields the equivalent of twelve ATP molecules. (Three NADH and one FADH$_2$ molecules oxidized in the respiratory chain provide nine and two molecules of ATP respectively, and one GTP molecule formed by substrate-level phosphorylation provides one molecule of ATP.) The activated fatty acid chain continues to be oxidized by the sequential removal of acetyl coenzyme A fragments, as in Equation 6, until the fatty acid is completely oxidized. (Fatty acid chains containing an odd number of carbons are oxidized to propionyl coenzyme A, a three-carbon fragment; this is carboxylated to succinyl coenzyme A that enters the Krebs cycle.)

The net result of the oxidation of fats is an energy yield well over twice that of the oxidation of a comparable weight of carbohydrates. For example, the complete oxidation of stearic acid ($C_{17}H_{35}COOH$) yields the equivalent of 146 molecules of ATP. If the yield is calculated on the basis of weight, one gram of glucose yields 0.21 mole of ATP, whereas one gram of stearic acid yields 0.58 mole of ATP.

OXIDATION OF PROTEINS

The cell may also extract energy from proteins after they have been hydrolyzed into amino acids by cellular proteolytic enzymes. Most amino acids are converted either to acetyl coenzyme A, or, by removal of the amino groups, to intermediates of the Embden-

The Regulation of Aerobic Respiration

Meyerhof pathway or of the Krebs cycle. For example, alanine, threonine, and cysteine are converted into pyruvic acid and then into acetyl coenzyme A; aspartic acid is converted into oxaloacetic acid; and glutamic acid is converted into α–ketoglutaric acid.

SYNTHESIS OF CELLULAR CONSTITUENTS

The Krebs cycle and the respiratory chain are clearly involved in the oxidation of fats and proteins as well as of carbohydrates (Fig. 15–5). However, the Krebs cycle, like the Embden-Meyerhof pathway, is not merely a degradative pathway; it also plays an important role in the synthesis of cellular constituents. As may be seen in Figure 15–6, the Krebs cycle contains intermediate compounds that are important precursors to porphyrins and amino acids. Some syntheses of cellular components occur by pathways that are essentially the reverse of degradative pathways, but this is not true in the majority of cases. Figure 15–6 also shows that instead of being oxidized by the Krebs cycle, acetyl coenzyme A may be used for the synthesis of fats and related compounds, under conditions in which ATP supplies are adequate to meet the immediate energy demands of the cell.

THE REGULATION OF AEROBIC RESPIRATION

In Chapter 14, we noted that the Embden-Meyerhof pathway is controlled primarily by the ratio of NAD^+ to NADH; in order to keep the Embden-Meyerhof pathway running, NAD^+ must be continually regenerated from NADH by the reduction of pyruvic acid. However, in aerobic respiration, NADH is oxidized by the respiratory chain, and pyruvic acid is converted to acetyl coenzyme A, which enters

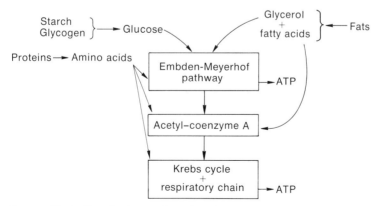

Figure 15–5 Schematic diagram illustrating the integration of pathways of carbohydrate, fat, and protein oxidation.

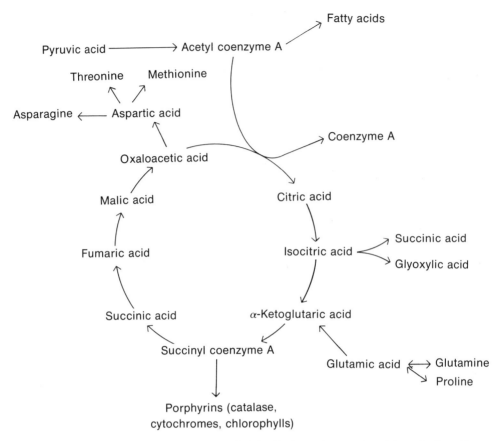

Figure 15–6 Illustration of the relationship of the Krebs cycle to various biosynthetic pathways of the cell. The diagram is by no means inclusive. (Adapted from J. Levy et al. [1973]. *Introductory Microbiology.* New York, John Wiley and Sons, Inc.)

the Krebs cycle. Under normal intracellular conditions, the operation of the Krebs cycle and respiratory chain is dependent primarily on the ratio of ADP to ATP. As mentioned previously, mitochondrial respiration (including substrate oxidation, oxygen utilization, and phosphorylation) is inhibited by low levels of ADP and inorganic phosphate and stimulated by high levels of these materials. This phenomenon, termed *respiratory control,* appears to be the principal control mechanism regulating the operation of the Krebs cycle and respiratory chain; it is called a *positive feedback system,* since *high* levels of ADP and phosphate *stimulate* respiratory activity (Fig. 15–7).

Another feedback control of the Krebs cycle involves oxaloacetic acid. Oxaloacetic acid is a competitive inhibitor (see Chapter 5) of succinic dehydrogenase, the enzyme that catalyzes the oxidation of succinic acid to fumaric acid. Consequently, the buildup of a high concentration of oxaloacetic acid (because of a lack of acetyl coenzyme A, or for some other reason) inhibits the formation of additional oxaloacetic acid. In contrast to respiratory control, this is a

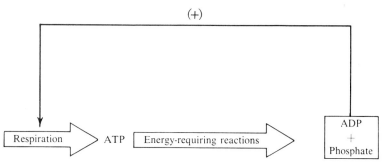

Figure 15-7

A schematic diagram illustrating respiratory control. Notice that ATP levels as such do not influence the rate of respiration. (*From* J.L. Howland [1968], *Introduction to Cell Physiology: Information and Control.* New York, Macmillan, Inc. Copyright © 1968, Macmillan, Inc.)

negative feedback system, because *high* levels of oxaloacetic acid *depress* the synthesis of oxaloacetic acid.

In addition to these fairly specific controls, a great number of less direct factors influence the rate of aerobic respiration. For example, we have noted that oxygen deprivation, which results in the depletion of oxidized coenzymes needed for the operation of the Krebs cycle, will bring the cycle to a halt. Competition between metabolic pathways for common intermediates plays a role in determining the rate of respiratory activity, just as it affects the operation of the Embden-Meyerhof pathway. This is particularly true of acetyl coenzyme A, which is an intermediate in numerous metabolic pathways in aerobic organisms. Another type of metabolic control, which we shall discuss in detail in Chapter 18, is the regulation of enzyme synthesis. Since enzymes are required for all metabolic reactions, any process that stimulates or inhibits the synthesis of a particular enzyme may greatly affect the rate of metabolic activity.

LOCALIZATION OF AEROBIC RESPIRATION

During the 1970s, researchers have correlated various existing evidence to reach the tentative conclusion that the plasma membrane *in prokaryotes* carries enzymes involved in electron transport and oxidative phosphorylation; enzymes of the Krebs cycle and fatty acid oxidation are presumably dispersed in the cytoplasm of prokaryotes, or loosely bound to structures within the cytoplasm. In general, however, the localization of oxidative enzymes within the prokaryotic cell has not been clearly delineated.

We have already mentioned that the plasma membranes of some aerobic prokaryotes possess small projections on their inner surfaces. In mitochondria, structures with similar morphology attached to the inner mitochondrial membrane are known to be involved in the phosphorylation of ADP (see below). Consequently, some workers have suggested that the projections seen on the plasma membranes of some aerobic prokaryotes may have a similar function.

In eukaryotes, aerobic oxidation of nutrients takes place within mitochondria. The mitochondrion, like the cell itself, is compartmentalized, and certain regions of the mitochondrion are associated with specific functions. The outer mitochondrial membrane, which is about 50 per cent protein and 50 per cent lipid, strongly resembles the plasma membrane in its chemical composition and its permeability properties. In fact, the outer mitochondrial membrane is considerably more permeable than the inner membrane. Associated with the outer membrane are the fatty acid thiokinases, the enzymes that catalyze the transfer of fatty acid to coenzyme A during fatty acid oxidation (Equation 4).

The outer membrane is easily removed by treating mitochondrial preparations with certain detergents. This treatment disrupts the outer membrane while leaving the inner membrane intact, and releases two enzymes that are present in the outer mitochondrial compartment. These are adenylate kinase, which catalyzes the reaction 2 ADP $\rightleftharpoons$ ATP + AMP, and *nucleoside diphosphokinase*, which catalyzes the transfer of a high-energy phosphate group from GTP to ADP (Equation 3).

Isolation and chemical analysis of the inner mitochondrial membrane show that it is about 75 per cent protein and 25 per cent lipid, although its precise molecular composition is not known. Indeed, debate has for some time centered on the question of how much, if any, of the mitochondrial membrane consists of "structural" protein lacking enzymatic activity. The inner membrane has been shown to contain the components of the respiratory chain, including succinic dehydrogenase as well as ATPase and other proteins responsible for the phosphorylation of ADP. In addition, the inner membrane also contains two enzymes of the Krebs cycle—pyruvate dehydrogenase and α–ketoglutarate dehydrogenase. The exact structural organization of the components of the respiratory chain has not been determined, although biochemical evidence suggests that these components may be arranged in clusters within the inner membrane. Such an arrangement would facilitate the passage of electrons between members of the chain.

The 8 to 9 nm spherical structures attached to the inside of the inner mitochondrial membrane appear to consist of ATPase (also called F_1 or *coupling factor*), a large protein with a molecular weight of about 285,000 (see Figure 11–21, p. 294). This was shown in a series of elegant experiments by Efraim Racker and his colleagues. Racker demonstrated that the spheres contained ATPase activity, and that membranes stripped of the spherical structures could oxidize coenzymes but were unable to phosphorylate ADP. When solutions containing ATPase activity were added back to the stripped inner membrane preparations, phosphorylating activity was again detected. Under the electron microscope, such reconstituted membranes were seen to possess spherical particles similar to those on the untreated inner membranes (Figure 15–8). Racker has also shown that preparations of purified mitochondrial ATPase appear as 8 to 9 nm spherical particles when observed by negative staining

Localization of Aerobic Respiration

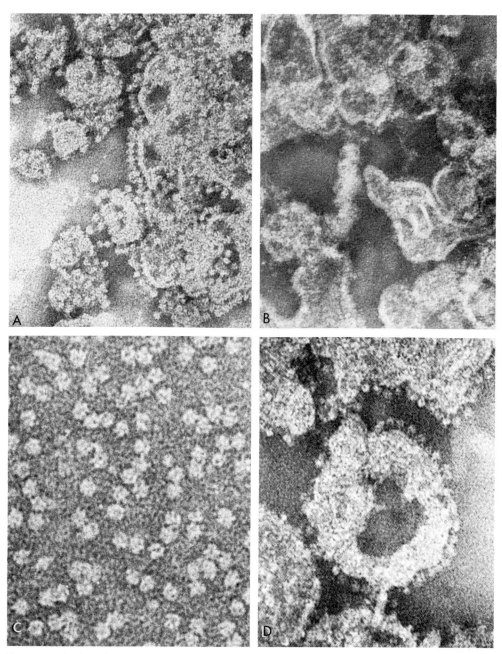

The reconstitution of inner mitochondrial membranes is shown in this series of electron micrographs. *a*, Untreated inner membranes ($\times$ 300,000). *b*, Stripped inner membranes from which spherical particles have been removed ($\times$ 300,000). *c*, Spherical particles isolated from inner membranes ($\times$ 600,000). *d*, Addition of isolated spherical particles to stripped membrane preparations. (Reproduced by permission of Efraim Racker, Cornell University, Ithaca, New York 14853.)

Figure 15-8

under the electron microscope. Because these spherical particles consist of ATPase, they are regarded as the site of oxidative phosphorylation.

When the mitochondrial inner membrane is disrupted, the contents of the matrix are released. In the matrix are located the soluble enzymes of the Krebs cycle, the enzymes of fatty acid oxidation, and the enzymes and other components (e.g., DNA, RNA, ribosomes) associated with mitochondrial protein synthesis.

The compartmentalization observed within the mitochondrion is achieved by the differential permeabilities of the two mitochondrial membranes. The outer membrane is highly permeable, and the composition of the outer compartment is not unlike the composition of the cytoplasm as a whole. The inner membrane, with its more selective permeability, maintains high concentrations of substrates and oxidative enzymes within the mitochondrial matrix.

THE DYNAMIC NATURE OF METABOLISM

The foregoing discussion should make it clear that an appreciation of the nature of cellular metabolism depends on an understanding of the integration of degradative and biosynthetic pathways, of energy capture and energy utilization. The metabolic system of the cell is continually making delicate adjustments to help maintain the integrity of the cell and the organism by counteracting, as much as possible, the effects of a hostile environment. Thus, if a cell requires energy, the cellular control mechanisms will drive metabolic reactions in the direction of energy capture by the oxidation of reserve material or, in a crisis, even of structural components of the cell. If the cell must synthesize material, the metabolism of the cell will favor reactions leading to biosynthesis: for instance, those generating reducing power (NADPH) needed for biosyntheses. If the cell has an ample supply of ATP and is undergoing biosynthesis at a low rate such that nutrient input more than compensates for the energy expenditure required for maintaining all cellular and organismal activities, nutrients will be stored in the form of carbohydrates or fats. There are, of course, many other possible metabolic adjustments; together these exemplify the dynamic nature of cellular metabolism.

REFERENCES

Azzone, G. F. (1972). Oxidative phosphorylation, a history of unsuccessful attempts: is it only an experimental problem? J. Bioenergetics 3:95.
Bronk, J. R. (1973). *Chemical Biology*. Macmillan, Inc., New York.
Giese, A. C. (1973). *Cell Physiology*. W. B. Saunders Co., Philadelphia.
Green, D. E., and Sungchul Ji (1972). The electromechanical model of mitochondrial structure and function. J. Bioenergetics 3:159.
Green, D. E., and J. H. Young (1971). Energy transduction in membrane systems. Am. Scientist 59:92.
Greville, G. D. (1969). Mitchell's chemiosmotic hypothesis. Current Topics in Bioenergetics 4:1.
Howland, J. H. (1968). *Introduction to Cell Physiology*. Macmillan, Inc., New York.

References

Levy, J., J. J. R. Campbell, and T. H. Blackburn (1973). *Introductory Microbiology.* John Wiley and Sons, Inc., New York.

Mitchell, P. (1965). Chemiosmotic coupling in oxidative and photosynthetic phosphorylations. *Biol. Revs. 41*:445.

Mitchell, P. (1968). *Chemiosmotic Coupling and Energy Transduction.* Glynn Research, Ltd., Bodmin, England.

Mitchell, P. (1972). Chemiosmotic coupling in energy transduction: a logical development of biochemical knowledge. J. Bioenergetics 3:5.

Morton, R. A. (1971). Ubiquinones, plastoquinones and vitamin K. Biol. Revs. *46*:47.

Racker, E. (1968). The membrane of the mitochondrion. Sci. Amer. *218*:32.

Racker, E. (1970). The two faces of the inner mitochondrial membrane. Essays in Biochem. *6*:1.

Slater, E. C. (1972). The mechanism of energy conservation in the mitochondrial respiratory chain. Harvey Lectures *66*:19. Academic Press, Inc., New York.

Wainio, W. W. (1970). *The Mammalian Mitochondrial Respiratory Chain.* Academic Press, Inc., New York.

White, A., P. Handler, and E. L. Smith (1973). *Principles of Biochemistry.* McGraw-Hill, Inc., New York.

Wolfe, S. L. (1972). *Biology of the Cell.* Wadsworth Publishing Co., Inc., Belmont, California.

Young, J. H. (1972). An enzymological approach to mitochondrial energy transduction. J. Bioenergetics *3*:137.

16

Membrane Structure and the Transport of Materials

Discussions in previous chapters have emphasized the importance of cellular membranes to the normal functioning of the cell. Of all cellular membranes, the plasma membrane is clearly the most important. It is present in all cell types and was undoubtedly the first membrane to develop during the evolution of living systems. In present-day cells, the plasma membrane is essential for the maintenance of cellular integrity. It not only forms a structural barrier surrounding the cell but determines the metabolic activity of the cell in a very fundamental way by controlling the passage of substances into and out of the cell. Thus, for example, the plasma membrane permits the entrance of water, nutrients, certain salts, and other essential materials into the cell while excluding unneeded or potentially harmful substances; it hinders the loss of metabolically useful substances and encourages the release of toxic or useless metabolic byproducts; and it controls the concentration of intracellular substances at acceptable levels.

The function of the plasma membrane is not exclusively that of a permeability barrier; in most cell types it serves additional functions, as well. In prokaryotic cells, the plasma membrane is involved in a number of metabolic processes. There is good evidence, for instance, that the DNA of bacteria is attached to a specific site on the cytoplasmic side of the plasma membrane and that this attachment is important in the process of cell division. In some bacteria the attachment of DNA appears to occur at mesosomes, complex infoldings of the plasma membrane. Furthermore, at least some of the enzymes associated with respiratory metabolism in aerobic prokaryotes may be attached to the inner side of the plasma membrane, and in some of these organisms the plasma membrane may be folded to form chondroids (Chapter 10). It has also been assumed that thylakoid membranes in prokaryotes are derived from the plasma membrane. Indeed, the thylakoid membranes in photosynthetic bacteria appear in many cases to be continuous with the plasma membrane, although this is not often seen in blue-green algae.

Membrane Structure and the Transport of Materials

In multicellular eukaryotes the plasma membrane, besides functioning as a permeability barrier, is important in interactions between cells. The differentiated junctional structures (desmosomes, septate desmosomes, and gap junctions) are believed to function not only as strong points of intercellular attachment but also in the intercellular transfer of ions and small organic molecules. There is now also good evidence that the plasma membrane of each cell type in a multicellular organism possesses specific molecular components termed *recognition molecules,* by which one cell type may distinguish between similar or different types. For example, embryonic development is characterized by complex cellular movements, in which cells of similar type (e.g., kidney) associate specifically to form tissues; the recognition molecules on plasma membranes are thought to play an important role in this process.

Indeed, it has been postulated that tumor cells differ from normal cells in possessing altered recognition molecules on their plasma membranes. This idea stems from the observation that normal eukaryotic cells and tumor cells show very different growth patterns in tissue culture. When two normal cells grown on a glass surface make contact with one another, they will stop dividing, with the result that only single layers (monolayers) of cells are formed. This phenomenon is called *contact inhibition.* Tumor cells, on the other hand, do not exhibit contact inhibition toward one another and consequently will pile up in multilayered areas. Thus it is thought that tumors may arise by changes in recognition molecules that destroy the normal control mechanisms regulating cell division and adhesion.

Plasma membranes in eukaryotes also possess *receptor sites* to which certain drugs, hormones, and other substances must attach in order to influence the cell. Receptor sites are specific, and there is a different receptor site for each substance or group of substances that is capable of binding to the membrane. Attachment of a substance to its receptor site depends upon specific chemical interactions between the two, much as an enzyme binds to its substrate. It has been known for some time that cellular resistance to infection by a particular virus is often due to the fact that the cell does not possess a specific receptor site to which the virus (or surface component of the virus) is capable of attaching; without attaching to the membrane, the virus cannot enter the cell. Some hormones (e.g., many of the pituitary hormones) and drugs have been shown to cause their characteristic effects merely by attaching to their receptor sites; that is, they do not have to enter the cell in order to trigger a physiological response. Thus, cells which lack receptor sites for such a substance are not affected by the substance.

The plasma membrane in multicellular eukaryotes is also important to cell function in another respect. Plasma membranes of different cell types within an organism often have different properties, and the specialized functions of many cell types are associated with these specific properties. Thus, for example, the conduction of nerve impulses is dependent upon the special permeability characteristics of the nerve cell plasma membrane, the plasma membrane is important in cell motility, and so forth.

Of course, it would be wholly incorrect to view the plasma membrane as the only important cellular membrane in eukaryotes. Eukaryotic cell structure is characterized by many internal membrane systems which divide the cell into a number of subcellular compartments responsible for carrying out specific biochemical functions. Presumably, such compartmentalization developed because it represented a particularly efficient way of carrying out certain cellular functions. It is clear, however, that integration of intracellular functions must require communication between compartments and that this communication must occur through the membranes that form the compartments.

Although not a great deal is known about the transfer of metabolites between compartments, evidence is accumulating that each of the membranes that forms an intracellular compartment has specific permeability characteristics. For instance, we know that all compartments are not equally permeable to every type of substance and that certain substances tend to be concentrated in specific locations within the cell. This differential distribution of materials must be a function of the permeability properties of the membranes that form the various compartments.

Certain intracellular membranes not only serve as selective barriers but have specific biochemical functions uniquely associated with them. We know that the components of the mitochondrial electron transport chain are integral constituents of the inner mitochondrial membrane; the pigments, enzymes, and other substances required for the light reactions of photosynthesis are tightly bound to or embedded within the thylakoid membranes. In both mitochondrial electron transport and the light reactions of photosynthesis, it is clear that the localization of enzymes and other components on membranes is crucial to the efficient operation of these processes. Many other enzymes may also be more or less specifically associated with membranes in a noncovalent (and reversible) fashion; such associations are important in the subcellular localization of enzymes and in enzyme regulation.

Thus, membranes play essential and varied roles in the living cell. In order to understand how membranes can display such diverse functions, we need to understand the molecular organization of membranes.

STRUCTURE OF CELL MEMBRANES

CHEMICAL COMPOSITION

All cellular membranes, including the plasma membranes as well as those in the cell interior, are composed primarily of protein and lipid, although carbohydrate is often present in association with

Structure of Cell Membranes

Table 16-1 APPROXIMATE COMPOSITION OF MEMBRANES FROM VARIOUS SOURCES (in Per Cent)

	MYELIN SHEATH[a]	ERYTHROCYTE PLASMA MEMBRANE	MITOCHONDRIAL INNER MEMBRANE	CHLOROPLAST LAMELLAE	ESCHERICHIA COLI
Protein + glycoprotein	20	70	75	50	75
Lipids	80	30	25	50	25
Cholesterol	20	8	1	0	0
Phospholipids	30	21	24	6	25
Glycolipids	21	0	0	20	0
Other	9	1	0	24	0

[a] The myelin sheath is the insulating membrane surrounding some nerve fibers.

protein (as glycoprotein) or lipid (as glycolipid). However, the relative proportions of protein and lipid vary considerably in membranes from different sources, as do the types of specific molecules actually present (Table 16-1). This variation in molecular composition extends to different kinds of membranes in the same cell, as well as to the same kinds of membranes in different cell types. Such molecular variation is hardly surprising in view of the great diversity in function that membranes display.

Lipids

There are three major types of membrane lipids: *phospholipids, cholesterol,* and *glycolipids.*

The phospholipids found in cellular membranes are of two general types. The most common type, the *phosphoglycerides,* includes the phosphatidic acids and their derivatives. Phosphatidic acids, as you recall, are composed of glycerol to which two fatty acid molecules and one molecule of phosphoric acid are esterified (see page 176). Only small amounts of free phosphatidic acids are, however, present in cell membranes. Instead, the great majority of membrane phosphoglycerides are derivatives of phosphatidic acids. In these derivatives choline, ethanolamine, serine, inositol, and sometimes other compounds are linked to one of the phosphate hydroxyl groups of phosphatidic acids (see Figure 7-26).

The only membrane phospholipids that are not derived from phosphatidic acids are the *sphingomyelins,* which are present primarily in plasma membranes of animal cells, particularly nerve cells and erythrocytes. The sphingomyelins are derived from *sphingosine* (Fig. 16-1), a molecule that has a polar base on one end and a nonpolar hydrocarbon chain on the other. In the sphingomyelins, a fatty acid is linked by an amide bond to the polar amino group of sphingosine, and phosphoric acid and choline are linked to the terminal hydroxyl group of sphingosine (Fig. 16-2). Because the composition of the fatty acid in sphingomyelin molecules may vary, the

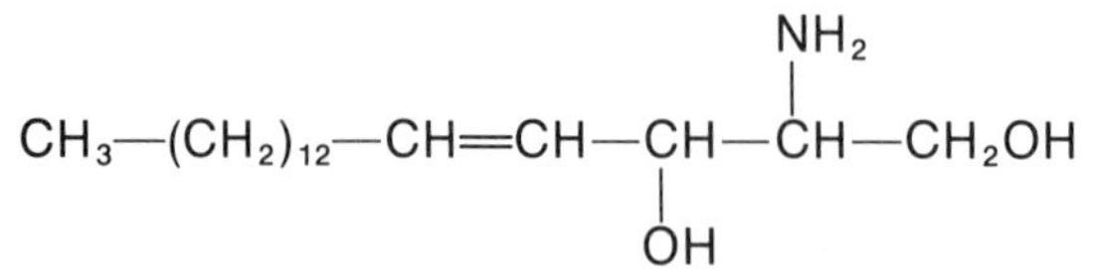

$$CH_3-(CH_2)_{12}-CH=CH-\underset{\underset{OH}{|}}{CH}-\underset{\underset{}{|}}{\overset{\overset{NH_2}{|}}{CH}}-CH_2OH$$

Figure 16-1 Chemical structure of sphingosine, which possesses a nonpolar hydrocarbon chain on one end and a polar group on the other.

sphingomyelins, like the phosphatidic acids, are a *class* of compounds.

It is important to note that all the phospholipids in membranes, whether phosphoglycerides or sphingomyelins, are similar in general conformation and charge; that is, they all contain two nonpolar hydrocarbon chains and a polar end.

Cholesterol is found in many animal membranes and is particularly abundant in the plasma membranes of some animal cells, where it may constitute up to about 30 per cent of the total lipids. The cholesterol molecule (see Figure 7-30) is largely nonpolar, and it is believed that cholesterol, when present in membranes, occurs in association with the nonpolar hydrocarbon chains of phospholipid molecules, forming a complex stabilized by hydrophobic bonds. The idea of such an association between cholesterol and phospholipid molecules is supported by studies of erythrocyte plasma membranes from various mammalian species, which show that cholesterol and phospholipid occur in a molecular ratio of 1:1.

The glycolipids represent a third group of lipids found in many membranes. As their name implies, the glycolipids are carbohydrate-containing lipids. They are all derivatives of sphingosine, and they differ in structure from sphingomyelins only with regard to the component that is attached to the terminal hydroxyl group of sphingosine. Whereas the sphingomyelins have phosphoric acid and choline linked to the hydroxyl group of sphingosine, the glycolipids have a hexose (either glucose or galactose) or a more complex oligosaccharide linked to the hydroxyl group of sphingosine (Fig. 16-3).

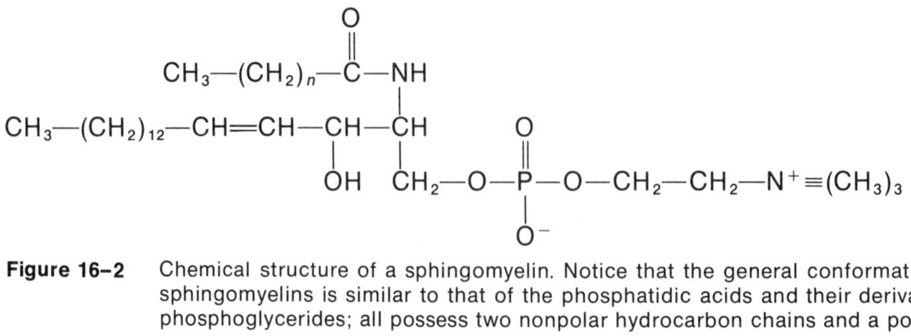

Figure 16-2 Chemical structure of a sphingomyelin. Notice that the general conformation of the sphingomyelins is similar to that of the phosphatidic acids and their derivatives, the phosphoglycerides; all possess two nonpolar hydrocarbon chains and a polar group.

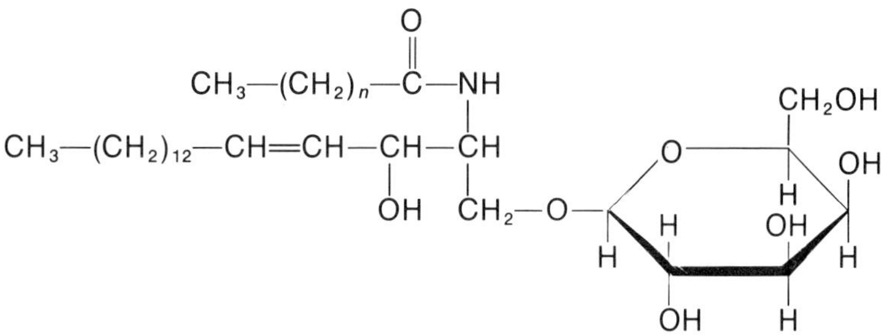

Structure of Cell Membranes

$$CH_3-(CH_2)_{12}-CH=CH-CH-CH \begin{array}{c} CH_3-(CH_2)_n-\overset{O}{\overset{\|}{C}}-NH \\ | \\ OH \quad CH_2-O- \end{array}$$

Chemical structure of a glycolipid. The hexose sugar shown here is galactose, but the carbohydrate moiety varies in different glycolipids.

Figure 16-3

Proteins

All membranes contain some protein; indeed, there is a rough relationship between the protein content of a membrane and its functional activity. The myelin sheath, the membrane surrounding some nerve axons, serves primarily an insulating function and is composed of about 80 per cent lipid and 20 per cent protein. Because lipid is an excellent insulator, the high lipid content of myelin is appropriate to its function. Plasma membranes of eukaryotes, which serve primarily as permeability barriers, possess about 50 per cent protein and 50 per cent lipid. Membranes that are actively involved in energy transfer, such as the inner membranes of mitochondria and chloroplasts as well as the plasma membrane of aerobic prokaryotes, have large amounts of protein, often as much as 75 per cent. The correlation between protein content and functional activity provided the first suggestion that proteins may perform specific membrane functions, whereas lipids may form the structural components of membranes.

Although it is clear that a single type of membrane contains a large number of different proteins, it has proved exceedingly difficult to isolate and characterize the protein components of membranes. Consequently, we still know very little about the types of proteins found in membranes, or about their functional roles. The main problem in characterizing membrane proteins is that many of them are very insoluble and cannot readily be extracted from membranes except by using drastic chemical procedures. For example, by treating membranes with detergents, which disrupt many protein–protein and protein–lipid interactions, it is possible to obtain a preparation of solubilized membrane proteins. However, even when a membrane protein can be purified from such a preparation, it is often difficult to determine its intramembrane function. One problem in assigning a functional role to a purified membrane protein is that many membrane proteins are active only when they are present in membranes. For instance, some membrane proteins are thought to act as carriers for the transport of substances across membranes, but this function cannot be tested once the protein has been removed from the membrane. In addition, many enzymes and other proteins are known to be denatured by treatment with detergents, and hence the lack of functional activity in a protein purified from detergent solutions does not necessarily mean that the protein is inactive within the membrane.

The use of detergents to study membrane proteins has, however, yielded an important piece of information about membrane structure. The fact that membrane proteins can be solubilized in detergents shows that the *individual protein and lipid molecules in membranes are held together by noncovalent interactions,* primarily hydrophobic interactions that are disrupted by treatment with detergents.

For some time, debate has centered on the question of whether membrane proteins play important structural roles in membranes or are present only as functional proteins (enzymes, carrier proteins), attached to, or embedded within, the membrane. David Green and his collaborators have isolated from various membranes a protein material that they contend represents "structural protein." However, many other workers claim that Green's "structural protein" is merely a mixture of denatured enzymes, and they suggest that proteins play no significant structural role in membranes. This suggestion is supported by the observation that phospholipids in aqueous solution form stable bilayers (Chapter 11) that have many properties similar to natural membranes, even though they are devoid of protein.

A few proteins in some membranes have relatively large amounts of carbohydrate associated with their polypeptide chains; these proteins are known as *glycoproteins*. It has recently been proposed that plasma membrane glycoproteins and glycolipids may serve as recognition molecules, which are thought to function in intercellular communication (as discussed on page 407).

THE MOLECULAR ORGANIZATION OF CELL MEMBRANES

Although we know that cell membranes are composed of lipids and proteins, and, in some cases, carbohydrates, we still are not certain how these molecular components are organized to form the three-dimensional structure of cell membranes. Our ideas about the molecular organization of membranes have evolved gradually as a result of research activity that was begun at the end of the nineteenth century and continues even today. Initially, biologists thought that all membranes were rigid structures with identical molecular organization; recent research on membrane structure has increasingly led us to the conclusion that membranes are dynamic and flexible entities and that the organization of membranes from different sources varies.

In general, the approach to the elucidation of membrane architecture has involved the construction of models of membrane structure that are able to account for observed functional characteristics of a particular membrane. Early research concentrated on the

Structure of Cell Membranes

plasma membrane of various cell types; indeed, it was not until the development and use of the electron microscope that biologists became aware of the existence and the importance of intracellular membranes. As information about the distribution, composition, and function of cellular membranes has accumulated, our views of membrane structure have been modified accordingly. It is for this reason that it is particularly appropriate to examine the major historical developments that led to our current views of membrane structure.

Cell Membranes as Lipid Bilayers

Probably the first study that bears upon the problem of the organization of cell membranes was that of E. Overton, published in 1895. Overton studied the permeability of the plasma membranes of plant cells to solutes and showed that substances that were soluble in lipid penetrated the membrane readily, whereas those that were insoluble in lipid did not. On the basis of these observations, Overton concluded that the plasma membrane was constructed of a very thin film of lipid, which would account for the penetration of lipid-soluble substances.

Subsequent work demonstrated that although Overton's observations were generally valid, they were inaccurate for very small molecules such as water and methanol, which penetrate cells much more readily than would be expected on the basis of their lipid solubilities. Consequently, Overton's ideas were modified to include the presence of small pores that allow the penetration of small nonpolar substances.

In 1925, work by E. Gorter and F. Grendel provided additional support for Overton's ideas of cell membrane structure. Gorter and Grendel demonstrated that lipids extracted from mammalian erythrocytes form a single layer (monolayer) that occupies twice the surface area of the cells from which the lipids have been extracted. From this observation they concluded that the surface lipids form a layer two molecules thick. Subsequent research has shown that Gorter and Grendel not only *underestimated* the surface area of the erythrocyte but did *not* extract all the lipids from the membranes. However, these two errors in technique cancelled one another, and their conclusion was essentially correct.

The Danielli-Davson Model

The first comprehensive model of membrane structure was formulated in an attempt to explain the observation by Harvey, Cole, and others in the early 1930s that the surface tension of the plasma membrane of a mammalian egg cell in aqueous medium is considerably lower than that of intracellular oil droplets. When it was subsequently shown by Harvey and Danielli that protein adsorbed to the surface of lipid droplets had the effect of reducing the surface tension of the droplets, it seemed reasonable to conclude that plasma membranes consist not only of lipid, as proposed by Overton and by Gorter and Grendel, but also of adsorbed protein. This conclusion, reached by Danielli and Davson in 1934, formed the basis of what is now known as the *Danielli-Davson model* or *bimolecular leaflet model* of membrane structure.

In their model, Danielli and Davson proposed that the plasma membrane consists of two layers of lipid molecules—a bimolecular leaflet—formed in such a way that the polar parts of the lipid molecules are in contact with the surrounding aqueous phase, whereas the nonpolar hydrocarbon portions are associated in the central region of the leaflet, removed from the aqueous phase. They further envisioned the polar ends of the lipid molecules as being associated with a monomolecular layer of polar globular protein molecules. The entire structure would thus consist of a double layer of lipid molecules sandwiched between two essentially continuous layers of protein (Fig. 16–4). Subsequently, the model was modified to include polar regions, presumably coated with protein, extending through the lipid bilayer and forming small pores 0.7 to 1.0 nm thick (Fig. 16–5).

EXTERIOR

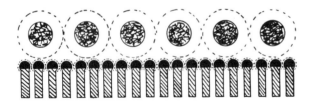

LIPOID

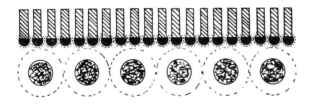

INTERIOR

Figure 16–4 The original Danielli-Davson model of membrane structure. The bimolecular layer of lipid molecules is of undefined thickness and is covered on each side by a continuous layer of globular proteins (From Danielli, J.F. and H. Davson [1935]. J. Cellular Physiol. 5:498.)

Structure of Cell Membranes

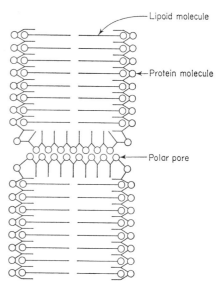

Figure 16-5

A modification of the original Danielli-Davson model, showing pores lined by polar protein molecules extending through the lipid bilayer. (From Stein, W.D. and J.F. Danielli [1956] Discuss. Faraday Soc. *21*:238).

It is interesting from a historical standpoint that although subsequent studies using X-ray diffraction, electron microscopy, and chemical analyses of membranes have indeed supported the idea that protein is associated with lipid in membranes, other experiments have shown that it is not necessary to assume a coating of protein on cell membranes in order to account for the reduced surface tension of the plasma membrane relative to that of intracellular oil droplets. The oil droplets on which the original surface tension data was based consisted of triglycerides and other storage lipids that are not usual components of membranes. We now know that phospholipids—the major lipid components of cell membranes—form artificial membranes that have a surface tension in water similar to that of cell membranes. Thus, the Danielli-Davson model, which appears to describe the structure of some cell membranes, was derived by drawing a reasonable conclusion from misleading experimental data.

The Unit Membrane Hypothesis

Independent evidence concerning the structure of cell membranes began to accumulate in the 1950s. The development of the electron microscope and improvements in techniques of fixing, staining, and sectioning of tissues led to a detailed microscopic examination of plasma membranes and intracellular membranes in a variety of different cell types. In the late 1950s, J.D. Robertson summarized a large amount of ultrastructural data obtained by himself and other workers and concluded that all cell membranes have a common molecular organization. Robertson noticed that plasma membranes and intracellular membranes from a wide variety of cells appear under the electron microscope as a three-layered (trilaminar) structure approximately 7.5 to 10.0 nm in thickness, composed of two dark electron-dense lines, each 2.0 to 2.5 nm thick, separated by a light electron-transparent region of 3.5 to 5.0 nm (Fig. 16–6a and b). Robertson termed this structure a "unit membrane," and he concluded that all biological membranes had unit membrane construction—a generalization that became known as the *unit membrane hypothesis.*

Robertson assumed that the dark lines of the trilaminar structure were proteins and polar groups of lipids and that the light central region consisted of nonpolar groups (Fig. 16–6 c). This molecular organization is similar to that proposed by Danielli and Davson. Studies on artificial phospholipid membranes tend to support Robertson's interpretation of the molecular basis of the unit membrane structure. You recall that phospholipid bilayers and micelles form in aqueous solution with the polar ends of the phospholipid molecules in intimate contact with the aqueous phase and the nonpolar ends buried in the central region of the structure, removed from the aqueous phase (Chapter 11). Such structures are presumably stabilized by hydrophobic bonds between the nonpolar hydrocarbon chains and by hydrogen bonding and other favorable electrostatic interactions between the polar lipid groups and water. When pure phospholipid bilayers, or phospholipid bilayers coated with water-soluble protein, are fixed with potassium permanganate or osmium tetroxide (two common fixatives used in electron microscopy), they have dimensions and appearance in the electron microscope similar to those of natural membranes fixed under similar conditions (see Figure 7–27). Results with phospholipid bilayers are thus consistent with the interpretation that osmium tetroxide and potassium permanganate stain the polar, hydrophilic regions of phospholipid and protein molecules, and they suggest that the electron-dense layers of the unit membrane are hydrophilic in nature.

Robertson's unit membrane hypothesis provided strong independent support of the Danielli-Davson model of membrane structure. In fact, the unit membrane hypothesis appeared so convincing

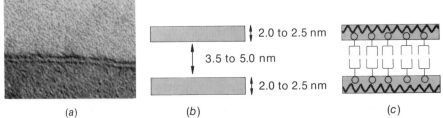

Figure 16–6 (a) Electron micrograph of the plasma membrane of a human red blood cell, illustrating unit membrane construction (×280,000) (From Robertson, J.D. in *Cellular Membranes in Development,* Locke, M. (editor), Academic Press, 1964, p. 3). (b) Dimensions of the unit membrane configuration. (c) Robertson's interpretation of the molecular basis of unit membrane structure. Each zig-zagged line represents a layer of protein, and circles represent hydrophilic ends of lipid molecules (Adapted from Robertson, J.D. [1959]. Biochemical Symp. *16*:24).

Structure of Cell Membranes

as an explanation for membrane structure that Danielli was prompted to make the following remarks about the structure of the plasma membrane at a meeting of the Biochemical Society in 1962:

> It now seems to be agreed that its basic structure is that which I suggested in 1934, and it is also highly probable that the same structure is present in many other intercellular [sic] membranes. So far as it is possible to predict at the present time, it is unlikely that this general picture will be substantially disturbed, and the focus of attention is likely to shift to other fields.

Questioning of the Danielli-Davson Model and the Unit Membrane Hypothesis

Danielli's prediction proved wrong. In spite of the evidence supporting both the Danielli-Davson model and the unit membrane hypothesis, a number of workers were disturbed by the simplicity and rigidity of these ideas. How, they wondered, could the diversity of membrane composition and function be reconciled with models that proposed a universally similar structure for all cellular membranes? Indeed, it is difficult to understand how the same structural organization could occur in myelin, which is largely metabolically inactive and functions primarily as an insulator; in plasma membranes, which function as the primary permeability barrier of the cell; and in membranes that carry out energy transfer activities and have enzymes and other specific components associated with them, probably in precise spatial orientations. Furthermore, the unit membrane hypothesis was contradicted by additional ultrastructural studies.

In 1963, F.S. Sjöstrand published electron micrographs of ultrathin tissue sections, which showed what appeared to be globular subunits forming the structure of mitochondrial membranes and membranes of smooth endoplasmic reticulum (Fig. 16-7). Sjöstrand also presented evidence which demonstrated that the plasma membrane of kidney cells, although not possessing a globular substructure, is geometrically asymmetric, with a thicker electron-dense layer at its cytoplasmic surface than at its outer surface. In addition, he was able to show a marked difference in the thickness of plasma membranes and mitochondrial membranes seen in the same electron micrograph.

Sjöstrand's observations were inconsistent with the unit membrane hypothesis in a number of respects. First, the globular substructure seen by Sjöstrand in membranes of mitochondria and smooth endoplasmic reticulum is difficult to reconcile with a continuous lipid bilayer model of membrane structure. Second, the unit membrane hypothesis does not account for the geometrical asymmetry that Sjöstrand observed in kidney cell plasma membranes. Third, Robertson claimed that differences in the measured thickness of cell membranes reported by various workers were produced by differences in fixation, staining, and sectioning techniques; he assumed that all types of membranes, because they have identical unit

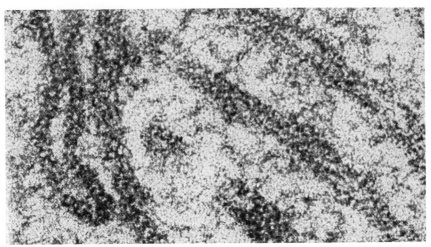

Figure 16-7 Electron micrograph of smooth endoplasmic reticulum (left) lying near the surface of a mitochondrion (right). Membranes of both structures appear to be composed of globular subunits (From Sjöstrand, F.S. [1963]. J. Ultrastructure Res. 9:357).

membrane structure, would have the same dimensions if they were studied under identical conditions. This assumption is, however, inconsistent with Sjöstrand's observation that two different membranes seen *in the same electron micrograph* have different thicknesses.

Sjöstrand has tentatively interpreted the globular subunits seen in mitochondrial membranes and smooth endoplasmic reticulum as representing lipids in the form of small micelles that are covered on their surfaces by protein. He envisions that a vast array of these particles form the membrane. Because protein covers the surface of the lipid particles, the lipid particles would not fuse to form a bimolecular layer, and hence the membrane would appear granular rather than smooth. In mitochondrial membranes, some of the protein molecules covering the lipid micelles would undoubtedly represent elements of the respiratory chain and phosphorylating system.

The work of Sjöstrand and others provided strong evidence against the idea that all cellular membranes are identical in molecular organization. "It seems justifiable at present," Sjöstrand concluded in 1963, "to consider the possibility that various types of cellular membranes may differ considerably with respect to their molecular architecture, and not only with respect to their chemical composition. It is, for instance, impossible to refer all the observed differences to slight variations in the molecular architecture still with a bimolecular leaflet as one basic component of the membranes." Thus began emphasis on the dynamic aspect of membrane structure based upon the realization that membranes involved in different functions most probably possess different structures.

Structure of Cell Membranes

Cell Membranes as Observed by Freeze-Etching

In the middle 1960s, then, the molecular organization of cell membranes was very much an open question, and there were no models of membrane structure that were versatile enough to account adequately for the functional and compositional variation of cellular membranes. Much of the ultrastructural data on membranes was criticized during this period because of the possibility that methods of tissue preparation caused changes in the natural membrane structure. In order to overcome this criticism, researchers turned to methods that allowed the examination of cellular membranes without treatment by the relatively harsh fixation and staining methods required for observing sectioned material in the electron microscope.

One method that avoids chemical fixatives is *freeze-etching*, a technique which is described in Chapter 11. In this technique, cells or pieces of tissue are frozen rapidly in liquid nitrogen to very low temperatures, the cell is fractured by a sharp blow with a microtome knife, and the exposed surfaces are replicated. The cleavage plane often splits a membrane in such a way that its inner and outer surfaces are separated, exposing the interior part of the membrane, which is not normally visible (Fig. 16–8). Since the membrane is only frozen and is not chemically treated in any way, it is assumed that structural alterations are minimized.

Daniel Branton has used the freeze-etching technique to examine membrane structure in a number of different cell types and has published electron micrographs that show the two outer surfaces of membranes, as well as their internal surfaces, as revealed by freeze-etching. On many of the exposed inner surfaces of membranes, Branton observed small spherical particles averaging about 8.5 nm in diameter. He also showed that the number, density, and grouping of the particles are dependent upon the particular type of membrane studied. For example, the inner surfaces of myelin membranes are

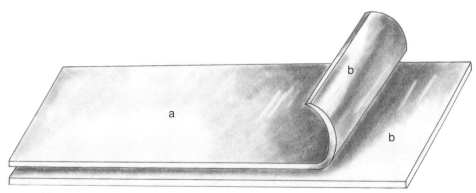

Figure 16–8 In freeze-etching, the cleavage plane often splits a membrane, revealing the inner membrane surfaces which are not otherwise visible. (a) Outer surface of membrane and (b) the two inner surfaces.

essentially free of such particles (Fig. 16-9a), whereas the inner surfaces of chloroplast thylakoid membranes appear highly particulate (Fig. 16-9b). Other types of membranes show intermediate degrees of particulate structure.

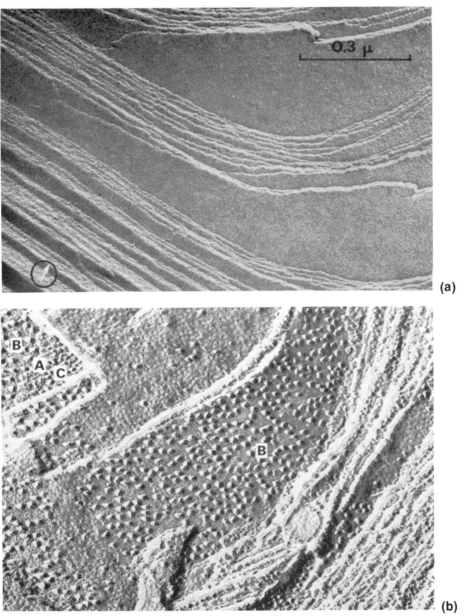

Figure 16-9 Inner surfaces of (a) myelin membranes and (b) chloroplast thylakoids exposed by freeze-etching. Notice the lack of globular particles on the inner surfaces of myelin membranes as compared with the particulate nature of the inner surfaces of chloroplast thylakoids. (a) From Branton, D. (1967). Exp. Cell Res. 45:705. (b) From Branton, D. and R.B. Park (1967). J. Ultrastructure Res. 19:289.

Structure of Cell Membranes

A number of recent experiments suggest that the 8.5 nm particles observed by Branton are specific membrane proteins, and some of these have been tentatively identified. There is also evidence that some globular particles may span the entire membrane. In these cases, the particles may penetrate both surfaces of the membrane, and thus they appear not to be restricted to internal positions within the membrane.

The Protein Crystal Model

The demonstration of globular particles, presumably proteins, as components of a number of cell membranes has led to new models of membrane structure. David Green and his coworkers have proposed a *protein crystal model* which suggests that membrane proteins polymerize to form arrays of two layers of loosely packed globular proteins, 3.0 to 4.0 nm in diameter. The proteins are envisioned as having extensive nonpolar, as well as polar, regions on their surfaces. These nonpolar regions may exhibit hydrophobic bonding with nonpolar groups of phospholipid molecules that fill cavities between the globular protein units; the polar heads of phospholipid molecules remain at the membrane's surface (Fig. 16–10).

The protein crystal model accounts for some of the observed chemical and functional characteristics of membranes. The existence of nonpolar regions on the surfaces of the globular membrane proteins would explain the observed insolubility of membrane proteins in aqueous solution in the absence of detergents. On the other hand, the existence of polar groups as well as nonpolar groups on the surfaces of the globular proteins could account for the permeability of many membranes to polar molecules. In addition, such a membrane would be expected to show trilaminar structure in electron micrographs, as does a bilayer of phospholipid molecules, because both structures essentially have a nonpolar core and a polar surface.

Green has recently modified the protein crystal model toward a more dynamic model in which the proteins and lipids are considered to move in the plane of the membrane. Green points out that a

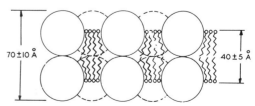

Figure 16–10 Diagrammatic representation of the protein crystal model of membrane structure. Phospholipid molecules forming short regions of lipid bilayer fill cavities between protein molecules (large circles). (From Vanderkooi, G. and D.E. Green [1970]. Proc. Nat. Acad. Sci. *66*:616.)

consequence of this modification is that the lipid bilayer forms essentially a continuum, and any protein-protein interactions are temporary.

The Fluid Mosaic Model

Another model of membrane structure was proposed by S.J. Singer in 1971 and has been widely accepted. Called the *fluid mosaic model*, it is similar in many respects to Green's modified version of the protein crystal model, although it proposes a less uniform structure.

Singer envisions cell membranes as mosaics of lipids and proteins. The lipids are thought to be arranged primarily in a bilayer in which proteins are embedded to varying degrees (Fig. 16–11). The extent of interaction between the proteins and the lipids would depend upon thermodynamic considerations. On the basis of the extent of their interaction with lipids, Singer classifies membrane proteins as *peripheral* or *integral.* Peripheral proteins require only mild chemical methods to dissociate them from the membrane, and hence they are considered to be held to the membrane only by weak electrostatic interactions. Integral proteins, on the other hand, are considered to form strong hydrophobic interactions with the lipids of the membrane and require detergents or other harsh treatments to

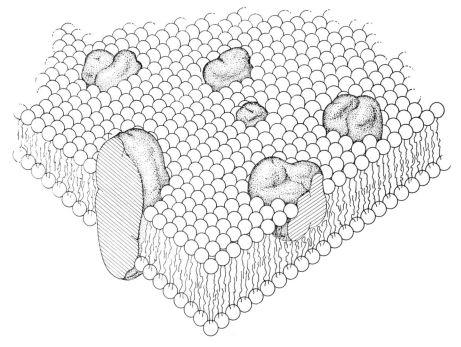

Figure 16–11 Diagram of Singer's fluid mosaic model of membrane structure. The large bodies are globular protein molecules which are embedded to varying degrees in a lipid bilayer. (From Singer, S.J. and G.L. Nicolson [1972]. Science *175*:723. Copyright © 1972 by The American Association for the Advancement of Science.)

dissociate them from the membrane. Thus, the overall structure of the protein would determine its molecular orientation and degree of interaction with the lipids of the membrane. The proteins, varied in size and dissolved to varying degrees in the lipid matrix, are envisioned as being able to diffuse laterally in the plane of the membrane, and the entire structure is hence dynamic.

The fluid mosaic model can be used to describe the structures of different membranes, even though the membranes may differ significantly in chemical composition (ratio of protein to lipid). The lipid bilayer forms the structural matrix which serves as the primary permeability barrier of the membrane. In membranes with high lipid content, the lipid bilayer is extensive and interrupted only occasionally by protein molecules; in membranes with high protein content, the extent of the lipid bilayer is reduced.

Thus, the fluid mosaic model can account for the molecular organization and ultrastructure of membranes in terms of their chemical composition. Membranes such as myelin with little or no enzymatic activity would exist primarily as a lipid bilayer with attached peripheral protein and would have a typical "unit membrane" configuration. On the other hand, in mitochondria and chloroplast membranes, which have a high protein content and vast arrays of enzymes associated with them, the protein components would be so extensive that the trilaminar framework would be almost obscured. Such membranes would appear highly granular in the electron microscope.

MEMBRANE STRUCTURE SUMMARIZED

We may summarize our present knowledge of membrane structure as follows: The proteins and lipids in membranes are held together by noncovalent interactions, including electrostatic forces and hydrophobic bonding. The precise structural organization of proteins and lipids in membranes is not well understood and undoubtedly varies in different membranes depending upon their composition and function. The natural tendency of phospholipids to form micelles and bilayers in aqueous solution suggests that this type of association is important in the organization of lipids in membranes. Whatever the arrangement of lipids, the permeability characteristics of membranes may be largely explained on the basis of a lipid barrier. The membrane proteins appear to serve specific functions as transport carriers, receptor sites, recognition molecules (in the form of glycoproteins), and enzymes in energy transfer reactions.

MEMBRANE TRANSPORT

Membranes clearly have a variety of important functions, but their most basic function is that of a permeability barrier. Whether forming the plasma membrane or various subcellular compartments, membranes serve as selective barriers that control the passage of substances from one region to another.

The movement of substances across membranes is known as *membrane transport*, and it involves two aspects: (1) membrane permeability and (2) transport mechanism.

Membrane permeability refers to the ability of a membrane to be permeated by a substance; in other words, its ability to allow a substance to pass through it. Membranes are permeable to a wide variety of substances—large and small; water soluble and lipid soluble; charged and uncharged; polar and nonpolar. As yet, we do not completely understand the factors that determine the permeability characteristics of cellular membranes. However, we do know that the permeability of different cellular membranes varies; this means that the same substances do not necessarily pass through all types of membranes with equal ease. Indeed, subcellular compartmentalization of function depends largely on the fact that each of the membranes forming subcellular compartments has unique permeability characteristics.

In view of the diverse range of molecules which pass through membranes, it is not at all surprising that there are a number of mechanisms by which such movement occurs. We are only beginning to understand how some of these transport mechanisms operate, but it is clear that there are two general types of transport mechanisms, which can be distinguished on the basis of their energy requirements. One of these, *passive transport*, is the movement of a substance across a membrane in response to a concentration gradient; that is, from one region to another region that possesses a *lower* concentration of the substance (Fig. 16–12a). Passive transport is a spontaneous process which, in order to occur, does *not* require the expenditure of energy by the cell itself. Other substances cross membranes from one region to another region that possesses a *higher* concentration of the substance. Such transport occurring against a concentration gradient is called *active transport*

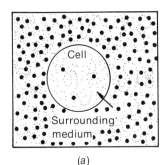
(a)

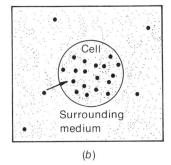
(b)

Figure 16–12 (a) In passive transport, a substance passes through the plasma membrane in response to a concentration gradient; (b) in active transport, a substance passes through the plasma membrane against a concentration gradient in a process that requires metabolic energy.

Membrane Transport

because it is driven by cellular metabolic energy in the form of ATP or other high energy compounds (Fig. 16–12b).

It should be stressed, however, that *all transport processes, whether passive or active, require energy* in one form or another. Thus, in passive transport, the required energy is supplied in the form of a concentration gradient. The existence of differential concentrations of a substance on two sides of a membrane should be considered as a situation potentially capable of yielding energy, and the collapse of a concentration gradient is a spontaneous, energy-yielding process that may be thought of as driving the passage of molecules across the membrane.

Most research on membrane transport has focused on the plasma membrane, and it is only recently that transport across intracellular membranes has been studied in any detail at all. Consequently, our discussion of membrane transport will concentrate on the plasma membrane, but we must keep in mind that intracellular membranes are involved in similar transport processes.

Membrane Permeability

Before beginning a detailed discussion of membrane permeability, it would be well to clarify what is meant by the term "permeable." In the first place, we speak of a membrane being permeable to a substance only in the sense that the substance actually passes *through* the membrane. Thus, for example, substances that enter the cell only by endocytosis (see Chapter 11) or leave the cell only by exocytosis do not actually pass through the plasma membrane, and the plasma membrane is not considered to be permeable to these substances.

Second, permeability is a rate phenomenon, and describes the rate at which a given substance passes through a membrane under defined conditions. Some molecules can pass through the plasma membrane at such fast rates that the membrane is said to be freely permeable to them; other molecules may pass through at such slow rates that the membrane is said to be only slightly permeable to them; still others may not pass through the membrane at all, and the membrane is said to be completely *impermeable* to them. All living membranes are *selectively permeable*: they allow some substances to pass through them more readily than others. With respect to their permeability characteristics, membranes of living systems differ from many artificial membranes that are *semipermeable* (allowing only the passage of solvent).

Third, when we speak of membrane permeability, we do not usually specify the mechanism of transport (as long as it is understood that the substance passes *through* the membrane). We therefore consider a membrane permeable to a substance whether it is normally transported through the membrane by a passive or by an active transport mechanism.

Permeability of the Plasma Membrane to Water

Of all the substances that pass through the plasma membrane, only carbon dioxide, oxygen, and a few others can enter the cell as quickly as can water. We discussed the dynamics of the passage of water through the plasma membrane in Chapter 9 (see especially Figure 9–5) and noted that cells exhibit osmotic properties because water passes through the plasma membrane at a much faster rate than do most solute molecules. Consequently, animal cells shrink when placed in a hypertonic medium and swell when placed in a hypotonic medium.

We can observe a similar phenomenon in plant cells. However, because of the presence of a rigid cell wall which is not deformed by the loss or gain of water by the cell, the plasma membrane of the plant cell placed in hypertonic solution will shrink away from the cell wall when the volume of the cytoplasm is reduced as a consequence of water loss. This phenomenon is known as *plasmolysis* (Fig. 16–13). If the solute is able to penetrate the plasma membrane (albeit at a slower rate than water), plasmolyzed cells will eventually regain their original volume if left in the hypertonic solution. This process is called *deplasmolysis* and results from the fact that the penetration of solute into the cell will, in turn, result in the passage of water into the cell in order to compensate for the entry of solute (Fig. 16–13). The rate of deplasmolysis is slow when the solute penetrates the membrane slowly and rapid when the solute penetrates the membrane rapidly. In fact, the speed of deplasmolysis was considered by many early workers to be an indication of the permeability of the plasma membrane to many substances.

The flow of water into a cell in response to a concentration gradient creates an internal pressure against the plasma membrane (and cell wall, if present). This pressure is termed an *osmotic pressure*. Osmotic pressures build up in animal cells because of the restricted elasticity of the plasma membrane. In extreme cases, the vol-

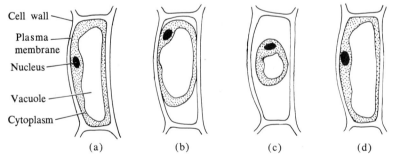

Figure 16–13 A diagram illustrating plasmolysis and deplasmolysis. Plasmolysis occurs when a normal plant cell (a) is placed in a hypertonic solution. Water leaves the cell and the plasma membrane shrinks away from the cell wall (b and c). If the solute can penetrate the plasma membrane, the cell will eventually regain water—a process termed deplasmolysis (d). (From Dowben, R.M. [1971], *Cell Biology*. New York, Harper & Row, Inc.).

ume of the cytoplasm swells until the osmotic pressure against the plasma membrane is so great that the cell bursts. In plant cells, an internal osmotic pressure pushes the plasma membrane firmly against the cell wall, causing the cell to become increasingly rigid as the pressure increases.

Osmotic pressure appears to play an important role in providing rigidity both to cells themselves and to intracellular organelles, thus preventing them from collapsing. Plants wilt, for example, when enough water is lost from their cells to cause a drastic decrease in osmotic pressure. Under normal conditions, the osmotic pressure in both animal and plant cells is maintained within a narrow range. The maintenance of osmotic pressure is important to numerous physiological processes, particularly those relating to the transfer of materials between cells.

The Permeability of Cells to Solutes

Plasma membranes, and other cellular membranes as well, are very selective in terms of their permeability to various solutes. This selectivity is, of course, a function of the mechanisms by which solutes enter cells—membranes themselves do not possess some magic ability that allows them to decide which substances should pass through the membrane and which substances should not. Rather, the chemical characteristics of both the permeant molecule and the membrane, the concentration of substances on both sides of the membrane, the rate of metabolic activity, and other physiological factors determine whether a particular substance will cross the membrane, and at what rate.

It is possible to make some generalizations about the passage of substances through the plasma membrane, although we must realize that there are many special cases in which such general rules do not hold. First, the greater the lipid solubility of a given molecule, the more readily it traverses the plasma membrane. This is perhaps not surprising, since lipid is a major constituent of cell membranes, and it would seem a priori that favorable interactions between the membrane and the permeant molecule would facilitate the passage of the molecule through the membrane. Because the lipid solubility of a molecule is determined by its molecular structure, and because nonpolar substances are more soluble in lipid than polar substances, it follows that the plasma membrane is more permeable to nonpolar substances than to polar ones.

Second, small molecules pass through the plasma membrane more readily than large ones; the membrane is essentially impermeable to large molecules such as proteins. In general, however, the lipid solubility of a molecule is more important in determining the rate of solute penetration than is molecular size. Thus, a molecule with high lipid solubility will usually traverse the plasma membrane more rapidly than a smaller molecule of low lipid solubility.

Third, the plasma membrane is more permeable to uncharged

particles than it is to charged ones. Electrolytes (substances whose solutions give positive and negative ions) pass through the plasma membrane more slowly than nonelectrolytes of similar dimensions; strong electrolytes (those that ionize completely in solution) pass through more slowly than weak electrolytes. The greater the charge on an ion, the slower its rate of penetration. Hence, trivalent ions penetrate more slowly than divalent ions of the same charge, and divalent ions penetrate more slowly than monovalent ones. In addition, anions generally pass through the plasma membrane more readily than do cations, although there are many exceptions. The reduced permeability of charged particles appears to be due both to the fact that ions possess hydration shells which increase their effective size and to the fact that ions are generally less soluble in lipid than are uncharged particles.

It should also be emphasized that the permeability of a membrane is related to its physiological state and the environmental conditions. Narcotics and anesthetics often alter the permeability of membranes, and injury to the cell by heat, radiation, pH change, salt imbalance and numerous other means increases permeability. Even among normal, healthy cells, permeability varies with functional state. It has been reported, for example, that active muscle cells are very permeable to nutrients and other materials, whereas inactive ones are not. Stimulation of a muscle or a nerve cell increases its permeability to ions; indeed, the excitability of muscle and nerve cells is closely connected with the passage of ions across the plasma membrane.

In addition to these somewhat general rules of membrane permeability, there are special situations in which the passage of a particular molecule through a membrane is determined by the molecular configuration of the molecule or the presence in the molecule of specific chemical groups. In these cases, transport is envisioned as occurring by virtue of the attachment of the molecule to specific membrane transport molecules (carrier molecules). Carrier molecules are presumably proteinaceous components of the membrane, and they somehow facilitate the transport of a molecule from one side of the membrane to the other. Such carrier processes are similar in certain respects to enzymatic processes, in which activity is dependent upon specific configurations or chemical groups in the substrate molecule(s). However, carrier processes are, in general, less specific than enzymatic processes, and one particular carrier molecule often transports a number of molecules of similar structure.

TRANSPORT MECHANISMS

The cellular transport of molecules across membranes occurs by a number of different mechanisms. Although many of these transport mechanisms are poorly understood, we are beginning to ac-

Membrane Transport

cumulate enough information to suggest how some of them may operate.

It is important to stress that a particular molecule may normally be transported across a membrane by more than one mechanism, depending upon physiological conditions. In addition, we do not know the mechanism or mechanisms by which most substances traverse a membrane.

Passive Diffusion

A great many substances are able to pass through the plasma membrane by *passive diffusion;* that is, from one region to another containing a lower concentration of the substance. Passive diffusion is the simplest type of transport mechanism. It requires no energy expenditure on the part of the cell, nor does it require any specific transport carrier molecules. As long as a favorable concentration gradient across the membrane is maintained, transport by passive diffusion may occur; when the concentration gradient is dissipated, passive diffusion stops. The rate of transport in passive diffusion is strictly proportional to the concentration difference across the membrane; the greater the concentration difference, the greater the rate of passage.

A concentration gradient may be maintained by the cell in such a way that a substance will pass through a membrane essentially in one direction. For example, if molecules of a particular nutrient, entering the cell by passive diffusion, are utilized in some metabolic process, additional molecules may continue to enter as long as an ample supply is readily available outside the cell. This occurs because the utilization of the nutrient keeps its intracellular concentration low and hence maintains the concentration gradient necessary for transport. In the same way, waste products may continue to leave the cell as long as they are washed away from the medium surrounding the outer surface of the plasma membrane.

It is clear that water, oxygen, and carbon dioxide pass through the plasma membrane by passive diffusion and that many other substances penetrate membranes by this mechanism at least to some extent. It has been postulated that molecules traversing membranes by passive diffusion must dissolve in the lipid component of the membrane, and that this accounts for the observed relationship between permeability and lipid solubility. However, some substances that pass through cell membranes at relatively rapid rates are quite insoluble in lipids, as mentioned previously. The inconsistency between observation and hypothesis has resulted in the suggestion that small, lipid-insoluble substances penetrate the membrane through hydrophilic pores approximately 0.8 to 1.0 nm in diameter. But the existence of such pores has not been confirmed, and furthermore, not all small, lipid-insoluble substances penetrate the plasma membrane at equal rates—an observation that is difficult to rationalize with the presence of pores, which presumably could not display a great deal of selectivity. Many workers have concluded that if pores in fact exist, it is quite probable that only water normally penetrates the membrane through them. This conclusion would account for the extremely rapid penetration of water relative to that of most other substances.

Facilitated Diffusion

Glucose, certain amino acids, and some other compounds enter the cell by diffusion, but the characteristics of their passage across the plasma membrane differ somewhat from those of passive diffusion. For although passage occurs in response to a concentration gradient, as it does with passive diffusion, the rate of transport is proportional to the concentration difference across the membrane only up to a particular concentration level of the permeant molecule. After that concentration level is reached, the rate of transport does not increase, regardless of the magnitude of the concentration differential.

This observation suggests that in these cases the passage of molecules across membranes occurs by some mechanism that becomes saturated at high concentrations. There is good evidence that the permeant molecule combines with specific carrier molecules in the membrane and that these carrier molecules facilitate the transport of the permeant molecule across the membrane. Because there are only a limited number of carrier molecules specific for a particular substance (or group of substances), and because the transport of a substance across the membrane by a carrier mechanism requires a given amount of time under defined conditions, it follows that the carrier process will become saturated at a certain concentration of permeant molecules. At the saturation level, all the specific carrier molecules will be involved in transporting permeant molecules, and hence the rate of passage will not increase as the concentration of permeant molecule rises about the saturation level (Fig. 16–14). Transport of this type is termed *facilitated diffusion* because combination of the permeant molecule with a carrier molecule in the membrane facilitates the passage of a molecule to which the membrane would otherwise be impermeable or only slowly permeable. It must be stressed that facilitated diffusion, like passive diffusion, is driven by a concentration gradient and does not require metabolic energy: it is the carrier alone that facilitates the process.

Active Transport

Many molecules are transported into and out of cells against a concentration gradient; that is, from one region to another of greater concentration. It is clear that the energy which drives such

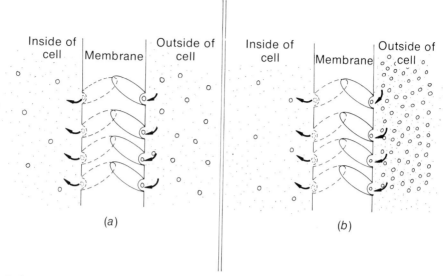

Illustration of the saturation phenomenon in carrier-mediated transport. In the hypothetical carrier mechanism shown here, the carrier transports a particular molecule (open circles) in the direction of the arrows by forming a complex with the molecule on the outside of the cell and oscillating to the other side of the membrane where the molecule is released. In the two cases illustrated, the molecule will be transported at the same rate, even though its external concentration in (b) is much higher than in (a). This is because all carriers in the membrane are involved in transporting molecules and the transport process requires a finite amount of time. In other words, the carrier process is operating as fast as it can in both cases.

Figure 16-14

transport cannot be derived from the concentration gradient across the membrane. In fact, such transport is driven by energy supplied by metabolic processes and is termed *active* or *metabolically linked transport*.

It is not always a simple matter to distinguish active transport from passive transport, and although there is no general consensus about how the term active transport should be used, we will define active transport as necessarily including two aspects:

(1) Active transport is transport against a concentration gradient.

(2) Active transport is driven by energy-yielding, spontaneous reactions of the cell; most frequently the immediate energy source for active transport is ATP.

There are, however, inherent problems in deciding whether a process meets these two criteria. In the first place, it is necessary to demonstrate that the process really *does* operate against a concentration gradient. This is more difficult than might be immediately apparent. For example, even though a substance may normally be present in much higher concentrations within the cell than without, it does not necessarily follow that the substance enters the cell by active transport. We must be able to demonstrate that the substance is *osmotically active* within the cell by being present in the free, uncombined form and soluble in the aqueous phase of the cytoplasm. A substance may be osmotically inactive within the cell, hence not contributing to the effective concentration gradient, if it combines with other substances to form insoluble compounds, if it attaches to binding sites on membranes shortly after entering the cell, if it dissolves in lipid and is thus removed from the aqueous phase, and so forth. Phosphate ions, for example, often precipitate as magnesium phosphate within the cell; glucose may be polymerized into glycogen or starch. In the former case, the intracellular concentration of *soluble* phosphate may be very low even though there is considerable phosphate within the cell in the form of insoluble magnesium phosphate. In the latter example, there may be much glucose in the form of glycogen or starch but little in the form of free, osmotically active glucose. Thus, in order to determine whether a substance contributes to a concentration gradient, we must know the chemical nature of the substance after it enters the cell.

It also is not always easy to determine whether a given transport process requires metabolic energy. It is possible that a transport process which operates against a concentration gradient may be linked to, and hence driven by, a spontaneous, passive transport process that operates in the opposite direction. Any transport process that is inhibited by certain respiratory inhibitor compounds is probably active transport. Thus, for example, active transport in aerobic cells is greatly inhibited by dinitrophenol (DNP) or cyanide. Both these substances block oxidative phosphorylation and thus prevent ATP production.* DNP acts to uncouple mitochondrial electron transfer in the respiratory chain from phosphorylation, with the result that electron transfer continues, often at accelerated rates, but ATP is not formed. Cyanide inhibits ATP formation by preventing the transfer of electrons from cytochrome oxidase to oxygen and hence blocking the transfer of electrons along the respiratory chain. Active transport is inhibited in anaerobic cells by compounds that block glycolysis; in photosynthetic cells, compounds that block both photophosphorylation and phosphorylation associated with respiration (either anaerobic or aerobic) often block active transport.

The inhibition of active transport by respiratory inhibitor compounds does not necessarily prove that a transport process occurs by an active transport mechanism, nor does the lack of inhibition prove that the process occurs by a passive transport mechanism. It is possible that the inhibitor is not, in fact, preventing ATP production, that the inhibitor is interfering with a diffusion-linked process, that some high-energy compound other than ATP is driving the process, and so on.

It should be obvious, then, that there are significant problems in distinguishing between active and passive transport processes. Nevertheless, by accumulating sufficient experimental data and using care in their analysis, it is possible in most cases to determine with

---

*In aerobic cells, some small amount of ATP may be generated by substrate level phosphorylation in the Embden-Meyerhof pathway; hence, active transport may not be *completely* inhibited by compounds that block oxidative phosphorylation, although the degree of inhibition is substantial.

Mechanisms of Carrier-Mediated Transport

Active transport, like facilitated diffusion, displays the characteristics of carrier-mediated transport. These are:

(1) The transport depends upon the structure and configuration of the permeant molecule.

(2) The transport can often be blocked by compounds which, because of close structural relationships to the permeant molecule, presumably compete for binding sites on the carrier protein.

(3) The rate of transport is not strictly proportional to the concentration of the permeant molecule but instead reaches a maximal rate at high concentration levels.

A number of investigators have concentrated their research efforts on the identification and isolation of carrier molecules responsible for transport, and some information has accumulated about the chemical characteristics of several substances presumed to be carriers. In addition, we now have enough data on carrier-mediated transport to suggest how the process may operate.

In the 1950s, bacterial geneticists discovered mutants of *Escherichia coli* that were *not* capable of transporting β-galactosides* into the bacterial cell. They assumed that these mutants lacked a membrane carrier protein that they termed *β-galactoside permease.* Subsequently, Eugene Kennedy and C. Fred Fox isolated from *E. coli* cells a protein that is presumably involved in β-galactoside transport. This protein, which they named the *M (membrane) protein,* has been shown to have a molecular weight of about 30,000. Because the M protein is not present in *E. coli* mutants unable to transport β-galactosides, and because the M protein binds β-galactosides in vitro, it is thought to be identical with β-galactoside permease. However, even though the M protein binds β-galactosides in vitro, this does not necessarily mean that it functions as a carrier protein in vivo. Furthermore, additional components may be involved in β-galactoside transport. In any case, the absence of the M protein in mutants incapable of transporting β-galactosides shows that the M protein is in some way necessary for the transport of β-galactosides, and it is generally assumed that the M protein is, in fact, the β-galactoside carrier (permease).

More recently, researchers have isolated other membrane proteins that are capable of binding specific small molecules, and these also have been implicated as carrier proteins. It thus seems reasonable to assume that some, if not all, carrier molecules are proteins.

It is one thing to say that carrier molecules facilitate transport

---

*A β-galactoside is a compound in which β-galactose is linked to another molecule by a glycosidic linkage (see page 168).

and quite another to explain how this occurs. Unfortunately, we do not have any good information to indicate the precise mechanism (or mechanisms) involved in carrier-mediated transport. However, many models have been devised to explain the process. These models are based upon the assumption that carrier molecules, if they are indeed proteins embedded within membranes, might be expected to display certain types of behavior, by analogy to enzymes and other proteins whose structures and functions have been carefully studied.

One such model of carrier-mediated transport envisions that a carrier protein changes its configuration when it attaches to the particular molecule that it normally transports. This change in configuration is seen as resulting in the transport of the molecule across the membrane (Fig. 16–15a). The idea of a configurational change of this type is not new; in fact, we have already mentioned that such changes occur in some enzymes during catalysis.

Another model of carrier-mediated transport suggests that carrier molecules may have a diameter approximately equal to the thickness of the membrane. Once the substance to be transported has become bound to the carrier molecule, transport across the membrane may occur by a rotation of the carrier-substrate complex through 180 degrees (Fig. 16–15b).

A third possible mechanism involves intramembrane shuttling, by which a small carrier molecule, once having bound a molecule on one side of the membrane, diffuses through the membrane to the other side and releases the transported molecule (Fig. 16–15c).

It is quite clear that numerous other mechanisms of carrier-mediated transport are possible, and there is no reliable evidence that any of the mechanisms described above actually occur. Indeed, there is no reason why we need attempt to settle on one particular mechanism at this time: different molecules may well be transported by different carrier mechanisms.

It is important to note that the main difference between the carrier-mediated mechanisms of active transport and facilitated diffusion is that the former require the expenditure of metabolic energy in order to operate, and the latter do not. Thus, any of the carrier mechanisms we have discussed could operate in either active transport processes or facilitated diffusion processes.

THE REGULATION OF INTRACELLULAR CONCENTRATIONS OF SODIUM AND POTASSIUM IONS

There is one carrier-mediated active transport system about which we have accumulated a significant amount of information. That is the process by which cells regulate their intracellular concentrations of potassium ions (K^+) and sodium ions (Na^+). Because the process is important in virtually all animal cells and provides an

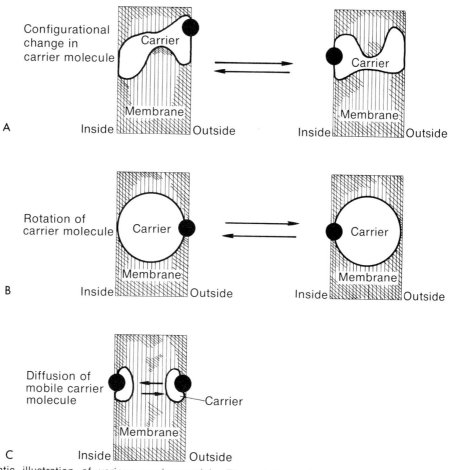

Diagrammatic illustration of various carrier models. The substance transported is represented by a dark circle.   **Figure 16-15**

example of how one active transport process operates, we will discuss it in some detail.

In general, cells maintain a high internal concentration of K^+ and exclude Na^+ from the interior. Indeed, cells maintain these levels in the face of an external environment that is usually high in Na^+ and low in K^+ (Table 16-2). It is not at all apparent why most cells maintain high levels of K^+ and low levels of Na^+, although it is clear that the maintenance of these ion concentrations has great physiologic significance, since over one third of the ATP utilized by an average "resting" cell serves to maintain K^+ and Na^+ gradients.

We do know that K^+ ions are required for the activity of many enzymes and for protein synthesis and thus may play a significant role in the regulation of metabolism. In addition, we know of several specific physiologic processes in which ion gradients are important. The conduction of impulses by nerve and muscle cells—electrical excitability—is largely dependent upon the imbalance of ion con-

Table 16-2 CONCENTRATIONS OF Na^+ AND K^+ (in millimoles/l) IN VARIOUS CELL TYPES*

Cell	INTRACELLULAR		FLUID SURROUNDING CELLS	
	Na^+	K^+	Na^+	K^+
Human erythrocyte	11	91	138	4.2
Rat erythrocyte	12	100	151	5.9
Frog muscle	16	127	106	2.6
Rat muscle	8	160	147	7.3
Chara cell sap (marine alga)	66	65	460	10
Valonia cell sap (marine alga)	35	576	460	10

*All data calculated from H.B. Steinbach, in *Comparative Biochemistry* (M. Florkin and H.S. Mason, Eds.), Volume 4, Part B, pp 677–720, Academic Press, Inc., New York, 1963.

centrations between the inside and the outside of the plasma membrane, and, of course, communication between cells (as determined by nerve impulses) and control of muscular activity are of great importance to multicellular organisms. In fact, cell excitability may also play an important role in the response of unicellular organisms to their environment.

It is likely that the requirement of many enzymes for K^+ is not a result but rather a consequence of the relatively high intracellular concentrations of K^+. Enzymes presumably underwent much of their evolution after the plasma membrane developed and probably after the establishment of high intracellular levels of K^+. If this was the case, then enzymes would have evolved configurations consistent with their ionic environment and thus acquired a dependence on the K^+ ion.

It has also been shown that in certain animal cells the Na^+ gradient across the plasma membrane may be used for the transport of some sugars and amino acids. For example, glucose may be pumped into the cell against a concentration gradient by a mechanism in which both Na^+ and glucose bind to the same carrier molecule and the two substances enter the cell together. The diffusion of Na^+ from a region of high concentration (outside the cell) to one of low concentration (inside the cell) thus drives the intracellular accumulation of glucose. In this case, the Na^+ gradient serves as a source of readily available energy. Some workers have postulated that the development of mechanisms for the maintenance of K^+ gradients and Na^+ gradients occurred originally because they provided cells with a "storage battery system" of readily available free energy.

Membrane Transport

The Sodium-Potassium Pump

The carrier-mediated transport system by which cells maintain high levels of K^+ and low levels of Na^+ relative to the external medium has been called the *sodium pump* or, more recently, the *sodium–potassium pump* (Na^+–K^+ pump), since we now realize that transport of the two ions is linked. The mechanism by which the sodium–potassium pump operates has been under investigation for many years. Much of this research has concentrated on the pump mechanism in red blood cells, simply because they are readily available and easy to work with.

The first real breakthrough in our understanding of the pump mechanism was the discovery of J. Skou in 1957 of an enzyme that was capable of hydrolyzing ATP to ADP and inorganic phosphate only in the presence of Na^+, K^+ and Mg^{++}:

$$ATP + H_2O \xrightarrow[\text{Enzyme}]{Na^+, K^+, Mg^{++}} ADP + H_3PO_4$$
$$(Na^+\text{–}K^+ \text{ ATPase})$$

This enzyme was termed the *sodium–potassium ATPase* (Na^+–K^+ *ATPase*), and its identification led to the idea that the exergonic hydrolysis of ATP might provide the energy for the active transport of Na^+ out of the cell and K^+ into the cell.

A great deal of evidence has accumulated which shows that the Na^+–K^+ ATPase is an integral part of, or perhaps identical to, the Na^+–K^+ pump. Some of this evidence includes the following:

(1) The Na^+–K^+ ATPase is associated with plasma membranes that actively transport Na^+ and K^+.

(2) There is greater Na^+–K^+ ATPase activity in plasma membranes of those cells which pump quantitatively more Na^+ and K^+ than others.

(3) The hydrolysis of ATP and the transport of Na^+ and K^+ are closely linked, and ATP is not hydrolyzed unless Na^+ and K^+ are transported.

(4) Both the Na^+–K^+ ATPase and the Na^+–K^+ pump are specifically inhibited to similar extents by a compound called ouabain (pronounced *wah-bane*).

(5) It is possible under certain conditions to phosphorylate ADP by allowing the Na^+ and K^+ gradients to collapse (reversing the pump).

These results taken together indicate that the hydrolysis of ATP by the Na^+–K^+ ATPase is linked to the transport of K^+ into the cell and Na^+ out of the cell.

Other experiments have provided evidence concerning the orientation of the Na^+–K^+ ATPase and Na^+–K^+ pump in the membrane. It has been shown that the Na^+ ion is not transported unless it is located in the aqueous medium within the cell, and K^+ is not transported unless it is located in the aqueous medium surrounding the cell. Furthermore, the Na⁺–K⁺ ATPase can only use as substrate ATP that is located within the cell, and the ATPase and pump are inhibited by ouabain only when ouabain is located in the aqueous medium outside the cell. This evidence suggests that the ATPase and pump are located within the membrane in such a way that they are accessible only to K⁺ and ouabain from one side and only to Na⁺ and ATP from the other (Fig. 16–16).

A further understanding of the pump mechanism was obtained by the finding that the ATPase reaction could be separated into two steps, one step requiring Na⁺ and Mg⁺⁺, and the other step requiring K⁺. The first step results in the formation of a high-energy phosphorylated enzyme intermediate:

$$\text{ATP} + \text{Enzyme} \xrightarrow{\text{Na}^+, \text{Mg}^{++}} \text{Enzyme} \sim \overset{\overset{O}{\|}}{\underset{OH}{P}} - OH + \text{ADP}^{\dagger}$$
(ATPase)

The phosphate group in the enzyme has been shown to be attached to the side chain of one glutamic acid residue in the ATPase enzyme (Fig. 16–17).

The next step in the reaction, the dephosphorylation of the enzyme, requires K⁺ and occurs as follows:

$$\text{ADP} + \text{Enzyme} \sim \overset{\overset{O}{\|}}{\underset{OH}{P}} - OH + H_2O \xrightarrow{\text{K}^+} \text{Enzyme} + H_3PO_4$$

---

†Mg⁺⁺ is required in the ATPase reaction as a cofactor; the transport of Mg⁺⁺ is not linked to this reaction.

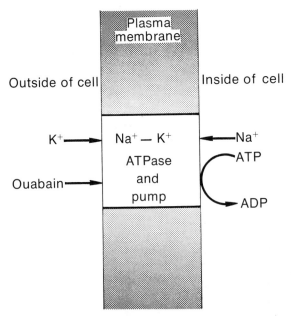

Figure 16–16 Orientation of the Na⁺–K⁺ ATPase and pump in the plasma membrane.

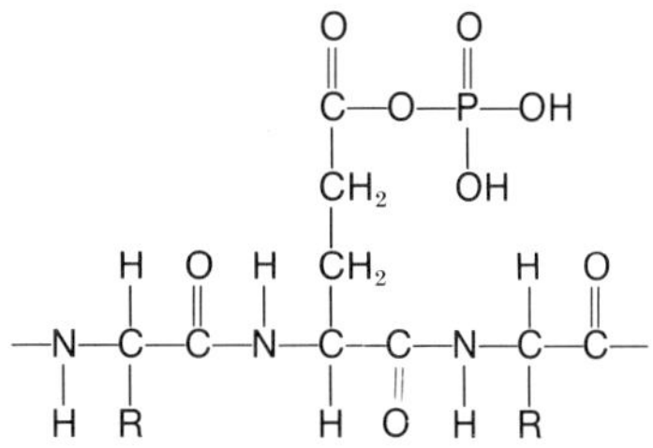

Membrane Transport

[chemical structure diagram]

In the phosphorylated form of the Na$^+$–K$^+$ ATPase, the phosphate group is bonded to the carboxyl side chain of a particular glutamic acid residue in the ATPase molecule.

Figure 16-17

We also know something about the stoichiometry of Na$^+$ and K$^+$ transport. Not only are the transport of Na$^+$ and the transport of K$^+$ closely linked but for each molecule of ATP hydrolyzed in the ATPase reaction, three sodium ions are pumped out of the cell and two potassium ions are pumped into the cell.

Na–K$^+$ ATPase has been isolated from plasma membranes of several cell types. The most recent evidence shows that the Na$^+$–K$^+$ ATPase from at least one cell type consists of two subunits: a catalytic subunit of 97,000 molecular weight and a glycoprotein subunit of 55,000 molecular weight. The larger subunit is thought to be the catalytic portion of the enzyme; the function of the glycoprotein subunit is not known.

Many models have been proposed to explain the molecular mechanism of the Na$^+$-K$^+$ pump, based upon the information discussed above. One of these models is diagrammed in Figure 16–18. The Na$^+$–K$^+$ ATPase, located within the membrane, is considered to exist in at least two different configurations, depending upon whether the enzyme is phosphorylated or not. It is postulated that the configuration of the dephosphorylated enzyme (E_1) is such that binding sites specific for Na$^+$ transport face inward toward the cell's cytoplasm. The binding of Na$^+$ to these sites is followed by the phosphorylation of the ATPase by ATP. The phosphorylation of the enzyme, in turn, is thought to convert the enzyme into a second configurational form (E_2). The configurational change from E_1 to E_2 carries the bound Na$^+$ ions to the outside surface of the membrane, where they are released. The second configurational form now has K$^+$–specific binding sites on the surface facing the exterior of the cell. Following binding of K$^+$ ions to E_2, dephosphorylation occurs, converting the ATPase back to its original configuration (E_1). The result of this conversion is to carry K$^+$ ions to the inside surface of the membrane, where they are released into the cytoplasm; at the same time, the new configuration again has Na$^+$–specific binding sites on its surface facing the cell's interior. The binding of Na$^+$ to these sites initiates another cycle. In order to account for the

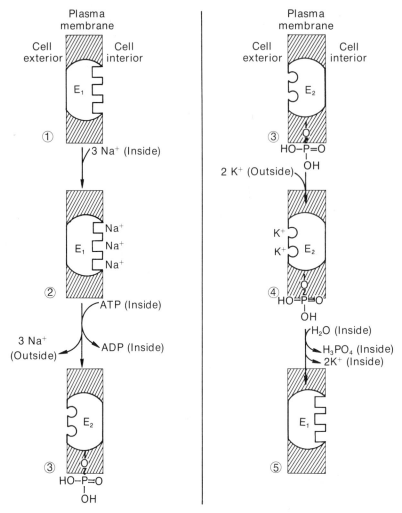

Figure 16-18 Diagram of one model illustrating the molecular mechanism of the sodium-potassium pump. See text for details.

stoichiometry of transport, it is assumed that each ATPase molecule possesses two sites that bind Na^+ and three sites that bind K^+.

The Calcium Pump

Sodium and potassium are not the only ions that are transported by a carrier-mediated active transport mechanism. A calcium (Ca^{++}) pump has been shown to be operative in skeletal muscle, and a Ca^{++}-activated ATPase that is an integral part of the Ca^{++} pump has been isolated. In skeletal muscle cells, an extensive membrane system—the sarcoplasmic reticulum—is involved in the control of Ca^{++} levels surrounding the contractile fibers of muscle.

In an uncontracted muscle fiber, Ca^{++} is pumped into the sarcoplasmic reticulum, and the Ca^{++} concentration surrounding the contractile fibers is consequently very low. When the membrane of the sarcoplasmic reticulum is stimulated by a nerve impulse, Ca^{++} is released from the membrane system, and the increased concentration of Ca^{++} brings about contraction of muscle fibers. It is thus apparent that the concentration of Ca^{++} around muscle fibers is an essential feature in the control of muscular contraction.

There are undoubtedly a variety of intramembrane carrier mechanisms that serve to regulate the intracellular levels of other ions as well as other substances. Some of these undoubtedly operate by a membrane-bound ATPase pump mechanism, whereas others may involve different mechanisms. Considerable research is now being concentrated on active transport processes, and this work should eventually improve our understanding of these processes.

ENDOCYTOSIS AND EXOCYTOSIS

A discussion of how substances get in and out of cells would be incomplete without mention of *endocytosis* and *exocytosis*. These processes do not involve the passage of substances *through* membranes and hence are not membrane transport processes in the strict sense.

We discussed endocytosis and exocytosis in detail in Chapter 11. Endocytosis appears to be a general mechanism by which macromolecules and other large substances enter cells; exocytosis is important in the secretion of cellular products such as hormones, as well as in the removal of undigestible material. Both processes require metabolic energy and are inhibited by substances that interfere with cellular energy production, but they do not necessarily transport substances against a concentration gradient. For example, any substance that is present in the fluid surrounding a cell may be brought into the cell by endocytosis, whether or not that substance is present at higher concentration outside the cell than within. Indeed, endocytosis and exocytosis do not have inherent specificity, and virtually any substance can move in or out of cells by these mechanisms. It is becoming clear, however, that some chemical substances induce endocytotic vesicles in certain cell types.

THE ORIGIN OF CELLULAR MEMBRANES

In previous chapters, we have mentioned a number of physicochemical processes that may have led to the formation of membranous structures during the origin of the first living systems. In particular, a great deal of research has concentrated on the formation of fatty acid and phospholipid bilayers as models for the formation of cell membranes. Not only have phospholipids and fatty acids been shown to orient themselves spontaneously in aqueous solutions to form micelles or bilayers but phospholipid bilayers have been shown to bind water-soluble protein to form artificial lipoprotein membranes that are similar in dimensions and patterns of electron density to cell membranes (see Chapters 7 and 9).

On the basis of these experiments and our admittedly limited knowledge of membrane structure and function, it is possible to speculate on the origin and subsequent development of cellular membranes. It is probable that the first membranes were simple phospholipid bilayers. Such membranes must have been relatively impermeable to polar substances, and the subsequent incorporation of protein into the lipid bilayer increased their permeability to water and other polar substances, thus introducing some specificity of function. Eventually, membranes developed that contained large amounts of protein and that possessed important metabolic functions. This trend toward the incorporation of increasing amounts of protein into membranes presumably would have occurred because of the advantages of localizing energy transfer reactions and other metabolic processes on membranes, as we have discussed in Chapters 14 and 15.

REFERENCES

Berlin, R.D. (1970). Specificities of Transport Systems and Enzymes. *Science, 168*:1539.
Branton, D. (1967). Fracture faces of frozen myelin. *Exp. Cell Res., 45*:703.
Branton, D. and R.B. Park (1967). Subunits in chloroplast lamellae. *J. Ultrastructure Res. 19*:283.
Branton, D. and R.B. Park (1968). *Papers on Biological Membrane Structure.* Little, Brown and Company, Boston.
Chapman, D., Ed. (1968). *Biological Membranes: Physical Fact and Function*, Academic Press, Inc., New York.
Culliton, B.J. (1972). Cell membranes: a new look at how they work. *Science, 175*:1348.
Dahl, J.L. and L.E. Hokin (1974). The sodium-potassium adenosinetriphosphatase. *Ann. Rev. of Biochem., 43*:327.
Davies, M. (1973). *Functions of Biological Membranes.* John Wiley and Sons, New York.
Fox, C.F. (1972). *The Structure of Cell Membranes. Scientific American*, February, 1973, p. 30.
Giese, A.C. (1973). *Cell Physiology.* W.B. Saunders Co., Philadelphia.
Hokin, L.E. (1972). *Metabolic Transport.* Academic Press, Inc., New York.
Roth, S. (1973). A molecular model for cell interactions. *Quart. Rev. Biol., 48*:541.
Sharon, N. (1974). Glycoproteins. *Scientific American,* May 1974, p. 74.
Singer, S.J. and G.L. Nicolson (1972). The fluid mosaic model of the structure of cell membranes. *Science 175*:720.
Sjöstrand, F.S. (1963). A new ultrastructural element of the membranes in mitochondria and of some cytoplasmic membranes. *J. Ultrastructure Res., 9*:340.
Sjöstrand, F.S. (1963). A comparison of plasma membrane, cytomembranes, and mitochondrial membrane elements with respect to ultrastructural features. *J. Ultrastructure Res., 9*:561.
Stein, W.D. (1967). *The Movement of Molecules Across Cell Membranes.* Academic Press, Inc., New York.
Stryer, L. (1975). *Biochemistry.* W.H. Freeman and Company, San Francisco.
Tedeschi, H. (1974). *Cell Physiology: Molecular Dynamics.* Academic Press, Inc., New York.
Vanderkooi, G. and D.E. Green (1970). Biological membrane structure. I. The protein crystal model for membranes. *Proc. Nat. Acad. Sci., 66*:615.

CHAPTER 17

Cell Division

One of the most critical steps in the evolution of living systems must have been the development of mechanisms for growth and reproduction. The perpetuation of life depends upon the fact that cells divide; without some type of cell division, living organisms would be unable to reproduce. In addition, growth of living systems without cell division would give rise to large cells whose plasma membranes would be easily disrupted. The increase in the ratio of cell volume to surface area that would occur during growth in the absence of cell division would also create a situation in which the exchange of materials across the plasma membrane would become increasingly difficult as cell growth continued.

It is obvious, however, that cell division is no answer to the problems of growth and reproduction unless it occurs in such a way that essential cell constituents are distributed to daughter cells. In particular, the viability of daughter cells depends upon the operation of a mechanism that ensures the distribution of hereditary information, which is required to direct most of the cell's activities.

Historically, the process of cell division was first studied in eukaryotes, simply because prokaryotes are too small to be observed in any detail under the light microscope. The division of eukaryotic cells was first described shortly after the formulation of the cell theory. It soon became apparent that nuclear division always precedes cell division, and this observation suggested to mid-nineteenth century biologists that the nucleus must play a central role in cell division.

In 1875, Eduard Strasburger, a German botanist, published a detailed description of *mitosis*—the process of nuclear division—in plant cells, and a few years later, Walther Flemming described mitosis in animal cells. It was Flemming who first observed the longitudinal division of nuclear structures that appeared only during nuclear division. These structures were named *chromosomes* by W. Waldeyer in 1888.

By the start of the twentieth century, biologists were beginning to realize that the nucleus was involved not only in the process of cell division per se but also in determining the hereditary characteristics of cells. In 1866, an Austrian monk named Gregor Mendel reported the results of a number of breeding experiments that he had performed on garden peas. Mendel noticed that certain characteristics in peas, such as the length of the stem, the color of the flowers, and so on, were transmitted in a predictable fashion from one generation to the next. Mendel assumed that hereditary traits were transmitted by physical particles: the hereditary units we now call *genes*. He correctly deduced from the results of his experiments that each characteristic he examined was determined by the interaction of two hereditary particles, one inherited from each parent.

Mendel's experiments were largely ignored until 1900, when they were rediscovered independently by three botanists, Hugo de Vries, Erich von Tschermak, and Carl Correns. By that time, chromosomes had been described, as had their behavior during nuclear division, and several biologists suggested that chromosomes might somehow be responsible for the transmission of the hereditary units. In 1903, Walter Sutton advanced the hypothesis that hereditary units were transmitted by chromosomes. Sutton's hypothesis, which was an important conceptual breakthrough, was based on the observation that there were many parallels between the behavior of chromosomes during cell division and the pattern of transmission of hereditary characteristics.

Continued research during the twentieth century has confirmed Sutton's ideas and has demonstrated the importance of the nucleus as a storage site for the hereditary information of the eukaryotic cell. We now know that hereditary information is encoded in DNA molecules in all cell types—eukaryotic DNA is associated with proteins and some RNA to form chromosomes, and prokaryotic DNA exists within the cell as a single unassociated macromolecule.

This chapter describes experiments that led to the discovery of DNA as the hereditary material, discusses how DNA is replicated and distributed to daughter cells during the process of cell division in both prokaryotes and eukaryotes, and examines what is known about the control of cell division. Chapter 18 describes how the hereditary information encoded in DNA molecules is decoded and thus utilized by the cell.

DNA: THE MOLECULAR BASIS OF HEREDITY

It has been known since the discovery of nucleic acids by Frederich Miescher in 1871 that nuclei contain large amounts of deoxyribonucleic acid (DNA) and protein. Until the 1940s, most biologists assumed that the proteins in the nucleus probably functioned as the hereditary material of the cell. This seemed like a reasonable assumption because proteins, composed of twenty different amino acid building blocks, are structurally more complex than nucleic acids, which consist of long chain polymers of only four dif-

DNA: The Molecular Basis of Heredity

ferent nucleotides (see Chapter 7 for a review of the primary structure of proteins and nucleic acids).

TRANSFORMATION

In 1928, Fred Griffith performed an experiment that set the stage for the discovery of the hereditary role of DNA. Griffith was studying pneumococcal pneumonia, a form of pneumonia caused by the bacterium *Diplococcus pneumoniae* (also known as pneumococcus). There are a number of strains of pneumococcus, some of which are pathogenic (cause disease) and some of which are not. The cells of pathogenic strains are surrounded by a slimy polysaccharide capsule, and when these strains are grown on the surface of an agar plate, they form large, smooth colonies. Other forms of pneumococcus are devoid of a polysaccharide coat and are nonpathogenic. These latter forms produce small, rough colonies on agar plates. Because of their characteristic appearances when grown on agar, the pathogenic pneumococcal strains are referred to as S (smooth) forms and the nonpathogenic variants are known as R (rough) forms. The presence or absence of a polysaccharide coat is a hereditary characteristic, and the pathogenicity of the S form is apparently due to the fact the the smooth, slippery pneumococci cannot be engulfed, and hence destroyed, by white blood cells.

Griffith knew that if S form pneumococci were injected into mice, the bacteria caused a fatal infection. In addition, he was aware that mice were not harmed when they were injected with S form pneumococci that had been killed by heat treatment. Much to his surprise, however, Griffith found that if he injected mice with a *mixture* of live R cells (nonpathogenic) and heat-killed S cells, the mice died of pneumonia (Fig. 17–1). Indeed, Griffith was able to isolate living S cells from the dead animals and to show that these bacteria were permanently of the S form; that is, all progeny of the bacteria were of the smooth type. Thus it seemed that dead S cells were somehow able to change living R cells into the S form. This change has been called *transformation*.

Subsequently, it was shown that transformation from R to S form can occur in vitro. For example, viable S cells could be isolated following the addition of a cell-free homogenate of S cells to a growing culture of R cells. Thus, intact dead cells were shown not to be necessary for transformation, and it appeared that some component of the homogenate was capable of causing the genetic change. This component became known as the "transforming principle."

In 1944, Oswald Avery, Colin MacLeod, and Maclyn McCarty published a paper entitled "Studies on the Chemical Nature of the Substance Inducing Transformation of Pneumococcal Types," which identified DNA as the "transforming principle." Avery, MacLeod, and McCarty based this conclusion on the finding that highly purified DNA extracted from S form pneumococci caused transformation when added to viable R cells in an in vitro system. They

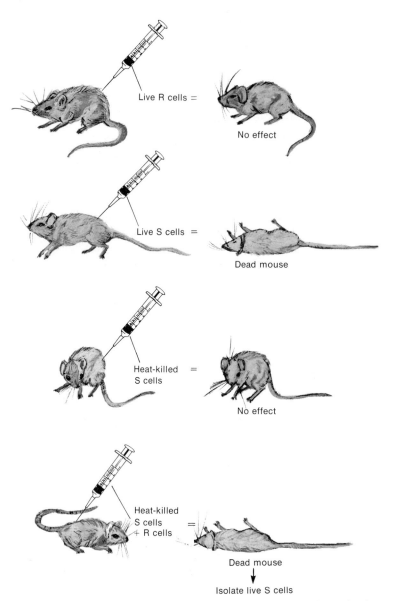

Figure 17-1 A summary of Griffith's experiments on transformation in pneumococcus.

found that the transforming activity of a crude DNA preparation was not diminished upon extraction of protein, lipid, or polysaccharide from the preparation, suggesting that transformation observed when using purified DNA was not due to a minor contaminant in the preparation. They also demonstrated that the transforming activity of a DNA preparation was abolished by incubating it with deoxyribonuclease (an enzyme that hydrolyzes DNA) but was not affected by ribonuclease (an enzyme that hydrolyzes RNA) or by the proteolytic enzymes trypsin and chymotrypsin (Fig. 17-2).

DNA: The Molecular Basis of Heredity

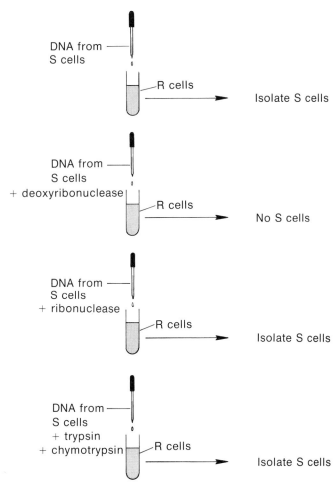

Summary of experiments by Avery, MacLeod, and McCarty, proving the genetic role of DNA. See text for details.   **Figure 17-2**

The conclusions of Avery, MacLeod, and McCarty were not immediately accepted, and it was argued by some that contaminants in the DNA preparation were responsible for transformation or that the DNA caused R cells to mutate into the S form. Although we still do not completely understand all the precise details of bacterial transformation, it can be shown that DNA from the S form bacteria is taken up by the recipient cell (R form) and is permanently incorporated as part of the genetic material of the R cell.

THE BLENDOR EXPERIMENT

Additional evidence that DNA is the genetic material came from research done by Alfred Hershey and Martha Chase in the early 1950s. Hershey and Chase were interested in discovering the mechanism by which T2, a bacterial virus, infects *Escherichia coli.* It was known at the time that T2 and a group of closely related bacterial viruses (also called *bacteriophages,* or *phages,* for short) are composed exclusively of DNA and protein; in the intact T2 particle, the viral DNA is surrounded by a protein coat of rather complex morphology (Fig. 17–3). A great deal of evidence had shown that T2 infection is initiated when T2 attaches by its tail to the bacterial wall. Approximately 30 minutes after this attachment occurs, the infected bacterium lyses (breaks apart), releasing several hundred newly synthesized T2 particles (Fig. 17–4).

Hershey and Chase wondered what molecular events occur between the time of T2 attachment to *E. coli* and the release of multiple copies of T2 from the infected cell. In order to investigate these events, they used radioactive isotopes to label, or tag, T2 molecules so that they could easily distinguish between viral DNA and viral protein. By using these labelled T2 particles to infect *E. coli,* they were able to trace viral DNA and protein during the infection process.

The first step in Hershey and Chase's experiment was to obtain specifically labeled T2 particles. They did this by growing *E. coli* in a medium that contained a simple carbon source, a nitrogen source, and essential minerals, and to which they had added radioactive sulfur (S^{35}) and radioactive phosphorus (P^{32}). They then infected these bacteria with T2, and after the bacteria had lysed, they harvested the newly formed T2 particles. These T2 particles, produced in *E. coli* cells growing in radioactive medium, had incorporated P^{32} into their DNA and S^{35} into their protein coats. (Recall that phosphorus is

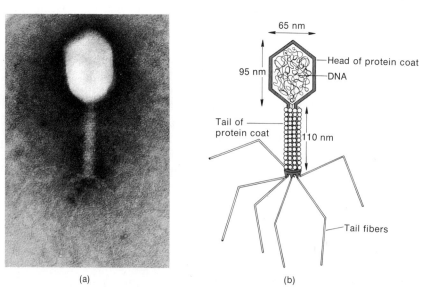

Figure 17–3 Detailed structure of bacteriophage T2. (a) Electron micrograph of T2, and (b) diagrammatic interpretation of cross-section of T2 particle. (a) From Stent, G.S. (1971), *Molecular Genetics,* San Francisco, © 1971 by W.H. Freeman and Company, page 305.

DNA: The Molecular Basis of Heredity

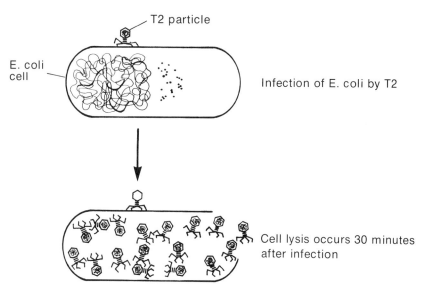

The attachment of bacteriophage T2 to *E. coli* is followed in about 30 minutes by lysis of the cell and the release of several hundred T2 particles.

Figure 17–4

present in DNA, but not in most proteins;* sulfur is present in some amino acids, but not in DNA). Thus, T2 DNA was labeled specifically with P^{32}, and T2 protein was labeled specifically with S^{35} (Fig. 17–5).

The second step was to infect a nonradioactive culture of *E. coli* with the radioactively labeled T2. After allowing about 2 minutes for the attachment of T2 bacteria, Hershey and Chase placed the liquid medium containing the infected bacteria in a Waring Blendor for a few minutes. Following blending, they pelleted the bacterial cells by centrifugation and then analyzed both the pellet and the supernatant (the remaining liquid) for P^{32} and for S^{35}.

The results of this experiment showed that (1) most of the P^{32} label was present inside the bacteria, (2) most of the S^{35} label was present in the supernatant, and (3) the process of infection was not influenced by the blending treatment, since new progeny virus were released from blended cells within about 30 minutes after T2 attachment.

Hershey and Chase interpreted these results in the following way: Attachment of T2 to the bacteria was immediately followed by injection of viral DNA (P^{32}-labeled) into the cell, but the S^{35}-labeled protein coats of the viral particles remained on the surface of the bacteria. Subsequently, the blending treatment, by developing strong shearing forces, tore the empty protein coats from the surface of the bacteria. Because the protein coats of the virus particles

---

*There are some conjugated proteins, termed *phosphoproteins*, that have prosthetic groups containing phosphorus, but the phosphoproteins constitute such a small proportion of cellular proteins that P^{32} is usually not incorporated into proteins to any significant extent.

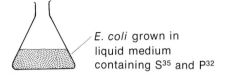

Figure 17–5   Diagram illustrating how Hershey and Chase obtained radioactively-labelled T2 particles.

are vastly smaller than the bacterial cells, the coats remained in the supernatant fluid when the bacteria were pelleted by centrifugation (Fig. 17–6).

Hershey and Chase concluded that:

> The sulfur-containing protein of resting phage particles is confined to a protective coat that is responsible for the adsorption [of the T2 particles] to bacteria, and functions as an instrument for the injection of phage DNA into the cell. This protein probably has no function in the growth of intracellular phage. The DNA has some function. Further chemical inferences should not be drawn from the experiments presented.

In spite of the guarded tone of these conclusions, it soon was recognized that the DNA of the virus is solely responsible for the replication of the virus; that is, the viral DNA is endowed with the hereditary information that specifies the production of new virus particles. Thus Hershey and Chase confirmed the conclusions of Avery, MacLeod, and McCarty, and DNA was indisputably recognized as the hereditary material.*

---

*Some viruses use RNA instead of DNA as their hereditary material, but DNA is the hereditary material in all prokaryotic and eukaryotic cells.

The Three-Dimensional Structure of DNA

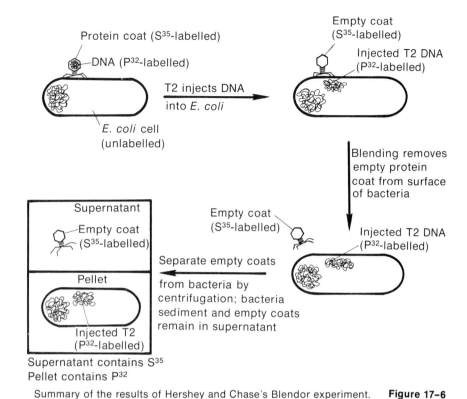

Summary of the results of Hershey and Chase's Blendor experiment. **Figure 17-6**

THE THREE-DIMENSIONAL STRUCTURE OF DNA

You will remember that DNA is composed of polynucleotide chains formed from nucleotides joined by phosphodiester linkages between the 3′ and 5′ carbon atoms of adjacent sugar residues (see Figure 7–35). The nucleotides in DNA are normally of four types, in which the nitrogen base is either adenine, guanine, cytosine, or thymine (See Figure 7–33).

The three-dimensional structure of the polynucleotide chains in DNA was elucidated in 1953 by James Watson* and Francis Crick. They analyzed x-ray diffraction photographs taken by Rosalind Franklin and Maurice Wilkins of DNA crystals and deduced a three-dimensional structure for DNA that was consistent with the observed diffraction patterns. Watson and Crick proposed that the DNA molecule is formed of two polynucleotide chains running in opposite directions and wound around one another in the form of a double helix (Fig. 17–7). The phosphoric acid and deoxyribose units, which

---

*Watson has written an extremely entertaining book entitled *The Double Helix*, which gives an account of the events leading to the discovery of the structure of DNA. A somewhat different view of the same events is presented in *Rosalind Franklin and DNA* by Anne Sayre.

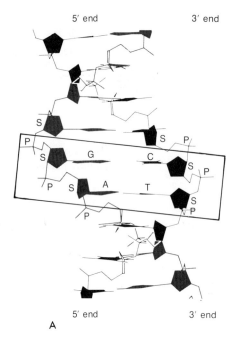

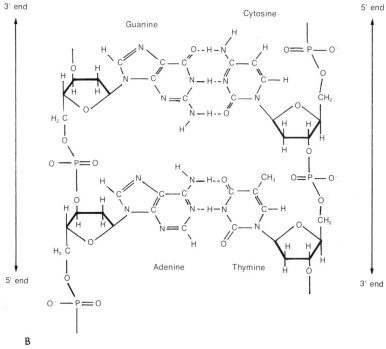

Figure 17-7 (a) A model of the three-dimensional structure of DNA. Two polynucleotide chains are wound around each other in the form of a double helix. Deoxyribose sugar (S) and phosphoric acid (P) units form the backbone of each chain; the purine and pyrimidine bases (A, G, C and T) lie perpendicular to the axis of the helix. (b) Primary structure of a portion of the DNA double helix. Notice that the two polynucleotide chains run in opposite directions, and that guanine is hydrogen bonded to cytosine, and adenine is hydrogen bonded to thymine. (a) From Stryer, L. (1975) *Biochemistry,* San Francisco, W.H. Freeman and Company. © 1975 by W.H. Freeman and Company.

form the backbone of the polynucleotide chain, are located on the outside of the helix; the purine and pyrimidine bases are on the inside and lie perpendicular to the axis of the helix. The double helical structure is stabilized by specific hydrogen bonding between the nitrogenous bases in the polynucleotide chains. With rare exceptions, adenine always hydrogen bonds with thymine, and guanine always hydrogen bonds with cytosine (see Figure 8–11). Because the adenine-thymine base pair is exactly as wide as the guanine-cytosine pair, this specific pattern of hydrogen bonding permits the polynucleotide chains in DNA to form a double helix of regular dimensions, with a diameter of 2.0 nm.

CELL DIVISION IN PROKARYOTES

Cell division in prokaryotes involves two aspects: (1) replication and division of cellular DNA so that each daughter cell receives a complete copy of the hereditary material of the original parent cell, and (2) cytokinesis (cyto = cell; kinesis = division), or the division of the cytoplasm into two approximately equal parts.

REPLICATION OF DNA IN PROKARYOTES

The hereditary material in prokaryotes usually consists of a single, very long molecule of DNA.* In some cases, the DNA has been shown to be circular, in the sense that it has no free ends. *Escherichia coli*, for example, possesses a circular DNA molecule with a molecular weight of 2.3×10^9 daltons. This corresponds to a contour length of 1.2 mm (1200 μm), which is extremely long considering that the *E. coli* cell is normally only 2 to 3 μm in length!

Watson and Crick recognized that the double helical structure that they proposed for DNA (and which has since been verified) suggests a simple mechanism for DNA replication. Because of the specific base pairing that occurs between the two polynucleotide strands of DNA, one strand is, in a sense, the complement of the other. That is, the sequence of bases on one strand determines the sequence of bases on the other strand, simply because only certain base pairings are allowed. For example, if one polynucleotide strand in a DNA molecule consists exclusively of deoxyadenylic acid residues linked together, then we know that the other strand consists exclusively of thymidylic acid residues. It was this characteristic

---

*In some bacteria, small molecules of DNA are present in the cytoplasm in addition to the large DNA molecule. These smaller pieces of DNA, called *cytoplasmic DNA* or *plasmids,* determine some specific characteristics such as resistance to drugs, the ability to synthesize certain antibiotics, and so on. The presence of cytoplasmic DNA in a bacterial cell is, however, the exception rather than the rule.

base pairing that Watson and Crick believed could account for the duplication of the DNA molecule:

> We imagine that prior to duplication the hydrogen bonds are broken, and the two chains unwind and separate. Each chain then acts as a template for the formation on to itself of a new companion chain, where we only had one before. Moreover, the sequence of the pairs of bases will have duplicated exactly.

Watson and Crick's hypothesis of DNA replication could be tested experimentally, since it would result in DNA molecules that are composed of one polynucleotide strand from the original parental DNA and a complementary, newly synthesized polynucleotide strand. This type of replication is termed *semiconservative,* since one of two parental strands is conserved in each replicated DNA molecule (Fig. 17-8a). Alternatively, it is possible that DNA replication is *conservative,* in which case the two newly synthesized polynucleotide strands, one the complement of one of the parental strands and one the complement of the other, unite to form a new DNA molecule. Conservative replication would thus result in the conservation of the original parental molecule and the synthesis of an entirely new molecule (Fig. 17-8b).*

These two alternatives were tested in an experiment designed by Matthew Meselson and Frank Stahl in 1958. Meselson and Stahl used a method which allowed them to distinguish newly synthesized

---

*Other types of replication in addition to semiconservative and conservative can be envisioned.

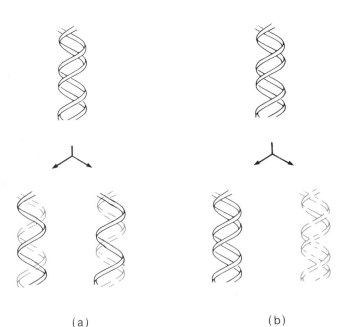

(a)   (b)

Figure 17-8   Two theoretical possibilities for the replication of DNA. Dashed lines represent newly synthesized polynucleotide chains. (a) Semiconservative and (b) conservative replication.

Cell Division in Prokaryotes

polynucleotide strands from existing (parental) ones. It depended on the fact that polynucleotide strands synthesized by bacteria grown in a medium containing nitrogen as N^{14} (the common isotope of nitrogen) possess a lighter density than polynucleotide strands synthesized by bacteria in a medium containing nitrogen in the form of N^{15}, the heavy isotope of nitrogen. (Nitrogen in the medium is incorporated into nitrogen-containing compounds, including the purine and pyrimidine bases of DNA.)

In their experiment, Meselson and Stahl initially grew *E. coli* in a medium containing N^{15} (in the form of $N^{15}H_4Cl$) as the sole nitrogen source. After many generations of growth, all of the *E. coli* DNA was labeled uniformly with N^{15}. The cells were then transferred to a growth medium containing nitrogen in the form of N^{14}. At intervals during the subsequent growth period, samples of cells were removed, and their DNA was extracted and purified.

The density of DNA in each of the purified DNA samples was determined by a technique known as *cesium chloride density gradient equilibrium sedimentation*. In this technique, purified DNA, dissolved in an aqueous solution of 8.8 M cesium chloride (CsCl), is centrifuged in a transparent quartz tube at extremely high speeds. CsCl is a very dense salt, and the centrifugal force will bring about a redistribution of CsCl in the aqueous solution. In fact, a gradient of CsCl concentration will be established in the centrifuge tube, with the highest concentration of CsCl in the solution near the bottom of the tube and the lowest concentration of CsCl in the solution at the top of the tube. Because of the high density of CsCl, the density of the solution at the bottom of the tube will be very high, and the density of the solution will gradually decrease toward the top of the tube, forming a density gradient. DNA molecules centrifuged along with the CsCl solution will congregate in a region of the density gradient that is equal to their own density, and they will form a sharp band. By using ultraviolet light to photograph the quartz centrifuge tube during centrifugation, it is possible to detect the exact position of the DNA band in the density gradient, since DNA absorbs ultraviolet light and CsCl does not (Fig. 17–9). Once the position of the DNA band is known, the density of the DNA molecules comprising the band can be calculated.

The results of Meselson and Stahl's experiment showed a single band of DNA in cells that had been transferred from a medium containing N^{15} and grown for one generation in an N^{14}-containing medium. The density of this band was exactly halfway between the density expected for DNA fully labeled with N^{15} and for DNA fully labeled with N^{14} (Fig. 17–10 *a, b,* and *c*). After two generations of growth in the N^{14}-containing medium, two DNA bands of equal intensities (and hence representing equal amounts) were obtained in a CsCl gradient. One of these bands had a density similar to that expected for DNA fully labeled with N^{14}; the density of the other band was similar to that obtained for DNA after only one generation of growth (Fig. 17–10*d*). These results and those of related experiments proved conclusively that DNA replication in *E. coli* is semiconservative. The

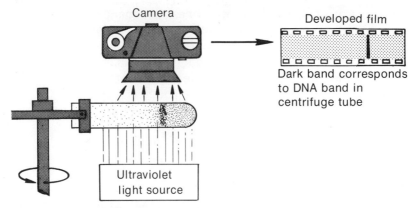

Figure 17-9 Diagram illustrating cesium chloride density gradient equilibrium centrifugation. See text for details.

parental DNA is not conserved intact during replication. Rather, one strand of DNA is always derived from the original parental DNA molecule and the other strand is its newly synthesized complement.

Subsequent research has shown that the DNA of other prokaryotes and of bacterial viruses is also replicated in a semiconservative fashion. In no case has DNA replication been found to occur by an alternative mechanism.

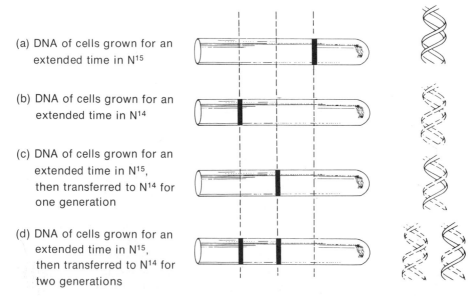

(a) DNA of cells grown for an extended time in N^{15}

(b) DNA of cells grown for an extended time in N^{14}

(c) DNA of cells grown for an extended time in N^{15}, then transferred to N^{14} for one generation

(d) DNA of cells grown for an extended time in N^{15}, then transferred to N^{14} for two generations

Figure 17-10 Summary of Meselson and Stahl's experiment demonstrating semiconservative DNA replication. The DNA helices at the right of the diagram show the composition of the DNA comprising each band; solid helical lines represent N^{15} label, and dotted lines represent N^{14} label.

Cell Division in Prokaryotes

MOLECULAR EVENTS IN DNA REPLICATION IN PROKARYOTES

DNA Polymerase Enzymes

In 1955, Arthur Kornberg discovered an enzyme that is capable of catalyzing the replication of DNA. The enzyme, now called DNA polymerase I, catalyzes the formation of a new polynucleotide strand complementary to a template strand by adding nucleotides sequentially to the 3'-hydroxyl end of an existing "primer" polynucleotide strand. In order to be added to a polynucleotide chain, the nucleotide must exist in the form of the nucleoside–5'–triphosphate; the terminal two phosphate groups are then hydrolyzed from the triphosphate at the time the nucleotide is added to the chain (Fig. 17–11).

The overall reaction catalyzed by DNA polymerase I may thus be summarized as follows

$$\left(\begin{array}{c}\text{"Primer"}\\ \text{polynucleotide}\end{array}\right)_n \text{residues} + \begin{array}{c}\text{dATP}\\ \text{dGTP}\\ \text{dCTP}\\ \text{TTP}\end{array} \xrightarrow[\text{Mg}^{++}]{\text{DNA polymerase I}} \left(\begin{array}{c}\text{"Primer"}\\ \text{polynucleotide}\end{array}\right)_{n+1}$$

$$+ \; {}^-\text{O}-\underset{\underset{\text{O}^-}{|}}{\overset{\overset{\text{O}}{\|}}{\text{P}}}-\text{O}-\underset{\underset{\text{O}^-}{|}}{\overset{\overset{\text{O}}{\|}}{\text{P}}}-\text{O}^- \qquad \Delta G^\circ = +0.5 \text{ kcal/mole} \quad (1)$$

(Pyrophosphate)

$${}^-\text{O}-\underset{\underset{\text{O}^-}{|}}{\overset{\overset{\text{O}}{\|}}{\text{P}}}-\text{O}-\underset{\underset{\text{O}^-}{|}}{\overset{\overset{\text{O}}{\|}}{\text{P}}}-\text{O}^- + \text{H}_2\text{O} \rightarrow 2 \; {}^-\text{O}-\underset{\underset{\text{O}^-}{|}}{\overset{\overset{\text{O}}{\|}}{\text{P}}}-\text{O}^- \qquad \Delta G^\circ = -8 \text{ kcal/mole} \quad (2)$$

Although the polymerization reaction (Step 1) is endergonic, it is driven by the exergonic hydrolysis of pyrophosphate (Step 2). Hence, hydrolysis of the pyrophosphate, derived from the nucleoside triphosphate, provides the energy that drives the reaction. This is why the nucleoside triphosphates are required for the polymerization of nucleotides into polynucleotide chains.

One important characteristic of DNA polymerase I is that it catalyzes the addition of nucleotides *only* to the 3'-hydroxyl end of an existing polynucleotide chain; that is, the elongation of the polynucleotide strand occurs in the 5' to 3' direction (5' → 3'). Also note that DNA polymerase I is template directed—it will only catalyze the addition of a nucleotide whose base is complementary to the base on the template strand (Fig. 17–11).

Under certain conditions, DNA polymerase I is also able to

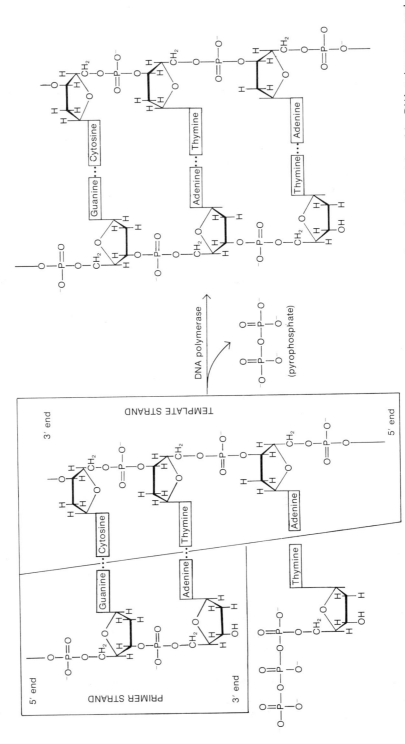

Figure 17-11 The synthesis of a DNA polynucleotide strand complementary to an existing template strand is catalyzed by DNA polymerase, and occurs by the addition of a nucleotide (initially in the form of a nucleoside triphosphate) to the 3'-hydroxyl end of a primer strand.

Cell Division in Prokaryotes

catalyze the *hydrolysis* of DNA progressively from the 3'-hydroxyl end of the DNA, thus removing nucleotides one by one. This activity is called 3' → 5' *exonuclease* (exo = outside, end) activity. A nucleotide may be removed in this way only if it is unpaired and has a free 3'-hydroxyl group. Experiments have also demonstrated that the hydrolytic activity of DNA polymerase I is limited primarily to mismatched bases and that there is little or no hydrolysis of a terminal nucleotide if it is properly matched. Hence it appears that the 3' → 5' exonuclease activity of DNA polymerase I has an important function in the prevention of errors during the replication process. Normally, only the proper nucleotide—the complement of the nucleotide on the template strand—is added to the existing chain, but if a mismatched nucleotide is added by mistake, it will be removed before the next nucleotide is added. This high specificity in the action of DNA polymerase I helps ensure the accuracy of DNA replication.

Since the discovery of DNA polymerase I, two other DNA polymerases (DNA polymerases II and III) have been discovered in *E. coli*. These enzymes, like DNA polymerase I, catalyze a template-directed synthesis of DNA from deoxyribonucleoside triphosphate precursors, adding nucleotides only to the free 3'-hydroxyl end of an existing polynucleotide chain. They also possess 3' → 5' exonuclease activity by which unpaired nucleotides are hydrolyzed from the 3' terminus. DNA polymerases II and III prefer a template of double-stranded DNA with short gaps; polymerase I, on the other hand, works optimally when the DNA template has extensive single-stranded regions. The specific functions of each of the polymerases is not known, however.

Mechanism of DNA Replication

How does DNA replication actually occur? Numerous experiments with *E. coli* have demonstrated that replication begins at a unique site on the circular DNA of *E. coli* and proceeds simultaneously in two opposite directions. Although the resolution of autoradiographic* and electron microscopic techniques used to observe DNA is too low to demonstrate single-strand breaks in the polynucleotide chains, it is nevertheless clear that the DNA of *E. coli* maintains its circular form while being replicated. Furthermore, long stretches of single-stranded areas in the DNA molecule are absent

---

*In *autoradiography* a living cell is incubated with a radioactive compound which will be incorporated specifically into a particular molecule. (For example, radioactive thymidine is incorporated into DNA.) The cell, or fragments of it, is then placed near a photographic emulsion in the dark. The high-energy particles emitted from the decaying radioisotope will expose the emulsion when they strike it, producing small dots that are visible when the emulsion is developed. The net result is to pinpoint the areas of a cell where the radioisotope has been incorporated. Although the resolution of autoradiography is low, it is a valuable technique because of its high specificity.

during replication. The circular nature of replicating DNA and the lack of single-stranded stretches demonstrates that parental DNA does not unwind completely before serving as a template. Instead, the DNA appears to unwind in small areas prior to replication, and then two strands, one of which is complementary to each of the parental strands, are synthesized. The sites where DNA is unwinding

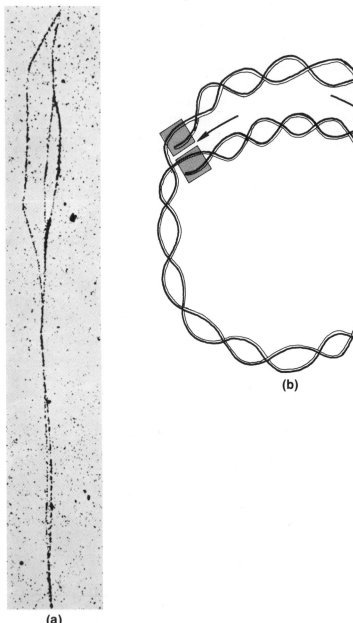

Figure 17–12 (a) Replicating DNA of *E. coli* as seen by autoradiography. This configuration, called a theta structure, is formed by the replication of a circular DNA molecule; replication begins at one point and proceeds in two opposite directions. (From Stryer, L. [1975], *Biochemistry*, San Francisco, Courtesy of Dr. John Cairns. © 1975 by W.H. Freeman and Company, p. 583.) (b) A diagrammatic model of a theta structure, showing replicating forks (arrows) and position of DNA polymerase enzymes (boxes).

Cell Division in Prokaryotes

and new synthesis is occurring are called *replicating forks*. Because the *E. coli* DNA molecule remains circular during replication, replicating DNA takes on the appearance of a large circle (the parental DNA) with an inner loop (newly synthesized DNA). Such replicating structures are called *theta structures* because of their resemblance to the Greek letter θ (Fig. 17–12).

The unwinding of DNA during replication is apparently a complex process that is facilitated by several enzymes and other proteins. It probably requires an *endonuclease* (endo = within) that is capable of making internal single strand breaks in at least one template strand and hence allows unraveling of the DNA without requiring the whole DNA molecule to twist. The opening of local regions following the production of single-strand breaks is apparently aided by specific *unwinding proteins,* which bind preferentially to single-stranded DNA and bring about the separation of the strands. Indeed, the binding of one unwinding protein to a DNA strand apparently promotes the binding of a second, so that the helix continues to unwind. Following replication, the break in a polynucleotide strand is repaired with the help of another enzyme, termed *DNA ligase,* which is capable of joining two strands end to end by catalyzing the formation of a phosphodiester bond between them (Fig. 17–13). This repair reaction is endergonic and requires an energy source in one form or another.

We have yet to explain one paradox in the process of DNA replication, related to the direction of DNA synthesis. We noted that all three polymerases synthesize DNA in the $5' \rightarrow 3'$ direction; we also mentioned earlier that the polynucleotide strands in a DNA double helix run in opposite directions. Hence, at a replicating fork, the *overall* direction of DNA synthesis must be $5' \rightarrow 3'$ for one daughter strand, but $3' \rightarrow 5'$ for the other (Fig. 17–14). Since we know of no enzyme that polymerizes DNA in the $3' \rightarrow 5'$ direction, how is it possible for one of the daughter strands to grow in the $3' \rightarrow 5'$ direction?

Work by Reiji Okazaki has helped to answer this question. Okazaki has shown that newly synthesized polynucleotides in the vicinity of replicating forks consist of small fragments about 1000 nucleotides long. These fragments, known as *Okazaki fragments,* become joined as replication proceeds. Thus, all fragments may be synthesized in the $5' \rightarrow 3'$ direction, and newly synthesized fragments base-paired to the same template strand are then joined end to end by a DNA ligase. This mechanism would mean that both strands are, in fact, synthesized in the $5' \rightarrow 3'$ direction; one of these strands only *appears* to be synthesized in the other direction (Fig. 17–15).

Initiation of DNA Synthesis

How does DNA synthesis begin? We have mentioned that all the known polymerase enzymes require a polynucleotide with a free 3'-hydroxyl group to serve as a primer to which nucleotides are added.

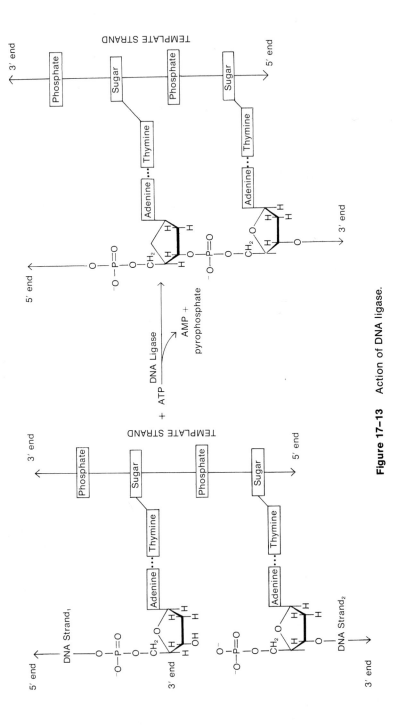

Figure 17-13 Action of DNA ligase.

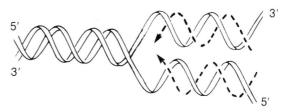

The *apparent* direction of DNA synthesis at a replicating fork.   **Figure 17-14**

What serves as the primer molecule in vivo is not certain. However, recent experiments have shown that RNA synthesis is required for the initiation of DNA synthesis and that a short piece of RNA is covalently bonded to one end of newly synthesized polynucleotide fragments of DNA. These results suggest that RNA is the primer for DNA synthesis, and the following mechanism has been proposed to explain how this might occur: An *RNA polymerase,* an enzyme which does not require a primer for the initiation of synthesis, catalyzes the synthesis of a short polynucleotide fragment of *RNA* (see Figure 7–35) that is complementary to one of the polynucleotide strands in DNA. The free 3'-hydroxyl group of the RNA then serves as a primer for the DNA polymerase-catalyzed elongation of the strand. It has been proposed that DNA polymerase III is the enzyme involved in this initial DNA synthesis. The RNA fragment is later hydrolyzed from the newly synthesized polynucleotide strand,

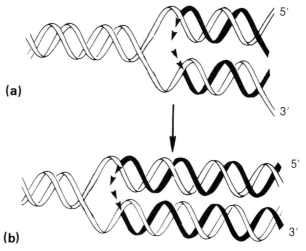

(a) At replicating forks, DNA is probably synthesized in small fragments (arrows). Fragments base paired to the same template strand are then joined end-to-end by DNA ligase, as shown in (b). Thus, DNA synthesis proceeds in the 5' to 3' direction for both strands.   **Figure 17-15**

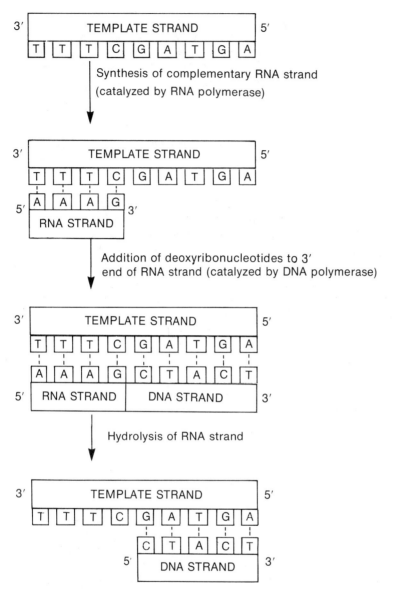

Figure 17-16 Postulated mechanism for the initiation of DNA synthesis by the use of an RNA primer strand.

leaving large single-stranded gaps which are filled in with the help of DNA polymerase I. The polynucleotide strand filling in the gap is then joined end to end with the polynucleotide strands on either side of the gap. This latter reaction is catalyzed by DNA ligase. The entire process is summarized in Figure 17-16.

CYTOKINESIS IN PROKARYOTES

Considerable evidence shows that cytokinesis in prokaryotes does not begin until a round of DNA replication is completed. In addition, cytokinesis must be preceded by growth of the cell; if it were not, it would result in cells of increasingly smaller size. Growth, in turn, requires the synthesis of structural proteins and of enzymes needed for the production of various cellular components. However, the required structural proteins and enzymes must be synthesized in appropriate amounts, so that the accumulation of various cellular materials occurs at a balanced rate. In other words, growth must be controlled. Because instructions for the synthesis of structural proteins and enzymes are contained in the cell's hereditary material, it is clear that balanced growth requires the regulation of gene expression—a topic discussed in Chapter 18.

Cytokinesis in prokaryotes first becomes evident with the inward growth of the cell wall and plasma membrane, forming a structure known as a *septum* (Fig. 17–17). In bacteria that possess mesosomes, septum growth often occurs at the site where a mesosome arises from the plasma membrane, and the cell's DNA is attached to the mesosome. In bacteria that do not regularly form mesosomes, the DNA is attached to a specific site on the plasma membrane, and septum formation appears to occur at or near this site. In both cases, the attachment of DNA to membranes is apparently important in the distribution of DNA to daughter cells: the two daughter DNA molecules are separated as new plasma membrane is synthesized between them.

As the cell wall and plasma membrane continue to grow inward, a complete septum is eventually formed, dividing the cell into two roughly equal parts. The septum then splits longitudinally in half, resulting in the separation of the two daughter cells. In some organisms, septum cleavage regularly does not occur, resulting in the formation of long chains of cells, as seen, for example, in certain filamentous strains of blue-green algae.

CONTROL OF CELL DIVISION IN PROKARYOTES

If cytokinesis is triggered by the completion of a round of DNA replication, then the control of cell division would appear to lie in those events that initiate DNA replication.

The initiation of DNA replication appears to occur when the ratio of cell weight to DNA content reaches a critical value. This observation has been interpreted as implying that the concentration of some cytoplasmic factor is important in the initiation of DNA replication. Some workers have postulated that a substance which stimulates initiation may have to accumulate in the cytoplasm; others contend that the increase in cell mass dilutes a substance which, at high concentrations, inhibits the initiation of DNA replication. There is some evidence supporting the former conclusion. *E. coli* mutants have been isolated that are unable to initiate DNA replication when grown at relatively high temperatures (at which they are viable) but are normal in this respect when grown at low temperatures. Such

Figure 17-17 Septum formation in the blue-green alga *Anacystis nidulans* (From Allen, M.M. [1968], J. Bact. *96*:845.)

sensitivity to temperature is known to be due to an altered protein which is inactivated at high temperatures (for example, 40° C) but is active at low temperatures (for example, 25° C). Consequently, the existence of mutants that display temperature sensitivity in the initiation of DNA replication indicates that one or more proteins are required for initiation.

It is obvious that we have only a vague outline of the events controlling cell division in prokaryotes. However, the field is currently a particularly active one and rapid progress is being made.

CELL DIVISION IN EUKARYOTES

Cell division in eukaryotes is similar in principle to that in prokaryotes, although it is more complex. It involves (1) replication of the hereditary material and division of the nucleus (mitosis), and (2) cytokinesis. Before we discuss the first of these topics, we need to know something about the composition and structure of chromatin, the DNA-protein complex in eukaryotes.

COMPOSITION OF EUKARYOTIC CHROMATIN

The DNA in eukaryotes is contained in a number of separate molecules which, together with a large amount of protein and a small amount of RNA (sometimes referred to as chromosomal RNA), form the *chromatin strands* of the interphase nucleus (the nucleus of a cell not in the process of division).

Proteins of Chromatin

Histones. The proteins in chromatin consist primarily of small, basic proteins called *histones*. In most eukaryotic cells, there are five classes of histones,[*] which can be separated by ion-exchange chromatography on the basis of their charge. All histones consist of a single polypeptide chain and have molecular weights between about 11,000 and 21,000, depending upon the particular histone. In addition, histones have a high content of the basic amino acids lysine and arginine.

The function of histones is not clearly understood, but they have been implicated in two processes. First, it seems quite likely that they aid in the packing of DNA within the nucleus. The basic amino acids of histones are positively charged at cellular pH and would act to neutralize the repulsive forces created by the negative charges on the phosphate groups of DNA. This neutralization of charge would

---

[*]A sixth histone class is present in amphibian and avian red blood cells.

presumably allow DNA molecules to pack closely—an important requirement, since a very large amount of DNA must be able to pack within the restricted volume of the nucleus.

Second, there is experimental evidence that the binding of histones to DNA serves to prevent the use (expression) of hereditary information. We know that all the hereditary information present in any one cell is not expressed at all times; instead, at any one moment, only a very small proportion of the total information is expressed. For example, all cells of a human being possess the information that determines the development of the eye, but this information is expressed only during a short period of embryonic life, and then only by cells that are involved in eye development. Somehow, then, there must be mechanisms that regulate gene expression so that at any given time the expression of most of the hereditary information is repressed while the expression of other information is induced.

Because of the similarity of histones in all eukaryotic cells, it does not seem likely that histones can play a specific role in the regulation of gene expression. Instead, histones appear to be *nonspecific* repressors of gene expression; that is, the binding of histones to DNA appears to prevent the expression of information, no matter what that information may be.

Whatever the function of histones, it seems clear that it is a critical one, for the amino acid sequences of the histones have changed very little during evolution. For example, the amino acid sequence of one histone—histone IV—of pea seedlings differs at only 2 out of 102 amino acid sites from the histone IV of calf thymus. Thus, only 2 per cent (2/102) of the sequence of histone IV has changed during the estimated 1200 million years since plants and animals diverged. This is a remarkable constancy of structure. In comparison, the sequence of cytochrome *c* has changed by 2 per cent every 40 million years, and hemoglobin has changed by the same amount every 12 million years.

Other histones are not as invariable from species to species as is histone IV, but the conservation of their amino acid sequences throughout evolution is nevertheless much greater than that of other well-investigated proteins. Because the amino acid sequences of histones are so invariable, it has been assumed that a precise sequence is important to the function of histones and that their amino acid sequences cannot be altered substantially without altering their functions.

Nonhistone Proteins

In addition to the histone proteins, there are nonhistone proteins that are also components of chromatin. Unlike histones, most of the nonhistone proteins are acidic, and they vary qualitatively in different cell types of the same organism. A number of experiments have shown that nonhistone proteins are complexed to areas of DNA whose information is being expressed, and hence it has been suggested that nonhistone proteins, along with chromosomal RNA which also binds to certain active portions of DNA, may somehow be involved in the specific control of gene expression. However, if nonhistone proteins do regulate gene expression, we do not know how this occurs.

DNA of Chromatin

Evidence is accumulating that the DNA in a chromatin strand consists of a single molecule of double-helical DNA. Bruno Zimm used a sensitive physical method to calculate the length of DNA molecules in *Drosophila* nuclei disrupted by gentle lysis so as to minimize breakage. He determined that the molecular weight of the largest molecule in the nuclear lysate was $41 \pm 3 \times 10^9$. If the known DNA content of the largest chromosome in *Drosophila* were present in a single molecule, it would have a molecular weight of 43×10^9 — a value which agrees closely with the molecular weight that Zimm determined for the largest DNA molecule he obtained. Zimm also observed *Drosophila* DNA by autoradiography and showed that it consists of linear, unbranched molecules.

Repetitive DNA

A number of experiments have revealed that the DNA of most eukaryotic cells, if not all, contains repeated sequences (repetitive DNA), whereas the DNA of prokaryotes does not. The ratio of repetitive to nonrepetitive DNA varies from one species to another, but it can be very high. It has been reported, for example, that 30 per cent of human DNA consists of sequences repeated at least twenty times.

One of the best known examples of repetitive DNA is that of genes coding for 28s, 18s, and 5s ribosomal RNA (rRNA) molecules in *Xenopus laevis*, the African clawed toad. The genes for 18s and 28s rRNA are located in the nucleolus and are tandemly repeated about 450 times (Fig. 17–18). The repeating unit consists of (1) a gene determining the synthesis of a 40s precursor RNA, which is later cleaved within the cell to yield 28s and 18s ribosomal RNA, and (2) a spacer region of about 5000 base pairs, whose function is not known (Fig. 17–19). Genes coding for 5s rRNA are tandemly repeated in clusters containing 100 to 1000 repetitive gene sequences at the ends of most of the 18 chromosomes of *Xenopus*. There is also a spacer region between each of the 5s rRNA genes.

During the maturation of egg cells (oocytes) in *Xenopus*, the number of nucleoli in each oocyte increases by about a thousand-fold and so do the genes determining 18s and 28s rRNA. Indeed, the genes for these ribosomal RNA's form about 75 per cent of the oocyte DNA. The replication of 18s and 28s genes during oocyte

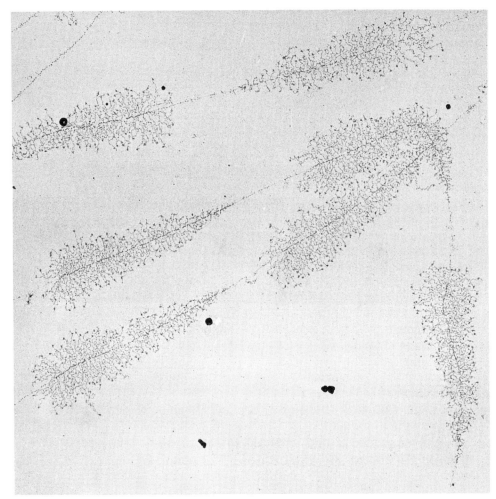

Figure 17–18 Electron micrograph of nucleolar DNA in the oocyte of *Triturus viridescens*. The units shaped like arrowheads consist of an axial fiber of DNA, corresponding to the 40s rRNA precursor gene; the lateral fibers are newly-synthesized 40s RNA. Numerous RNA molecules are sequentially initiated at the tip of each arrowhead before synthesis of the first RNA molecule is completed, and consequently RNA molecules are visible at all stages of completion. The portion of the DNA fiber between arrowhead regions corresponds to spacer regions. (From Miller, O.L. and Beatty, B.R. [1969], Science *164*:956. Copyright © 1969 by The American Association for the Advancement of Science.)

maturation is termed *gene amplification* and is apparently one mechanism by which cells can produce additional amounts of a needed product. In this case, the oocyte is able to produce large amounts of ribosomes needed for protein synthesis after fertilization (during embryonic development).

This is not to say that all genes are repeated, and in fact it is known that there is only one copy of many genes. For example, it has been shown that there is but one copy of the gene that determines silk fiber production in the silkworm, *Bombyx mori,* although the larval stage of this organism synthesizes large amounts of silk. This suggests that large amounts of a single gene product can be

Cell Division in Eukaryotes

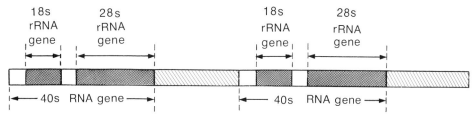

Diagram showing the organization of genes determining the 40s RNA precursor of 28s and 18s ribosomal RNAs.

Figure 17-19

synthesized by an alternative mechanism that does not involve tandem repetition of genes or gene amplification. We will return to this point in Chapter 18.

CHROMATIN ORGANIZATION

Our ideas about the organization of chromatin are presently in a state of flux. A variety of models describing the organization of chromatin have been proposed, but there is as yet no clear consensus as to which of these most clearly describes intranuclear chromatin.

Hans Ris and coworkers have examined densely packed chromatin (heterochromatin) by electron microscopy, and they contend that *in the native state* it consists of fibers 20 to 30 nm thick (Fig. 17-20). In the presence of chelating agents that bind metal ions,

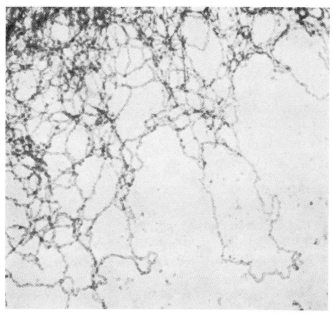

Electron micrograph of chromatin fibers, approximately 20 nm thick, obtained from frog erythrocyte nuclei (×36,000). (From Ris, H. [1975]. *The Structure and Function of Chromatin*, Ciba Foundation Symposium 28 [new series], Elsevier, p. 12.)

Figure 17-20

these fibers unravel into smaller fibers with diameters of about 10 nm. If this latter fiber is treated with urea, which disrupts noncovalent interactions (presumably, in this case, between histones), the fiber unravels into a unit fibril 2 to 4 nm thick, which Ris contends represents the single DNA double helix with histones wrapped around it.

Although Ris and others visualize the 20 to 30 nm fiber as the intranuclear form of chromatin, some workers argue that native (intranuclear) chromatin consists of 10 nm fibers and that 20 to 30 nm fibers are, for the most part, artifacts of preparation.

Recent work published by Ada Olins and Donald Olins has provided yet another model of chromatin organization. They suggest that native chromatin in rat thymus, rat liver, and chicken erythrocytes does not consist of supercoiled fibers of regular diameter but instead resembles a "string of beads." The "string of beads" structure, which they have observed in electron micrographs of gently lysed nuclei, consists of spherical particles (called ν-bodies or nucleosomes) about 7 nm in diameter, connected by thin filaments, about 1.5 to 2.0 nm thick (Fig. 17–21). Subsequent work has suggested that the thin filament consists of DNA thinly covered by histone and that the "beads" consist of about 190 base pairs of DNA supercoiled in some way with 8 histone molecules. Additional supercoiling of this basic "string of beads" fiber might give rise to a folded or helical packing of ν-bodies.

As yet, we do not know which, if any, of these models accurately describes the structure of intranuclear chromatin. It is currently fashionable to support the "string of beads" structure, but this preference may well change as future discoveries are made.

CHROMATIN STRUCTURE AND GENE EXPRESSION

There is evidence that the densely staining heterochromatin in interphase nuclei is genetically inactive and that the more lightly staining euchromatin represents areas of chromatin whose information is being expressed. However, although heterochromatin is presumably more tightly coiled than euchromatin, it is not at all clear

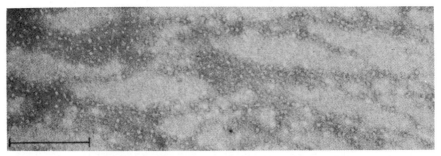

Figure 17–21 Electron micrograph showing "string of beads" structure of chromatin. ν-bodies and connecting strands are evident. (From Olins, A.L. and D.E. Olins [1974], Science *183*:330. Copyright © 1974 by The American Association for the Advancement of Science.)

Cell Division in Eukaryotes

how the various models of chromatin structure relate to the structure of heterochromatin and euchromatin in the interphase nucleus.

CHROMOSOME ORGANIZATION

During cell division the chromatin strands in the nucleus condense by coiling and folding in a predictable manner to form compact structures called *chromosomes*. Because of its specific folding pattern, each chromosome has a characteristic morphology when observed under the light microscope. It is important to remember that the *information content* (the DNA) of each chromosome is identical to that of the chromatin strand from which it is derived by coiling, although the proteins associated with the two forms may vary to some small extent. In a general sense, then, chromosomes are merely condensed chromatin strands, and the difference between the two is primarily one of three-dimensional configuration and not one of composition.

When whole chromosomes of the great majority of cells are observed in the electron microscope, they appear not to have any regular folding pattern but to consist of fibers going every which way in

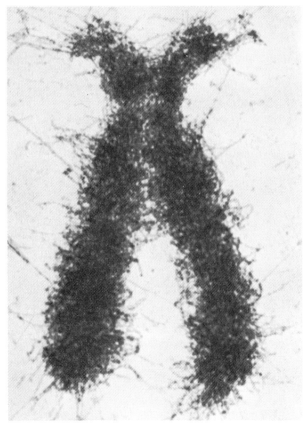

Figure 17-22 Electron micrograph of a whole-mounted human chromosome, which appears to consist of a tangle of fibers approximately 30 nm in diameter. (From Macleod, A.G. [1973], *Cytology*, The Upjohn Company, Kalamazoo, Michigan, p. 101.)

a dense configuration (Fig. 17–22). The irregular, jumbled appearance of most chromosomes in electron micrographs has prompted Hewson Swift to describe chromosome morphology as resembling "something like a bad day at a macaroni factory."

In some preparations, the fibers that form chromosomes appear to be 20 to 30 nm in diameter; in other preparations, the diameter of the fibers appears more variable, with some as small as 2 nm, some as large as 30 nm, and some with intermediate values. David Comings and Tadashi Okada have some evidence suggesting that the fibers forming chromosomes are actually 3 to 5 nm thick within the cell, and that this fiber forms loops and folds when the nucleus is disrupted, resulting in an artificially supercoiled fiber of 20 to 30 nm in diameter. The "string of beads" structure has not been described for the chromatin of chromosomes, and it is not apparent whether the structure of chromatin in the interphase nucleus is similar to the structure of chromatin in chromosomes.

In addition to the usual chromosome structure seen in most dividing cells, there are a few cell types that have chromosomes of unusual configurations. These are discussed in Chapter 18.

REPLICATION OF EUKARYOTIC DNA

Eukaryotic DNA is replicated in a semiconservative fashion, as is prokaryotic DNA. This was first demonstrated in 1957 by J.H. Taylor, who performed an autoradiographic study of dividing root-tip cells in the broad bean, *Vicia faba*. Taylor incubated dividing root-tip cells with radioactive thymidine and then allowed the cells to divide without radioactivity—a protocol similar, in principle, to that used by Meselson and Stahl to demonstrate semiconservative replication in *E. coli*. Although there are complications (which are not important here) in interpreting Taylor's results, he was able to show that upon replication of a chromatin strand, each resulting strand consisted of one parental strand and a newly synthesized strand.*

Electron microscope studies have also shown that eukaryotic DNA is replicated bidirectionally, as is prokaryotic DNA, but from many origins, rather than one (Fig. 17–23). Each of these replicating areas is called a *replicon,* and two complete daughter molecules result when all replicons of the parental molecule have replicated. Apparently, any particular replicon, once activated to replicate, cannot be reactivated until the entire DNA molecule is replicated. Replication from many origins seems to be necessary because of the large size of eukaryotic DNA. It has been calculated, for example, that replication of the DNA in the largest *Drosophila* chromosome would take more than 16 days if there were only one

---

*Although this is an oversimplified interpretation of Taylor's results, it is nevertheless an accurate one.

Mitosis: Nuclear Division in Somatic Cells

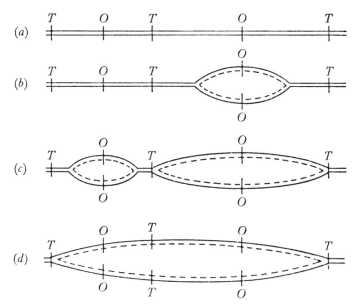

Figure 17-23 Diagram illustrating bidirectional DNA replication from many origins, as occurs in eukaryotes. O and T indicate, respectively, the sites of origin and termination of replication. (From Huberman, J.A. and A.D. Riggs [1968], J. Mol. Biol. 32:340.)

initiation site for replication. In fact, the DNA in this chromosome replicates in about 3 minutes.

DNA polymerases have been isolated from a number of eukaryotic cells of different types, and these polymerases are similar in most respects to DNA polymerases in prokaryotes. For example, nucleotides are added only to the 3′–hydroxyl group of a "primer" polynucleotide strand. Thus, at the molecular level, the replication of eukaryotic DNA is probably very similar to DNA replication in prokaryotes.

MITOSIS: NUCLEAR DIVISION IN SOMATIC CELLS

The distribution of hereditary material during cell division in eukaryotes presents a problem considerably more complex than that faced by prokaryotes. The genetic material of prokaryotes is present as a single molecule of "naked" DNA. The DNA is anchored to an attachment site on the plasma membrane, and once the DNA is replicated, a copy is distributed to each daughter cell by growth of the plasma membrane between the region where the two DNA molecules are attached. This results in the separation of the two DNA molecules into different halves of the dividing cell.

Eukaryotic cells contain much more DNA than do prokaryotic ones. Human cells, for example, have more than one thousand times as much DNA as *E. coli!* Probably because it is present in such large amounts, the total nuclear DNA in eukaryotes is not in the form of a single molecule. Instead, it is broken up into a number of separate molecules which, with attached proteins, form the chromatin strands of the nucleus. During cell division, the chromatin strands condense to form chromosomes. Each chromosome has a characteristic morphology, and cells of each eukaryotic species have a characteristic number of chromosomes. For instance, the somatic cells (all cells other than sperm cells and egg cells) of humans have 46 chromosomes (Fig. 17–24). These 46 chromosomes are actually present as 23 pairs of chromosomes—one chromosome of each pair is derived from the mother (as represented by the egg), and the other member of each pair is derived from the father (as represented by the sperm). The two chromosomes of a pair are called *homologous chromosomes.* Homologous chromosomes are similar in the sense that genes which affect the same traits are present at the same position on each chromosome. Cells which, like human somatic cells, have one pair of each chromosome type are called *diploid* cells (di = two; ploid = in the form of).

When a somatic cell divides, each daughter cell nucleus must receive a copy of each of the chromosomes in the parental cell

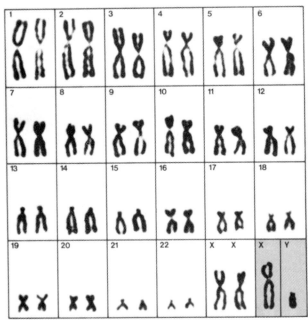

Figure 17–24 Human karyotype. The female has 22 pairs of chromosomes plus a pair of X chromosomes; the male has the same 22 pairs of chromosomes plus one X and one Y chromosome. (From Macleod, A.G. [1973], *Cytology,* The Upjohn Company, Kalamazoo, Michigan, p. 78.)

Mitosis: Nuclear Division in Somatic Cells

nucleus. If it did not receive a complete copy of the genetic information in the parent cell, genetic abnormality would result. Thus, each daughter cell of a human diploid cell also has 23 pairs of chromosomes.

A complex mechanism has evolved in eukaryotes to ensure the even distribution of hereditary material during cell division. The mechanism is called *mitosis*. Since mitosis occurs in a similar way in all eukaryotes, it undoubtedly evolved early in the evolution of eukaryotic organisms. Mitosis is a very efficient process, and errors occur very rarely. The efficiency and accuracy of mitosis is not surprising, considering the central importance of hereditary material to the cell: faulty mitosis is usually selectively disadvantageous.

The definitive study of mitosis in animal cells was published in 1879 by Walther Flemming, who observed the mitotic process in erythroblasts and in the large clear epithelial cells of the salamander. Flemming not only described mitosis in accurate detail but he recognized the central features of the process: prior to nuclear division, each chromosome appears double along its length. Then "the threads [chromosomes] divide themselves in half longitudinally," each half ending up in one of the two daughter nuclei. The end result is that each daughter cell obtains a copy of every chromosome of the parental cell.

Although the electron microscope has allowed us to describe the process of mitosis in greater detail than did Flemming, we still do not know much about the molecular events that initiate and control the mitotic process.

In describing mitosis, it is convenient to divide the process into four phases: *prophase, metaphase, anaphase,* and *telophase*. The period between successive mitoses is called *interphase*. It must be emphasized that the division of mitosis into stages is wholly artificial. Mitosis does not actually occur in discrete steps but is a continuous process, and hence it is not always clear when one stage ends and another begins.

The main events of mitosis are shown in Figure 17–25 and summarized in diagrammatic form in Figure 17–26. Details of mitosis are described below. Although all aspects of mitosis are not absolutely identical in all cell types, the following description nevertheless applies to the great majority of cells.

INTERPHASE

Most DNA, RNA, and protein synthesis required for cell division occurs during the interphase period. Thus interphase is a time of high metabolic activity and not a time of "rest" between cell divisions, as was originally believed.

On the basis of observations of cell division in cultured mammalian cells, interphase has been divided into three stages: G_1 (first gap), S (synthesis), and G_2 (second gap). The G_2 period immediately

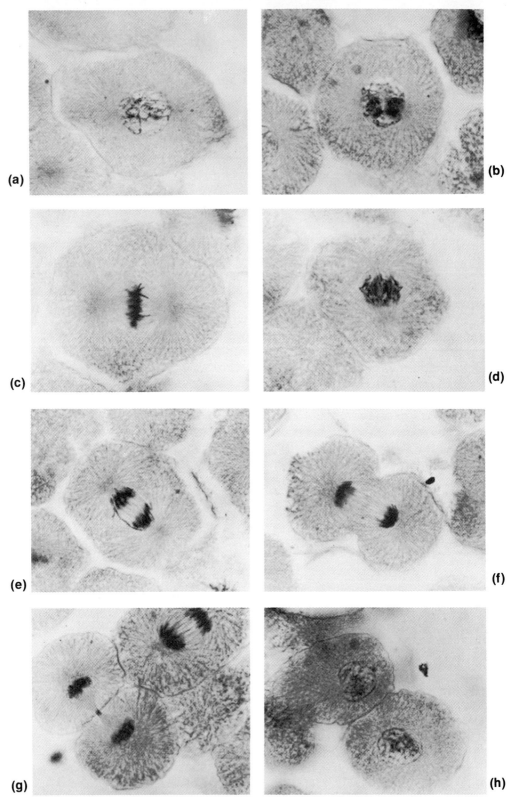

Figure 17-25 Mitosis in the developing whitefish embryo. (a) early prophase, (b) late prophase, (c) metaphase, (d) early anaphase, (e) late anaphase, (f) early telophase, (g) late telophase, and (h) interphase. (From Dowben, R.M. [1969], *General Physiology: A Molecular Approach*, New York, Harper & Row, Inc., p. 547.)

Mitosis: Nuclear Division in Somatic Cells

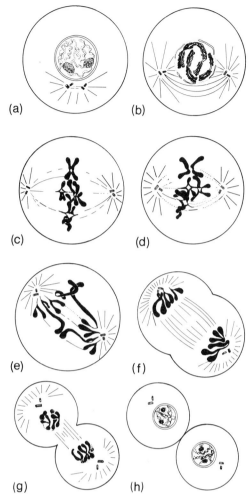

A diagrammatic interpretation of the mitotic events shown in Figure 17–25. (a) Early prophase, (b) late prophase, (c) metaphase, (d) early anaphase, (e) late anaphase, (f) early telophase, (g) late telophase and (h) interphase.

Figure 17–26

precedes mitosis (M) and cytokinesis (D) (see Figure 17–27). RNA synthesis, protein synthesis, and general mass increase of the cell occur during G_1, S, and G_2. DNA synthesis is limited to the S stage, and once a cell enters the S stage, it usually will proceed through the other stages of cell division until it again reaches the G_1 stage. Typically, an animal cell in tissue culture repeats this cycle approximately every 24 hours. The G_1 phase normally lasts about 10 hours, S lasts about 8 hours, G_2 lasts about 5 hours, and M and D together take about 30 minutes. If the division cycle is longer than about 24 hours, cells are commonly arrested in G_1, which consequently becomes prolonged.

Thus, the replication of DNA and the synthesis of histone proteins occur during interphase, resulting in the duplication of the

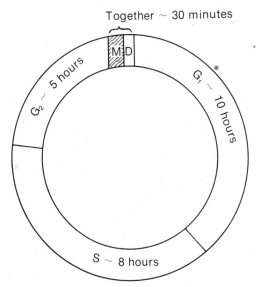

Figure 17-27 The cell division cycle of the "typical" mammalian cell grown in tissue culture.

chromatin strands. During this time, the nucleus shows little internal differentiation; in fact, it was the lack of morphologic change in the interphase nucleus that prompted early microscopists to describe interphase as a "resting" state.

PROPHASE

The nuclear chromatin, which has by now duplicated, begins to condense by coiling and folding, and chromosomes first become visible as very faint threads in the nucleus. (The term mitosis comes from the Greek word *mitos,* meaning "thread".) Shortly after it becomes visible under the light microscope, each chromosome appears to consist of two chromatin strands. Each strand of a chromosome is termed a *chromatid,* and the two identical strands of a single chromosome are called *sister chromatids.* Sister chromatids are attached firmly to each other at a specific point on each chromatid called a *centromere (kinetochore* or *primary constriction).* As condensation continues throughout prophase, the chromosomes become short and thickened, and the centromere of each chromosome is visible as a constriction or poorly staining gap along the length of the chromosome.*

Although the structure of centromeres has not been completely resolved, it is clear that the chromatin strand is continuous through

---

*The centromere is actually double, since each sister chromatid has a centromere. But because the sister chromatids are firmly attached at their centromeres, the centromere region does not appear double.

Mitosis: Nuclear Division in Somatic Cells

the centromere and that there is some protein, as yet uncharacterized, which is also present in the centromere region. In addition, it has been shown that the DNA in the centromeres is composed of highly repetitive DNA sequences. For example, centromeres of mouse chromosomes consist of many thousands of copies of a repeating sequence of about 300 base pairs.

Chromosomes in late prophase may also show characteristic constrictions other than centromeres. These constrictions are called *secondary constrictions.* Indeed, the positions of the centromere and secondary constrictions, along with differences in length and general shape, allow individual chromosomes to be distinguished from one another. The complete set of chromosomes, visible in late prophase or early metaphase, is known as the *karyotype* of the cell. The karyotype of human somatic cells is shown in Figure 17–24.

Another change that occurs during prophase is the degeneration of the nucleolus. The nucleolus becomes progressively smaller as prophase proceeds; by late prophase, it has broken down completely and has disappeared.

The *spindle apparatus,* a system of microtubules that is involved in the separation of sister chromatids, is also assembled during prophase. The formation of the spindle apparatus in animal cells, and in plant cells with motile germ cells, is apparently directed by *centrioles* (see Chapter 11). The interphase nucleus in these cells usually has a single pair of centrioles lying near the nuclear envelope. Another pair of centrioles is assembled in late interphase prior to the beginning of prophase. During prophase, one pair of centrioles migrates to the opposite side of the nucleus, so that the two pairs of centrioles eventually come to lie directly opposite one another. As centriole migration occurs, short microtubules can be seen radiating outward in all directions from each centriole pair, forming starlike structures called *asters* (aster = star). Some microtubules run between the two pairs of centrioles and appear to lengthen as the centriole pairs separate. Consequently, by the time the centrioles have migrated to their final positions, there is a system of microtubules—the spindle apparatus—running between the two pairs of centrioles. The final positions of the centrioles determine the two *poles* of the spindle apparatus.

There is evidence to suggest that most of the protein subunits forming the spindle microtubules are reassembled from subunits already present in the cytoplasm. Many cells undergo a change to a spherical shape during mitosis, and some workers contend that this change in shape is due to the depolymerization of the microtubule "cytoskeleton" (see Chapter 11) into its component protein subunits. The subunits presumably are then repolymerized to form spindle microtubules.

Although centrioles in animal cells thus appear to direct in some way the formation of the spindle apparatus, most plant cells, including cells of all higher plants, do not contain centrioles, and asters do not form. Nevertheless, these organisms form a spindle apparatus and undergo mitosis without difficulty.

Concomitant with the formation of the spindle apparatus is the breakdown of the nuclear envelope, which has disintegrated by late prophase. There is some evidence that the nuclear envelope comes apart in discrete units which may be reassembled during the formation of the envelopes of daughter nuclei.

METAPHASE

After breakdown of the nuclear envelope, the chromosomes migrate to an imaginary plane that lies perpendicular to the middle of the spindle apparatus. This plane is called the *metaphase plate*. By the time the chromosomes reach the metaphase plate, a bundle of four to eight or more spindle microtubules has attached to the centromere of each sister chromatid. This attachment occurs in such a way that one of the centromeres of each chromosome is attached to one pole and the other centromere is attached to the other pole (Fig. 17-28).

Many spindle microtubules do not make connections with chromosomes, remaining extended from pole to pole. The function of these spindle microtubules is not known.

ANAPHASE

During anaphase, the sister chromatids of each metaphase chromosome begin to separate and move to opposite poles of the

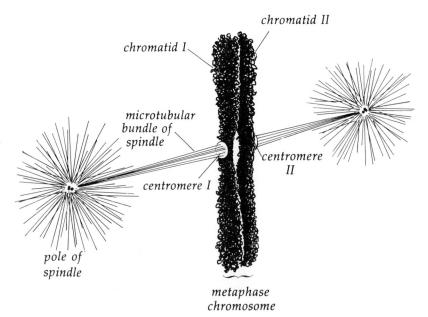

Figure 17-28 Illustration of the attachment of spindle microtubules to metaphase chromosomes during mitosis. (From Wolfe, S.L. [1972], *Biology of the Cell*, Belmont, California, Wadsworth Publishing Company, Inc., p. 299. © 1972 by Wadsworth Publishing Company, Inc. Reprinted by permission of the publisher.)

Mitosis: Nuclear Division in Somatic Cells

spindle, eventually aggregating at the opposite poles. We do not know how the centromeres hold two sister strands together prior to anaphase, nor do we understand the mechanism by which centromeres separate during anaphase.

We also do not know the driving force that causes the chromosomes to move to opposite poles during anaphase. It is thought that microtubules may be directly responsible for chromosome movement, both because of their attachment to chromosomes and because the disruption of the spindle apparatus with antimitotic chemicals or low temperature results in the failure of chromosomes to move to opposite poles.

Several theories have been proposed to explain how microtubules may bring about chromosome movement. One of these suggests that chromosomes move because force is generated by the depolymerization of microtubules into their protein subunits. This theory contends that the shortening of spindle microtubules resulting from the removal of microtubule subunits at the regions of both poles pulls the chromosomes apart. Another theory suggests that a sliding motion between microtubules running between the poles and those attached to chromosomes pulls the chromosomes apart, in a manner similar to the mechanism that has been proposed to account for flagellar motion (see Chapter 10). As yet, there is no indication as to which, if any, of these mechanisms is correct.

TELOPHASE

During telophase, the chromosomes aggregate at the poles. The spindle apparatus is disassembled, and a new nuclear envelope forms around each set of chromosomes. The new nuclear envelopes are apparently assembled, at least in part, from fragments that are derived from the disassembled nuclear envelope of the parent cell. The chromatin mass within each daughter nucleus becomes diffuse as the chromosomes uncoil, probably by reversing the condensation process. Finally, a nucleolus is reorganized in each daughter nucleus at the chromosomal nucleolar organizer site.

CYTOKINESIS IN SOMATIC CELLS

In most cell types, the initial events of cytokinesis take place around mid-anaphase. However, coordination between cytokinesis and mitosis does not occur in all cells. In many insects, for example, the development of fertilized eggs proceeds by repeated nuclear division until several thousand nuclei are formed; then the cytoplasm undergoes multiple cytokinesis, which encloses each nucleus in a separate cell. Some plants, particularly many fungi, undergo nuclear division without cytoplasmic division and normally exist as single-celled, multinucleate organisms called *coenocytes*.

In the most frequently observed situation, in which cytokinesis is coordinated with nuclear division, the plane of cytoplasmic division normally corresponds with the plane occupied by the metaphase plate. The position of the cytoplasmic division plane seems to be determined rather early in the process of nuclear division, since displacement of the spindle apparatus from its normal position by centrifugation of a dividing cell after late metaphase does not usually alter the position of the division plane. If, however, the spindle is displaced before this time, the position of the cytoplasmic division plane is altered.

Cytokinesis in animal cells is initially visible at the cell surface as a narrow groove or indentation, termed a *furrow,* which extends in a ring around the surface of the cell and which indicates the future division plane (Fig. 17–29). The furrow deepens progressively until the cell is divided into roughly equal parts.

In Chapter 11, we noted that cytoplasmic microfilaments appear to be involved in furrow formation. Furrowing and cytokinesis stop if the cell is treated with cytochalasin B, an antibiotic which disrupts microfilaments, and both processes resume upon the reappearance

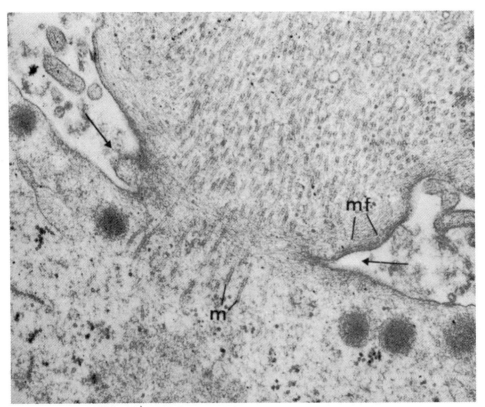

Figure 17–29 Electron micrograph showing a furrow (arrows) in a dividing rat cell. Microfilaments (mf) are visible at the advancing edge of the furrow; remnants of spindle microtubules (m) are visible in the cytoplasmic area connecting the two daughter cells. (From DeRobertis, Saez and DeRobertis [1975], *Cell Biology,* Philadelphia, W. B. Saunders Company, p. 289.)

Mitosis: Nuclear Division in Somatic Cells

of microtubules following removal of cytochalasin. However, it is not known whether or not microfilaments are directly responsible for furrowing.

In plants, cytokinesis occurs not by furrowing but by a somewhat different mechanism. In late anaphase, at the position previously occupied by the metaphase plate, a double membrane called a *cell plate* begins to form by the fusion of small vesicles and other membrane fragments. In some cells, the vesicles and fragments forming the cell plate appear to be derived from the Golgi apparatus. When the formation of the cell plate is complete, the two membranes composing it eventually fuse with the plasma membrane of the parent cell. When this occurs, each daughter cell becomes surrounded by a continuous plasma membrane (Fig. 17–30). New cellulose walls are then laid down between the two membranes that separate the daughter cells. We have already described the process of cell wall formation in Chapter 11.

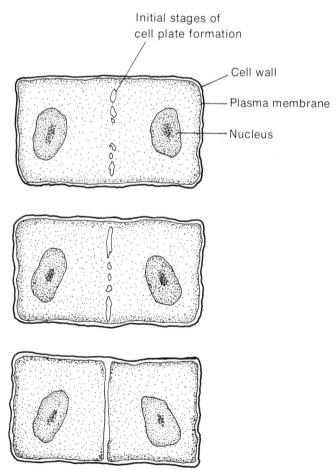

Diagram illustrating the formation of the cell plate in a dividing plant cell. **Figure 17-30**

MEIOSIS: NUCLEAR DIVISION IN GERM CELLS

The process of mitosis, described in the foregoing section, results in the production of two daughter cells, each possessing the same number and types of chromosomes as are present in the original cell. Thus, the chromosome number remains constant from one generation of somatic cells to the next.

In the late 1800s, it was discovered that a fertilized egg, from which a complete organism develops, results from the fusion of an egg cell and a sperm cell. This discovery was confirmed and expanded by Edouard van Beneden, who reported in 1883 that germ cells* (eggs and sperm) in the roundworm *Ascaris* contain half as many chromosomes as the somatic cells, and that following fertilization, all cells again contain the normal somatic number of chromosomes. Van Beneden, however, did not fully understand the significance of his observations, and it was not until about four years later that August Weismann predicted that the chromosome number must *always* be reduced to half during the formation of germ cells. Weismann reasoned that if the chromosome number were not reduced during germ cell formation, then the production of a new individual by the fusion of an egg and sperm would necessarily result in a doubling of the chromosome number every generation—a situation which does not occur. Instead, the fusion of an egg and a sperm, each possessing half as many chromosomes as a somatic cell, would reestablish the characteristic somatic number of chromosomes.

Subsequent work has confirmed Weismann's prediction, but we now know that the halving of chromosome number during germ cell formation occurs in a highly specific way. Weismann initially believed that individual chromosomes were qualitatively similar, and hence he did not suggest that any special mechanism was necessary for the reduction of chromosome number, as long as 50 per cent of the chromosomes present in a somatic cell ended up in a sex cell. However, when it was discovered that chromosomes existed in somatic cells as qualitatively distinct homologous pairs, it became clear that during germ cell formation, the chromosomes must be distributed so that each germ cell receives *one chromosome of each homologous pair.* Cells that possess one chromosome of each homologous pair are called *haploid.* The fusion of two haploid cells (an egg and a sperm) would thus result in a fertilized egg that possessed not only the *somatic number of chromosomes* but also *one pair of each chromosome type*: a diploid cell.

The specific distribution of chromosomes to germ cells occurs by a process called *meiosis* (Fig. 17–31). Whereas mitosis involves *one division* of the nucleus following a single duplication of the somatic number of chromosomes, meiosis involves *two divisions* of the nucleus following a single duplication of the somatic number of chromosomes.

---

*Germ cells are also known as sex cells.

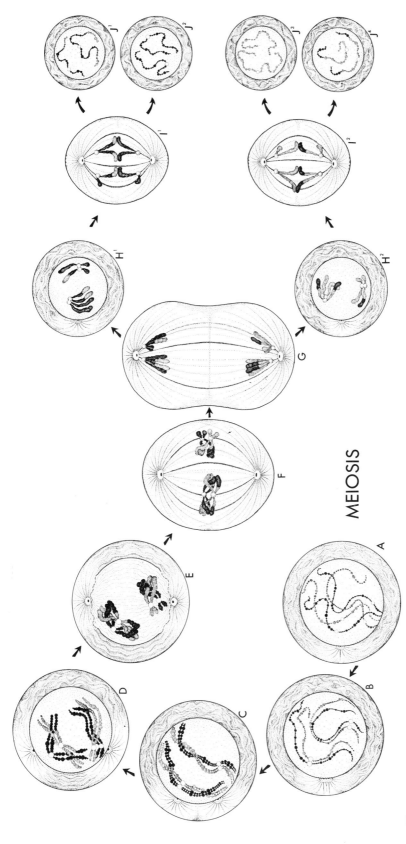

Diagram summarizing the events in meiosis. (A) Early prophase I, (B), (C), and (D) middle prophase I, (E) late prophase I, (F) metaphase I, (G) late anaphase I, (H) telophase I, (I) early anaphase II, (J) late telophase II. (From King, R.C. [1965], *Genetics*, New York, Oxford University Press, Inc.)

Figure 17-31

THE FIRST MEIOTIC DIVISION (MEIOSIS I)

Interphase I

The events occurring in interphase preceding the first meiotic division (interphase I) are similar to those of mitotic interphase. The chromatin strands exist in the extended form, and are duplicated during the S phase of the cell cycle. Research has shown that the "decision" to undergo a meiotic, rather than a mitotic, division occurs at some time in early G_2 immediately following the S phase. We do not know the mechanism for this control.

Prophase I

During prophase of the first meiotic division, the chromatin strands begin to condense, just as they do in prophase of mitosis. Shortly after the chromosomes first become visible, homologous chromosomes come to lie adjacent to one another and begin to pair closely along their lengths. The pairing of homologous chromosomes during meiosis is called *synapsis*. As condensation of the chromosomes continues, it becomes apparent that each chromosome consists of two chromatids (as a result of doubling of chromatin strands in the previous interphase). The pairing of homologous chromosomes thus forms a four-stranded structure called a *tetrad*. During synapsis in the tetrad stage, material from one chromosome may be reciprocally exchanged for similar material on the homologous chromosome, a process termed *crossing over*. Crossing over

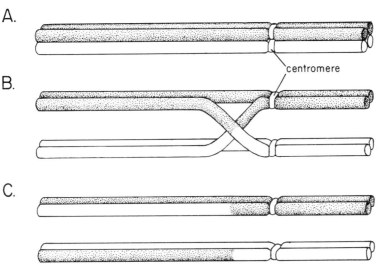

Figure 17-32   Illustration of the process of crossing over, which occurs during synapsis in the tetrad stage. Crossing over results in the reciprocal exchange of material between homologous chromosomes; this most probably occurs by breakage and reunion of chromatids. (From Moody, P.A. [1967], *Genetics of Man,* New York, W.W. Norton, Inc., p. 33.)

Meiosis: Nuclear Division in Germ Cells

may result in an alteration of gene types on chromosomes and is extremely important in determining variability during sexual reproduction (Fig. 17–32).

Late prophase I is characterized by disintegration of the nuclear envelope, disappearance of the nucleolus, and formation of the spindle apparatus. All these events take place in the same way as they do in prophase of mitosis. As in mitosis, centrioles are involved in the formation of the meiotic spindle apparatus in animal cells and in plants with motile sex cells; in most other plants, the meiotic spindle apparatus assembles without the presence of centrioles.

Metaphase I

The tetrads come to lie at the metaphase plate, and a bundle of spindle microtubules attaches to each centromere. This attachment occurs in such a way that *both chromatids of a single chromosome are attached to one pole, and both chromatids of the homologous chromosome are attached to the other pole* (Fig. 17–33).

Anaphase I

Because of the characteristic pattern of attachment of spindle microtubules, one chromosome of each homologous pair migrates to opposite poles; sister chromatids are not separated from one another in this division.

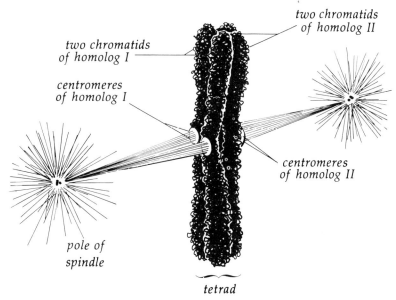

Figure 17–33
Attachment of spindle microtubules to the centromeres of a metaphase chromosome during meiosis (From Wolfe, S.L. [1972], *Biology of the Cell,* Belmont, California, Wadsworth Publishing Company, Inc., © 1972 by the Wadsworth Publishing Company, Inc. Reprinted by permission of the publisher.)

Telophase I

In some organisms telophase I results in the formation of a nuclear envelope and in the uncoiling of the chromosomes; in other organisms no nuclear envelope is formed and the chromosomes do not uncoil. In any case, a nucleolus does not ordinarily re-form during telophase I.

THE SECOND MEIOTIC DIVISION (MEIOSIS II)

The second meiotic division results in the separation of sister chromatids and the production of four haploid cells.

Interphase II

The length of the interphase period following the first meiotic division may vary, depending upon the cell type. If the chromosomes did not uncoil during telophase I, then there is no interphase II period. The important point to note is that *no DNA replication occurs during interphase II,* whether the interphase period is long or short.

Prophase II

If the chromosomes have uncoiled during telophase I, prophase II will be characterized by condensation of the chromosomes. If, on the other hand, there has been little uncoiling of the chromosomes during telophase I, prophase II will be very short. In both cases, a spindle apparatus begins to form in late prophase II at each of the two areas (poles) where the chromosomes aggregated in telophase I. The nuclear envelope, if it was formed during telophase I, also disintegrates in late prophase.

Metaphase II

During metaphase II, each of the two groups of chromosomes becomes aligned at the center of its spindle. Spindle microtubules attach to each chromosome so that sister chromatids are connected to opposite poles.

Anaphase II

Sister chromatids are separated from one another and move to opposite poles during anaphase II. Thus, at the end of anaphase II one complete haploid set of chromosomes has aggregated at each pole (of both spindles).

Telophase II

In telophase II, nuclear envelopes form around each haploid set of chromosomes, the chromosomes uncoil and become diffuse, and a nucleolus may re-form within each nucleus.

CYTOKINESIS DURING GERM CELL FORMATION

Cytokinesis during germ cell formation occurs by the same mechanisms as does cytokinesis following mitosis; that is, by furrowing in animal cells and by the formation of cell plates (and cell walls) in plant cells.

In some organisms, cytokinesis occurs at both telophase I and telophase II. This pattern of cytokinesis gives rise to two cells at telophase I, and then each of these cells divides again at telophase II, yielding a total of four haploid cells. In other organisms, cytokinesis is postponed until telophase II. In this latter case, four haploid cells are produced simultaneously, with two cytoplasmic division planes forming roughly at right angles to one another.

GAMETOGENESIS: THE FORMATION OF SEX CELLS (GAMETES)

The formation of sperm cells (*spermatogenesis*) and egg cells (*oogenesis*) in animals is summarized in Figure 17–34. The main difference between the two processes is that spermatogenesis results in the production of four functional sperm cells, whereas oogenesis gives rise to one functional egg. In the latter case, the other three products of meiotic division—the polar bodies—are nonfunctional and eventually disintegrate. Because the cytoplasm is divided unequally in the two meiotic divisions of oogenesis, an egg cell is produced which possesses large amounts of cytoplasm containing stored nutrients and other materials needed during the early development of the fertilized egg.

Meiosis also is involved in the formation of reproductive cells in plants. In most plants, the diploid organism (called the *sporophyte*) produces male and female haploid spores by meiosis. The male and female spores germinate, respectively, into multicellular male and female structures (called *gametophytes*) which, in turn, produce male and female haploid germ cells (pollen and eggs) by mitosis. Fusion of a pollen cell and an egg cell again produces the diploid sporophyte. The alternation of sporophyte and gametophyte organisms, which occurs in the life cycles of most plants, is referred to as *alternation of generations* (Fig. 17–35).

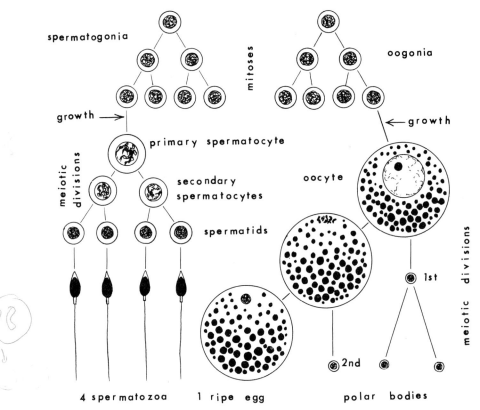

Figure 17-34 Summary of spermatogenesis and oogenesis in animals (From Bolinsky, B.I. [1975], *An Introduction to Embryology*, 4th Ed., Philadelphia, W. B. Saunders Company, p. 63.)

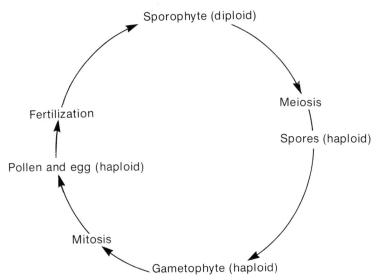

Figure 17-35 Summary of the alteration of generations which occurs in most plants.

CONTROL OF CELL DIVISION IN EUKARYOTES

Primarily because of interest in determining the cause of uncontrolled proliferation of benign and cancerous tumor cells, a great deal of research is now being done in attempts to elucidate the mechanism or mechanisms that control cell division in eukaryotes. Unfortunately, we have made very little progress towards this goal, and it has become increasingly clear that many factors influence cell proliferation. Some of these factors are discussed below.

Cell Size

Some experiments have shown that cell division in eukaryotes usually occurs when a particular cell weight is reached. Consequently, the rate of cell division is dependent upon growth conditions—the faster the rate of mass accumulation, the faster the rate of cell division. The correlation between cell weight and initiation of cell division implies the need to accumulate a substance that promotes division or to dilute a substance that inhibits division.

Enzyme Levels

Some evidence suggests that the availability of certain enzymes may serve to regulate cell division. Some workers, for example, have speculated that genetically regulated fluctuation in DNA polymerase levels may control cell division by determining the timing of DNA replication. The trouble with this idea is that although the level of DNA polymerase in many cells does rise just before DNA synthesis and drop shortly thereafter, other cells normally have continuously high levels of DNA polymerase throughout their growth cycles.

Chalones

In 1960, W.S. Bullough and E.B. Laurence isolated from epidermal cells a substance that acts as a mitotic inhibitor both in vivo and in vitro. They termed this substance a *chalone* (chalon = slacken), and chalones have now been isolated from virtually every tissue that has been studied. Chalones are tissue specific and their effects are reversible. Some are glycoproteins, whereas others appear to be simple polypeptides. It has been shown that chalones are released from receptor sites on the plasma membranes of damaged cells; the cells, now freed of the antimitotic effects of chalones, undergo division. This results in the repair of damaged tissue.

Cyclic AMP

Cyclic AMP (see page 188) is present in most, if not all, cells and has been shown to be involved in the regulation of many metabolic processes. It is required for the activation of a number of enzymes and also serves as an intracellular mediator for the effects of a number of hormones. Because of its virtually ubiquitous distribution in cells, some workers have suggested that cyclic AMP may control cell proliferation. The division rate of some cells grown in tissue culture has been shown to be inversely proportional to intracellular cyclic AMP levels, and indeed, the addition of cyclic AMP to cultures of some tumor cells slows their growth.

Hormones

Many animal and plant hormones cause the proliferation of cells. For example, in vertebrates, growth hormone from the pituitary gland (somatotrophin) stimulates proliferation of a wide variety of cell types. Androgens (e.g., testosterone) and estrogens also promote cellular proliferation of accessory sex tissues and of other tissues, as well. In fact, estrogens have been shown to encourage the formation of cancer in some individuals.

In plants, the auxins (indoleacetic acid and its derivatives), the kinens (derivatives of adenine), the gibberellins (derivatives of gibberellic acid), and other hormone-like chemical substances have been shown to affect cell growth and proliferation.

Contact Inhibition

In Chapter 16, we mentioned that membrane glycoproteins and glycolipids may be responsible for the phenomenon of contact inhibition, whereby normal cells cease dividing when they come into contact with other cells. It is thought that tumor cells may possess altered glycoprotein and glycolipid membrane components and hence are not subject to the normal mechanisms that control cell proliferation.

EVOLUTION OF CELL DIVISION

It is quite likely that division of the most primitive cells was a purely physical phenomenon, dependent on forces external to the cell. For example, we can envision a process analogous to that which occurs in the division of microspheres and coacervates (see Chapter 9). Thus, we might assume that as a cell increased in size by the accumulation of cellular material as a result of primitive metabolic processes, it must eventually have reached a size at which it tended to be broken up into smaller particles by the action of tides and winds.

Initially, then, division was probably a very crude process. However, as cellular metabolism and structure increased in complexity,

Evolution of Cell Division

and particularly as a primitive system for storing and retrieving hereditary information was developed, individual cells that acquired mechanisms for the replication and distribution of hereditary material to daughter cells would have been selected for. These mechanisms, evolving in organisms ancestral to present-day prokaryotes, were undoubtedly relatively simple ones. Because the amount of hereditary material was small, it could have been distributed to daughter cells by a mechanism involving replication of the hereditary material followed by binary fission, much as cell division occurs in prokaryotes today.

The development of eukaryotic cells, which are considerably more complex than prokaryotic ones, was accompanied by a large increase in the amount of hereditary material. There was presumably too much hereditary information to be replicated efficiently as a single molecule (there is about 300 times as much DNA in the unicellular eukaryote *Euglena gracilis* as in *E. coli*), and consequently the DNA became distributed into a number of smaller molecules. Increase in the number of DNA molecules must have required, in turn, a distribution mechanism more elaborate than that in prokaryotes. This requirement may have led to the evolution of mitosis, and eventually, meiosis. Determination of the division plane by the position of the spindle apparatus was an important step following the evolution of mitosis and meiosis, since it guaranteed that cytokinesis would result in the separation of the two sets of daughter chromosomes.

We do not know how mitosis and meiosis evolved, but there are currently two theories—the same two theories that have been suggested to account for the origin of mitochondria and chloroplasts (see Chapter 11). The *classical theory* suggests that mitosis (and meiosis) must have arisen by the gradual, step by step accumulation of selectively advantageous mutations. The *symbiotic theory,* proposed by Lynn Margulis, is based upon the ultrastructural similarities among flagella, cilia, centrioles and basal bodies. You will recall from Chapter 11 that the microtubules in flagella and cilia are arranged in a $(9 + 2)$ configuration, consisting of nine peripheral sets of two microtubules surrounding a central pair; centrioles and basal bodies have a $(9 + 0)$ microtubular configuration, consisting of nine peripheral sets of three microtubules and no central ones.

Margulis has suggested that a primitive ameboid heterotroph containing mitochondria acquired a population of spirochaete-like, motile symbionts that possessed a $(9 + 2)$ microtubular configuration. Presumably, such a symbiotic relationship would have been selectively advantageous to both organisms, since it permitted the motile symbiont to exist in a relatively stable and secure environment, and it conferred motility on the host organism. By theory, the motile symbionts gave rise to flagella, cilia, and, eventually, to basal bodies, centrioles, and spindle microtubules; the process of mitosis itself evolved by a long series of steps which may have taken as long as one billion years.

As evidence to support her theory, Margulis points out that symbiotic relationships of the proposed type exist in several present-day organisms. For example, a surface spirochaete is responsible for movement in *Myxotrichia,* a flagellate found in the guts of termites. In addition, there have been some reports that centrioles and basal bodies contain DNA, which Margulis thinks may be a relic of the genetic material of the ancestral symbiont organism. Furthermore, centrioles and basal bodies are somehow able to direct their own assembly—another characteristic that suggests an autonomous origin.

Critics of the symbiotic theory point out that the initial reports of DNA in centrioles and basal bodies have been disputed, and it now appears unlikely that centrioles and basal bodies contain DNA. Opponents also emphasize that there is no present-day prokaryote that shows a (9 + 2) microtubular structure.

On balance, the evidence supporting the symbiotic theory for the origin of (9 + 2) and related structures is considered very weak. In fact, many biologists who support the idea of a symbiotic origin of chloroplasts and mitochondria do not accept the symbiotic origin of (9 + 2) structures.

Even if centrioles and spindle microtubules did evolve from (9 + 2) symbiont organisms, it is not at all clear how mitosis itself evolved. Consequently, the symbiotic theory does not, in fact, provide us with any special insights into the evolution of mitosis; it merely contends that part of the mitotic apparatus may have arisen as a result of a sudden symbiotic event.

Thus, although the selective advantage of accurate cell division seem obvious to us, it is not immediately apparent by what route the mechanisms of cell division, particularly mitosis, evolved. We simply know that they did evolve, and we must depend on future research to clarify the problem.

REFERENCES

Avery, T.O., C.M. MacLeod, and M. McCarty (1944). Studies on the chemical nature of the substance inducing transformation of pneumococcal types. *J. Exp. Med.* 79:137.

Bryan, J. (1974). Microtubules. *Bioscience* 24:701.

Cold Spring Harbor Symposium on Quantitative Biology (1968), Vol. 33, *Replication of DNA in Microorganisms,* Cold Spring Harbor Laboratory Quantitative Biology, New York.

Cold Spring Harbor Symposium on Quantitative Biology (1973), Vol. 38, *Chromosome Structure and Function,* Cold Spring Harbor Laboratory Quantitative Biology, New York.

DeRobertis, E.D.P., F.A. Saez, and E.M.F. DeRobertis (1975). *Cell Biology,* 6th Edition. W. B. Saunders Co., Philadelphia.

Felsenfeld, G. (1975). String of pearls. *Nature* (News and Views) 257:177.

Fitzsimons, D.W. and G.E.W. Wolstenholme (1975). *The Structure and Function of Chromatin,* American Elsevier Publishing Co., New York.

Inoué, S. and H. Sato (1967). Cell motility by labile association of molecules. *J. Gen. Physiol.* 50:259.

Macleod, A.G. (1973). *Cytology.* The Upjohn Company, Kalamazoo, Michigan.

Margulis, L. (1970). *Origin of Eukaryotic Cells.* Yale University Press, New Haven.

References

Meselson, M. and F.W. Stahl (1958). The replication of DNA in *Escherichia coli. Proc. Nat. Acad. Sci., 44*:671.

Miller, O.L. Jr. and B.R. Beatty (1969). Visualization of Nucleolar Genes. *Science 164*:955.

Olins, A.L. and D.E. Olins (1974). Spheroid chromatin units (μ-Bodies), *Science 183*: 330.

Strickberger, M.W. (1968). *Genetics.* Macmillan, Inc., New York.

Stryer, L. (1975). *Biochemistry,* W. H. Freeman, San Francisco.

Voeller, B.R., Ed. (1968). *The Chromosome Theory of Inheritance.* Appleton-Century-Crofts, New York.

Watson, J.D. and F.H.C. Crick (1953). Genetical Implications of the Structure of Deoxyribonucleic Acid. *Nature 171*:964.

Watson, J. D. and F. H. C. Crick (1953). The structure of DNA. Cold Spring Harbor Symposium Quantitative Biology *18*:123.

White, M.J.D. (1973). *The Chromosomes,* 6th Edition. Chapman and Hall, London.

Wolfe, S.L. (1972). *Biology of the Cell.* Wadsworth Publishing Co., Inc., Belmont, California.

Zeuthen, E. (1964). *Synchrony in Cell Division and Growth.* Interscience Publishers, New York.

CHAPTER 18

The Genetic Code: Function and Evolution

Throughout our discussions of cell structure and function we have repeatedly emphasized the central role of DNA, the cell's hereditary (genetic) material. We have noted that the sequence of nucleotides in DNA forms a code which is selectively decoded (expressed) by the cell as specific patterns of protein synthesis. These proteins, possessing enzymatic, structural, and other functions, ultimately account not only for the structural and metabolic characteristics of individual cells but also, in multicellular organisms, for the structural and metabolic characteristics of the entire organism. A cell's DNA, then, is the repository of information accumulated throughout evolution, and the decoding process is essential to the utilization of this information.

The most impressive breakthrough of modern biochemical research has been the "cracking" of the genetic code. As a result of this accomplishment, we now know much about how information is stored in DNA and how it is used by the cell for the synthesis of protein—processes that we will discuss at the beginning of this chapter. Once we understand how this system of information storage and retrieval operates, we will be in a good position to discuss its control and to speculate on its evolution.

PROTEIN SYNTHESIS: A BRIEF SUMMARY

In all normal cells, whether prokaryotic or eukaryotic, decoding of the hereditary information involves two general aspects: *transcription* and *translation.* In transcription, DNA is used as a template for the synthesis of a single-stranded molecule of RNA, called *messenger RNA (mRNA).* (See Chapter 7 for a review of the primary structure of RNA.) This template-directed synthesis of RNA occurs in much the same way as does the template-directed synthesis of polynucleotide strands of DNA. The net result of the transcription process is the transfer of the coded information (nucleotide sequence) in DNA to a nucleotide sequence in messenger RNA. Following transcription, the DNA is no longer directly involved in the steps that follow.

The nucleotide sequence in messenger RNA is now translated, or decoded, into specific sequences of amino acids in proteins. Translation occurs on ribosomes, small cytoplasmic structures composed of ribosomal RNA and structural protein. Another class of RNA, called *transfer RNA (tRNA)* transports specific amino acids to the ribosome in response to specific nucleotide sequences in messenger RNA, and the amino acids are joined together to form proteins.

The process of protein synthesis is basically similar in both prokaryotic and eukaryotic cells, although some minor differences do exist. Many of these differences are, in fact, a function of the complexity of the eukaryotic cell relative to the complexity of the prokaryotic one.

PROTEIN SYNTHESIS IN PROKARYOTES

TRANSCRIPTION: THE SYNTHESIS OF MESSENGER RNA

The Messenger Hypothesis

In 1961, Francois Jacob and Jacques Monod, two French microbiologists, published a paper in the *Journal of Molecular Biology* in which they postulated the existence of a chemical "messenger" that acts as an intermediate in protein synthesis by transporting the coded information in DNA to the ribosome. The messenger concept was originally based primarily on the observation that DNA did not appear to be directly involved in protein synthesis. It was known that DNA in eukaryotic cells is confined to the nucleus and that protein synthesis occurs in the cytoplasm. It thus seemed reasonable to assume that some compound must serve to transfer the information in DNA to the ribosomes.

Jacob and Monod accumulated a large amount of genetic data on the control of protein synthesis in *Escherichia coli,* and this data was consistent with the messenger hypothesis. They found that certain enzymes, normally produced in very small amounts, are synthesized in very large amounts after specific compounds called *inducers* are added to the growth medium. This switch from minimal to maximal synthesis occurred within a few minutes after the addition of an inducer, and maximal synthesis terminated within an equally short time following the removal of the inducer. Jacob and Monod explained these observations by assuming the existence of a short-lived messenger compound whose synthesis was promoted by the presence of inducer and terminated in the absence of inducer. The transitory existence of the messenger would explain the rapid response of protein synthesis to the presence and absence of inducer.

Jacob and Monod predicted that the messenger should have the following characteristics:

(1) It should be a polynucleotide, presumably RNA.

(2) The messenger RNA should occur in various lengths, reflecting the lengths of individual genes or groups of genes on DNA.

(3) Its nucleotide sequence should reflect the nucleotide sequence of DNA.

(4) It should be short-lived.

(5) It should be associated with the ribosomes, at least transiently.

In the same year that Jacob and Monod proposed the messenger hypothesis, Brenner, Jacob, and Meselson reported the appearance of a short-lived RNA fraction a few minutes after infection of *E. coli* with bacteriophage T2. The RNA fraction, presumably the messenger carrying information from the T2 DNA, became associated with preexisting ribosomes. This and other experiments provided strong support for the messenger hypothesis, and the existence of messenger RNA was soon universally accepted.

RNA Polymerase

At about the same time that the concept of messenger RNA was first proposed, researchers in several laboratories reported the discovery of an enzyme, called *RNA polymerase,* which was capable of catalyzing the synthesis of RNA. (You will recall that RNA is similar in primary structure to DNA, with the exceptions that in RNA, uracil substitutes for thymine and ribose substitutes for deoxyribose. In addition, RNA is usually single-stranded, whereas DNA is double-stranded.)

Synthesis of RNA, catalyzed by RNA polymerase (Fig. 18–1), is similar in many respects to DNA polymerase-catalyzed synthesis of DNA (see Figure 17–11). RNA polymerase requires a DNA template, all four ribonucleoside triphosphates, and either Mg^{++} or Mn^{++} ions. The synthesis of RNA, like that of DNA, is driven by the hydrolysis of pyrophosphate, and chain elongation occurs by addition of ribonucleotides to the 3'-hydroxyl end of the growing chain. Thus, synthesis of an RNA strand complementary to a DNA template strand occurs in the 5' → 3' direction, as it does in DNA synthesis. Unlike DNA polymerase, RNA polymerase does not require a primer and does not show nuclease activity.

Since DNA is a double-stranded molecule, we might ask whether RNA polymerase uses both DNA strands, or only one, as a template for RNA synthesis in vivo. A priori, it would seem that only one of the strands in DNA would be transcribed into RNA, for the nucleotide sequence of one DNA strand specifies the sequence of the other. Hence, the information on one strand determines the information on the complementary strand.

Protein Synthesis in Prokaryotes

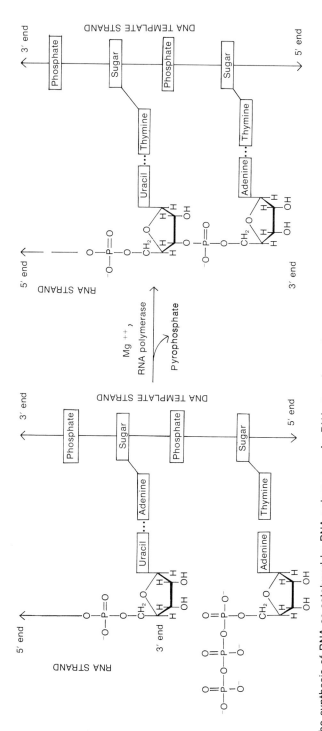

Figure 18–1

The synthesis of RNA as catalyzed by RNA polymerase. An RNA strand complementary to an existing DNA template strand is synthesized by the addition of a nucleotide, initially in the form of a nucleoside triphosphate, to the 3'-hydroxyl end of the growing chain. Unlike DNA polymerase, RNA polymerase does not require an existing primer strand.

In fact, this reasoning is correct. However, it has been shown both in *E. coli* and in several bacteriophages that although only one DNA strand serves as a template for the transcription of RNA in vivo, the *same* strand is not always transcribed. Instead, some sequences are transcribed from one DNA strand, and some sequences are transcribed from the other. Because the two DNA strands run in opposite directions, and because RNA is synthesized only in the 5' → 3' direction, it follows that RNA transcription must occur in one direction on one DNA template strand and in the other direction on the other template strand.

Initiation and Termination of RNA Transcription

How is RNA transcription initiated and terminated? A great deal of research has shown that RNA polymerase in *E. coli* possesses a protein subunit, called sigma (σ), which is involved in the initiation of transcription, at least during in vitro RNA synthesis. The sigma subunit is able to recognize on a DNA template strand specific nucleotide base sequences, which serve as "start" or initiation signals for transcription. These base sequences are called *promoter regions*. The binding of the sigma subunit of RNA polymerase to a promoter region leads to local unwinding of the DNA helix and the beginning of transcription. After the initiation of transcription, the sigma subunit dissociates from the rest of the enzyme, which continues to catalyze the template-directed synthesis of RNA. The sigma subunit, released into the cytoplasm, may then associate with another RNA polymerase molecule.

RNA transcription proceeds along a DNA template strand which becomes unwound from its complementary strand only at the region of transcription. Thus, small regions of DNA are transiently unwound prior to transcription and then re-form their normal double-helical configuration following transcription (Fig. 18–2). Consequently, transcription is not accompanied by the unwinding of extensive regions of DNA.

In addition to initiation signals, there are also specific base sequences on DNA that signify the termination of transcription. Some of these base sequences are recognized by RNA polymerase itself, whereas others are apparently recognized, at least in vitro, by a specific protein called *rho* (ρ), which has a molecular weight of about 200,000 and binds to RNA polymerase. Somehow, rho interacts with the termination regions to prevent further transcription by RNA polymerase. At the termination of transcription, the RNA polymerase and the RNA transcript separate completely from the DNA.

The net result of transcription, then, is the production of an RNA polynucleotide strand whose base sequence is complementary to that of the DNA template strand from which the RNA was synthesized. Hence the information in DNA is preserved in its RNA transcript.

Protein Synthesis in Prokaryotes

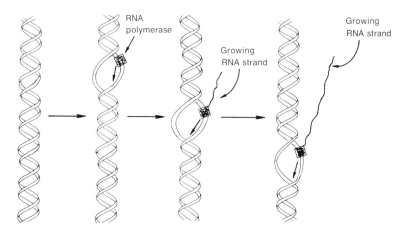

Transcription of a DNA strand occurs by transient unwinding of the DNA double helix. **Figure 18-2**

THE GENETIC CODE

Codons

The nucleotide sequence in messenger RNA contains a code. A large number of genetic and biochemical experiments have shown that each specific sequence of three nucleotide bases (a triplet) constitutes a code word, or *codon,* which specifies one amino acid. Because there are four possible bases for each position in a codon, there are sixty-four possible codons ($4 \times 4 \times 4 = 64$).

Marshall Nirenberg was the first to crack the genetic code. Nirenberg showed that if polyuridylic acid (a polyribonucleotide of uridylic acid) is added to an in vitro, cell-free system capable of sustaining protein synthesis, a polypeptide consisting exclusively of phenylalanine residues is synthesized. Because the polyuridylic acid serves as messenger RNA for the in vitro protein synthesis, Nirenberg's results demonstrated that the triplet UUU codes for the amino acid phenylalanine. A number of similar experiments by Nirenberg and others soon yielded codon assignments for all amino acids.

All sixty-four codons have now been deciphered. Sixty-one of these code for amino acids, and three (UAA, UAG, and UGA) code for the termination of a polypeptide chain. Because there are only twenty amino acids in naturally occurring proteins, it is evident that the code is *degenerate;* that is, many amino acids are designated by more than one codon. But the code is not normally *ambiguous*; instead, one codon normally designates only one amino acid.

Different codons specifying the same amino acid are called *synonyms,* and it has been shown that most synonyms differ only in the last base of their triplets. For example, UUU and UUC are synonyms for phenylalanine. The characteristic sequence of synonymous codons has led Francis Crick to propose the *wobble hypothesis,* which assumes that during translation of messenger RNA, recognition of the third base in a codon is less discriminating than the recognition of the first two. Thus, the mechanism that translates the mRNA code recognizes the first two bases with fidelity, but may "wobble" in its recognition of the third.

Reading of the Code

The code is read sequentially, three bases at a time, from an initiator codon near the 5' end of messenger RNA. There are no "spacer" regions in the code; that is, there are no bases between codons that are not read. The absence of spacer regions can be demonstrated by genetic experiments, which show that the deletion or addition of a single base causes improper reading of the message by shifting the reading frame (Fig. 18–3).

Universality of the Code

The genetic code is the same in all organisms. Thus, the code is said to be *universal.* Indeed, the universality of the genetic code provides strong support for the idea that all organisms are related to one another. The fact that the code has remained unchanged over

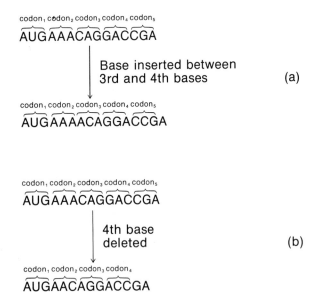

Figure 18–3 (a) Addition of a base or (b) deletion of a base in a DNA template strand causes shifting of the reading frame in the messenger RNA.

Protein Synthesis in Prokaryotes

several billion years of evolution is not surprising, since mutations that alter the code would, in turn, change the amino acid sequences of most, or all, of the proteins synthesized by an organism. Because many of these changes in proteins would undoubtedly be deleterious, any organism that possessed such a mutation would be subjected to strong negative selection pressures.

TRANSLATION: THE SYNTHESIS OF PROTEINS USING INFORMATION SPECIFIED BY MESSENGER RNA

Messenger RNA, which contains a transcript of the DNA's genetic code, is now decoded at the ribosomes into the sequence of amino acids in a protein.

Ribosomes

As you will recall, ribosomes of prokaryotes are roughly spherical particles, about 20 nm in diameter. The intact ribosome, which has a sedimentation coefficient of 70, is formed by the aggregation of a 30s and a 50s subunit. Both subunits are composed of *ribosomal RNA (rRNA)* and structural proteins. In *E. coli,* the 30s subunit is formed from a single molecule of 16s rRNA and 21 different proteins, all arranged in a precise three-dimensional configuration; the 50s subunit is composed of one molecule of 23s rRNA, one molecule of 5s rRNA and about 35 proteins, also arranged in a defined configuration. Like mRNA, the rRNA molecules are transcribed from DNA, as we discussed in Chapter 17.

We do not completely understand the function of rRNA in ribosomes. However, because the rRNA in all ribosomes of a particular cell type is similar, it does not seem likely that rRNA has any messenger function. Studies on the reassembly of the *E. coli* 30s ribosomal subunit from a mixture of its 21 characteristic proteins and 16s rRNA have shown that the 16s rRNA is essential for both the assembly and function of *E. coli* ribosomes, and that most of the 21 proteins are required for activity of the subunit. In fact, the reconstitution of a 30s subunit (or a 50s subunit) from its constituent RNA and proteins occurs spontaneously. Thus, the information required for assembly of the subunits is contained in the structure of their components, much as the amino acid sequence of a protein determines the protein's three-dimensional configuration.

Transfer RNA

Transfer RNA (tRNA) molecules are involved in transporting amino acids to the ribosomes for assembly into proteins. Each amino acid is transported by a tRNA molecule of specific structure, and tRNA molecules are named for the amino acid they carry. For example, the tRNA that carries alanine is called alanyl-tRNA, the tRNA that carries valine is called valyl-tRNA, and so on.

In 1965, Robert Holley culminated more than 7 years of research by reporting the nucleotide sequence of alanyl-tRNA from yeast cells—work for which Holley eventually received the Nobel prize. As shown in Figure 18-4, yeast alanyl-tRNA consists of 77 ribonucleotides, many of which contain "odd" bases other than G, C, A, and U. If the structure of alanyl-tRNA is written so that approximately half of the bases are base-paired, the "cloverleaf" configuration shown in Figure 18-4 is obtained.

In the last several years, sequences of many other tRNA molecules have been determined. Although each of these tRNA molecules differs in structure, all have many structural features in common. These include the following:

(1) A molecular weight of about 25,000.

(2) A high proportion of "odd" bases.

(3) The base sequence CCA at the 3' end of the tRNA molecule, to which the amino acid is attached when carried to the ribosome.

(4) The same general "cloverleaf" configuration, when drawn

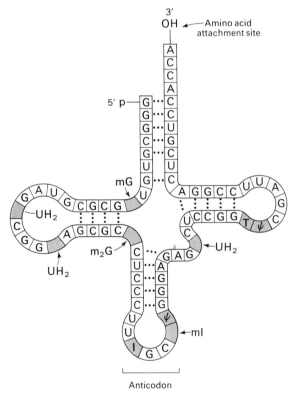

Figure 18-4 The "cloverleaf" configuration of yeast alanyl-tRNA. Notice that the tRNA molecule contains many "odd" bases, including inosine (I), methylinosine (mI), dihydrouridine (UH$_2$), ribothymidine (T), pseudouridine (ψ), methylguanosine (mG), and dimethylguanosine (m$_2$G). (From Stryer, L. [1975], *Biochemistry*, San Francisco, © 1975 by W.H. Freeman and Company, p. 651.)

Protein Synthesis in Prokaryotes

with about half of the nucleotides base-paired to form double helices.

(5) A loop of seven unpaired nucleotides, called the *anticodon loop.* The anticodon loop contains three bases, called the *anticodon,* which are specifically involved in the recognition of mRNA codons.

The in vivo three-dimensional structure of yeast phenylalanyl-tRNA has recently been determined by Alexander Rich and Aaron Klug. They have shown that the hydrogen bonding pattern assumed in the "cloverleaf" model is essentially correct. Phenylalanyl-tRNA is L-shaped, and hydrogen-bonded regions fall in two segments. The CCA terminus, to which the amino acid is attached, is associated with a hydrogen-bonded region at one end of the L, and the anticodon loop is associated with a hydrogen-bonded region at the other end of the L (Fig. 18–5).

Coupling of Amino Acids to tRNA Molecules

The linkage of an amino acid to a transfer RNA molecule (the *charging* of tRNA) occurs in a two-step reaction that is catalyzed by an enzyme called an aminoacyl-tRNA synthetase. There is at least

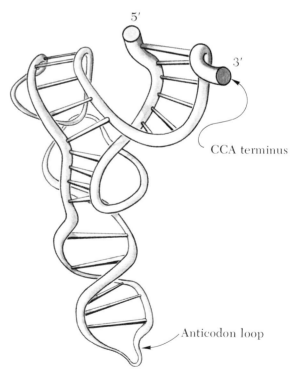

A diagram illustrating the three-dimensional configuration of yeast phenylalanyl-tRNA, as determined by X-ray diffraction.

Figure 18–5

one aminoacyl-tRNA synthetase for each amino acid, and thus each synthetase is specific both for a particular amino acid and for a particular tRNA.

In the initial step of the charging reaction, an aminoacyl-tRNA synthetase catalyzes the addition of an amino acid to AMP, forming a high-energy intermediate called an *aminoacyl-AMP* or an *activated amino acid*. The reaction may be summarized as follows:

$$^{+}H_3N-\underset{H}{\overset{R}{C}}-\overset{O}{\underset{}{C}}-O^- + {}^-O-\overset{O}{\underset{O^-}{P}}-O\sim\overset{O}{\underset{O^-}{P}}-O\sim\overset{O}{\underset{O^-}{P}}-O-\text{ribose-adenine} \xrightarrow{\text{An aminoacyl-tRNA synthetase enzyme}} \rightleftarrows$$

(Amino acid)   (ATP)

$$\rightleftarrows \text{Enzyme} ({}^{+}H_3N-\underset{H}{\overset{R}{C}}-\overset{O}{\underset{}{C}}-O\sim\overset{O}{\underset{O^-}{P}}-O-\text{ribose-adenine}) + {}^-O-\overset{O}{\underset{O^-}{P}}-O\sim\overset{O}{\underset{O^-}{P}}-O^- \quad (1)$$

(Aminoacyl-AMP-Enzyme Complex)   (Pyrophosphate)

The energy to drive amino acid activation is thus supplied by the hydrolysis of ATP.

The aminoacyl-AMP actually remains bound to the synthetase enzyme, and in the next step of the reaction, the aminoacyl group of aminoacyl-AMP is transferred to the amino acid attachment site (see Figure 18–4) of a tRNA molecule:

$$\text{Enzyme} ({}^{+}H_3N-\underset{H}{\overset{R}{C}}-\overset{O}{\underset{}{C}}-O\sim\overset{}{\underset{O^-}{P}}-O-\text{ribose-adenine}) + \text{tRNA} \rightleftarrows$$

(Aminoacyl-AMP-Enzyme Complex)

$$\rightleftarrows {}^{+}H_3N-\underset{H}{\overset{R}{C}}-\overset{O}{\underset{}{C}}-O\sim\text{tRNA} + {}^-O-\overset{O}{\underset{O^-}{P}}-O-\text{ribose-adenine} \quad (2)$$

(Aminoacyl-tRNA)   (AMP)

The net result of both reactions is the formation of an aminoacyl-tRNA: a high-energy complex in which the amino acid is linked by a high-energy phosphate bond to the 3′-hydroxyl group of the terminal adenylic acid residue of tRNA (Fig. 18–6). The overall reaction (the sum of Reactions 1 and 2) is slightly *endergonic*, and the energy to drive it is supplied by the highly exergonic hydrolysis of pyrophosphate formed in Reaction 1. Thus, the net energy expen-

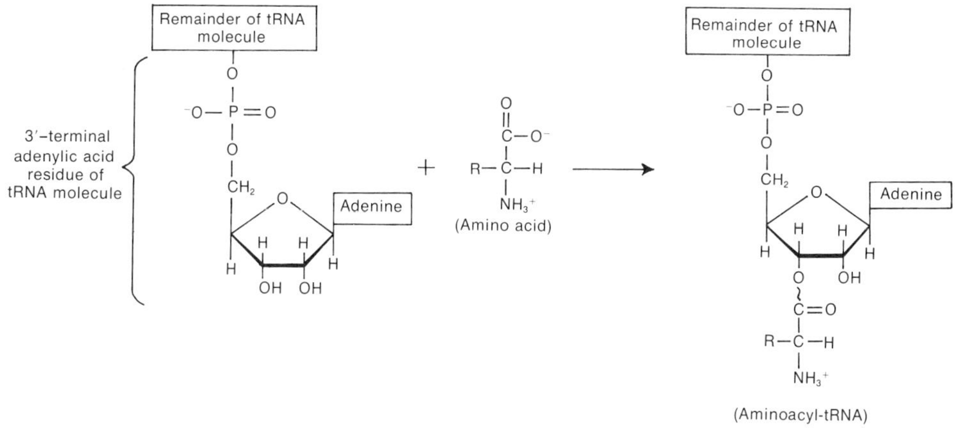

Protein Synthesis in Prokaryotes

In the formation of an aminoacyl-tRNA, an amino acid is linked by a high energy phosphate bond to the 3'-hydroxyl group of the terminal adenylic acid residue of tRNA.

Figure 18-6

diture for charging one molecule of tRNA is *two* high-energy phosphate bonds: one is utilized in the amino acid activation step (formation of aminoacyl-AMP) and another in the hydrolysis of pyrophosphate.

It is interesting to note that fatty acid oxidation (page 397) proceeds by the attachment of the fatty acid to coenzyme A to form the high-energy complex, fatty acyl-coenzyme A. This activation step occurs by a mechanism analogous to that involved in the charging of tRNA. The similarity in the mechanisms of the two reactions is another example illustrating the economy of metabolic design.

Recognition of mRNA Codons by tRNA

The three bases forming the anticodons of tRNA molecules "recognize" mRNA codons by base pairing. Thus, for example, a GCG codon in mRNA becomes base-paired with an aminoacyl-tRNA whose anticodon is CGC (Fig. 18–7). Such specific base pairing between anticodons of aminoacyl-tRNAs and mRNA codons occurs at

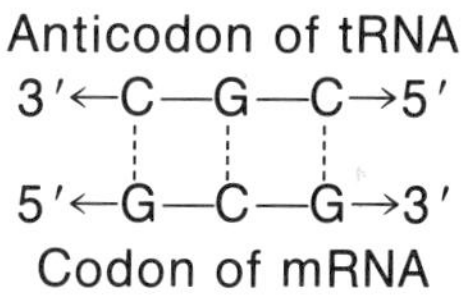

Anticodon of tRNA
3'←C—G—C→5'
5'←G—C—G→3'
Codon of mRNA

Recognition of mRNA codons by tRNA occurs by base pairing of the anticodon of tRNA to a complementary mRNA codon.

Figure 18–7

the ribosome and serves to bring specific amino acids to the ribosome where the amino acids are linked into proteins.

It has been shown unambiguously that codon recognition depends on the anticodon of a tRNA molecule and not on the amino acid that the tRNA molecule carries. This point was demonstrated in a brilliant experiment performed by F. Chapeville and others. Cysteine was attached to its tRNA, and then the sulfur atom in cysteine was removed by reacting the cysteinyl-tRNA with Raney nickel. The removal of the sulfur atom from cysteine converted cysteine to alanine. Thus, alanine was now attached to the tRNA specific for cysteine (Fig. 18–8). When this amino acid-tRNA complex was used in an in vitro protein synthesizing system, alanine was incorporated into the protein at positions where cysteine would normally have been.

Initiation of Translation

Translation of *E. coli* mRNA at the ribosome is initiated by the codon AUG, which is recognized by the anticodon of a special tRNA that carries the amino acid formylmethionine (Fig. 18–9), a derivative of methionine. There are actually two tRNAs with identical anticodons that are capable of recognizing AUG by base pairing. Both are capable of carrying the amino acid methionine; however, one of these tRNAs allows the methionine attached to it to be formylated (forming formylmethionine), and the other tRNA does not. Only formylmethionyl-tRNA binds to AUG codons that code for the initiation of protein synthesis, whereas only methionyl-tRNA binds to AUG

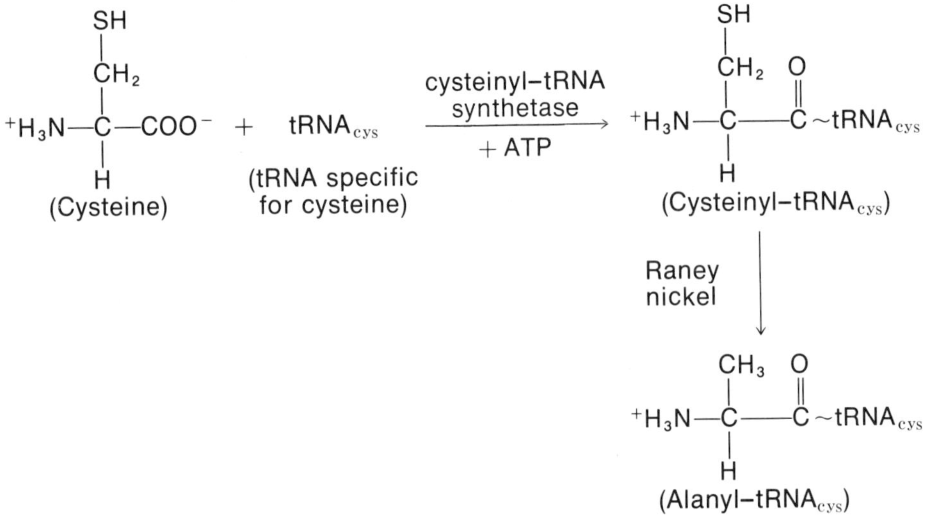

Figure 18–8   Treatment of cysteinyl-tRNA carrying cysteine (cysteinyl-tRNA$_{cys}$) with Raney nickel converts cysteine to alanine, forming a complex in which alanine is attached to the tRNA specific for cysteine (alanyl-tRNA$_{cys}$).

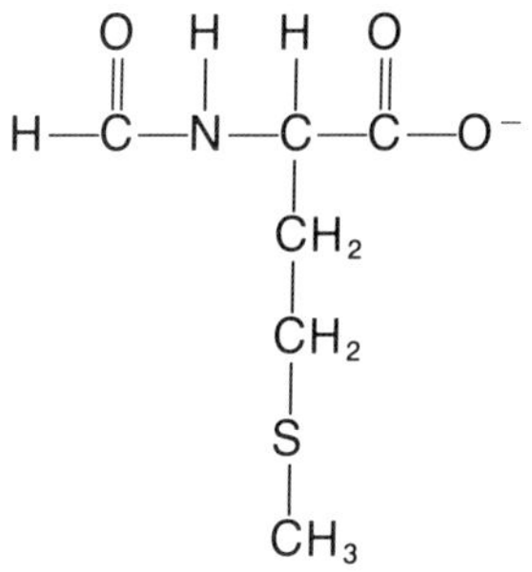

Protein Synthesis in Prokaryotes

$$\begin{array}{c} \text{O} \quad \text{H} \quad \text{H} \quad \text{O} \\ \| \quad | \quad | \quad \| \\ \text{H—C—N—C—C—O}^- \\ | \\ \text{CH}_2 \\ | \\ \text{CH}_2 \\ | \\ \text{S} \\ | \\ \text{CH}_3 \end{array}$$

The chemical structure of formylmethionine. **Figure 18-9**

triplets that code for internal methionines. It is not at all clear how these two types of tRNA can discriminate between initiator AUG codons and AUG codons that specify the insertion of methionine within a polypeptide chain.

The Initiator Complex

The first step in the initiation of translation is the formation of a 30s initiation complex from a 30s ribosomal subunit, mRNA, and formylmethionyl-tRNA. Formation of the initiation complex also requires GTP and at least three protein *initiation factors*. Following the formation of the 30s initiation complex, a 50s ribosomal subunit joins it to form a *70s initiation complex;* during this step, GTP is hydrolyzed to GDP.

In the 70s initiation complex, formylmethionyl-tRNA is positioned at the so-called P (peptidyl) site on the ribosome, and the anticodon of formylmethionyl-tRNA is base paired with the AUG initiator codon. Another site on the ribosome, called the A (aminoacyl) site, is next to the P site; the A site is capable of accepting another tRNA molecule, which carries the next amino acid to be incorporated into the chain.

The formation of the 70s initiator complex, and the position of the A and P sites, are illustrated schematically in Figure 18-10.

Chain Elongation

Elongation of the polypeptide chain (Fig. 18-11) begins with the insertion of a charged tRNA (a tRNA carrying an amino acid) into the A site. Of course, the tRNA that is inserted must have an anticodon that can base pair with the mRNA codon positioned at the A site. The binding of a charged tRNA to the ribosome requires GTP (which is hydrolyzed to GDP and phosphate) and a protein *elongation factor*.

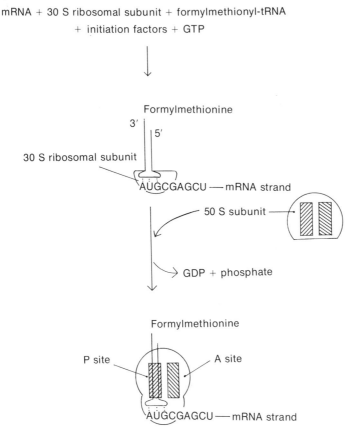

Figure 18-10 Diagrammatic illustration of the formation of the 70S initiation complex. (Adapted from Stryer, L. [1975], *Biochemistry*, San Francisco, W.H. Freeman and Company. p. 655.)

The next step in chain elongation is the formation of a peptide bond between amino acids carried on tRNAs that are positioned at the P and A sites. Peptide bond formation is catalyzed by an enzyme, *peptidyl transferase,* which is an integral part of the 50s ribosomal subunit. The first peptide bond is formed by the transfer of activated formylmethionine (the amino acid carried by the tRNA at the P site) to the amino acid carried by the tRNA at the A site (Fig. 18–12).

The tRNA at the P site is now uncharged (it carries no amino acid) and leaves the ribosome. The tRNA carrying the growing peptide chain (called the peptidyl-tRNA) then translocates to the empty P site, and the ribosome moves along the mRNA a distance of three nucleotides. Translocation requires GTP (which is hydrolyzed to GDP and phosphate) and another protein elongation factor, called a *translocase*.

The process just described continues as additional amino acids are added to the growing chain. The binding of each new tRNA to the ribosome requires the hydrolysis of a high-energy phosphate bond in GTP, as does each translocation.

Protein Synthesis in Prokaryotes

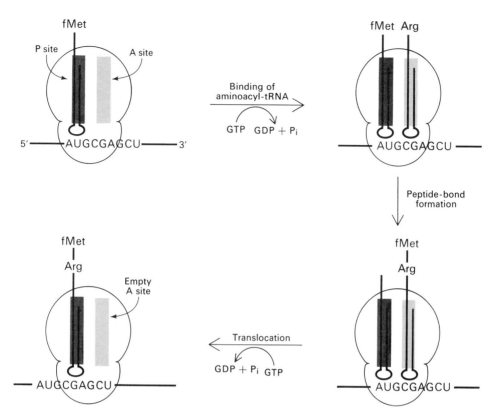

Diagram illustrating the events involved in chain elongation during protein synthesis. (From Stryer, L. [1975], *Biochemistry,* San Francisco, W.H. Freeman and Company, p. 666.)

Figure 18–11

As a polypeptide chain lengthens at the ribosome, the polypeptide begins to assume the characteristic three-dimensional shape determined by its amino acid sequence.

Chain Termination

Normal *E. coli* cells do not contain tRNAs that bind to the codons UAA, UGA, or UAG. Rather, these codons are recognized by protein *release factors.* The binding of a release factor to a termination codon at the A site of the ribosome apparently activates peptidyl transferase so that it catalyzes the hydrolysis of the growing polypeptide chain from the peptidyl-tRNA at the P site. The polypeptide chain then leaves the ribosome, and the ribosome dissociates into 30s and 50s subunits; the 30s subunit may now form a new 30s initiation complex.

Sometime after the initiation of protein synthesis in *E. coli,* the formyl group is enzymatically removed from the N-terminal methionine residue. The removal of the formyl group explains why completed *E. coli* proteins do not usually possess formyl groups at their

514 The Genetic Code: Function and Evolution

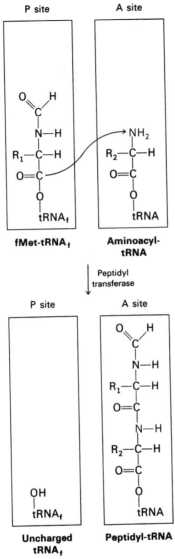

Figure 18-12 The mechanism of peptide bond formation during protein synthesis. (From Stryer, L. [1975], *Biochemistry,* San Francisco, © 1975 by W.H. Freeman and Company, p. 667.)

N-terminal ends. In some proteins, the N-terminal methionine itself, and sometimes several other amino acids, are also hydrolyzed from the N-terminal end.

Polyribosomes

Actually, more than one ribosome is able to translate one mRNA molecule simultaneously. When this occurs, a structure is formed that consists of a number of ribosomes (often two to eight) attached to the mRNA. This structure is called a *polyribosome,* or *polysome* (Fig. 18–13).

ENERGY REQUIREMENTS OF PROTEIN SYNTHESIS

The addition of each amino acid to a growing chain requires four high-energy phosphate bonds. One high-energy phosphate bond is utilized in amino acid activation, a second is utilized in linking the activated amino acid to tRNA, a third is utilized in the attachment of a charged tRNA to the ribosome, and a fourth is utilized in translocation.

PROTEIN SYNTHESIS IN EUKARYOTES

Protein synthesis in eukaryotes differs from that in prokaryotes in several respects.

RIBOSOMES

Eukaryotic ribosomes are somewhat larger than prokaryotic ribosomes and consist of subunits with sedimentation coefficients

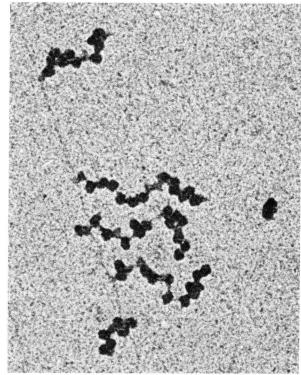

Figure 18–13 Polyribosomes in *E. coli.* (From Miller, O.L., Jr. and C.A. Thomas, Jr. [1970], Science *169*:394. Copyright © 1970 by The American Association for the Advancement of Science.)

of 40s and 60s. Together, these subunits form an 80s ribosome. The 40s subunit contains a strand of 18s RNA and a large, although undetermined, number of structural proteins; the 60s subunit contains a strand of 28s RNA, a strand of 5s RNA, and structural proteins.

LOCATION OF PROTEIN SYNTHESIS

In prokaryotes, protein synthesis occurs on ribosomes that are "free" in the cytoplasm; that is, they are not attached to other cellular structures. In most eukaryotic cells, some ribosomes are also located free in the cytoplasm, but others are bound to membranes of the rough endoplasmic reticulum and sometimes to the outer membrane of the nuclear envelope.

It is not known what determines whether a particular mRNA molecule is synthesized on membrane-bound ribosomes or on free cytoplasmic ribosomes. We do know that cells which secrete proteins, such as acinar cells of the pancreas, have a highly developed rough endoplasmic reticulum. In such cells, many, if not all, proteins synthesized on membrane-bound ribosomes end up in vesicles of endoplasmic reticulum. As we described in Chapter 11, these vesicles may fuse with membranes of the Golgi apparatus, where the proteins may be modified and concentrated before they are packaged into small vesicles which bud off the periphery of the Golgi bodies. These vesicles, in turn, fuse with the plasma membrane, releasing their products from the cell by exocytosis.

HETEROGENEOUS NUCLEAR RNA

Large amounts of RNA are continually synthesized in the nuclei of most eukaryotic cells, at least in those cells grown in tissue culture. Most of this RNA, termed *heterogeneous nuclear RNA* (heterogeneous because it varies in molecular weight), remains in the nucleus and is broken down shortly after it is synthesized. Many workers have suggested that heterogeneous nuclear RNA is a precursor to messenger RNA. They contend that although the cell transcribes large amounts of DNA into RNA, only a small proportion of the total RNA synthesized—the messenger RNA—remains intact and is transported to the cytoplasm. It is not at all clear why such an energetically expensive method for synthesizing mRNA would evolve, and the function of heterogeneous nuclear RNA remains obscure.

PROTEIN SYNTHESIS IN MITOCHONDRIA AND CHLOROPLASTS

You will recall that protein synthesis in prokaryotes is similar in many respects to protein synthesis in chloroplasts and mitochondria. For example, the size of chloroplast ribosomes is the same as ribosomes of blue-green algae, and ribosomes in both mitochondria and chloroplasts more closely resemble prokaryotic ribosomes in antibiotic sensitivity than they do eukaryotic ribosomes. These similarities have been interpreted by some biologists as suggesting that chloroplasts and mitochondria may have been derived from prokaryotes that originally existed as symbionts of an ancestral eukaryotic cell; however, this conclusion has been disputed by others. (See Chapter 11 for a review of this material).

CONTROL OF PROTEIN SYNTHESIS (REGULATION OF GENE EXPRESSION)

It is obvious that the control of protein synthesis—the regulation of those processes by which genetic information is expressed—is critical to all cells, whether prokaryotic or eukaryotic. Indeed, the structural and functional characteristics of any cell depend upon the expression of genetic information in DNA. Quite simply, a cell is what it is, and does what it does, because of the specific proteins it produces.

The regulation of gene expression is involved in two general processes. First, it is important in *cellular differentiation,* the process by which cells of both unicellular and multicellular organisms become specialized to perform specific functions. Sporulation in bacteria (Chapter 10), heterocyst formation in blue-green algae (Chapter 10), and the production of specialized cells, tissues, and organs during the development of a multicellular organism from a fertilized egg are all examples of processes involving differentiation. A large amount of information, including the observation that cellular specialization can be reversed in many cell types, has led to the conclusion that differentiation is achieved not by the loss or alteration of genetic information but rather by the selective expression of the genetic information that is present in identical form in all cell types.

The regulation of gene expression is also of paramount importance in the day-to-day functioning of the cell. If the organism is to survive and to reproduce, environmental challenges must be met by appropriate cellular responses. For example, the depletion of ATP supplies for any of a number of reasons must be countered by the stimulation of processes that generate ATP or by the inhibition of processes that utilize ATP. Such a response may, in turn, depend upon the synthesis of certain enzymes, termination of the synthesis of other enzymes, inactivation of still other enzymes (perhaps by the synthesis of a proteinaceous inactivator substance), and so forth.

Although we have some knowledge of specific mechanisms that control the expression of particular genes, or groups of genes, in certain cell types, we still do not have any comprehensive idea of how a cell manages to regulate the expression of all of its genetic information. If we rule out alterations in a cell's DNA as a mechanism of gene regulation, then we are apparently left with several other mechanisms. These include *transcriptional controls* involving the selective transcription of certain regions of DNA at specific times to produce various types of messenger RNAs; *translational controls*, involving the selective translation of certain messenger RNAs at specific times; or a combination of these two mechanisms.

TRANSCRIPTIONAL CONTROLS IN BACTERIA

THE OPERON MODEL

The synthesis of a number of enzymes in *Escherichia coli* is regulated at the transcriptional level. This was first demonstrated by Jacob and Monod in studies of lactose metabolism in *E. coli.*

The metabolism of lactose in *E. coli* requires the enzyme *β-galactosidase,* which catalyzes the hydrolysis of lactose to galactose and glucose (Fig. 18–14). If an *E. coli* cell is grown on a medium containing a carbon source other than lactose—for example, glucose—there are very few molecules of β-galactosidase (fewer than ten) in the cell. When lactose is added to the medium, however, the number of β-galactosidase molecules increases several hundredfold within a very short time, owing to the synthesis of new enzyme molecules.

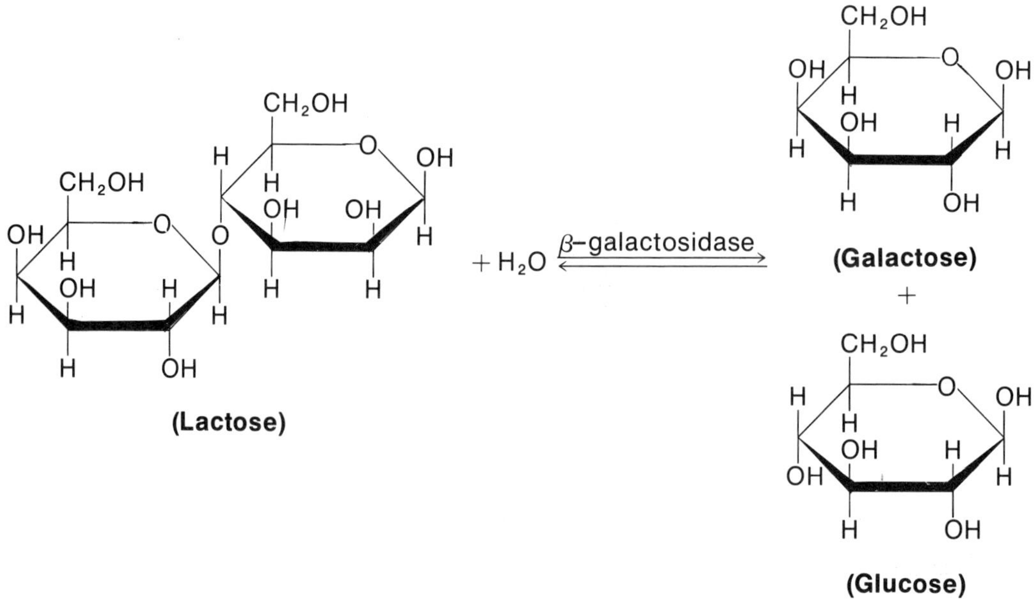

Figure 18–14 β-galactosidase catalyzes the hydrolysis of the disaccharide lactose to galactose and glucose.

Transcriptional Controls in Bacteria

Because β-galactosidase is synthesized when the cell is exposed to certain substances (e.g., lactose), β-galactosidase is termed an *inducible enzyme.* The induction of β-galactosidase synthesis in normal cells is accompanied by the synthesis of two other enzymes (*β-galactoside permease* and *transacetylase*) and the three enzymes are said to be *coordinately induced.* As noted in Chapter 16, β-galactoside permease is required for the transport of lactose into the cell; the role of transacetylase in lactose metabolism is not well understood. The genes coding for β-galactosidase, β-galactoside permease, and transacetylase, called the z, y, and a genes, respectively, are located immediately adjacent to one another on the *E. coli* DNA molecule.

In order to explain the results of a large number of genetic and biochemical experiments on the control of lactose metabolism in *E. coli,* Jacob and Monod proposed a mechanism known as the *operon model.* In its current form, the operon model suggests that transcription of the z, y, and a genes (also called *structural genes*) is controlled by three other genes: a *regulator gene (i),* a *promoter gene (p),* and an *operator gene (o).* The o, p, z, y, and a genes have been shown to be contiguous, in that order, and form what is known as the *lac operon.* The i gene is located a short distance away from the genes of the lac operon.

According to the operon model, the regulator gene codes for the synthesis of a protein, termed the *lac repressor,* which is capable of binding to the operator gene and thus preventing transcription of all three structural genes. Transcription is blocked because the binding of the lac repressor to the operator gene interferes with the passage of RNA polymerase from its binding site at the p gene to the region of the structural genes (Fig. 18–15a). However, in the presence of inducer, the lac repressor binds with inducer molecules, forming a complex that is unable to interact with the operator gene. Consequently, transcription of the structural genes occurs in the presence of inducer, producing one molecule of mRNA containing the transcript of all three structural genes (Fig. 18–15b). Central to the operon model is the concept of a short-lived mRNA that is available for translation only for short times after its production.

Many details of the lac operon model have been substantiated by experimental evidence. Both the proteinaceous lac repressor and the lac operator gene have been isolated, and it has been shown that the lac repressor does indeed bind to the operator gene and to inducer molecules.

Recently, it has also been shown that cyclic AMP (page 188) and another specific protein form a complex that binds to the promoter site and stimulates the initiation of transcription. Because the presence of glucose lowers the intracellular concentration of cyclic AMP, transcription of the structural genes is blocked by high levels of glucose. This mechanism has obvious adaptive value, since β-galactosidase, β-galactoside permease, and transacetylase are not needed if glucose is available as a nutrient source. Indeed, the utilization of a number of other sugars by *E. coli* is repressed in the

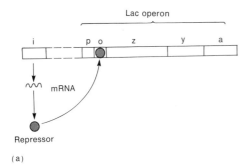

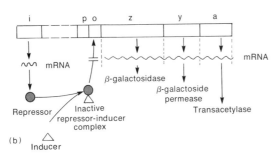

Figure 18-15 The Jacob and Monod operon model for the control of protein synthesis. (a) The lac repressor binds to the operator gene (o) and prevents transcription of the three structural genes (z, y and a). (b) The lac repressor binds with inducer to form an inactive complex which is incapable of binding to the o gene; consequently, transcription of the z, y and a genes occurs.

presence of glucose, a phenomenon known as *catabolite repression.*

Several other operons have been identified in *E. coli,* including those for enzymes involved in the synthesis of histidine, tryptophan, and arginine, and in the degradation of galactose and glycerol. Not all of these operons are regulated in precisely the same way as is the lac operon. For example, some of the regulatory mechanisms involve positive control, in which the protein product of the regulatory gene promotes, rather than inhibits, transcription.

ROLE OF RNA POLYMERASE

We have already mentioned the role of the sigma factor in initiating RNA transcription. In fact, the sigma factor increases the efficiency and alters the specificity of RNA polymerase. For example, the *E. coli* RNA polymerase and its sigma factor can only transcribe the genes of bacteriophage T4 that specify proteins needed early in phage replication. One result of this early gene activity is to bring about a modification of the RNA polymerase enzyme so that it is able to transcribe T4 genes whose products are needed during the final stages of T4 replication.

TRANSLATIONAL CONTROLS IN BACTERIA

Traditionally, the control of gene expression in prokaryotes has emphasized transcriptional control mechanisms, undoubtedly because of the early work of Jacob and Monod and others, which elucidated control mechanisms in the lac operon. Some workers have suggested the existence of translational controls in *Escherichia coli*, including mechanisms that regulate the availability of specific transfer RNAs, enzymes, cofactors, and other substances required for translation. Although such mechanisms may well be important in the control of translation, it is not easy to see how they would possess the required specificity.

TRANSCRIPTIONAL CONTROLS IN EUKARYOTES

Ample evidence exists that transcription in eukaryotes must be preceded by uncoiling of the chromatin and, consequently, that the processes which control chromatin condensation are important in the regulation of gene expression. Some of the evidence supporting this conclusion is discussed below.

CHROMOSOME PUFFS

Giant *polytene* (poly = many; tene = strands) *chromosomes* are found in certain cells of many dipteran larvae—for example, in the salivary glands of larval *Drosophila melanogaster*. Polytene chromosomes get their name from the fact that they are formed by many parallel chromatids, often more than a thousand strands, which do not separate from one another following duplication and which remain in perfect register. Along each chromatid strand some regions of chromatin are tightly coiled and other regions are less so, with the result that polytene chromosomes appear to consist of light and dark bands when observed under a microscope (Fig. 18–16). Each polytene chromosome of any one species possesses a characteristic banding pattern, and it is therefore possible to identify any polytene chromosome on the basis of this pattern.

During larval development, specific areas on polytene chromosomes become uncoiled, forming localized regions called "puffs" (Fig. 18–17). It has been shown that "puffs" represent regions of active RNA synthesis (transcription). A number of workers have reported that a particular area of a polytene chromosome becomes puffed only at a specific stage in the development of the larva. Transitory puffing of various chromosomal regions at specific developmental stages has been interpreted as demonstrating the sequential expression of different genetic material. However, some recent evidence has suggested that the puffing pattern of polytene chromosomes may not be as specific as was once supposed. In any case,

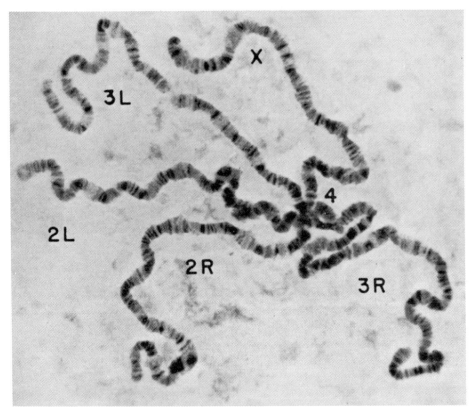

Figure 18-16 Light micrograph of the polytene chromosomes from the salivary glands of *Drosophila melanogaster*. (From Kaufman, B.P. [1939], Journal of Heredity 30:5, Frontispiece.)

observations of polytene chromosomes support the conclusion that RNA transcription occurs primarily in areas of a chromosome which become uncoiled (and are presumably free of bound protein).

LAMPBRUSH CHROMOSOMES

Similar evidence linking the uncoiling of chromatin to transcriptional activity has been obtained from studies of oocyte development in amphibians. At certain stages, the chromosomes in amphibian oocytes show a "lampbrush" configuration; that is, the chromosomes consist of a central axis of condensed chromatin from which unwound loops of chromatin extend (Fig. 18–18). The loops have been shown to be sites of active transcription, presumably involved in the production of yolk and other egg materials.

HISTONES

We noted in Chapter 17 that the attachment of histone proteins to DNA inhibits transcription in experiments performed in vitro; transcription is promoted when histone proteins are removed.

Translational Controls in Eukaryotes

Figure 18-17
Electron micrograph of part of a polytene chromosome of *Chironomus*, showing two large puffs. (From Harris, H. [1974], *Nucleus and Cytoplasm*, 3rd Ed., Oxford, Clarendon Press.)

TRANSCRIPTIONAL CONTROL MECHANISMS

Although we have evidence, then, that transcription requires the uncoiling of condensed chromatin, we do not know what actually determines the coiling state of a particular region of chromatin. A number of investigators have suggested mechanisms by which regulator substances could repress or induce the condensation of chromatin, and it appears that some hormones may operate in this way. There is also evidence that acidic chromosomal proteins may somehow be directly involved in promoting transcription (Chapter 17).

TRANSLATIONAL CONTROLS IN EUKARYOTES

Many workers have proposed that translational controls of gene expression may be particularly important in eukaryotes, where there is evidence that at least some messenger RNAs exist in the cytoplasm for very long periods of time. Some of the following translational control mechanisms have been considered.

MASKING OF MESSENGER RNA

Messenger RNA in the cytoplasm may often be "masked" in some way, perhaps by forming a complex with proteins, so that

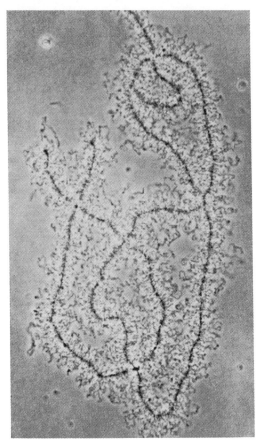

Figure 18-18 Light micrograph of lampbrush chromosomes in an oocyte of *Triturus viridescens*. (From Herskowitz, I.H. [1965], *Genetics,* Boston, Little, Brown & Company, Second Ed.)

translation is repressed; "unmasking" of the RNA will allow its translation. There is, in fact, good evidence that such a control mechanism may operate in certain cell types. For example, during early development of the fertilized egg, there is no RNA synthesis, even though this is an active metabolic period during which protein synthesis occurs on preexisting, maternal ribosomes. Consequently, mRNA used for protein synthesis during early development must be synthesized prior to this period and stored in the cytoplasm in some inactive (masked) form prior to its translation.

TRANSLATIONAL REPRESSORS

This model proposes that stimulation of enzyme synthesis may result from the decay of a labile translational repressor that is maintained by continuous transcription. Such a mechanism is supported by experiments which demonstrate that some enzymes are synthesized only when transcription is inhibited.

AVAILABILITY OF SUBSTANCES NEEDED FOR TRANSLATION

The availability of tRNA, free ribosomes, enzymes, and other necessary substances may control the rate of translation. As mentioned previously, however, it is not clear how control mechanisms of this type possess any inherent specificity.

TRANSPORT OF mRNA ACROSS THE NUCLEAR MEMBRANE

The passage of macromolecules across the nuclear membrane has been shown to fluctuate in response to changes in physiological conditions, and some workers have postulated that the nuclear membrane may exercise some control over which mRNA molecules pass into the cytoplasm. For instance, it has been proposed that some type of post-transcriptional modification of an mRNA molecule might be essential for its transfer from nucleus to cytoplasm. Many eukaryotic messenger RNA molecules have been shown to possess a long sequence of polyadenylic acid (50 to 200 adenylic acid residues) covalently bonded to the 3'-terminus. Because there is some evidence that polyadenylic acid sequences are added to mRNA after its transcription, it has been proposed that the polyadenylic acid sequences may be necessary for transport of the RNA to the cytoplasm, perhaps by facilitating the binding of a carrier protein.

THE MOLECULAR BASIS OF MUTATION

Now that we understand the chemical nature of the genetic code and the process by which the code is decoded into specific patterns of protein synthesis, we are in a good position to understand the molecular basis of hereditary variation, which is central to the evolutionary process. To put it simply: hereditary variation is due to changes, which we also call *mutations,* that occur in the base sequence of DNA. It is clear that if one base is permanently substituted for another, or if one or more bases are deleted from, or inserted into, a DNA molecule, the information contained in the DNA will probably be altered. Because such changes occur purely by chance, some mutations will be deleterious. This is particularly true of insertions or deletions, which may shift the reading frame of a sequence of DNA bases and thus may change much of the sequence of a protein (see Figure 18–3).

In multicellular organisms, we must distinguish between two types of mutations: (1) those that occur in somatic cells (*somatic mutations*) and, although inherited by descendants of the *mutated* somatic cell, are not passed on to the next generation; and (2) those that occur in germ cells (*germ cell mutations*) and *are* passed on to

526   The Genetic Code: Function and Evolution

the next generation. For the most part, somatic mutations are unimportant from an evolutionary standpoint because they are not transmitted to future generations. Of course, in some instances, somatic mutations may have severe, even fatal, effects on individual organisms in a population, and hence may adversely influence the reproductive capacity of these organisms.

SPONTANEOUS MUTATIONS

Mutations are often classified as either *spontaneous* or *induced*. Spontaneous mutations—those that are the result of inherent inaccuracies in replication—occur in organisms of a given species at a predictable frequency. For example, it has been estimated in *E. coli* that the spontaneous mutation rate is 4×10^{-10} per replicated base. This means that, per generation, the chance of a spontaneous mutation in any one base is 1 in 40 billion.

How do spontaneous mutations occur at the molecular level? One mechanism that has been proposed is the occurrence of faulty base pairing during replication, as a result of the tendency of bases to exist to a very small extent in alternative molecular forms. These

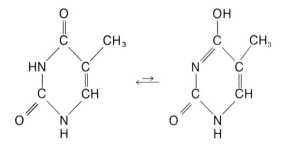

Figure 18–19   Chemical structures of adenine and thymine tautomers.

The Molecular Basis of Mutation

alternative forms occur because hydrogen atoms on each of the bases can change their locations. The alternative forms of a single base are called *tautomers* (Fig. 18–19). A rare tautomer of adenine can form hydrogen bonds with cytosine, and a rare tautomer of thymine can hydrogen bond with guanine (Fig. 18–20). Should such improper base pairing occur, it is likely to result in the permanent substitution of one base for another in the next generation. For example, hydrogen bonding of the rare form of adenine with cytosine, or the rare form of thymine with guanine, will result in the substitution of an A–T pair for a G–C pair in the next generation (Fig. 18–21).

In other cases, substitution of bases may occur because of defective DNA polymerases which permit incorrect base pairing during replication.

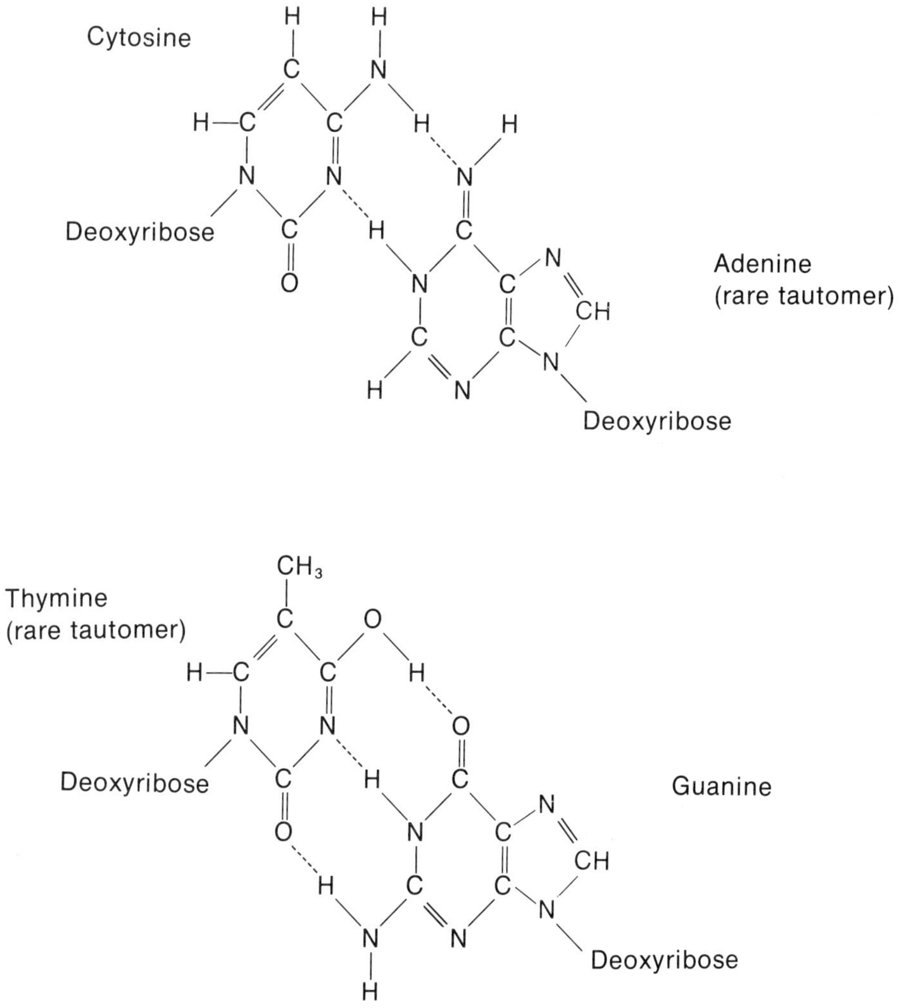

Figure 18–20 Hydrogen bonding between (a) cytosine and a rare tautomer of adenine, and (b) guanine and a rare tautomer of thymine.

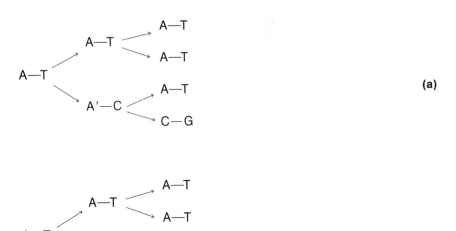

Figure 18-21 Hydrogen bonding of (a) a rare tautomer of adenine with cytosine or (b) a rare tautomer of thymine with guanine results in the substitution of a G–C pair for an A–T pair in the next generation.

INDUCED MUTATIONS

By far the greatest proportion of mutations are of the induced type; that is, they are brought about by chemicals or other agents called *mutagens*. Different mutagens may cause mutations in different ways.

Chemical Modification of Bases

Some chemicals cause mutations by chemically modifying the DNA bases. One effect of nitrous acid, a potent chemical mutagen, is to deaminate cytosine to uracil (Fig. 18–22). In the next round of replication, uracil will pair with adenine, resulting in the substitution of a U–A pair for a G–C one.

Intercalation of Chemicals in the DNA Helix

Other mutagens, in particular the acridine dyes, cause mutations by slipping in between bases (intercalating in the DNA molecule) and distorting the geometry of the double helix. This often leads to the insertion or deletion of bases during replication.

Ionizing Radiation

Ionizing radiation (x–rays, α–rays, β–rays and γ–rays) may also cause mutations. The cause of the mutagenic effect of ionizing radi-

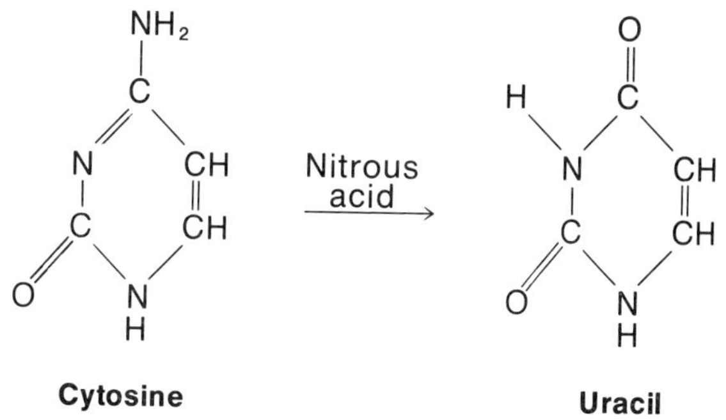

The Molecular Basis of Mutation

Cytosine —Nitrous acid→ **Uracil**

One effect of nitrous acid on DNA is to deaminate cytosine to uracil.   **Figure 18-22**

ation is not well understood, but such radiation may destroy or alter nitrogenous bases, or may produce double-strand breaks in DNA molecules which can lead to deletions or insertions of DNA fragments.

Ultraviolet Light

Another mutagen that has been well studied is ultraviolet light. Ultraviolet light is absorbed strongly by all bases, particularly the pyrimidine ones, and often causes the formation of *thymine dimers*—the covalent linking of thymines that lie adjacent to one another on the *same* polynucleotide chain (Fig. 18–23). Dimerization reduces the distance between nucleotides and produces a distortion in the DNA double helix that may block or cause errors in replication.

REPAIR PROCESSES

It should be emphasized that not all changes in DNA bases result in permanent effects. There is considerable evidence that altered DNA in *E. coli* is often repaired before replication. For instance, it has been shown that thymine dimers are often excised from a DNA strand by a process believed to occur as illustrated in

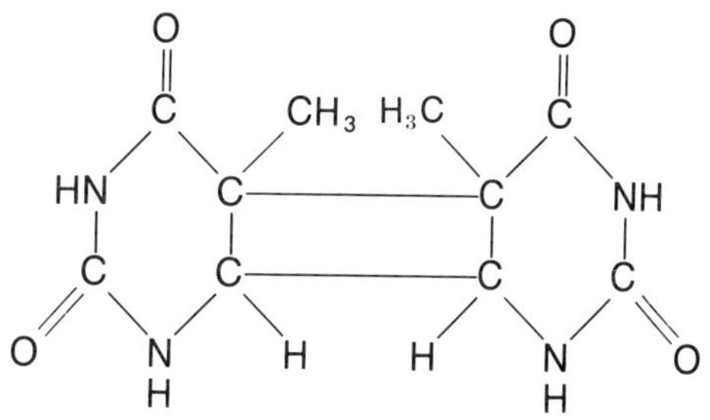

Chemical structure of a thymine dimer.   **Figure 18-23**

530 The Genetic Code: Function and Evolution

Figure 18-24. An endonuclease detects the distorted region and makes a nick at or near the dimer. The dimer, or a larger region including the dimer, is then excised in a reaction catalyzed by the 5'→3' exonuclease activity of DNA polymerase. DNA polymerase then catalyzes new synthesis that fills in the gap produced by excision of the dimer, and the newly synthesized fragment and the original polynucleotide strands are joined by DNA ligase.

Evidence is now accumulating that repair processes may be very important in all organisms in eliminating potential mutations.

ORIGIN OF THE HEREDITARY APPARATUS

Any acceptable theory that attempts to describe the origin of living systems must include an explanation of how the cell's hereditary

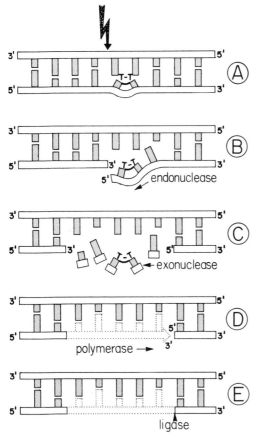

Figure 18-24 Diagram illustrating the mechanism by which thymine dimers are excised from DNA. (A) The production of a thymine dimer by ultraviolet light; (B) an endonuclease catalyzes a break in the polynucleotide strand containing the thymine dimer; (C) the dimer region is excised with aid of the 5'——→3' exonuclease activity of DNA polymerase; (D) DNA polymerase catalyzes repair of the excised region; and (E), DNA ligase catalyzes the end-to-end joining of the newly synthesized segment with the original DNA strand. (From DeRobertis, E.D.P., F.A. Saez and E.M.F. DeRobertis, Jr. [1975]. *Cell Biology*, 6th Ed., Philadelphia, W. B. Saunders Company.)

Origin of the Hereditary Apparatus

apparatus developed. Indeed, it is clear that the development of a system for storing and utilizing hereditary information is perhaps the most crucial step in the evolution of life, for evolution through natural selection requires some type of hereditary apparatus.

It is reasonable to assume that the first mechanisms of information storage and retrieval were very different from the highly complex mechanism we encounter in modern cells. Most workers have assumed that the earliest hereditary systems must necessarily have been rather inaccurate and relatively simple, and that accuracy (and complexity) were later evolutionary developments.

In discussing the development of the most primitive hereditary systems, we are referring to a time at the borderline between nonliving and living systems, when the process of biological evolution first became operative in its most rudimentary form. Previous to that time, chemical evolution may have given rise to reasonably stable molecular aggregates, of which coacervates and microspheres are prototypes. These molecular aggregates may have been subject to selection pressures of a primitive sort. For example, those aggregates that in some way acquired the ability to accumulate materials from their surroundings faster than others would grow in mass more rapidly; these rapidly growing aggregates would, in turn, more quickly reach a size at which they become unstable and undergo a primitive type of division. Eventually, as this process continued, we might expect that the fastest-growing aggregates would tend to predominate in the primitive broth, replacing more slowly growing forms.

It should be obvious that the selection process just described differs fundamentally from that of Darwinian natural selection. In the former process, there is no specified hereditary component, no defined *mechanism* by which an aggregate can acquire selectively advantageous characteristics (variation) and transmit them to its descendants.

A number of models have been proposed to account for the development of a hereditary system. These models fall into two general categories. One category of models suggests that small polynucleotides synthesized abiotically in the primitive broth were randomly incorporated into molecular aggregates. A few of these polynucleotides, because of their composition and base sequences, may have contained information that could be used to direct the synthesis of peptides whose functions were selectively advantageous to the aggregate. For example, a particular peptide may have endowed an aggregate with the ability to increase the rate at which it can accumulate material from the surroundings and with the ability to divide faster. Thus, those aggregates would be selected for which by chance had incorporated a polynucleotide that specified the synthesis of a useful peptide. Models of this type generally assume that the polynucleotide was capable of replication and served as a direct template for polypeptide synthesis; that is, the side chain group of a specific amino acid could bind directly only to certain areas of the polynucleotide template. Once attached to the

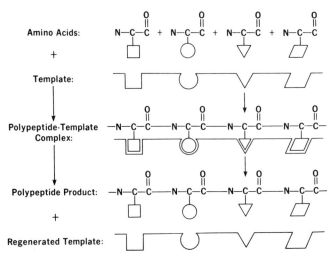

Figure 18-25 Schematic diagram of a direct coding mechanism. (From Black, S. [1973], Advances in Enzymology 38:210.)

template, the amino acids may have formed peptide bonds between them (Fig. 18–25).

Other selectively advantageous polynucleotides could be accumulated by the same mechanism, and gradually the amount of hereditary material in a primitive cell would increase. It is assumed that the primitive direct coding mechanism originally specified only classes of related amino acids (e.g., strongly hydrophobic amino acids) rather than individual ones. Eventually, however, increasing specificity would have evolved, and a change would have occurred from a direct coding system to the indirect one found in modern cells.

Another type of model assumes that the development of the hereditary system began with the incorporation into molecular aggregates of peptides that by chance possessed selectively advantageous functions. These peptides, in turn, served as templates for the formation of polynucleotides of specific sequence. The polynucleotides presumably could both replicate and serve as templates for the direct synthesis of more peptide. Additional polynucleotides (hereditary material) could have accumulated in the primitive cell by a repetition of this mechanism.

REFERENCES

Black, S. (1973). A theory on the origin of life. *Advances in Enzymology,* 38:193.
Brawerman, G. (1974). Eukaryotic messenger RNA. *Ann. Rev. Biochem.,* 43:621.
Brenner, S., F. Jacob, and M. Meselson (1961). An unstable intermediate carrying information from genes to ribosomes in protein synthesis. *Nature,* 190:576.
Britten, R.J. and E.H. Davidson (1969). Gene regulation for higher cells: a theory. *Science,* 165:349.

References

Chamberlin, M.J. (1974). The selectivity of transcription. *Ann. Rev. Biochem., 43*:721.

Cold Spring Harbor Symposium on Quantitative Biology (1969), Vol. 34, *Mechanisms of Protein Biosynthesis,* Cold Spring Harbor Laboratory, New York.

Cold Spring Harbor Symposium on Quantitative Biology (1970), Vol. 35, *Transcription of Genetic Material,* Cold Spring Harbor Laboratory, New York.

Grossman, L., A. Braun, R. Feldberg, and I. Mahler (1975). Enzymatic repair of DNA. *Ann. Rev. Biochem., 44*:19.

Harris, H. (1974). *Nucleus and Cytoplasm.* Clarendon Press, Oxford.

Haselkorn, R. and L.B. Rothman-Denes (1973). Protein synthesis. *Ann. Rev. Biochem., 42*:397.

Jacob, F. and J. Monod (1961). Genetic regulatory mechanisms in the synthesis of proteins. *J. Mol. Biol., 3*:318.

Losick, R. (1972). In vitro transcription. *Ann. Rev. Biochem., 41,* 409.

Miller, S.L. and L.E. Orgel (1974). *The Origins of Life on the Earth.* Prentice-Hall, Inc. Englewood Cliffs, New Jersey.

Nirenberg, M. (1968). The genetic code. *Nobel Lectures: Physiology or Medicine* (1963–1970), p. 372, American Elsevier Publishing Co., New York.

Orgel, L.E. (1973). *The Origins of Life: Molecules and Natural Selection.* John Wiley and Sons, Inc., New York.

Stryer, L. (1975). *Biochemistry.* W.H. Freeman and Co., San Francisco.

Weinberg, T.A. (1973). Nuclear RNA metabolism. *Ann. Rev. Biochem. 42*:329.

Woese, C.R. (1967). *The Genetic Code.* Harper and Row, Inc., New York.

CHAPTER 19

Epilogue

All attempts to make things comprehensible require the medium of theories, mythologies and lies, and a self-respecting author should not omit, at the close of an exposition, to dissipate these lies so far as may be in his power.

—Herman Hesse

The theory of evolution and the cell theory run as parallel themes throughout this book. Each complements the other, and the end result of this dual approach is to provide some coherence to the study of cell biology. But if there are benefits to such an approach, so also there are dangers: The reader of this book may take its evolutionary perspectives and speculations as fact.

In the final analysis, we must not forget that evolution is a historical process of the kind that cannot be repeated; currently held ideas on the origin and subsequent development of living systems are supported not by definitive proof, but by a massive accumulation of circumstantial evidence. In time, some of these ideas will receive additional support, some will no doubt be substantially modified, and some will be discarded.

Thus, it would be a mistake to adhere rigidly to the evolutionary scheme outlined in the previous chapters. On the other hand, the speculative nature of the scheme does not necessarily detract from its value. By allowing us to organize vast amounts of data and to stress the integration of cellular structure and function, the evolutionary approach enhances our understanding and appreciation of the living cell.

APPENDIX I

Microscopy

The purpose of any microscope is to make visible an object that normally cannot be seen, or cannot be seen clearly, by the naked eye. There are a number of different types of microscopes, and these vary in their ability to clearly distinguish separate parts of an image—a property of a microscope termed its *resolving power.* The resolving power can be defined in practice as the smallest separation at which we can distinguish two objects rather than one. Thus, for example, if two parts of an image are 0.01 μm apart, they can be resolved as separate entities by an electron microscope, which has a resolving power of 0.5 nm (0.0005 μm), but not by a light microscope, which has a resolving power of 0.2 μm.

An object can be seen in a microscope because of an interaction between it and the electromagnetic radiation (visible light, ultraviolet light, electrons, etc.) used to illuminate it. For instance, when waves composing the electromagnetic radiation pass *through* a transparent region of an object, they may be bent and slowed in velocity, or some waves may be absorbed; if waves encounter an opaque region of an object, they may be scattered (diffracted). Microscopes operate by converting such disturbances in the propagation of waves into an image of the object.

It is important to note, however, that waves will only be disturbed by an object that is sufficiently large in relation to the wavelength of the waves. If the wavelength is large (long), and the object is very small, the waves will not be disturbed, and the object will consequently remain invisible. Therefore, we conclude that the resolving power of a microscope is dependent upon the wavelength of electromagnetic radiation used to illuminate the object; the shorter the wavelength utilized, the better will be the theoretical resolving power of the microscope. The resolving power of a light microscope, for example, is approximately equal to one half the wavelength of light used to illuminate the object.

THE COMPOUND LIGHT MICROSCOPE

The *compound light microscope* uses visible light for illuminating the object and contains lenses that magnify the image of the object and focus the light on the retina of the observer's eye. In its simplest form, the compound microscope consists of two lenses, one at each end of a hollow tube (Fig. I–1a). The lens closer to the eye is called the *eyepiece,* and the lens closer to the object being viewed is called the *objective.* The object, supported by a glass slide under the objective lens, is illuminated by light beneath it. In some microscopes, a third lens, called a *condenser* is located between the object and the light source and serves to focus the light on the object. In practice, the eyepiece and objective lenses are each com-

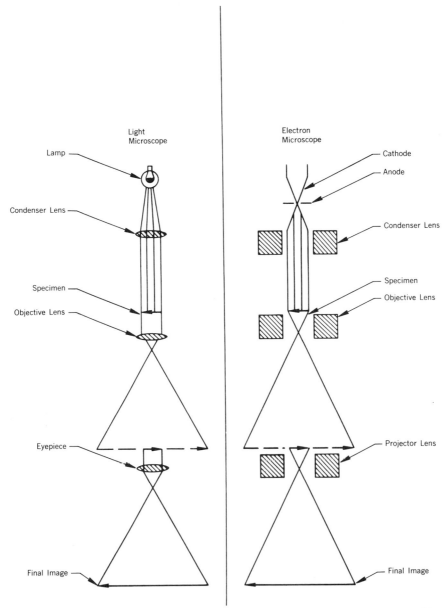

Figure I–1 Diagram illustrating the process by which thin sections are prepared for examination in the transmission electron microscope. (*From* Jensen, W.A. and R.B. Park [1967]. *Cell Ultrastructure,* Belmont, California, Wadsworth Publishing Company, Inc., © 1972 by Wadsworth Publishing Company, Inc. Reprinted by permission of the publisher.)

posed of a number of lenses that are combined in such a way as to overcome certain optical aberrations that occur when single lenses are used.

The limit of the resolving power of the best compound light microscope is about 0.2 μm. Thus, it will just resolve all but the smallest bacteria; however, the image will obviously not distinguish any internal detail in these cells.

In the compound light microscope, the quality of the image—its *definition* or *contrast*—depends primarily upon differences in the absorption of light by various regions of the object. Since, however, living cells are largely transparent, it is not always easy to see much internal structure when viewing even very large cells in the ordinary light microscope. Consequently, various staining methods have been developed that use colored dyes having affinity for specific cellular structures. Some of these dyes, or *stains,* do not kill the cells when added to the medium at low concentrations; such dyes are called *vital stains.* Two widely used vital stains are neutral red, which stains cytoplasm, and Janus green B, which selectively stains mitochondria.

Most staining methods require that, before staining, the specimen be treated with a preservative, also known as a *fixative,* that stabilizes the structures of the cell by precipitating proteins and other macromolecules and destroying hydrolytic enzymes. The fixative may be an acid such as acetic acid, an organic substance such as alcohol, or a water-soluble substance such as formaldehyde or glutaraldehyde. Fixation is generally followed by dehydration in an organic solvent. If the material to be examined consists of a block of tissue, it is usually embedded in wax after fixation. Because the wax has approximately the same hardness as the tissue, it is possible to cut sections of the tissue (usually 10 to 15 μm in thickness). The sections can be mounted on a glass slide and then stained with a nonvital stain to increase contrast.

Nonvital stains fall into two main classes: *acid stains* such as eosin, orange G, and fast green, which combine with basic molecules in the fixed cell, and *basic stains* such as hematoxylin, which combine with nucleic acids and other acidic molecules. Cellular structures that stain with acid stains are called *acidophilic* (acido = acid; philic = loving); those that stain with basic dyes are called *basophilic.*

THE PHASE CONTRAST MICROSCOPE

Vital dyes are not entirely harmless to cells, and often we desire to examine a cell without staining it, particularly if we wish to observe some physiological process like mitosis. Unstained living cells can be viewed in the *phase contrast microscope.*

The phase contrast microscope has the same resolving power as the ordinary light microscope but takes advantage of the fact that different parts of the cell may have slight differences in their *refractive indices,* defined as the ratio of the velocity of light in a vacuum to its velocity in a transmitting medium. Because light is transmitted through a structure at a velocity inversely proportional to the refractive index of the structure, light waves emerging from structures with differing refractive indices will be out of phase with one another. The phase contrast microscope is able to convert these differences in phase to differences in light intensity, producing an image with good contrast. In the ordinary light microscope, structures varying in refractive index are poorly contrasted because the transmitted light, although slowed in velocity by some regions of the object, is not diminished in intensity. Hence, all regions of the object appear to the eye to have equal brightness.

The phase contrast microscope has been successfully used for visualizing many physiological processes. For example, much of what we know about mitosis, meiosis, cell permeability, and endocytosis has been derived from the observation of living cells under the phase contrast microscope.

THE TRANSMISSION ELECTRON MICROSCOPE

The ordinary transmission electron microscope is similar in many respects to the light microscope (Fig. I–1b). Instead of using visible light to illuminate the object, the electron microscope uses a beam of accelerated electrons, and it focuses the electron beam with electromagnets (magnetic "lenses"). An image is formed when electrons strike a fluorescent screen or when they fall directly onto a photographic film, producing an electron micrograph.

Unlike the compound light microscope, in which image formation depends primarily upon differences in light absorption, the electron microscope forms images as a result of differences in the way electrons are scattered by various regions of the object. Electrons have a very low penetrating power; that is, they are easily scattered by objects in their paths. The degree to which electrons are scattered is determined by the thickness and atomic density of the object: Regions of high density (possessing, for example, atoms of high atomic number) scatter electrons more than regions of lesser density and consequently appear darker in the final image.

Because electrons are scattered so easily, the specimen used in electron microscopy must be extremely thin. Sections of material viewed in the transmission electron microscope are usually 10 to 50 nm thick (ultrathin sections). If the sections were not extremely thin, most of the electrons would be scattered and a uniform dark image would result. In fact, electrons are scattered even by gas molecules, and consequently the electron beam must travel through the electron microscope in a vacuum, and the samples must be dry (to remove water vapor) and otherwise nonvolatile.

Most biological material does not vary significantly in atomic density, and because of this, specimens must be prepared for elec-

The Transmission Electron Microscope

tron microscopy by the use of *electron stains.* These stains contain salts of heavy metals that react differentially with various cellular structures. Those areas of the cell that take up the heavy metal atoms become opaque to the electron beam and therefore appear dark in an electron micrograph.

Osmium tetroxide (OsO_4), alone or in combination with glutaraldehyde, is often used for both fixing and staining specimens. Once fixed or stained, tissues or cell pellets are embedded in plastic and sectioned; the sections are then supported on a fine wire mesh through which they can be viewed in the electron microscope (Fig. I-2). In some cases, the sections are also stained with uranyl ace-

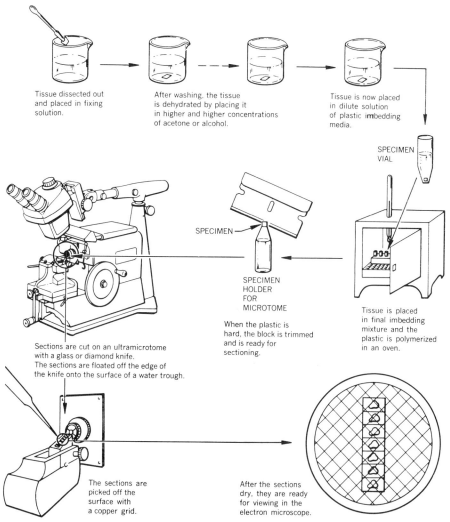

Figure I-2

Optical design of *(a)* the compound light microscope and *(b)* the transmission electron microscope. To emphasize the similarities between the two types of microscopes, the light microscope is shown inverted. (From Jensen, W.A. and R.B. Park [1967]. *Cell Ultrastructure,* Belmont, California, Wadsworth Publishing Company, Inc. © 1967 by Wadsworth Publishing Company. Reprinted by permission of the publisher.)

tate, lead citrate, or other salts of heavy metals to provide additional density.

In addition to thin sectioning, other preparative techniques may be used to prepare specimens for observation in the transmission electron microscope. These techniques include *freeze-etching, shadowing,* and *negative staining,* which are described in detail in the text.

Because accelerated electrons form an electron beam of very short wavelength, an electron microscope is capable of resolving structures as close together as 0.4 to 0.5 nm in biological specimens. Although resolution is good, however, the extensive preparative treatment needed for transmission electron microscopy casts some doubts on the reality of the final image. Many a "structural feature" observed in electron micrographs has later been shown to be an artifact of sample preparation.

THE SCANNING ELECTRON MICROSCOPE

The scanning electron microscope is similar in most respects to the transmission electron microscope. However, the former does not depend on the penetration of electrons through ultrathin sections to form an image. Rather, a beam of electrons is used to scan the specimen, and electrons that are scattered from the specimen are collected on a grid. When the signals from the grid are transferred to a cathode ray tube, a three-dimensional image is formed. The entire process is similar to the way an image is formed by the eye from light reflected from solid objects.

The resolving power of the scanning electron microscope is not as great as that of the transmission electron microscope, but because the former has an impressive depth of field, it has been used very successfully to observe the surface both of living cells and organisms.

REFERENCES

Culling, C.F.A. (1974). *Modern Microscopy.* Butterworth and Company (Publishers) Ltd., London.
Grimstone, A.V. (1968). *The Electron Microscope in Biology.* Edward Arnold (Publishers) Ltd., London.
Holwill, M.E. and N.R. Silvester (1972). *Introduction to Biological Physics.* John Wiley and Sons, London.
Oster, G., Ed. (1971). *Physical Techniques in Biological Research.* Volume IA, Optical Techniques. Academic Press, Inc., New York.
Slayter, E.M. (1970). *Optical Methods in Biology.* John Wiley and Sons, New York.

APPENDIX II

Some Practical Rules of Bonding

We have stated as a general rule that the orbitals which can be used in bonding must be those in the outer valence shell of an atom. We have seen, for example, that hydrogen has only one orbital available for bonding, the 1s orbital containing a single electron. Hydrogen is capable, therefore, of forming only one covalent bond with another atom. Carbon, with an atomic number of 6, has a filled 1s orbital (two electrons); its outer valence shell consists of one 2s orbital and three 2p orbitals, and hence it has a capacity of eight electrons. It possesses, however, only four electrons in this valence shell. Nitrogen (atomic number = 7) possesses five electrons in its outer valence shell, and oxygen (atomic number = 8) possesses six electrons in the outer valence shell.

Notice that in the case of carbon, nitrogen, and oxygen, the outer valence shell, consisting of the 2s and 2p orbitals, may hold up to eight electrons. This is also true for elements whose outer valence shell is formed by the 3s and 3p orbitals (see Figure 3–3). This observation forms the basis for an organizational rule, the *octet rule*, which states that *when an atom combines with another atom, it will share or transfer electrons until eight electrons, or four pairs, surround each atom (except, of course, for hydrogen, which can accommodate only two electrons).* There are numerous exceptions to the octet rule, but it works reasonably well for many elements that we encounter in living systems.

The number of electrons in the outer valence shell of a neutral atom is often represented by so-called *Lewis electron dot structures;* in this system, valence electrons are represented by dots surrounding the chemical abbreviation for each element. Lewis electron dot structures for the eleven most common elements found in living systems are shown in Table II–1. It must be remembered that these structures do *not* consider the orbital placement of electrons, but only the number of electrons present in the outer valence shell.

Once we know the number of electrons in the outer valence shell, we can use the octet rule to predict the chemical structure of molecules. To see how this is done, let us consider the chemical

Table II-1 LEWIS ELECTRON DOT STRUCTURES

H·	·Mg·	·C̈·	·N̈·	·Ö·	:C̈l·
Na·	·Ca·		·P̈·	·S̈·	
K·					

structure of methane (CH₄), a molecule whose bonding and geometry we have already mentioned.

Remember that carbon possesses four electrons in its outer valence shell (Table II–1) and thus requires an additional four electrons to fulfill the octet rule; hydrogen, an exception to the octet rule, must have two electrons surrounding it in the combined state. Both of these conditions are fulfilled in the structure of methane:

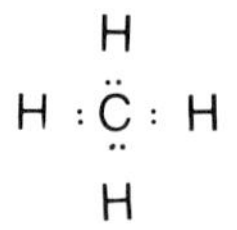

$$\begin{array}{c} H \\ H : \overset{..}{C} : H \\ H \end{array}$$

Each hydrogen shares its one electron with carbon, and each carbon shares its four electrons, one with each hydrogen. This is more obvious if we represent the electrons originally derived from hydrogen atoms by ×'s, and those derived from the carbon atom by circles, as in the following:

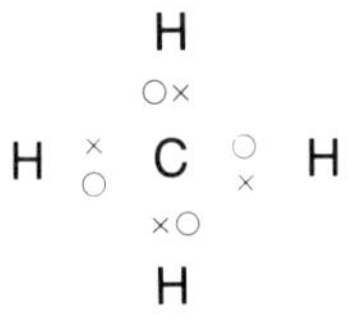

$$\begin{array}{c} H \\ {}^{O}_{\times} \\ H \,{}^{\times}_{O}\, C \,{}^{O}_{\times}\, H \\ {}^{\times}_{O} \\ H \end{array}$$

The end result of the electron sharing between the carbon and the four hydrogens in methane is that each hydrogen is surrounded by two electrons and the carbon is surrounded by eight.

Using the same procedure, we can draw a number of other molecules.

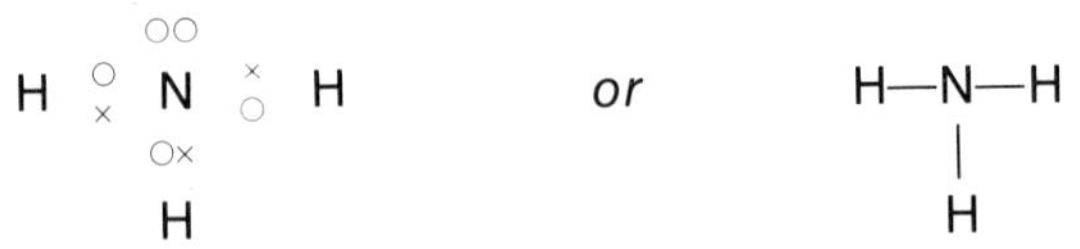

$$H \,{}^{O}_{\times}\, \overset{OO}{\underset{\overset{O\times}{H}}{N}} \,{}^{\times}_{O}\, H \qquad\text{or}\qquad \begin{array}{c} H-N-H \\ | \\ H \end{array}$$

Ammonia

$$H \,{}^{\times}_{O}\, \overset{OO}{\underset{OO}{O}} \,{}^{O}_{\times}\, H \qquad\text{or}\qquad H-O-H$$

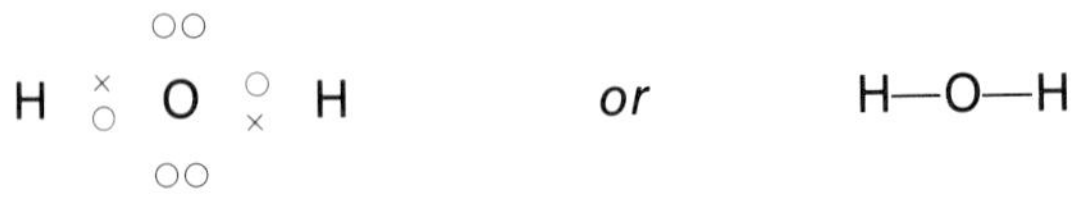

Water

Appendix II: Some Practical Rules of Bonding

$$O :\!\!:\!\! C :\!\!:\!\! O \quad\quad \text{or} \quad\quad O=C=O$$

Carbon dioxide

Notice how the double bonds are represented in the structure of carbon dioxide (CO_2). Since each double bond requires the sharing of *two* pairs of electrons, the octet rule is not violated. That is, each carbon and each oxygen atom is surrounded by eight electrons.

In considering the organic compounds in living systems, it is a useful general rule (but not without exceptions) that hydrogen forms one covalent bond, oxygen forms two, nitrogen forms three, and carbon forms four. The situation is more complicated with phosphorus and sulfur, since under certain circumstances phosphorus may form either three or five covalent bonds and sulfur may form two, four, or six.

The other five most abundant atoms in living systems—K, Na, Mg, Ca, and Cl—usually form ionic bonds involving the transfer of electrons, and these transfers obey the octet rule. For example chlorine, which possesses a high electron affinity, accepts one electron from another atom and thus is surrounded by eight electrons. Both potassium and sodium tend to lose the single electron that they contain in their outer valence shells to another atom, and magnesium and calcium tend to lose the two electrons in their outer valence shells. The result of these electron losses may be thought of as exposing the underlying valence shell, which is filled with eight electrons.

The composition of ionic compounds is determined by the charges carried on the constituent ions. To obtain electrical neutrality, the number of oppositely charged particles in a solid aggregate must be equal. The sodium ion, for example, has a charge of $+1$ (because it is formed by the loss of one electron from the neutral atom), and the chloride ion has a -1 charge (because it is formed by the gain of one electron). Therefore, solid sodium chloride must contain one sodium ion for each chloride ion and is written NaCl. The magnesium ion, formed by the loss of two electrons, carries a $+2$ charge; thus magnesium chloride ($MgCl_2$) must contain one magnesium ion and two chloride ions.

APPENDIX III

The pH Scale

The pH scale is based upon the hydronium ion concentration in pure water. Water, which acts as both a weak acid and a weak base, has the following equilibrium equation:

$$k = \frac{[H_3O^+] \times [OH^-]}{[H_2O]}$$

Since the concentration of H_2O does not vary appreciably in pure water or in dilute solutions, the equation can be rewritten

$$k \times [H_2O] = [H_3O^+] \times [OH^-] = k_w$$

k_w is called the ion product of water and is equal to 1×10^{-14}. Thus

$$[H_3O^+] \times [OH^-] = 1 \times 10^{-14}$$

In a neutral solution, in which the hydronium ion concentration equals the hydroxide ion concentration,

$$[OH^-] = [H_3O^+] = 1 \times 10^{-7} \text{ M}$$

and the pH of the solution is 7. This explains why 7 has been chosen as the midpoint of the pH scale.

Because of the constant ion product of water in dilute aqueous solutions, an increase in the hydronium ion concentration results in a decrease in the concentration of hydroxide ion. For example, if we add an acid to water in such a quantity that the final solution contains a hydronium ion concentration of 10^{-3} M, the hydroxide ion

Appendix III: pH Scale

concentration must equal 10^{-11} M. This is shown in the following calculations:

if $\quad [H_3O^+] \times [OH^-] = 1 \times 10^{-14}$

then $\quad [OH^-] = \dfrac{1 \times 10^{-14}}{[H_3O^+]}$

and $\quad [OH^-] = \dfrac{1 \times 10^{-14}}{1 \times 10^{-3}}$

$$[OH^-] = 1 \times 10^{-11} \text{ M}$$

Thus as the acidity of a solution increases, its basicity decreases, and vice versa.

APPENDIX IV

Mechanisms of Oxidative Phosphorylation

DIRECT CHEMICAL COUPLING HYPOTHESIS

It is not surprising that the first proposed mechanism of oxidative phosphorylation, formulated by E.C. Slater several decades ago, suggested a direct enzymatic coupling between phosphorylation and particular electron transfer steps in the respiratory chain—a mechanism that is used in several biochemical processes, including substrate-level phosphorylation.

According to the direct chemical coupling hypothesis, the coupling of electron transport to phosphorylation may occur in a number of steps as summarized by the following equations:

$$AH_2 + B + X + Y \rightleftharpoons A + BH_2 + X{\sim}Y \qquad (1)$$

$$X{\sim}Y + \text{Phosphate} \rightleftharpoons X{\sim}\text{Phosphate} + Y \qquad (2)$$

$$X{\sim}\text{Phosphate} + ADP \rightleftharpoons ATP + X \qquad (3)$$

In these equations A and B are oxidized components of the electron transport chain, AH_2 and BH_2 are the reduced forms, and X and Y are hypothetical intermediate compounds. Because there are apparently three phosphorylation sites at which coupling between electron transfer and phosphorylation must occur, the identity of A and B will vary, depending upon the particular phosphorylation site. Equations 1 through 3 may also be represented in diagrammatic form, as shown in Figure IV–1. This makes it somewhat simpler to visualize how the process is thought to operate.

Because there is experimental evidence that inorganic phosphate is not required for electron transport per se, it has been suggested that the common intermediate compound linking electron transport and phosphorylation does not contain phosphate. This

Direct Chemical Coupling Hypothesis

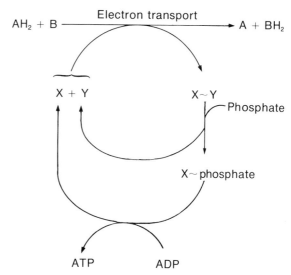

Figure IV-1

Diagrammatic summary of the direct chemical coupling hypothesis of oxidative phosphorylation. (Adapted from Howland, J.L. [1968], *Introduction of Cell Physiology*, New York, Macmillan, Inc., p. 119).

is why it has been postulated that there exist two substances, X and Y, which react to form the high energy intermediate $X \sim Y$, using the energy released in electron transport. It is thus $X \sim Y$ which is visualized as being intermediate to both the electron transport and phosphorylation processes.

The mechanism by which $X \sim Y$ is formed is not indicated in Equations 1 through 3 or in Figure IV-1, and it may occur in several steps as follows:

$$B + Y \rightleftharpoons B-Y \qquad (4)$$

$$B-Y + AH_2 \rightleftharpoons A + BH_2 \sim Y \qquad (5)$$

$$BH_2 \sim Y + X \rightleftharpoons BH_2 + X \sim Y \qquad (6)$$

Thus, B and Y are visualized as reacting to produce a low-energy compound (B—Y) which, as a result of reduction by AH_2 (Equation 5) acquires a high-energy conformation ($BH_2 \sim Y$). $X \sim Y$ is then formed by a group transfer of $\sim Y$ to X.

Once $X \sim Y$ is formed, the actual phosphorylation of ADP to ATP (Equations 2 and 3) occurs by the formation of a high-energy phosphorylated intermediate ($X \sim$ Phosphate) from $X \sim Y$, again by group transfer. In the final stage of the phosphorylation process, the high-energy phosphate group of $X \sim$ Phosphate is transferred to ADP to form ATP.

The chemical coupling hypothesis suffers from the fact that it has not been possible to isolate the proposed phosphorylated or nonphosphorylated chemical intermediates. This is not, however, as devastating to the hypothesis as it might at first appear, since such intermediates would be expected to be present in very low concentrations and to be very labile, and hence would probably be difficult to isolate. Indeed, isolation of such intermediates would be further complicated if they were bound to mitochondrial membranes, which would seem likely. One strong point in favor of the direct chemical coupling hypothesis is that it employs a mechanism which has been found to operate in other biological processes.

CHEMIOSMOTIC HYPOTHESIS

In 1961, Peter Mitchell proposed that oxidative phosphorylation is driven, not by chemical coupling between electron transport and the phosphorylation reactions, but by a proton gradient (differential H^+ concentration) that arises between the two sides of the inner mitochondrial membrane (cristae) during electron transport. Mitchell suggests that because the establishment of a proton gradient is an endergonic process, driven presumably by energy released during electron transport, it is not unreasonable to assume that the spontaneous collapse of such an improbable state would release energy that could be used to drive phosphorylation.

ESTABLISHMENT OF A PROTON GRADIENT

Mitchell proposes that the establishment of a proton gradient across the inner mitochondrial membrane occurs during electron transport because the components of the respiratory chain are so aligned in the inner membrane that the hydrogen carriers, as distinct from the pure electron carriers, pick up protons from the mitochondrial matrix and release them into the outer mitochondrial compartment. Such a hypothesis requires that hydrogen carriers in the chain alternate with pure electron carriers, as illustrated in Figure IV–2. It also requires that the membrane be relatively impermeable to H^+ and OH^- ions, so that any gradient that is established will be maintained. The net effect of such an arrangement is that respiratory chain activity will "pump" protons from the mitochondrial matrix to the outer mitochondrial compartment, creating a differential concentration of protons on either side of the inner membrane. (Since protons are derived from the ionization of water [$H_2O \rightleftarrows H^+ + OH^-$], it is clear that the translocation of protons from the mitochondrial matrix to the outer compartment will also create an OH^- gradient across the inner mitochondrial membrane).

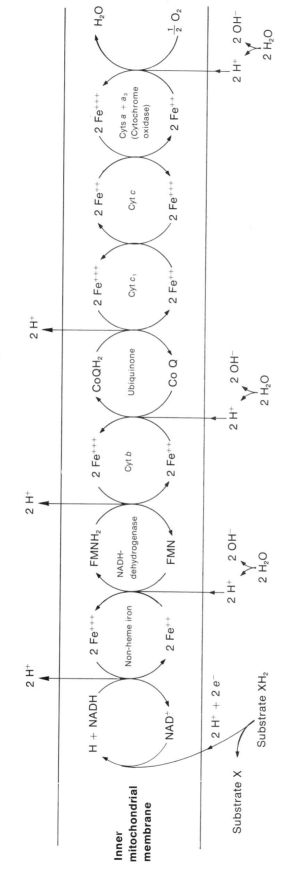

Figure IV-2

A possible arrangement of electron carriers in the inner mitochondrial membrane, as required by the chemiosmotic hypothesis. Because hydrogen carriers alternate with pure electron carriers, the net effect of electron transfer is to "pump" hydrogen ions (protons) from the mitochondrial matrix into the outer mitochondrial compartment, thus establishing a gradient of protons (or hydroxyl ions) on the two sides of the membrane. (Adapted from Wolfe, S.L. [1972]. *Biology of the Cell*, Belmont, California, Wadsworth Publishing Company, Inc., p. 117, © 1972 by Wadsworth Publishing Company, Inc. Reprinted by permission of the publisher.)

PHOSPHORYLATION

It is the proton gradient, according to Mitchell, that powers the phosphorylation of ADP to ATP. For this to occur, Mitchell assumes the existence of a membrane-bound reversible ATPase capable of catalyzing the reaction

$$ATP + H_2O \xrightleftharpoons{ATPase} ADP + H_3PO_4 \quad \Delta G° = -7.3 \text{ kcal/mole} \quad (7)$$

Since this reaction proceeds strongly in the direction of ATP hydrolysis, it is quite clear that a reduction in the concentration of water at the site of the enzyme would force the reaction in the reverse direction. Now Equation 7 can be rewritten as follows:

$$ATP + \underbrace{2H^+ + O^{-2}}_{\updownarrow \; H_2O} \xrightleftharpoons{ATPase} ADP + H_3PO_4$$

In this case, it is apparent that any process that lowers the concentration of H^+ and O^{-2} at the enzyme will favor ATP formation by reducing the concentration of H_2O, providing that the membrane itself is also poorly permeable to water, as is assumed.

ATP synthesis, then, is accomplished by the dehydration of ADP and H_3PO_4 and would be accompanied by the formation of protons and O^{-2} ions. Mitchell suggests that the ATPase that catalyzes the dehydration reaction is oriented in the inner membrane in such a manner that protons produced from the reaction are transferred to the mitochondrial matrix on the inside of the membrane and that O^{-2} ions are transferred to the outer compartment on the outside of the membrane. The synthesis of ATP would, therefore, reduce the proton gradient by the formation of water on both sides of the membrane, as shown in Figure IV–3. Thus, the synthesis of ATP can be thought of as being driven by the proton gradient. To put it another way, if the active center of the ATPase is inaccessible to water because of impermeability of the membrane, dehydration of ADP and H_3PO_4 at the enzyme may be thought of as occurring by the removal of H^+ from the membrane into the inner compartment and the removal of O^{-2} in the opposite direction. In a sense, then, *the formation of ATP is driven by the reduction of the charge difference on the two sides of the membrane.* Because three ATP molecules are phosphorylated for every six protons translocated across the membrane, it has been assumed that the translocation of two protons is required for the formation of one molecule of ATP.

Actually, Mitchell's hypothesis is somewhat more complicated than this description implies. It is assumed that the charge separation across the inner mitochondrial membrane gives rise to an

Chemiosmotic Hypothesis

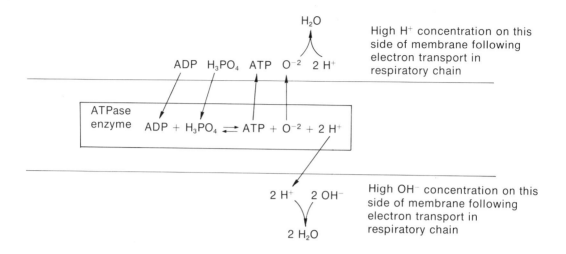

Figure IV-3. Schematic diagram illustrating the phosphorylation of ADP as described by the chemiosmotic hypothesis. See text for further details.

energy differential that produces the high-energy intermediate $X \sim Y$ by a complicated mechanism. It is then $X \sim Y$ that brings about the dehydration of ADP and H_3PO_4. Thus, although the chemiosmotic hypothesis does not propose a phosphorylated intermediate as does the direct chemical coupling hypothesis, both hypotheses postulate the formation of a high-energy intermediate. However, in the chemiosmotic hypothesis, the high-energy intermediate is used only in the final stages of ATP production and is not, as in the direct coupling hypothesis, the common intermediate by which electron transport is coupled to phosphorylation. Rather, the chemiosmotic hypothesis proposes that electron transport and phosphorylation are coupled by the generation of a proton gradient formed by the translocation of protons as a result of respiratory chain activity. This is shown in Figure IV-4, which should be compared with Figure IV-1 in order to demonstrate the main differences between the chemiosmotic and direct chemical coupling hypothesis.

Mitchell's chemiosmotic hypothesis is supported by experiments which show that respiration in mitochondria is associated with proton translocation out of mitochondria, and by other experiments which demonstrate that the formation of an artificial pH gradient across the mitochondrial inner membrane is accompanied by a short burst of ATP synthesis. As mentioned in Chapter 14, Jagendorf has also observed the phosphorylation of ADP by isolated thylakoid membranes as a result of an abrupt change in the pH of the medium, suggesting that a proton gradient may drive phos-

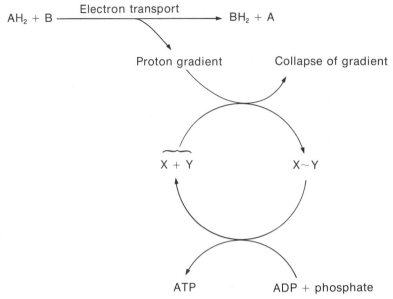

Figure IV-4 Diagrammatic summary of the chemiosmotic hypothesis of oxidative phosphorylation.

phorylation in chloroplasts as well as mitochondria. Another fact that supports Mitchell's hypothesis is the finding that the mitochondrial inner membrane is, in fact, relatively impermeable to protons and hydroxyl ions; this is, of course, essential if a proton gradient is to build up across the membrane. It has not been possible, however, to demonstrate that the arrangement of electron carriers shown in Figure IV-2 actually exists. In particular, it seems unlikely that coenzyme Q is located between cytochromes b and c_1, since there is no evidence that flavoproteins react directly with cytochrome b.

CONFORMATIONAL COUPLING HYPOTHESIS

A third hypothesis, proposed within the last few years by David Green and his associates, suggests that electron transport produces an excited state in macromolecular components of the mitochondrial inner membrane; the excited state presumably arises as a result of configurational changes in these components. It is proposed that the return of the excited molecules to their original low-energy configuration releases sufficient energy to drive the phosphorylation of ADP to ATP. In support of this idea, electron micrographs of isolated mitochondria show different configurations when they are inactive and when they are actively phosphorylating. However, it is not clear how the configurational changes in mitochondria are related to the configuration of the macromolecular components of the mitochondria, or how the dissipation of the energy of the excited state is coupled to ADP phosphorylation.

Index

Numbers in *italic* refer to illustrations; numbers followed by a (t) indicate tables.

Absorption, of light, 362–365. See also *Light, absorption of.*
Absorption spectrum, in photosynthetic pigments, 361, *372*
Accessory pigments, 363
Acetyl coenzyme A, 383–385, *385*, 386(t)
Acid(s), 76–79
　fatty. See *Fatty acid(s).*
　nucleic. See *Nucleic acids.*
Acid stains, 537
Adaptation, in metabolic development, 329
Adenine, 200
Adenine tautomer, *526*, *527*
Adenosine, prebiotic synthesis of, 206
Adenosine diphosphate. See *ADP.*
Adenosine triphosphate. See *ATP.*
ADP, 187
　formation of, 96, *97*
　in Embden-Meyerhof pathway, 351
　phosphorylation of, 320, *321*, 332, *376*, 550–552
　light-driven, 365–367, *366*
ADP-ATP system, in energy transfer, 94–97, *94*
Adsorption, of molecular aggregates, 231–232
Aerobe(s), anaerobic metabolism in, 394–395
　definition of, 323
　glucose oxidation in, 339
Aerobic autotrophs, in evolution, 38
Aerobic chemoautotrophs, *337*
Aerobic heterotrophs, 38
Aerobic metabolism, 383–405
　definition of, 324
Aerobic oxidation, in eukaryotes, 402
Aerobic respiration, definition of, 324
　localization of, 401–404
　pathways of, 335–338
　regulation of, 399–401
Akinetes, 255–257
D-Alanine, *146*
L-Alanine, *146*
Alcohol, ethyl, oxidation to acetaldehyde, 123–124
Aldehydes, in chemical evolution, 199
Alga(e), blue-green. See *Blue-green alga(e).*
　brown, evolutionary role of, 4
　green, in Bitter Springs fossils, 36
Alpha-carbon, 140

Alternation of generations, 491, *492*
Amino acid(s), 8–10, 139–152
　acidic, 143–144, *143*
　activated, 508
　as buffers, 80
　basic, 144–145, *144*
　coupling of, to tRNA, 507–509, *509*
　general structure of, 140
　in enzymes, 104
　in Murchison meteorite, 203(t)
　incorporation into proteinoids, 210
　neutral, 141–143, *142*
　nitrile-mediated condensations of, 212–213
　optical activity of, 145–149
　peptide bonds in, 149–152
　polymetaphosphate ethyl ether as condensing agent for, 211
　prebiotic synthesis of, 195–204
　production of, kinetics of, 197, *198*
　simulation experiments for, 195–204
　Strecker synthesis of, 198
　thermal condensation of, 207
　thermal polymerization of, 210
Amino acid residues, 149
Amino group, 139
Aminoacyl-AMP, 508
Aminoacyl-tRNA synthetase, 507
Ammonia, solubility of, 73, *73*
Ammonium cyanide, adenine synthesis from, 200
Amoeba(e), lysosomes in, 288
AMU, 24, 24(t)
Amylopectin, *172*
Amylose, *172*
Anabaena, gas vacuoles from, *252*
　heterocyst of, *257*
Anabolism, 6
　definition of, 317
　relation to catabolism, 317, 318, *319*
Anacystis nidulans, septum formation in, *466*
Anaerobe(s), definition of, 323
　facultative, 38
　　definition of, 323
　　in evolution, 38
　glucose oxidation in, 339
Anaerobic autotrophs, origin of, 37
Anaerobic bacteria, 333, *334*
Anaerobic carbohydrate metabolism, pentose phosphate pathway and, 351

Anaerobic chemoautotrophs, 331
Anaerobic heterotroph(s), ancestral organism as, 37
 evolution of, 37
 metabolism of, 340–357
Anaerobic heterotrophy, as fundamental metabolic mechanism, 328
 pathways of, 329–330, *330*
Anaerobic metabolism, definition of, 324
 in aerobes, 394–395
Anaerobic oxidation, of organic molecules, 321
Anaerobic photosynthetic bacteria, 331
Anaerobic respiration, 329, *330*
 definition of, 324
Anaphase, *479*, 482–483
Anaphase I, 489
Anaphase II, 490
Ancestral organism, as anaerobic heterotroph, 37
Anemia, sickle cell, 152, *154*
Animalia, evolutionary role of, *4*
Anion, 56
Anomer, 166
Anomeric carbon, 166
Antibodies, 139
Anticodon, 507
Anticodon loop, 507
Aqueous solution, ionic bonds in, 58
Archaeosphaeroides barbertonensis, 29–31, *30*
Arginase, pH optimum of, 106
Arginine, 144, *144*
Aristotle, 12
Asparagine, 143, *143*
Aspartic acid, 143, *143*
Asters, 481
Asymmetric carbon, 145
Atmosphere, primitive, 118–131, *135*
 chemical properties of, 122–128
 composition of, 16, 128–131
 escape of, 118–122
 free oxygen in, 126–127
 inert noble gases in, 118
 oxidation-reduction reactions in, 122–128
 oxygen accumulation in, 337, 358–382, 383
 reducing character of, 125–128
Atom(s), chemical properties of, effect on mass of, 24
 chemical reactivity of, 48
 composition of, 23–26, 24(t)
 electrons in, 23, 24(t)
 ground state of, 47
 mass number of, 24
 mass of, effect on chemical properties of, 24
 multiple bonding in, 50–51
 neutral, protons in, 23–24, 24(t)
 nucleus of, composition of, 23
 orbitals of. See *Atomic orbital(s).*
 structure of, 44–60
 valence shell of, 47
 weak interactions between; 57–60
Atomic mass unit (AMU), 24, 24(t)
Atomic nucleus, composition of, 23
Atomic number, 24
Atomic orbital(s), 44, *44, 45, 52*
ATP, 187
 catabolism and, 317
 chemical structure of, 96, *97*

ATP (*Continued*)
 high energy bonds in, 95–96
 hydrolysis of, 95, *97*
 in active transport, 431
 in ADP-ATP system, 94–97, *94*
 in anaerobic chemoautotrophs, 331
 in carbon dioxide fixation, 365
 in cell, function of, 10
 in Krebs cycle, 389
 in pentose phosphate pathway, 355
 synthesis of, 550–552
Autoradiography, 459
Autotrophic carbon dioxide fixation, 331
Autotroph(s), aerobic, 38
 anaerobic, origin of, 37
 chemosynthetic, 6
 photosynthetic, 6
 evolution of, 111
Avery, Oswald, 445

Bacillus cereus, 256
Bacteria, anaerobic photosynthetic, 331
 and *Eobacterium isolatum,* similarities in, 28, 30
 capsule of, 242
 differentiation from blue-green algae, 237
 in Bitter Springs formation, 36
 prokaryotic structure of, 19
 purple, absorption spectra of, *362*
 bacterial chlorophyll a in, 361
 transcriptional controls in, 518–520
 translational controls in, 521
Bacterial chlorophylls, 361
Bacterial photosynthesis, light reactions of, 360–369
Barghoorn, E.S., 28
Basal body(ies), 302
 of prokaryotic flagellum, 242, *245*
Base(s), 76–79
 chemical modification of, mutation from, 528
 chemical properties in water, 76
 strong, 77
 weak, 77
Basic stains, 537
Beck Springs dolomite, 35, *35*
Beneden, Edouard van, 486
Bilin, 251
Bimolecular leaflet model of cell membrane, 414
Biological evolution. See *Evolution, biological.*
Biological system(s), chemical nature of, 138–190
 origin and early development of, 37–38
Bioluminescence, 335, *336*
Bitter Springs formation, 31–32, 35–36, *36*
Black, Simon, 193
Blendor experiment, for DNA, 447–450, *451*
Blue-green alga(e), and *Archaeosphaeroides barbertonensis,* similarities of, 29–31, *30*
 and Bulawayan stromatolites, 33–34, *34*
 and Gunflint stromatolites, *34*, 35
 and Soudan stromatolites, 34
 carotenoids of, 251
 differentiation from bacteria of, 237
 electron transfer in, *374*
 evolutionary role of, *4*
 in Bitter Springs formation, 36, *36*

Index

Blue-green alga(e) (*Continued*)
  light reactions in, 369–377
  precursors of, 37
  prokaryotic structure of, 19
  sheath of, 242
  thylakoids in, 249, *249*
Bond, chemical, 48–60
  energy of, 51, 52(t)
  ionization and, 57
  covalent, 49–50
  coordinate, 50
  high energy, 95–96
  hydrogen, 58, *59*
  hydrophobic, 60
  ionic, 56–57
    in aqueous solution, 58
  isopeptide, 160
  multiple, 50–51
Bond angle, 53
Bond dipole moment, 54
Bonding, chemical, practical rules for, 541–543
Branton, Daniel, 419
Brooks, J., 31
Brown algae, evolutionary role of, 4
Buffer(s), 80–82
Bulawayan stromatolites, 33–34, *34*

Calcium pump, 440–441
Calvin cycle, 365, 378–380, *379*
Cambrian Period, 25(t)
Carbohydrate(s), 164–171
  glucose as, 10
  in living systems, function of, 10
Carbohydrate metabolism, 340
  anaerobic, pentose phosphate pathway and, 351
  relation to other metabolic processes, 396–399, *399*
Carbon, 61
  anomeric, 166
  asymmetric, 145
  stability of, 60
Carbon cycle, 75
Carbon dioxide, biological role of, 74–82
  double bonding and, 51
  hydrogen ion concentration and, 75
  passive diffusion of, 429
  reduction by anaerobic bacteria, 333, *334*
Carbon dioxide fixation, autotrophic, 331
  development of, 333
  in bacterial photosynthesis, ATP in, 365
Carbon tetrachloride, 55, *56*
Carbonaceous chondrite, chemical analysis of, 202–204
Carbonic acid–bicarbonate buffer system, 81–82
Carboniferous System, 25(t)
Carboxyl group, 139
Carotenes, 361
β-Carotene, chemical structure of, *182*
Carotenoid(s), 180, 333
  absorption spectrum of, *372*
  energy transfer in, 363
  in photosynthesis, 360
  of blue-green algae, 251
Carrier molecules, 428
  in facilitated diffusion, 430

Catabolic reactions, 94
Catabolism, 6
  definition of, 317
  relation to anabolism, 317, 318, *319*
    Embden-Meyerhof pathway and, 351, *352*
Catabolite repression, 520
Catalase, in aerobes, 335
Catalysts, 102–110, *103*
Cation, 56
Cell(s), ancestral, 192
  animal, size of, 18–19
  "average," molecular composition of, 40(t)
  bacterial, size of, 18–19
  carbohydrates in, glucose as, 10
  chemical constituents of, 39–43
  compartments of, metabolite transfer in, 408
  control of Embden-Meyerhof pathway by, 350–351
  design of, evolutionary significance of, 11
    universal features of, 11
  diploid, 476
  division of. See *Cell division*.
  dry weight of, 40, 40(t)
  elemental composition of, 40–43
  energy and, 83–112
  energy capture in, 320–321
  energy storage in, ATP in, 10
  energy transfer in, 94–97, *94*
  entropy preservation by, 111
  enzymes in, function of, 10
  eukaryotic, origin of, 38
    structures of, *20*, 21–22. See also names of specific structures.
  evolution from molecules, 217–233
  flexibility of, 217
  free energy coupling in, mechanism of, 97–99
  gap junction, *272*, 273
  glucose uptake by, 341
  haploid, 486
  ionic composition of, 42
  integration of, 5
  intermediate junction, *270*, 272
  interphase, chromatin in, 284
  loose junction, *270*, 272
  metabolic adaptability of, 235
  metabolic pathways in, 6–8
  metabolism of, design of, 7, 11
    evolution of, 317–339
    plasma membrane and, 406
    processes of, 7
  nucleus of, 283–286, *284*
    in Beck Springs fossils, 35, *35*
  organization of, 6
  oxidation in, 320
  permeability of, 225–226
    to solutes, 427–428
  plant, size of, 18–19
  plasma membrane of, metabolic activity and, 406
  primary and secondary elements in, 41–42
  prokaryotic, 19
    structures of, *22*, 22–23. See also names of specific structures.
  protein synthesis in, 10
  recycling of substances in, 8, *9*
  reduction in, 320
  refractive indices of, 537

Cell(s) (Continued)
  "resting," 235
  size of, in control of cell division, 493
  somatic cytokinesis in, 483–485
    nuclear division in, 475–485
  stability of, 217
  structure of, alteration by fixation and desiccation, 237
    study of, 234–237
    variability of, 235
  tight junction, 270, 273
  trace elements in, 41, 42
  tumor, 407
  "typical," 234
  water in, 39
Cell division, 443–497
  evolution of, 494–496
  in eukaryotes, 467–475
    control of, 493–494
  in prokaryotes, 453–467
  plasma membrane in, 406
Cell membrane, adsorbed protein in, 414
  bimolecular leaflet model of, 414
  chemical composition of, 408–412, 409(t)
  cholesterol in, 410
  Danielli-Davson model of, 413–415, 414, 415
  effect of environment on, 428
  facilitated diffusion and, 430
  fluid mosaic model of, 422–423, 422
  freeze-etching of, 419–421, 419, 420
  function of, 6
  glycolipids in, 410, 411
  glycoproteins in, 412
  integral proteins in, 422
  internal, size of, 19
  lipid bilayer model of, 413
  lipids in, 409–411
  molecular organization of, 412–423
  origin of, 441–442
  particulate structure of, 419, 420
  peripheral protein in, 422
  permeability of, 424–428
    lipid solubility and, 429
  phosphatidic acids in, 409
  phosphoglycerides in, 409
  phospholipids in, 409
  protein crystal model of, 421–422, 421
  protein in, 411–412
    study of, use of detergents, 411–412
  selectively permeable, 425
  sphingomyelins in, 409, 410
  sphingosine in, 409, 410
  structure of, 408–423
    summary of, 423
  transport in. See Membrane transport.
  unit membrane hypothesis of, 415–418, 416
  unit membrane of, 416
Cell plate, 485, 485
Cell theory, 1–2
  microscope technology and, 18–19, 19
Cell wall, in Eobacterium isolatum, 28
  in eukaryotic cells, 21
  in prokaryotes, 22, 239–242
Cellular differentiation, gene expression and, 517
Cellulose, 169–171, 170
Cenozoic era, 25(t), 27

Centrioles, 20, 302, 302, 481
Centromere, 286, 480
  structure of, 480–481
Cesium chloride density gradient equilibrium sedimentation, 455, 456
Chain elongation, in RNA translation, 511–513, 513
Chain termination, in RNA translation, 513–514
Chalones, in cell division control, 493
Chase, Martha, 447
Chemical bonding, 48–60
  practical rules for, 541–543
Chemical equilibrium, 77
  free energy and, 92–93
Chemical kinetics, 99–112
Chemical reaction, completion of, 79
Chemical selectivity, of living organisms, 42–43
Chemical synthesis, light absorption in, 132–135
Chemical thermodynamics, 83–99
Chemiosmotic hypothesis of oxidative phosphorylation, 548–552, 549, 551
Chemoautotrophs, aerobic, 337, 337
  anaerobic, 331
Chitin, 171
Chlamydomonas, 36
  flagellum of, 304
Chlorobium chlorophyll, 361
Chlorophyll(s), bacterial, 361
  Chlorobium, 361
  porphyrins as, 333
Chlorophyll a, absorption spectrum of, 372
  bacterial, chemical structure of, 250
  chemical structure of, 250
Chlorophyll b, absorption spectrum of, 372
Chloroplast(s), 21, 295, 297–298, 299
  protein synthesis in, 516–517
Chloroplast thylakoid, 420
Cholesterol, 180, 183
  in cell membrane, 410
Chondroids, of prokaryotic cells, 248
Chromatid(s), 286, 480
  sister, 480
Chromatin, 20
  eukaryotic, composition of, 467–471
  organization of, 471–472
  structure of, gene expression and, 472–473
Chromosome(s), 284
  homologous, 476
  in eukaryotic cells, 21
  lampbrush, 522, 524
  movement of, 483
  organization of, 473–474
  polytene, 521, 522
  secondary constrictions in, 481
Chromosome puffs, 521, 523
Cilia, 302–306, 303
  longitudinal structure of, 305, 305
  motion of, sliding filament model of, 306
Citric acid, in Krebs cycle, 387, 388
Citric acid cycle. See Krebs cycle.
Classical theory of eukaryotic cell origin, 312–315
Clays, molecular aggregates on, 232
Cleavage furrow, 300
Cloud, P.E., 35

Index

Coacervate(s), 227–231, *227, 228, 230*
Codon(s), 503–504
   mRNA, recognition by tRNA, 509–510, *509*
Coenocytes, 483
Coenzyme(s), 105
Coenzyme A, 383
   chemical structure of, 384, *385*
Coenzyme Q, 367, *367*
   in respiratory chain, 391
Cofactors, 105
   prosthetic group of, 105
Coiled bodies, 284
Cold accretion theory of formation of earth, 115–117, *116*
Collision theory of formation of earth, 115
Color, wavelength and, *361*
Compartmentalization, in eukaryotic cells, 262–264
Compound(s), chemical, definition of, 48
   covalent, in oxidation-reduction reactions, 123–125
   organic, complex, prebiotic synthesis of, 204–215
      in meteorites, 202–204
      prebiotic synthesis of, 191–216
      simple, prebiotic synthesis of, 195–204
Compound light microscope, 535–537, *536*
Concentration gradient, in passive diffusion, 429
Condensations, nitrile-mediated, in prebiotic synthesis of peptides, 212–213
Conformational coupling hypothesis of oxidative phosphorylation, 552
Conjugated proteins, 163–164
Contact inhibition, 407
   in cell division control, 494
Cork, structure of, *2*
Correns, Carl, 444
Cosmozoa, 15
Coupled reaction, 93
Covalent bond(s), 49–50, *49*
Covalent compounds, in oxidation-reduction reactions, 123–125
Cretaceous Period, 25(t)
Crick, Francis, 451
Crossing over, 488, *488*
C-terminal residue, 149
Cyanamide, in dehydration condensation reactions, 206
Cyanide, inhibition of active transport by, 432
Cyclic AMP, in cell division control, 493–494
   in transcription, 519
Cyclic photophosphorylation, 365–367, *366*
Cytochrome(s), 164, 333
   in bacterial cyclic photophosphorylation, 366, *366*
Cytochrome oxidase, 392
Cytokinesis, in germ cell formation, 491
   in prokaryotes, 464–465
   in somatic cells, 483–485
Cytoplasm, 19, *19*
   filamentous structures of, 296–301
   of prokaryotic cell, 246, *247*
Cytoplasmic NADH, oxidation of, 393–394, *394*
Cytoplasmic organelles
   in Beck Springs fossils, 35, *35*
   in eukaryotes, 21

Cytoplasmic streaming, 264
Cytosine, prebiotic synthesis of, 201

Danielli-Davson model of cell membrane, 413–415, *414, 415*
Darwin, Charles, 2–3
Dehydration condensation reaction(s), 149
   cyanamide in, 206
   in aqueous solutions, 205
   in prebiotic synthesis of organic compounds, 205–206, *206, 207*
   nitriles in, 205, *206*
Dehydrogenation, 123
Denaturation, of enzymes, 105–106
   of protein, 161
Deoxyadenosine, prebiotic synthesis of, 206
Deoxyribonucleic acid. See *DNA.*
Deoxyribose, 165
   in DNA, 451
Deplasmolysis, 426, *426*
Desmosome, *270*, 272
Devonian Period, 25(t)
de Vries, Hugo, 444
Dielectric constant, 72–74
Dictyosomes, 278
Differential reproduction, 3, *3*
Diffusion, 225
   facilitated, 430
   in hypertonic solution, 226, *226*
   in hypotonic solution, 226, *226*
   in isotonic solution, 226, *226*
   passive, 429–430
Digestive vacuole, 272
2,4-Dinitrophenol (DNP), 396
Dinitrophenol, inhibition of active transport by, 432
Diplococcus pneumoniae, transformation experiments with, 445, *446*
Direct chemical coupling hypothesis of oxidative phosphorylation, 546–548, *547*
Disaccharides, 168
Dissociation energy, 49
DNA, 444–496
   Avery, MacLeod, McCarty experiments of, 445–447, *447*
   Blendor experiment for, 447–450, *451*
   density of, 455
   effect of nitrous acid on, *529*
   enzymes and, 107
   eukaryotic, replication of, 474–475
   excision of thymine dimer from, *530*
   function of, 8–10
   in cell nucleus, 283
   in prokaryotic cell, 22, 244
   in stroma of chloroplast, 296
   mutation of, in hereditary variation, 10
   nucleotides of, 184, *184, 186,* 451
   of eukaryotic chromatin, 469
   purine and pyrimidine bases in, 453
   repetitive, 469–471
   replicating forks, *460*, 461
   replication of, catalysis of, 457
      conservative, 454, *454, 456*
      in *E. coli*, 455
      in prokaryotes, 453–464
         molecular events in, 457–464

DNA (*Continued*)
  replication of, mechanisms of, 459–461
    semiconservative, 454, *454, 456*
    unwinding in, 461
  self-replication of, 218–219
  synthesis of, initiation of, 461–464
    RNA in, 463, *463*
  theta structures in, *460,* 461
  three-dimensional structure of, 451–453, *452*
  transcription, *503*
DNA helix, intercalation of chemicals in, 528
DNA ligase, 461, *462*
DNA polymerase, isolation of, 218
DNA polymerase enzymes, 457–459
Double bond, 50
Dual pigment systems, 370–373
Duysens, Louis, 363

Early Precambrian Epoch, 25(t)
Earth, age of, 117–118
  formation of, 114–118, *116,* 120
  primitive, 113–137
    atmosphere of, 118–131
      contents of, 16
      escape of, 118–122
    free energy sources on, 131–135, 131(t)
Electric discharge, on primitive earth, 131(t), 132
Electron(s), electrical charge in, 23, 24(t)
  energies of, 45–48, *46*
  in atoms, chemical behavior and, 24
  in neutral atoms, mass of, 23, 24(t)
  number of, 23–24
Electron density, 44, *44*
Electron micrographs, interpretation of, 235, *236*
Electron microscope, development of, and cell theory, 18–19
Electron stains, 538
Electron transfer, in oxidation-reduction reactions, 124
Electron transport, inner mitochondrial membrane and, 408
  photosynthetic, 373–376
Electron transport chain, 365
Electronegativity, 54, 55(t)
Element(s), cell composition and, 40–43
  composition of, 23–26
  in cell, primary and secondary, 41–42
  light, formation of earth and, 120
  neutral atoms of, composition of, 23–26
  relative abundances of, 43(t)
    on earth and universe, 119(t)
  trace, in cell, 41, 42
Elongation factor, in RNA translation, 511
Embden-Meyerhof pathway, 340–351, *342*
  catabolic and anabolic processes and, 351, *352*
  cellular control of, 350–351
  evolutionary development of, 341
  reactions of, 341–351
Emerson, Robert, 370
Emerson enhancement effect, 70
Endergonic reactions, 91
Endocytosis, 269, *269,* 441
Endonuclease, 461

Endoplasmic reticulum, *20,* 21, 261, *262, 273–276, 274, 275*
  rough, *20,* 274, *274, 275*
    protein synthesis in, 274
  smooth, *20,* 274, *275*
    membrane of, 417, *418*
Endospores, 255, *256*
Endothermic process, 84
Energy, activation of, catalysts and, 103, *103*
  and first law of thermodynamics, 83–85
  cellular, 83–112
  conservation of, 84
  free, 90–99
    of activation, 100
    standard, 91–92
  metabolism and, 319
  transfer of, ADP-ATP system in, 94–97, *94*
Energy diagram, 101, *102*
Engel, A.E.J., 31
Enolase, in Embden-Meyerhof pathway, 348
Entropic doom, 87
Entropy, 87
  chemical evolution and, 193
  in cells, 111
  increase of, 88, *89*
  net, calculation of, 90
Environment, metabolic development and, 329
Enzyme(s), 104–110, 139
  absorption by coacervates, 229
  action of, 106–107
    "induced fit" hypothesis of, 107, *110*
    "lock and key" hypothesis of, 107, *108, 109*
    "orbital steering" hypothesis of, 107
  amino acid sequence of, 104
  as biological catalysts, 10
  chemical structure of, 104–105
  coenzymes and, 105
  cofactors in, 105
  denaturation of, 105–106
  DNA mutation and, 107
  DNA polymerase, 457–459
  effect of heat on, 69
  efficiency of, 106
  in cell division, 493
  in metabolic evolution, 326, *326*
  inducible, 519
  inhibition of, 107–109
  molecular configuration of, *148*
  mutase, 345
  nomenclature of, 104
  pH and, 106
  polynucleotide synthesis by, 229, *231*
  specificity of, 105
  starch synthesis by, 229, *229*
  temperature and, 105–106
  thermal inactivation of, 105
Eobacterium isolatum, 28, *30*
Eocene Epoch, 25(t)
Equilibrium constant, 78
Eras, geologic, 25(t), 27. See also name of specific era.
Escherichia coli, DNA polymerases in, 459
  DNA replication in, 455
  lactose metabolism in, 518–519
  M protein in, 433
  polyribosomes in, *515*

Index

Escherichia coli (*Continued*)
   protein synthesis in, control of, 499
   T2 infection of, 447–450
Esterification, 174
Ethyl alcohol, oxidation to acetaldehyde, 123–124
Euchromatin, 284, *285*
Eukaryote(s). See also *Eukaryotic cell(s)*.
   aerobic oxidation in, 402
   and Beck Springs fossils, 35, *35*
   cell division in, 467–475
      control of, 493–494
   chromatin in, composition of, 467–471
   DNA replication in, 474–475
   in Bitter Springs fossils, 36, *36*
   multicellular, plasma membrane in, 407
   photosynthetic, 369–377
   protein synthesis in, 515–516
   transcriptional controls in, 521–523
   translational controls in, 523–525
Eukaryotic cell(s), annular material in, 281
   brush border of, *267,* 269
   cell exterior of, 264–266, *265*
   cell wall of, 264–266, *265*
      primary, 266
      secondary, 266
   compartmentalization in, 262–264
      communication between compartments, 264
   endoplasmic reticulum of, 273–276, *274, 275*
   "external compartment" of, membrane system of, 266–283
   internal compartment of, intracellular inclusions of, 283–306
   lysosomes in, 288
   microbodies in, 289–290
   microvilli in, *267,* 269
   middle lamella of, 266
   nuclear envelope of, 279–283
      pore complex in, 281, *282*
      pores in, 280, *281*
   nucleolus of, 286–287
   origin of, 307–315, *309*
   perinuclear space, 279, *280*
   pits in, 266
   plasma membrane of, 267–273, *267, 268*
      function of, 268
   properties of, 261
   ribosomes of, 288
   structure of, 261–307
   surface area to volume ratio in, 261–263
   vacuoles of, 290
Eutectic point, definition of, 200
Evolution, bacteria in, *4*
   biological, as origin of living systems, 2–5
      metabolic development in, 327–339
   blue-green algae in, *4*
   chemical, 191–216, *193*
      aldehydes in, 199
      entropy and, 193
      nitriles in, 199
      thermodynamics and, 193
   chemical kinetics and, 110–112
   differential reproduction in, 3, *3*
   facultative anaerobes in, 38
   from molecules to cells, 217–233
   heritable variation in, 37
   history of, phylogeny as, 4, *4*

Evolution (*Continued*)
   individual variation in, 3
   materialistic theory of, 16
   metabolic pathways in, 37
   of cellular metabolism, 317–339
   of photosynthetic autotrophs, 111
   prokaryotes in, 4, *4,* 22–23
   simulation experiments and, 113–114
   theory of, 2–5
   thermodynamics and, 110–112
   unicellular organisms in, 4, *4*
Excited state, behavior of molecules in, 133
   definition of, 132
Exergonic reactions, 91
Exocytosis, 272, 441
Exothermic process, 84

Facilitated diffusion, 430
Facultative anaerobe(s), 38
   definition of, 323
   in evolution, 38
FAD (flavin adenine dinucleotide), 367, *368*
$FADH_2$, 367, *368*
Fat(s), 173–176
   formation of, *176*
   oxidation of, 397–398
   prebiotic synthesis of, 215
Fatty acid(s), 173–176, *175*
   in aqueous systems, *174*
   in sphingomyelin, 409
   polyunsaturated, 173
   prebiotic synthesis of, 201–202
   saturated, 173
   unsaturated, 173
Fermentation, 340
Ferredoxin, 367
Ferredoxin reducing substance (FRS), 375
Fig-Tree formation, 27–33
   age of, 28
Filament(s), 296
   of cell, *20*
   of prokaryotic flagellum, 242, *243,* 245
First law of thermodynamics, 83–85
Fixatives, 537
Flagellin, 242
   synthesis of, 242
Flagellum(a), 302–306, *304*
   longitudinal structure of, 305, *305*
   motion of, sliding filament model of, *306*
   of prokaryotic cells, 242, *243*
   polar, of prokaryotic cells, 242
Flavin adenine dinucleotide (FAD), 367, *368*
Flavin mononucleotide (FMN), 367, *368*
Flemming, Walther, 443, 477
Fluid mosaic model of cell membrane, 422–423, *422*
Fluorescence, 364
FMN (flavin mononucleotide), 367, *368*
$FMNH_2$, 367, *368*
Formylmethionine, 510, *511*
Fossil record, 23–36
   dating of, 23–27
      radioisotopes in, 23–27
Fossil(s), in sedimentary rocks, dating of, 26–27
   Precambrian, 27–33
   radioisotope dating of, 23–26

Fox, C. Fred, 433
Franklin, Rosalind, 451
Free energy, 90–99
    chemical equilibrium and, 92–93
    of activation, 100
    standard, 91–92
Free energy coupling, 93–99
    mechanism of, 97–99
Free energy diagram, 102, *103*
Free oxygen, in primitive atmosphere, 126–127
Free radicals, 133
Freeze-etching, of cell membrane, 419–421, *419, 420*
Fructose, 165, 167
Fructose-6-phosphate, phosphorylation of, 345
Fumarase, in Krebs cycle, 389
Fumaric acid, in Krebs cycle, 388, 389
Functional organization, as characteristic of living systems, 8–11
    of nucleic acids, 8–10
Fungus(i), evolutionary role of, *4*
Furrow, in cytokinesis, 484, *484*
Fusion, heat of, in water, 70

β-Galactosidase, in *E. coli* lactose metabolism, 518–519, *518*
Gametes, formation of, 491
Gametogenesis, 491
Gametophytes, 491
Gas vacuoles, in prokaryotic cells, *247, 252–253, 252*
    functions of, 253
Gases, inert noble, in primitive atmosphere, 118
GDP (guanosine diphosphate), in Krebs cycle, 388
Gene(s), expression of, chromatin structure and, 472–473
    regulation of, 517–518
Gene amplification, 470
Generation, spontaneous. See *Spontaneous generation.*
Genetic code, 498–533
    degeneracy of, 503
    direct coding mechanism, *532*
    origin of, 530–532
    reading of, 504, *504*
    universality of, 504–505
Geologic time scale, 25(t), 27
Germ cells, formation of, cytokinesis in, 491
    nuclear division in, 486–494, *487*
Germ cell mutation, 525
Gibbs, J. Willard, 90–91
Glauconite, in rock dating, 27
Glucose, 165
    "activation" of, 344
    aerobic oxidation of, efficiency of, 395, 396(t)
    anaerobic oxidation of, 340–356
        similarity in all organisms, 349–350
    cellular uptake of, 341
    facilitated diffusion of, 430
    in Calvin cycle, 378
    in cell, function of, 10
    in Embden-Meyerhof pathway, 351
    membrane transport of, 436
    oxidation of, in aerobes, 339

Glucose (*Continued*)
    oxidation of, in anaerobes, 339
    oxidation to acetyl coenzyme A, 384, 386(t)
    phosphorylation of, 343
    ring structure of, 165, *166*
    stored, breakdown of, 344
Glucose phosphate isomerase, in Embden-Meyerhof pathway, 345
Glucose units, 167–168
Glutamic acid, 143, *143*
Glutamine, 143, *143*
Glyceraldehyde phosphate dehydrogenase, in Embden-Meyerhof pathway, 346
Glyceraldehyde-3-phosphate, oxidation of, 346
Glycerides, in simulation experiments, 215
Glycerol, prebiotic synthesis of, 201–202
Glycogen, 171
    metabolism of, in smooth endoplasmic reticulum, 275
Glycogen granules, *20*
    of prokaryotic cells, 253, *254*
Glycolipids, in cell membrane, 410, *411*
Glycolysis, 340
Glycoproteins, in cell membrane, 412
Glycosidic linkage, 168
Glyoxysomes, 290
Golgi apparatus, *20, 276–279, 277*
    function of, 278
    secretion by, 278, *279*
Golgi bodies, in eukaryotic cells, 21
Gorten, E., 413
Gram-negative, definition of, 225
Gram-positive, definition of, 225
Grendel, F., 413
Green, David, 412, 421
Green algae, in Bitter Springs fossils, 36
Griffith, Fred, 445
Ground state, 47
GTP (guanosine triphosphate), in Krebs cycle, 388
Guanosine diphosphate, in Krebs cycle, 388
Gunflint Iron formation, *34, 35*

Haeckel, Ernst, 15–16
Haldane, J.B.S., materialistic theory of, 16
Haploid cells, 486
Heme, 163, *163*
Hemoglobin, primary structure of, 152
    quaternary structure of, *161*, 162
Hereditary apparatus, origin of, 530–532
    models of, 531–532
Hereditary variation, DNA mutation in, 10
Hershey, Alfred, 447
Heterochromatin, 284, *285*, 471, *471*
Heterocyst(s), 257–258, *257*
Heterogeneous nuclear RNA, 516
Heterotroph(s), 7
    aerobic, 38
    anaerobic, evolution of, 37
    formation of, 111
Hill, Robert, 377
Hill reaction, 377
Histidine, 144, *144*

Index

Histone(s), 467–468
    in eukaryotic transcription, 522
Holland, H. D., 129–130
Holley, Robert, 506
Holocene Epoch, 25(t)
Hook, of prokaryotic flagellum, 242, *245*
Hooke, Robert, 1
Hormone(s), 139
    in cell division control, 494
Horowitz, H.N., 325
Hydration, 57, 72, *73*
Hydrocarbon chain, in fatty acids, 173
Hydrogen, in primitive atmosphere, 128
    stability of, 60
Hydrogen bond, in ice, 66
    in tautomers, 527, *527, 528*
    in water, 65, *66*
Hydrogen bonding, 58, *59*, 158
    in nucleic acids, 214, *214*
Hydrogen ion(s), hydronium ions and, 79–80
    in respiratory chain, 392
Hydrogen peroxide, anaerobic bacteria and, 335
Hydrogenation, 123
Hydrolysis, of ATP, 95, *97*
Hydronium ions, hydrogen ions and, 79–80
Hydrophobic bond, 60
Hypotonic solution, diffusion in, 226, *226*

Ice, hydrogen bond in, 66
    structure of, 65–67, *65*
Igneous rocks, dating of, 26
Imino acid, 141
Individual variation, 3
"Induced fit" hypothesis of enzyme action, 107, *110*
Inducers, 499
Inductive resonance, 363
Initial photochemical event, 363, 364–365
Initiation factors, in RNA translation, 511
Initiator complex, 511
Inorganic phosphate, in Embden-Meyerhof pathway, 351
Insulin, bovine, primary structure of, 152, *153*
Interchromatin granules, 284
Interphase, 477–480, *478*
Interphase I, 488
Interphase II, 490
Intramitochondrial granules, *292*, 294
Intrathylakoidal space, 296
Ion-dipole interactions, 58
Ionic character, 57
Ionic compounds, composition of, 543
Ionization, 57
Ionizing radiation, mutation from, 528–529
Isocitric acid, in Krebs cycle, *387*, 388
Isomerases, 345
Isomers, D and L, 147
    optical, 145
    D and L forms of, 145
Isopeptide bonds, 160
Isoprenoid compounds, synthesis by prokaryotes, 333
Isoprenoid lipids, 180, *181*
Isotonic solution, diffusion in, 226, *226*
Isotope(s), 24
    radioactive, 24–27

Jacob, Francois, 499
Jagendorf, Andre, 376
Jupiter, atmosphere of, 129
Jurassic Period, 25(t)

Kant, Immanuel, 114–115
Kant-LaPlace theory of formation of earth, 114–115
Kaolin, in prebiotic synthesis of disaccharides, 215
Karyotype, *476*, 481
Kennedy, Eugene, 433
Kinases, 344
Kinetics, chemical, 99–112
    evolution and, 110–112
Kornberg, Arthur, 457
Krebs, Hans, 385
Krebs cycle, 385–390, *387*
    primary function of, 389
    synthesis of cell constituents by, 399, *400*

Lac operon, 519
Lac repressor, 519
Lampbrush chromosomes, 522, *524*
LaPlace, Pierre de, 114–115
Late Precambrian Epoch, 25(t)
Lecithin, 177
Leukocytes, lysosomes in, 288
Lewis electron dot structures, 541, 542(t)
Life, origin of, materialistic theory of, 16, 191
    mechanistic theory of, 15–16
    Oparin-Haldane hypothesis of, 191
    panspermia theory of, 15
    spontaneous generation theory of, 12–15
Light, absorption of, 362–365
    in chemical synthesis, 132–135
    polarized, 145, *146*
    relationship of wavelength and color, *361*
    ultraviolet, on primitive earth, 132, 133, 134
Light microscope, development of, and cell theory, 18–19
Linoleic acid, *175*
Lipid(s), 172–181
    function of, 172
    in cell, *20*
    in cell membrane, 409–411
    isoprenoid, 180
    metabolism of, by smooth endoplasmic reticulum, 275
    solubility of, membrane permeability and, 429
Lipid bilayer model of cell membrane, 413
Lipoic acid, 384
Lipoprotein vesicles, formation of, *221*
Living systems, cells in, 5, 8
    characteristics of, 6–12
    origin of, 1–17
        anaerobic heterotrophs in, 37
        ancestral organism in, 18–38
        fossil record in, 23–37
    unicellular and multicellular organisms in, 5–12
Living versus non-living, 11–12
"Lock and key" hypothesis of enzyme action, 107, *108, 109*

Luciferin, 335
Lysine, 144, *144*
Lysosome(s), *20*, 21, 288–289
  secondary, 272
Lysozyme, 160
  tertiary structure of, 160, *161*

M protein, 433
MacLeod, Colin, 445
Magnesium chloride, hydration of, 72, *73*
Malic acid, in Krebs cycle, 389
Margulis, Lynn, 495–496
Mass number, 24
Materialistic theory, 16
McCarty, Maclyn, 445
Mechanistic theory, 15–16
Meiosis, 286, 486–494, *487*
  crossing over in, 488, *488*
  first meiotic division, 488–490
  second meiotic division, 490–491
  synapsis in, 488
Meiosis I, 488–490
Meiosis II, 490–491
Membrane(s), cell. See *Cell membrane.*
  in molecular aggregates, 221–222, *221*
  intracellular, biochemical functions of, 408
  selectively permeable, 225
  semipermeable, 225, 425
Membrane transport, 423–441
  active, 424, *424*, 430–433
    ATP in, 431
    concentration gradient in, 431
    inhibition of, 432
  carrier-mediated, intramembrane shuffling, 434
    mechanisms of, 433–434
    models of, 434, *435*
  membrane permeability and, 424
  metabolic energy needs of, 432
  metabolically linked, 430–433
  passive, 424, *424*
  passive diffusion, 429–430
  transport mechanisms in, 428–434
Membrane transport molecules, 428
Mendel, Gregor, 444
Meselson, Matthew, 454
Mesosome(s), of prokaryotic cells, 247–248
  function of, 22
Mesozoic Era, 25(t), 27
Messenger hypothesis, 499–500
Messenger RNA, masking of, 523–524
  synthesis of, in prokaryotes, 499–500
  transport across nuclear membrane, 525
Metabolic pathway(s), 6–8, *7, 9*, 37
Metabolism, aerobic, 324, 383–405
  anaerobic, definition of, 324
  in aerobes, 394–395
  as characteristic of living systems, 6–8
  carbohydrate, 340
    relation to other metabolic processes, 396–399, *399*
  cellular, evolution of, 317–339
    processes of, *7*
  development of, adaptation and, 329
    in biological evolution, 327–339
    in changing environment, 329

Metabolism (*Continued*)
  development of, selection pressure and, 325
  dynamic nature of, 404
  energy and, 319
  first pathways of, 324–327
  of metabolic pathways, 328
  multistep nature of, 326, *326*
  of anaerobic heterotrophs, 340–357
  principles of, 319–324
  stepwise nature of, 322–323, *323*
Metabolites, transfer between cell compartments, 408
Metamorphic rocks, dating of, 26
Metaphase, *479*, 482
Metaphase I, 489, *489*
Metaphase II, 490
Metaphase plate, 482
Meteorite(s), 117
  as origin of living systems, 15
  chemical analysis of, 202–204
  Murchison, *202*
    amino acids in, 203(t)
  organic compounds in, 202–204
Methane, structure of, 54, *54*
Micelles, 173
Microtubules, *20*, 300–301, *301*
  chromosome movement and, 483
Microbodies, 289–290
Microfilaments, 296, *300*
Micrometer, 18
Microscope, compound light, 535–537, *536*
  condenser of, 536
  contrast of, 537
  electron, development of, and cell theory, 18–19
  eyepiece of, 536
  light, development of, and cell theory, 18–19
  objective of, 536
  phase contrast, 537–538
  scanning electron, 540
  transmission electron, *536*, 538–540
    preparation of thin sections for, *539*
Microscopy, 535–540
Microsphere(s), 222–227, *222, 223, 224, 226*
Middle Precambrian Epoch, 25(t)
Miescher, Frederick, 444
Miller, Stanley, 195–199
Minerals, Precambrian, 127
Miocene Epoch, 25(t)
Mississippian Period, 25(t)
Mitchell, Peter, 548
Mitochondrion(a), *20*, 291–295, *292, 293*
  compartments of, 291
  elementary particles of, 293
  in eukaryotes, 21
  inner membrane of, composition of, 402
  reconstitution of, *403*
  matrix of, 291
  membranes of, 291, *293*
  membrane structure of, 417, *418*
  protein synthesis in, 516–517
  respiratory chain in, 390, *391*
    NAD^+ in, 391
Mitosis, 21, 286, 475–485
  first description of, 443
Molarity, 72

Index

Molecular aggregates, 219-232
  adsorption to surfaces, 231-232
  membranous structures in, 221-222, *221*
Molecular orbital, 49
Molecule(s), aggregation of. See *Molecular aggregation.*
  carrier, 428
  D and L isomers of, 147
  evolution to cells, 217-233
  excited state of, 132
  geometry of, 52-54, *52, 53*
  organic, anaerobic oxidation of, 321
  polar, 54-55
  recognition, 407
  self-replicating, 218-219
    polynucleotides as, 218-219
    polypeptides as, 219
  structure of, 44-60
Monera. See *Prokaryotes.*
Monod, Jacques, 499
Monosaccharide(s), 164-168
  classification of, 164
  prebiotic synthesis of, 201
Muir, M., 31
Murein hydrolase, 241
Murchison meteorite, *202*
  amino acids in, 203(t)
Mutagens, 528-529
Mutase enzymes, 345
Mutation, 10, 525-530
Multiple bonding, 50-51
Myelin membrane, *420*

NAD, in Embden-Meyerhof pathway, 346-347, *347, 349,* 351
NAD^+, in mitochondrial respiratory chain, 391
  photoreduction of, 367-369, *369*
NADH, 346, *347*
  cytoplasmic, oxidation of, 393-394, *394*
  in carbon dioxide fixation, 365
NADP, 353
$NADP^+$, photoreduction of, 373-376
NADPH, formation of, 375
Nanometer, 18
Natural selection, in evolution of metabolic pathways, 37
  in metabolism development, 325
Neutrons, 23-24
Nicotinamide adenine dinucleotide. See *NAD.*
Nicotinamide adenine dinucleotide phosphate. See *NADP.*
Nirenberg, Marshall, 503
Nitrates, reduction by anaerobic bacteria, 333, *334*
Nitrile(s), as condensing agents, 205, *206*
  in chemical evolution, 199
  in prebiotic synthesis of peptides, 212-213
Nitrogen, stability of, 60
  utilization pathways of, 334-335
Nitrogen fixation, 334
Nitrous acid, effect on DNA, *529*
Nostoc, 36, 254
N-terminal residue, 149
Nuclear envelope, *20*
  in eukaryotic cells, 21

Nucleic acid(s), 182-190
  chemical composition of, 8-10
  hydrogen bonding in, 214, *214*
  in living systems, functional organization of, 9-10
Nucleoid, in prokaryotes, 22, 244, *247*
Nucleolar organizer region, 286
Nucleolus, *20,* 286-287
  ribosomal RNA synthesis by, 286
Nucleoplasm, 284
Nucleoside(s), 184
  nomenclature of, 187(t)
  prebiotic synthesis of, 206-207, *207*
Nucleosomes, 472, *472*
Nucleotide(s), 182-189
  as subunits of molecule, 8
  cellular functions of, 187
  formation of, *185*
  nomenclature of, 187(t)
  of DNA, 184, *184, 186*
  of RNA, 184, *184, 186*
  prebiotic synthesis of, 207
  subunits of, 182
Nucleus, atomic, 23-24
Nucleus, cell, 21, 283-286, *284*

Ocean, primitive, 135-136, *135*
Octet rule, of chemical bonding, 541-543
Okazaki, Reiji, 461
Okazaki fragments, 461
Oleic acid, *175*
  coacervate formation in, 227
Oligocene Epoch, 25(t)
Oligonucleotides, prebiotic synthesis of, 213-215
Oligosaccharide(s), 168-169
  prebiotic synthesis of, 215
On the Origin of Species, 2-3
Onverwacht formation, 27-33, *32*
Oogenesis, 491, *492*
Oparin, A.I., 16
Oparin-Haldane hypothesis of origin of life, 191
Operon model, 518-520, *520*
Optical activity, in "primitive broth," 147
  of amino acids, 145-149
Optical isomers, 145
"Orbital steering" hypothesis of enzyme action, 107
Ordovician Period, 25(t)
Organ(s), 5
Organelles, cytoplasmic, in eukaryotes, 21
  intracellular, osmotic pressure and, 427
Organism(s), ancestral, 18-22
  cell organization in, 18-20
  fossil record of, 23-36
  evolutionary relationship between, *4*
  multicellular, in living systems, 5
  unicellular, evolutionary role of, *4*
  in living systems, 5
Oro, Juan, 200
Oscillatoria, 36
Osmosis, in microspheres, 225, *226*
Osmotic pressure, effect on intracellular organelles, 427
  in cells, 426

Ouabain, 437
Overton, E., 413
Oxaloacetic acid, in Krebs cycle, 389, 400
Oxidation, aerobic, in eukaryotes, 402
    and phosphorylation of ADP, 320, *321*
    definition of, 122
    in cell energy capture, 320–321
    of glucose, anaerobic, 340–356
Oxidation-reduction potential, 320
Oxidation-reduction reaction(s), electron transfer in, 124
    in primitive atmosphere, 122–128
    with covalent compounds, 123–125
Oxidation states, 125, *125*
Oxidative phosphorylation, 393
    mechanisms of, 395–396, 546–552
Oxidizing agent, 123
Oxygen, accumulation in primitive atmosphere, 37–38, 337, 358–382, 383
    elimination of, 330–334
    free, in primitive atmosphere, 126–127
    passive diffusion of, 429
    stability of, 60
Ozone, in primitive atmosphere, 133–134
    production of, 134

P890, 363
Paleocene Epoch, 25(t)
Paleozoic Era, 25(t), 27
Palmitic acid, 173, *175*
Panspermia, 15
Passive diffusion, 429–430
Pasteur, Louis, 13–15
Pathways, metabolic. See *Metabolic pathways*.
Pectin, in eukaryotic cell, 266
    synthesis of, 278
Penicillin, antibacterial action of, 241
Pennsylvanian Period, 25(t)
Pentose phosphate pathway, 351–356, *354*
    ATP yield of, 355
    Calvin cycle and, 380
    evolution and, 356
    reductive. See *Calvin cycle*.
Pentose phosphate shunt, 353
Pepsin, pH optimum of, 106
Peptide(s), 149
    prebiotic synthesis of, 207–211
        nitrile-mediated condensations in, 212–213
Peptide bonds, 149–152
    formation of, *149*
Peptidyl transferase, 512
Perichromatin fibers, 284
Perichromatin granules, 284
Peroxisomes, 289, *289, 290*
Permian Period, 25(t)
pH, 79–80, *80*
    enzymes and, 106
    pepsin and, 106
pH scale, 544–545
Phagocytosis, 269
Phagocytotic vesicles, 269
Phanerozoic Era, 27
Phase contrast microscope, 537–538
Phosphate, inorganic, in Embden-Meyerhof pathway, 351

Phosphatidic acid(s), 176, *177*
    derivatives of, 178
    in cell membrane, 409
Phosphodiester bond, 189
Phosphoenolpyruvic acid, formation of, 348
Phosphoglycerate kinase, in Embden-Meyerhof pathway, 348
2-Phosphoglyceric acid, dehydration of, 348
Phosphoglycerides, in cell membrane, 409
Phosphoglyceromutase, in Embden-Meyerhof pathway, 348
Phospholipid(s), 176–180
    in aqueous systems, 178
    in cell membrane, 409
    membranes and, 221
Phosphoribulokinase, in Calvin cycle, 380
Phosphoric acid, ADP formation and, 96
    in DNA, 451
Phosphorus, metabolic role of, 61
Phosphorylation, oxidative, chemiosmotic hypothesis of, 548–552, *549, 551*
    conformational coupling hypothesis of, 552
    direct chemical coupling hypothesis of, 546–548, *547*
    mechanisms of, 546–552
    substrate-level, 348
Photon, 132
Photo-oxidation, 333
Photophosphorylation, 373–376
    cyclic, 365–367, *366*
Photoreduction, of NAD^+, 367–369, *369*
    of $NADP^+$, 373–376
Photosynthesis, 21, *331*, 358–382
    absorption spectra in, 361
    bacterial, light reactions of, 360–369
    chemical mechanisms of, 358
    dark reactions of, 359, 377–380, 381–382
    efficiency of, 380
    development of, 37, 330–334
    dual aspect of, 359
    dual pigment systems in, 370–373
    electron transport in, 373–376
    first photosynthetic organisms, 332
    in *A. barbertonensis,* 31
    in bacteria, blue-green algae, and eukaryotes, comparison of, 381–382
    in prokaryotic cells, 251
    light reactions of, 359, 381–382
        efficiency of, 376–377
        electron transfer in, *374*
        thylakoid membranes and, 408
    localization of, 381
    multistep nature of, 359
    origin of, 37
    oxygen-eliminating, development of, 369
    pigments in, energy transfer in, 372–373, *373*
Photosynthetic unit, 364
Phycobiliproteins, 251
Phycobilisomes, 252
Phycocyanin, 251
    absorption spectrum of, *372*
Phycocyanobilin, chemical structure of, *251*
Phycoerythrin, 251
    absorption spectrum of, *372*
Phylogeny, 4
Phytane, 180, *181*

Index

Pigment(s), accessory, 363
  photosynthetic, 360–362
    energy transfer in, 372–373, *373*
Pigment system, 364
Pigment system I, characteristics of, 371, 371(t)
Pigment system II, characteristics of, 371, 371(t)
F-pilus, 244
Pilus, of prokaryotic cells, 244
Pinocytosis, 269
Pinocytotic channels, *20*
Pinocytotic extensions, *20*
Pinocytotic vesicles, *20*, 269
Plantae, evolutionary role of, *4*
Plasma membrane, 6, *20*
  function of, 6
  in cell division, 406
  in eukaryotic cells, 21
  in multicellular eukaryotes, 407
  in prokaryotes, 22, 238, *239*
  osmotic pressure and, 426
  permeability to water, 426–427
  receptor sites of, 407
  sodium gradient across, 436
Plasmalemma, of prokaryotic cell, 238, *239*
Plasmodesmata, 241
  of eukaryotic cell, *265*, 266
Plasmolysis, 426, *426*
Plastid(s), 295–296
Plastid differentiation, 295, *295*
Platinum shadowing technique, 28, *29*
Pleistocene Epoch, 25(t)
Pliocene Epoch, 25(t)
PMPE. See *Polymetaphosphate ethyl ester.*
Polar molecules, 54–55
Polarized light, 145, *146*
Polyadenylic acid, enzymatic synthesis of, 230, *231*
Polymerization, thermal, of amino acids, 210
Polymetaphosphate ethyl ester (PMPE), 206, *207*
  in polypeptide synthesis, 211
Polynucleotide(s), 189–190
  as self-replicating molecules, 218–219
  enzymatic synthesis and degradation of, 229, *231*
  prebiotic synthesis of, 213–215
Polynucleotide chain, 189
Polypeptide(s), 149
  prebiotic synthesis of, 207–211
    polymetaphosphate ethyl ester in, 211
Polypeptide chain, chemical structure of, *150*
  rotation of amino acids in, *151*
Polyphosphate granules, of prokaryotic cells, *247*, 253
Polyribosome(s), 514–515, *515*
Polysaccharide(s), 169–171
  prebiotic synthesis of, 215
Polysomes, *20*
Polytene chromosomes, 521, *522*
Porphyrin(s), 162–163, *163*
  synthesis of, development of, 333
Positive feedback system, 400
Potassium ion(s), enzymes and, 435
  intracellular concentrations of, regulation of, 434–441, 436(t)
  protein synthesis and, 435
Potato phosphorylase, in starch formation, 229, *229*

Precambrian Era, 25(t), 27
Precambrian Epoch, 25(t)
Preliminary activation step, in fat oxidation, 398
"Primitive broth," 135–316
  optical activity in, 147
Primitive earth, atmosphere of, 16
Principal quantum number, 45–48
Pristane, 180, *181*
Prokaryote(s), 237–238. See also *Prokaryotic cell(s).*
  cytokinesis in, 464–465
  DNA replication in, 453–464
    molecular events in, 457–464
  evolutionary role of, *4*, 4, 22–23
  in Onverwacht formation, 31
  metabolic variability of, 238
  mRNA synthesis in, 499–500
  oxygen-producing, origin of, 37
  RNA transcription in, 499–502
  synthesis of isoprenoid compounds by, 333
Prokaryotic cell(s). See also *Prokaryotes.*
  cell division in, 453–467
    control of, 465–467
    plasma membrane in, 239
  cell wall of, 239, *242*
    chemical structure of, 240, *240*
    composition of, 22
    function of, 22
  characteristics of, 237
  chondroids of, 248
  cytoplasm of, 246, *247*
  cytoplasmic inclusions in, 252–255
  diversity of, 260
  DNA in, 22, 244
  exterior of, 238–242
  flagella of, 242, *243*
    movement of, 244
  gas vacuoles in, *247*, 252–253, *252*
  glycogen granules of, 253, *254*
  interior of, 244–247
  intracellular membranes of, 247–252
  lipid droplets in, 253
  mesosomes of, 247–248, *247*, *248*
  nucleoid of, 22, 244, *247*
  photosynthesis in, 251
  pili of, 244
  plasma membrane of, 238, *239*
  polyhedral bodies in, 253–255, *254*
  polyphosphate granules of, *247*, 253
  protein synthesis in, 246, 499–515
    energy requirements of, 515
  ribosomes in, 246
  slime, 242
  specialized, 255–258
  structure of, 234–259
  surface appendages of, 242–244
  thylakoids of, 249–251, *249*
Proline, 141
Prophase, *479*, 480–482
Prophase I, 488–489
Prophase II, 490
Proplastids, 295
Protein(s), 139–164
  adsorbed, in cell membrane, 414
  as buffers, 80
  as enzymes, 10

Protein(s) (*Continued*)
  conjugated, 163–164, *163*
    prosthetic group in, 164
  denaturation of, 161
  fibrous, secondary structure of, 155
  folding of, mechanism of, 161–162
  globular, 158–159
  in cell membrane, 411–412
    study of, with detergents, 412
  in living systems, 8–10
  in viruses, 218
  integral, 422
  M, 433
  nonhistone, in eukaroyotic chromatin, 468–469
  of eukaryotic chromatin, 467–468
  oxidation of, 398–399
  peripheral, 422
  physiological functions of, 139
  "structural," 412
  structure of, 152–163
    primary, 152–154, *153*
    quaternary, 162–163, *162*
    secondary, 154–155, *156*
    tertiary, 155–162
      x-ray diffraction and, 160
      stabilization of, 160, *160*
  synthesis of, 517–518
    in prokaryotes, 499–515
      energy requirements of, 515
    in rough endoplasmic reticulum, 274
  unwinding, 461
Protein crystal model of cell membrane, 421–422, *421*
Proteinoid(s), amino acids and, 210
  catalytic properties of, 208
  microsphere formation and, 222
  proteins and, 209
  structure of, 209, *209*
  synthesis of, 208
Protista, evolutionary role of, *4*
Protofilaments, 300
Proton gradient, in oxidative phosphorylation, 548
Proton(s), 23, 24(t)
Pseudomonas fluorescens, polar flagella of, *243*
Purine(s), 183
  prebiotic synthesis of, 200–201
Pyrimidine(s), 183
  prebiotic synthesis of, 200–201
Pyruvate dehydrogenase complex, 384
Pyruvate kinase, in Embden-Meyerhof pathway, 349
Pyruvic acid, in Embden-Meyerhof pathway, 349
  oxidative decarboxylation of, 384

Quantasomes, 296, *299*, 381
Quantum, definition of, 132
Quantum Theory, 44
Quaternary Period, 25(t)
Quinones, in bacterial cyclic photophosphorylation, 366, *366*

Racemic mixture, 147
Racker, Efraim, 402

Radioisotope(s), half-life of, 24–26
  in fossil dating, 24–27
Radioisotope dating, 23–26
Reaction(s), activation energy of, 100
  catalysts and, 103, *103*
  catabolic, 94
  coupled, 93
  endergonic, 91
  energy diagram of, 101, *102*
  exergonic, 91
  photochemical, 102
  rate of, 100–109
    collision theory of, 100–101, *101*
    factors in, 102–104
    transition state theory of, 101–102
Reaction center, in photosynthesis, 363
Receptor sites, of plasma membranes, 407
Recognition molecules, 407
Redi, Francesco, 12
Redox potential, 320
Redox reaction(s). See *Oxidation-reduction reaction(s)*.
Reducing agent, definition of, 123
Reductants, strong, 320
Reduction, definition of, 122
  in cell energy capture, 320–321
  in primitive atmosphere, 125–128
Reductive pentose phosphate pathway. See *Calvin cycle*.
Release factors, in RNA translation, 513
Repetitive DNA, 469–471
Replicating forks, in DNA, *460*, 461, *464*
Replication, conservative, 454, *454*
  semiconservative, 454, *454*, 456
Replicon, 474
Respiration, aerobic, definition of, 324
    localization of, 401–404
    pathways of, 335–338
    regulation of, 399–401
  anaerobic, 329, *330*
  definition of, 324
Respiratory chain, 390–394
  coenzyme Q in, 391
  hydrogen ions in, 392
  synthesis of cell constituents by, 399
Respiratory control, 400, *401*
Resonance, inductive, 363
Reticulum, endoplasmic, rough, 20, 274, *274*, *275*
  smooth, 20, 274, *275*, 417, *418*
  sarcoplasmic, 275, 440
Ribonucleic acid. See *RNA*.
Ribose, 165
Ribosome(s), 20, 288
  in eukaryotic protein synthesis, 515–516
  in prokaryotic cell, 246
  in RNA translation, 505
  in stroma of chloroplasts, 296
Ribulose diphosphate carboxylase, in Calvin cycle, 380
Ris, Hans, 471
mRNA. See *Messenger RNA*.
tRNA. See *Transfer RNA*.
RNA, function of, 10
  heterogeneous nuclear, 516
  in cell nucleus, 283
  in DNA synthesis, 463, *463*

RNA (Continued)
   in stroma of chloroplast, 296
   nucleotides of, 184, *184, 186*
   ribosomal, synthesis of, 286
RNA polymerase, 463, 500–502, *501*
   in bacterial transcription, 520
RNA transcription, in prokaryotes, 500–502
   initiation and termination of, 502
RNA translation, in prokaryotes, 505–515
      70s initiation complex, 511, *512*
Rocks, dating of, 26–27
   use of glauconite in, 27
Rough endoplasmic reticulum, *20, 274, 274, 275*
Rubidium-87, half-life of, 24
Rubidium-strontium system, in rock dating, 26

Saccharomyces, mitochondrial structure of, 294
Sanger, F., 152
Sarcoplasmic reticulum, 275
   calcium pump and, 440
Scanning electron microscope, 540
Schleiden, Matthias, 1
Schopf, J.W., 28
Schramm, Gerhard, 206
Schwann, Theodor, 1
Second law of thermodynamics, 85–90
Secondary constrictions, 481
Sedimentary rocks, dating of, 26–27
Selectivity, chemical, of living organisms, 42–43
Septum, in prokaryotic cytokinesis, 465, *466*
Sex cells, formation of, 491
Sickle cell anemia, 152, *154*
Side chain, 139
Sigma subunit, 502
Silurian Period, 25(t)
Simulation experiment(s), 195–216
   for purine and pyrimidine bases, 200–201
   of primitive earth, 113–114, 130–131
Singer, S.J., 422
Sjöstrand, F.S., 417
Sliding filament model, of ciliary and flagellar motion, *306*
Slime, of prokaryotic cell, 242
Sodium chloride, ionic bonding in, 56
Sodium ions, intracellular concentrations of, regulation of, 434–441, 436(t)
Sodium-potassium ATPase, 437–440
   phosphorylated, *439*
Sodium-potassium pump, 437–440, *438, 440*
Solar system, formation of, 117
Solutes, cell permeability and, 427–428
Solution(s), 71–72, *71*
Solvation, 72
Somatic cells, cytokinesis in, 483–485
Soudan Iron formation, 34
Spallanzani, Lazaro, 12–13
Spark discharge apparatus, 195, *195, 197*
Specific heat, of water, 68–69
Spectrum, absorption, in photosynthetic pigments, 361
   in purple and green bacteria, *362*
Spermatogenesis, 491, *492*
Spheroplast, 240
Sphingomyelin, fatty acid composition of, 409
   in cell membrane, 409, *410*

Sphingosine, in cell membrane, 409, *410*
Spindle apparatus, 481
   poles of, 481
Spontaneous generation, 12–15, *13, 14*
Sporophyte, 491
Stahl, Frank, 454
Staining, of microspheres, 225
Stain(s), acid, 537, 538
Standard free energy, 91–92
Standard state conditions, 92
Staphylococcus aureus, cell wall of, *241*
Starch, 171
   enzymatic synthesis and degradation of, 229, *229*
Stearic acid, 173, *175*
Steroids, 180
Strasburger, Eduard, 443
Strecker synthesis, of amino acids, 198
Stroma, 296
Stromatolites, Bulawayan, 33–34, *34*
Strong oxidants, 320
"Structural protein," 412
Substrate-level phosphorylation, 348
Succinic acid, in Krebs cycle, 388
   oxidation of, 392
Succinic dehydrogenase, 392
Succinyl coenzyme A, in Krebs cycle, *387*, 388
Sulfa drugs, competitive enzyme inhibition by, 109
Sulfate, reduction by anaerobic bacteria, 333, *334*
Sulfur, metabolic role of, 61
Surface replica technique, 28, *29*
Surface tension, of water, 70
Suspensions, 71, *71*
Sutton, Walter, 444
Symbiosis, intracellular, *311*
Symbiotic theory of eukaryotic cell origin, 308–312, *309*
Synapsis, 488
Synechococcus lividus, 249
Synonyms, of codons, 503
Systems, living. See *Living systems.*

T2 bacterial virus, 447–450, *448*
Tautomers, hydrogen bonding in, 527, *527, 528*
Taxis, 244
Telophase, *479, 483*
Telophase I, 490
Telophase II, 490
Tertiary Period, 25(t)
Tetrad, 488
Thermal condensation, in prebiotic synthesis of peptides and polypeptides, 207–211
Thermodynamics, chemical, 83–99, 110–112
   evolution and, 110–112
   first law of, 83–85
   second law of, 85–90
Theta structures, *460*, 461
Thiamine pyrophosphate, 384
Thiobacillus, polyhedral bodies in, 254
Thylakoid(s), 296
   arrangement of, 296, *297, 298*
   of prokaryotic cells, 249–251, *249*
      photosynthesis by, 251

Thylakoid membrane(s), chloroplast, *420*
  light reactions of, photosynthesis and, 408
  plasma membrane and, 406
  ultrastructure of, 296
Thymine dimer, chemical structure of, *529*
  excision from DNA, *530*
Thymine tautomer, *526, 527*
Tissues, 5
Transcription, RNA, in prokaryotes, 499–502
Transfer RNA, 505–507
  coupling of amino acids to, 507–509, *509*
  recognition of mRNA codons by, 509–510, *509*
Transformation, 445–447
Transition state, 101
Transition state theory of reaction rate, 101–102
Translation, in eukaryotes, availability of substances in, 525
  RNA, in prokaryotes, initiation of, 510–511
    initiator complex, 511
Translational repressors, in eukaryotes, 524
Translocase, 512
Transmission electron microscope, *536,* 538–540
  preparation of thin sections for, *539*
Transport, carrier-mediated, saturation phenomenon in, 430, *431*
Triassic Period, 25(t)
Tricarboxylic acid (TCA) cycle. See *Krebs cycle.*
Triose phosphate isomerase, 345
Tripalmitin, 174
Triple bond, 50
Triturus viridescens, oocyte of, nucleolar DNA in, *470*
Tschermak, Erich von, 444
Tubulin, 300
Tumor cells, 407

Ubiquinones, 367, *367*
Ultraviolet light, mutation from, 529
  on primitive earth, 132, 133, 134
Uncoupling agents, in phosphorylation, 395
Unit membrane, electron-dense layers of, 416
Unit membrane hypothesis of cell membrane structure, 415–418, *416*
Uracil, prebiotic synthesis of, 201
Urey, Harold, 128

Vacuole(s), *20,* 290
  contractile, 290
  digestive, 272

Vacuole(s) (*Continued*)
  gas, 247, 252–253, *252*
Valence shell, 47
van Beneden, Edouard, 486
van Leeuwenhoek, Antony, 12
Vaporization, heat of, 69–70
Virus(es), as borderline systems, 11
  characteristics of, 11
  composition of, 218
  non-living properties of, 11
  protein coats of, 449
Vital stains, 537
Vitamin(s), coenzyme production by, 105
  prosthetic group production by, 105
Vitamin A, chemical structure of, *182*
Volcanos, *121*
Volutin granules, *247,* 253
von Tschermak, Erich, 444
Vries, Hugo de, 444

Wallace, Alfred Russel, 2
Water, as cell constituent, 39
  atomic orbitals of, *52*
  biological role of, 63–74
  density of, 67–68
  heat of fusion of, 70
  heat of vaporization of, 69–70
  hydrogen bonding in, 65, *66*
  molecular structure of, 64–65, *64*
  passive diffusion of, 429
  plasma membrane permeability and, 426–427
  physical and chemical properties of, 67–74
  specific heat of, 68–69
    effect on environment, 69
  surface tension of, 70
Watson, James, 451
Weismann, August, 486
Wilkins, Maurice, 451

Xanthophylls, 361
Xenopus laevis, repetitive DNA in, 469
X-ray diffraction, in determination of tertiary protein structure, 160

Yeast alanyl-tRNA, nucleotide sequence of, 506, *506*

Zimm, Bruno, 469